HURST'S
THE HEART
COMPANION HANDBOOK

Editors

R. WAYNE ALEXANDER, M.D., Ph.D.

R. Bruce Logue Professor of Medicine
Director, Division of Cardiology
Department of Medicine
Emory University School of Medicine
Chief of Cardiology, Emory Hospital and Emory Clinic
Atlanta, Georgia

ROBERT C. SCHLANT, M.D.

Professor of Medicine (Cardiology)
Emory University School of Medicine
Atlanta, Georgia

VALENTIN FUSTER, M.D., Ph.D.

Director, The Zena and Michael A. Wiener
    Cardiovascular Institute
Richard Gorlin, M.D./Heart Research Foundation
    Professor of Cardiology
Dean for Academic Affairs
The Mount Sinai Medical Center
The Mount Sinai School of Medicine
New York, New York

ROBERT A. O'ROURKE, M.D.

Charles Conrad Brown Distinguished Professor
    in Cardiovascular Disease
University of Texas Health Science Center
San Antonio, Texas

ROBERT ROBERTS, M.D.

Don W. Chapman Professor of Medicine and Chief
    of Cardiology
Professor of Cell Biology
Director, Bugher Foundation Center for Molecular Biology
Baylor College of Medicine
Houston, Texas

EDMUND H. SONNENBLICK, M.D.

Edmond J. Safra Distinguished University Professor
    of Medicine
Chief Emeritus, Division of Cardiology
The Albert Einstein College of Medicine
Bronx, New York

NINTH EDITION

HURST'S
THE HEART
ARTERIES AND VEINS

COMPANION HANDBOOK

Editors

R. WAYNE ALEXANDER, M.D., Ph.D.

ROBERT C. SCHLANT, M.D.

VALENTIN FUSTER, M.D., Ph.D.

ROBERT A. O'ROURKE, M.D.

ROBERT ROBERTS, M.D.

EDMUND H. SONNENBLICK, M.D.

McGraw-Hill
HEALTH PROFESSIONS DIVISION

New York St. Louis San Francisco Auckland Bogotá
Caracas Lisbon London Madrid Mexico City Milan
Montreal New Delhi San Juan Singapore Sydney
Tokyo Toronto

McGraw-Hill

A Division of The McGraw·Hill Companies

HURST'S THE HEART, Ninth Edition,
COMPANION HANDBOOK

234567890 DOCDOC 99

ISBN 0-07-001024-2

This book was set in Times Roman by Better Graphics.
The editors were Joseph Hefta, Pamela Hanley, and Muza Navrozov.
The index was prepared by Barbara Littlewood.
The production supervisor was Helene G. Landers.
The text and cover design was done by Marsha Cohen / Parallelogram
Graphics.
R. R. Donnelley and Sons, Inc. was the printer and binder.
This book is printed on recycled, acid-free paper.

Cover Illustration reproduced with permission from S. B. King and J. S.
Douglas, "Coronary Arteriography and Angioplasty," Copyright © 1985,
McGraw-Hill, N.Y. Illustration by Michael Budowick, Medical Artist,
Emory University School of Medicine, Office of Medical Illustration.

**Congress Cataloging-in-Publication Data is on file for this title at the
Library of Congress.**

To

Jane, Kate, Melissa, and David
(R.W.A.)

Mia and Stephanie
(R.C.S.)

Maria, Silvia, and Pablo
(V.F.)

Suzann, Michael, Kevin, Sean, Katie, and Ryan
(R.A.O.)

Donna, Brandon, and Allison
(R.R.)

and

Linda, Emily and Charlotte
(E.H.S.)

NOTICE

Medicine is an ever-changing science. As new research and clinical experience broaden our knowledge, changes in treatment and drug therapy are required. The editors and the publisher of this work have checked with sources believed to be reliable in their efforts to provide information that is complete and generally in accord with the standards accepted at the time of publication. However, in view of the possibility of human error or changes in medical sciences, neither the editors nor the publisher nor any other party who has been involved in the preparation or publication of this work warrants that the information contained herein is in every respect accurate or complete, and they are not responsible for any errors or omissions or for the results obtained from use of such information. Readers are encouraged to confirm the information contained herein with other sources. For example and in particular, readers are advised to check the product information sheet included in the package of each drug they plan to administer to be certain that the information contained in this book is accurate and that changes have not been made in the recommended dose or in the contraindications for administration. This recommendation is of particular importance in connection with new or infrequently used drugs.

CONTENTS

CONTRIBUTORS*

JAMES K. ALEXANDER, M.D. [27]

Professor of Medicine (Cardiology), Baylor College of Medicine; Staff Physician, Houston Veterans Affairs Medical Center, Houston, Texas

R. WAYNE ALEXANDER, M.D., Ph.D. [8,10]

R. Bruce Logue Professor of Medicine; Director, Division of Cardiology, Department of Medicine, Emory University School of Medicine; Chief of Cardiology, Emory Hospital and Emory Clinic, Atlanta, Georgia

ROBERT C. ALLEN, M.D. [39]

Chief, Vascular and Endovascular Surgery, Carolinas Heart Institute, Carolinas Medical Center, Charlotte, North Carolina

JOSEPH S. ALPERT, M.D. [23]

Robert S. and Irene P. Flinn Chair of Medicine and Head, Department of Medicine; University of Arizona College of Medicine, Tucson, Arizona

HARISIOS BOUDOULAS, M.D. [4]

Professor of Medicine and Pharmacy, Division of Cardiology, The Ohio State University Medical Center; Director, Overstreet Teaching and Research Laboratory, Division of Cardiology, The Ohio State University College of Medicine, Columbus, Ohio

MICHAEL R. BRISTOW, M.D., Ph.D. [17]

Professor of Medicine; Head, Division of Cardiology; University of Colorado Health Sciences Center, Denver, Colorado

PETER M. BUTTRICK, M.D. [35]

Professor of Medicine and Physiology; Chief, Section of Cardiology; University of Illinois at Chicago, Chicago, Illinois

LOUIS R. CAPLAN, M.D. [37]

Neurologist, Beth Israel Deaconess Medical Center, Boston; Professor of Neurology, Harvard Medical School, Boston, Massachusetts

*The numbers in brackets refer to chapter(s) authored or co-authored by the contributor.

SIMON CHAKKO, M.D. [2]

Professor of Medicine, University of Miami School of Medicine; Chief of Cardiology Section, Veterans Affairs Medical Center, Miami, Florida

NISHA CHIBBER CHANDRA, M.D. [6]

Professor of Medicine, John Hopkins University School of Medicine; Director, Coronary Intensive Care Unit, John Hopkins Bayview Medical Center, Baltimore, Maryland

MELVIN D. CHEITLIN, M.D. [29]

Professor Emeritus (Medicine), University of California at San Francisco; Former Chief of Cardiology, San Francisco General Hospital, San Francisco, California

JAMES E. DALEN, M.D. [23]

Vice President for Health Sciences; Dean, College of Medicine, University of Arizona College of Medicine, Tucson, Arizona

LAURA A. DEMOPOULOS, M.D. [1]

Associate Director, Clinical Development Department, Merck and Company, Inc., West Point, Pennsylvania

DAVID T. DURACK, M.B., D. Phil. [21]

Worldwide Medical Director, Becton Dickinson Microbiology Systems, Baltimore, Maryland; Consulting Professor of Medicine, Duke University Medical Center, Durham, North Carolina

KIM A. EAGLE, M.D. [40]

Professor of Internal Medicine, Senior Associate Chair, Department of Internal Medicine; Chief of Clinical Cardiology, Division of Cardiology; Co-Director, Heart Care Program, University of Michigan, Ann Arbor, Michigan

ERICA DIANA ENGLESTEIN, M.D. [5]

Assistant Professor of Medicine, Indiana University School of Medicine; Director, Cardiac Electrophysiology, R. L. Roudebush Veterans Affairs Medicine Center, Indianapolis, Indiana

MICHAEL D. FREED, M.D. [22]

Associate Professor of Pediatrics, Harvard Medical School; Senior Associate in Cardiology, Chief In-Patient Cardiovascular Service, Childrens Hospital, Boston, Massachusetts

WILLIAM H. FRISHMAN, M.D. [47, 48]

Professor and Chair, Department of Medicine; Professor of Pharmacology, New York Medical College; Chief of Medicine, Westchester Medical Center, Valhalla, New York

VALENTIN FUSTER, M.D., Ph.D. [46]

Director, The Zena and Michael A. Wiener Cardiovascular Institute; Richard Gorlin, M.D./Heart Research Foundation Professor of Cardiology; Dean for Academic Affairs, The Mount Sinai Medical Center, The Mount Sinai School of Medicine, New York, New York

EDWARD M. GILBERT, M.D. [17]

Associate Professor of Medicine, Division of Cardiology; Director, Heart Failure Treatment Program, University of Utah School of Medicine, Salt Lake City, Utah

SCOTT M. GRUNDY, M.D., Ph.D. [7]

Professor of Internal Medicine, Director, Center for Human Nutrition; Chairman, Department of Clinical Nutrition, University of Texas Southwestern Medical Center at Dallas, Dallas, Texas

ROBERT J. HALL, M.D. [32]

Clinical Professor, Department of Medicine, Baylor College of Medicine, Houston, Texas; Clinical Professor of Medicine, The University of Texas Medical School at Houston; Director, Cardiology Education, St. Luke's Episcopal Hospital and Texas Heart Institute, Houston, Texas

W. DALLAS HALL, M.D. [11]

Professor Emeritus (Medicine), Emory University School of Medicine, Atlanta, Georgia

JUHA P. KOKKO, M.D., Ph.D. [42]

Asa G. Candler Professor of Medicine; Chairman, Department of Medicine, Emory University School of Medicine, Atlanta, Georgia

MARK J. KULBASKI, M.D. [39]

Fellow in Vascular Surgery, Emory University School of Medicine, Atlanta, Georgia

CHARLES L. LAHAM, M.D. [28]

Fellow in Cardiology, University of Medicine and Dentistry of New Jersey, New Jersey Medical School, Newark, New Jersey

THIERRY H. LEJEMTEL, M.D. [43]

Professor of Medicine, Albert Einstein College of Medicine, Bronx, New York

MARLO F. LEONEN, M.D. [16]

Assistant Professor in Medicine, Division of Cardiology, Wayne State University; Director of Echocardiography Laboratory, Veterans Affairs Medical Center, Detroit, Michigan

RICHARD P. LEWIS, M.D. [4]

Professor of Medicine, Ohio State University College of Medicine, Columbus, Ohio

JOSEPH LINDSAY, JR., M.D. [38]

Professor of Medicine, The George Washington University School of Health Care Sciences; Director, Section of Cardiology, Washington Hospital Center, Washington, D.C.

JAMES E. LOYD, M.D. [24]

Professor of Medicine, Division of Allergy, Pulmonary and Critical Care Medicine, Vanderbilt University Medical Center North, Nashville, Tennessee

BARRY J. MARON, M.D. [18]

Director, Cardiovascular Research Division, Minneapolis Heart Institute Foundations, Minneapolis, Minnesota

JOHN H. MCANULTY, M.D. [34]

Professor of Medicine and Head, Division of Cardiology; Director, Arrhythmia Service; Oregon Health Sciences University, Portland, Oregon

WILLIAM E. MITCH, M.D. [30]

Garland Herndon Professor of Medicine; Director, Renal Division, Department of Medicine, Emory University School of Medicine, Atlanta, Georgia

RAUL D. MITRANI, M.D. [3]

Assistant Professor of Medicine, University of Miami School of Medicine; Director, Arrhythmia and Pacemaker Center, Miami, Florida

DOUGLAS C. MORRIS, M.D. [41]

J. Willis Hurst Professor of Medicine (Cardiology), Emory University School of Medicine; Director, The Emory Heart Center; Director, Carlyle Fraser Heart Center, Crawford Long Hospital of Emory University; Director of Clinical Cardiology Services, The Emory Clinic, Emory University Healthcare Systems, Atlanta, Georgia

ROBERT J. MYERBURG, M.D. [2]

Professor of Medicine and Physiology; Director, Division of Cardiology, University of Miami School of Medicine, Jackson Memorial Medical Center, Miami, Florida

STEVEN D. NELSON, M.D. [4]

Associate Professor of Internal Medicine, The Ohio State University College of Medicine; Director, Cardiac Electrophysiology Laboratory; Director, Arrhythmia Monitoring Sevices, Columbus, Ohio

JOHN H. NEWMAN, M.D. [24]

Elsa S. Hanigan Chair in Pulmonary Medicine; Professor of Medicine, Vanderbilt University School of Medicine; Chief, Medicine Service, Nashville Veterans Administration Hospital, Nashville, Tennessee

JOHN B. O'CONNELL, M.D. [16]

Professor and Chairman, Department of Internal Medicine, Wayne State University School of Medicine; Physician-in-Chief, Detroit Medical Center, Detroit, Michigan

ROBERT A. O'ROURKE, M.D. [12,13,14,15]

Charles Conrad Brown Distinguished Professor in Cardiovascular Disease, University of Texas Health Science Center, San Antonio, Texas

STEPHEN O. PASTAN, M.D. [30]

Assistant Professor, Emory University School of Medicine; Medical Director, Gambro Peachtree Dialysis Center, Atlanta, Georgia

WILLIAM H. PLAUTH, JR., M.D. [22]

Cardiologist, Emory-Egleston Children's Heart Center; Professor of Pediatrics, Emory University School of Medicine, Atlanta, Georgia

TIMOTHY J. REGAN, M.D. [28]

Professor of Medicine, Division of Cardiology, University of Medicine and Dentistry of New Jersey, New Jersey Medical School, Newark, New Jersey

WILLIAM C. ROBERTS, M.D. [26]

Executive Director, Baylor Cardiovascular Institute; Dean, A. Webb Roberts Center for Continuing Education, Baylor University Medical Center, Dallas, Texas

STEPHEN F. SCHAAL, M.D. [4]

Professor of Internal Medicine, Ohio State University College of Medicine, Columbus, Ohio

JAMES SCHEUER, M.D. [35]

Baumritter Professor and Chairman, Department of Medicine, Albert Einstein College of Medicine, Bronx, New York

ROBERT C. SCHLANT, M.D. [8,26,40,44]

Professor of Medicine (Cardiology), Emory University School of Medicine, Atlanta, Georgia

STEVEN P. SCHULMAN, M.D. [36]

Associate Professor of Medicine, Johns Hopkins University School of Medicine; Director, Coronary Care Unit, Johns Hopkins Hospital, Baltimore, Maryland

ADAM SCHUSSHEIM, M.D. [46]

Fellow in Cardiology, Mount Sinai Medical Center, New York, New York

RALPH SHABETAI, M.D. [19,20]

Professor of Medicine, University of California San Diego School of Medicine; Chief, Cardiology Section, San Diego Veterans Affairs Medical Center, San Diego, California

ANDREW L. SMITH, M.D. [31]

Assistant Professor of Medicine (Cardiology), Department of Medicine; Medical Director, Heart Failure and Cardiac Transplantation, Emory University School of Medicine, Atlanta, Georgia

ROBERT B. SMITH III, M.D. [39]

John E. Skandalakis Professor of Surgery and Head, General Vascular Surgery, Emory University School of Medicine; Medical Director, Associate Chief of Surgery, Emory University Hospital, Atlanta, Georgia

EDMUND H. SONNENBLICK, M.D. [1,25,43]

Edmond J. Safra Distinguished University Professor of Medicine; Chief Emeritus, Division of Cardiology, The Albert Einstein College of Medicine, Bronx, New York

PANAGIOTIS N. SYMBAS, M.D. [33]

Professor of Cardiothoracic Surgery, Emory University School of Medicine; Director, Cardiothoracic Surgery, Grady Memorial Hospital, Atlanta, Georgia

PIERRE THÉROUX, M.D. [9]

Professor of Medicine, University of Montreal; Director, Coronary Care Unite, Montreal Heart Institute, Montreal, Quebec, Canada

RAYMOND L. WOOSLEY, M.D., Ph.D. [45]

Professor of Pharmacology and Medicine; Chairman, Department of Pharmacology; Interim Director, Institute for Cardiovascular Sciences, Georgetown University Medical Center, Washington, D.C.

DOUGLAS P. ZIPES, M.D. [5]

Distinguished Professor of Medicine, Pharmacology, and Toxicology; Director, Division of Cardiology and the Krannert Institute of Cardiology, Indiana University School of Medicine, Indianapolis, Indiana

JOEL ZONSZEIN, M.D. [25]

Associate Professor of Medicine, Albert Einstein College of Medicine; Director, Clinical Diabetes Center, The Jack D. Weiler Hospital of the Albert Einstein College of Medicine, Division of Montefiore Medical Center, Bronx, New York

PREFACE

This *Companion Handbook* to the 9th edition of *Hurst's The Heart* is the fourth such handbook. Like its predecessors, it was prepared to meet the needs of physicians and students for a concise, portable handbook that they could carry with them and use in the middle of the night or in circumstances when they did not have access to a larger reference textbook of cardiology. It is not intended to be used in place of more specialized textbooks of cardiology or medicine. While it is designed to be used in conjunction with the 9th edition of *Hurst's The Heart*, it can also be used by itself. Most of the chapters were prepared by one or more of the authors of the corresponding chapters in the big book. We express our very deep appreciation to all of our authors. We also would invite our readers to send us their suggestions of ways to improve the *Handbook* to make it more useful.

CHAPTER

1

THE RECOGNITION AND TREATMENT OF HEART FAILURE

Laura A. Demopoulos
Edmund H. Sonnenblick

INTRODUCTION

Congestive heart failure has become one of the most common syndromes afflicting our population, now accounting for more than 1 million hospitalizations annually in the United States. Those suffering from heart failure constitute one of the most rapidly expanding patient groups in this country, and with the increasing age of the population, the absolute number of patients is increasing progressively. There are now 4 million people with heart failure in the United States, and there will be 6 million by the year 2000.

Despite recent major advances in both diagnostic techniques to identify coronary and valve disease and therapeutic interventions to control hypertension, repair valvular defects, and alleviate ischemia, the incidence of heart failure has continued to increase, and the prognosis of patients with advanced congestive heart failure remains poor. The mortality of those with late-stage disease who are severely symptomatic approaches 50 percent at 1 year; even for those with mild symptoms, this 50 percent mortality is approached in 4 to 5 years. In the past decade, clinical investigation has shown that early identification and treatment of patients with left ventricular (LV) dysfunction may have a profound impact on their ultimate clinical course and prognosis. It is therefore vital that physicians be experienced in the recognition and management of this growing patient population.

DEFINITION

Congestive heart failure (CHF) describes a clinical syndrome, or a group of symptoms, generally the eventual result of abnor-

1

malities in the function of the heart. The term does not imply the presence of any specific underlying disease, since the possible causes of CHF are numerous. In general, the symptoms result from an imbalance between the output of blood from the heart and the body's demand for it and from associated salt and water retention, resulting in central and/or peripheral edema. This mismatch may be secondary to a poor cardiac output with normal or even low demand or to a normal-to-high cardiac output with abnormally elevated systemic demands.

Symptoms of CHF may be totally absent for prolonged periods despite markedly reduced ventricular function. Conversely, patients with little or no cardiac depression may be highly symptomatic and limited. This illustrates the fact that measures of systolic ventricular performance such as the ejection fraction do not correlate with the severity of a patient's symptoms. However, a reduced ejection fraction is a good predictor of mortality. In contrast, while symptoms do not predict mortality, they do reflect alterations in the peripheral circulation and skeletal musculature.

The symptoms that characterize CHF are generally of two types: (1) low flow, presenting as hypotension, mental confusion, intestinal ischemia, renal failure, and/or other symptoms of poor tissue perfusion, and (2) volume overload, including the congestive symptoms of pulmonary edema, paroxysmal nocturnal dyspnea, orthopnea, hepatic congestion, ascites, and peripheral edema. Patients whose disease is stable or compensated may experience these symptoms in mild forms, with exercise intolerance as an expression of low flow or only minimal signs of congestion. Acutely unstable or decompensated patients may present with a predominance of symptoms of one type or with a combination of symptoms of both types.

Possible etiologies of CHF may be grouped according to the supply/demand relationship between the heart and the body. They are characterized by the function of the heart as measured by ejection fraction (the ratio of stroke volume to end-diastolic volume).

Low Output/Normal Demand

SYSTOLIC (CONTRACTILE) DYSFUNCTION
Ischemic (from large or small vessel obstruction), late hypertensive, diabetic, toxic (including alcohol), infectious, valvular, idiopathic. These conditions are characterized by an increased diastolic left ventricular volume and a reduced ejection fraction.

DIASTOLIC (FILLING) DYSFUNCTION

Early hypertensive, aortic stenosis, familial hypertrophic, hypertrophic obstructive cardiomyopathy (idiopathic hypertrophic subaortic stenosis). These conditions are characterized by a normal or reduced diastolic ventricular volume and a normal or increased ejection fraction.

Most patients with poor systolic myocardial function also have associated reduced myocardial compliance, so that systolic and diastolic dysfunction often coexist.

High Output/Increased Demand

Sepsis, anemia, thyrotoxicosis, arteriovenous fistula, Paget's disease. Notably, although patients with these illnesses may present with signs and symptoms of heart failure, they generally have no structural cardiac abnormality. Therapy must therefore be directed primarily at the underlying disease rather than at the heart. Considerations of high-output heart failure is therefore excluded from the remainder of this discussion.

PATHOPHYSIOLOGY

The syndrome of heart failure can be divided conceptually into three phases: ventricular injury, ventricular remodeling, and peripheral remodeling. Ventricular injury and myocyte loss may occur either over time or abruptly. Diseases such as hypertension and valvular heart disease lead to gradual myocyte loss as a result of ventricular work overload, while a myocardial infarction or acute myocarditis lead to sudden loss of myocardial cells. When myocardial cells are lost, whether segmentally as in myocardial infarction or diffusely as in cardiomyopathy, they are replaced by fibrosis. The process of apoptosis, or programmed cell death, may contribute to ongoing myocyte loss even after the initial insult is complete. This transfers ventricular work to the remaining myocytes, creating a secondary work overload. The natural adaptation of the heart to a work overload is hypertrophy and then ventricular dilation. This process has been termed *ventricular remodeling*.

The process of ventricular remodeling contributes to hemodynamic compensation for the loss of myocardial cells. Additional circulatory compensation results from the activation of neurohumoral systems. These include the sympathetic nervous system, releasing norepinephrine, and the renin/angiotensin system. Antidiuretic hormone is also released in increased amounts. This neurohumoral activation creates peripheral vasoconstriction, which shunts the limited available blood flow to vital organs and produces salt and water retention, allowing the dilated heart to function on a more optimal portion of its Starling curve. In addition, the activity of the counterregulatory hormone atrial natriuretic peptide, which normally produces diuresis and vasodilation, is suppressed, probably via altered mechanics at the receptor level.

As a result of ventricular remodeling and neurohumoral activation, many patients subsequently experience an asymptomatic phase that is highly variable in duration but may last years. This comprises the period of asymptomatic left ventricular (LV) dysfunction. Most patients remain stable at some reduced level of function for prolonged periods, with repeated episodes of decompensation characterized by pulmonary congestion, after which they may stabilize at a new, lower level. Often, the reason for a progressive decline in clinical status cannot be determined, as decompensation frequently occurs in the absence of further measurable decreases in LV function.

Although vasoconstriction and fluid retention may provide needed short-term hemodynamic compensation in the setting of reduced cardiac output, long-term activation of these systems is ultimately detrimental. The increase in systemic vascular resistance serves as a further burden to the already impaired heart, and fluid retention produces symptoms of pulmonary and peripheral congestion. The inevitably maladaptive nature of neurohumoral activation leads to an increase in cardiac workload, with further ventricular dilation and perhaps damage to existing myocytes. This leads to greater compromise in tissue perfusion and may cause still further activation of neurohumoral systems and a downward spiraling clinical course.

Symptomatic heart failure ultimately develops when the hemodynamic compensation resulting from ventricular remodeling and neurohumoral activation fails. Changes in the structure and function of the peripheral vasculature and skeletal muscles, termed *peripheral remodeling*, contribute to the development of symptoms. The maximal dilatory capacity of the peripheral vasculature becomes blunted, and the skeletal muscles lose the ability to carry out oxidative metabolism effectively. These peripheral changes lead to premature muscle fatigue during exertion and the development of exercise intolerance, a classic symptom of heart failure. Treatment of CHF is directed at treating the initial etiologic factor when possible while inhibiting the effects of detrimental compensatory systems and preventing maladaptive changes in the periphery.

DIAGNOSIS

History

A careful history is an important part of the evaluation of a patient with CHF. This includes the patient's risk factors for or history of ischemic heart disease, hypertension, or diabetes and their control; known murmurs or other evidence of valvular heart disease; family history of cardiomyopathy; recent pregnancy; viral illness; risk factors for AIDS; alcohol consumption; thyroid disease; or other systemic illnesses.

The pattern of the patient's symptoms is also important. The history should include the timing of the onset of any symptoms relative to the probable time of cardiac damage, the frequency of deteriorations, likely precipitants, any past interventional therapy, and past and current functional status. The latter should be described according to the New York Heart Association (NYHA) functional classification system, with class I describing asymptomatic patients, class II describing those with symptoms only during strenuous activity, class III describing those with symptoms during activities of daily living, and class IV those with symptoms at rest. If the patient has coronary artery disease, anginal symptoms and their relation to symptoms of CHF are also important. Patients with valvular heart disease should be questioned carefully for evidence of subacute endocarditis as a cause of clinical deterioration. Associated symptoms of thromboembolic and arrhythmic events should also be assessed.

The causes of decompensations in a patient with chronic CHF should be evaluated, with attention given to dietary or medical noncompliance or to the increased stress of an ischemic episode, infection, or anemia. A period of tachycardia may precipitate ischemia and worsened CHF in a patient with coronary artery disease and may reduce ventricular filling time, thus compromising output in a patient with diastolic dysfunction.

Physical Examination

Findings on examination vary widely. Patients may appear acutely or chronically ill. Blood pressure may be elevated in patients with acute decompensations or poorly controlled hypertension, but those with advanced disease typically have low blood pressure with a narrow pulse pressure. Patients generally have tachycardia in the absence of coexisting conduction system abnormalities. The skin is usually cool and dry, reflecting decreased cardiac output and vasoconstriction. Neck vein distention may be prominent in those with right-sided heart disease, and jugular venous pulsations may reflect tricuspid valve regurgitation. Exaggerated venous pulsations may also reflect left ventricular hypertrophy and failure due to transmission across the intraventricular septum, even when right-sided heart failure is absent. Hepatojugular reflux is also noted in those with involvement of the right side of the heart. Carotid and peripheral pulsations are generally of low amplitude in patients with advanced disease, and bruits may reflect diffuse atherosclerosis. Lung examination may reveal pulmonary congestion with rales and wheezes as well as pleural effusions. The cardiac examination often demonstrates a displaced apex impulse, and gallops may be palpable. The heart sounds are often distant, and third or fourth sounds may be present, depending on the degree of disease and its cause. An S3 is often audible in those with poor systolic function, and an S4 may be heard in those with associated poor ventricular compliance. Murmurs may provide a key to the diagnosis of the underlying heart disease or may be secondary to ventricular dilatation affecting the AV valve annuli, producing functional tricuspid and mitral regurgitation. The abdominal examination may reveal ascites and hepatomegaly in those with right-sided heart involvement, and the liver may be pulsatile in patients with severe tricuspid regurgitation. The lower extremities or other dependent areas may develop pitting edema, and the overlying skin may show chronic pigmentation from long-standing hypoperfusion.

Laboratory Evaluation

Blood tests in patients with CHF are generally not specific, although cardiac enzymes may be elevated acutely in those with ischemia or myocarditis. In addition, serologic evaluations in those with suspected underlying infectious or inflammatory disorders may be useful.

The chest x-ray may show cardiomegaly, vascular redistribution, and the typical pattern of central pulmonary congestion in addition to the pleural effusions and Kerley B lines. The electrocardiogram (ECG) may reveal evidence of infarction or acute ischemia or the nonspecific findings of ST-segment and T-wave changes. Criteria for left ventricular hypertrophy or left bundle branch block are common.

The echocardiogram is very useful for determining overall cardiac function, although indices of systolic performance are far more sensitive and specific than indices of diastolic performance. The echocardiogram permits serial evaluation of chamber sizes, function, and hypertrophy; diagnosis and quantitation of valvular heart disease; and the identification of possible valvular vegetations, thrombi, and pericardial effusions. Nuclear imaging may be useful in identifying viable myocardium in regions of focal reduced contraction in patients with coronary artery disease; some regions may regain contractile function with a revascularization procedure. This has been termed *hibernating myocardium.* As many as 50 percent of defects that appear to be nonviable scars on standard thallium scans may prove to be viable myocardium with reinjection techniques or magnetic resonance imaging. In addition, gallium scanning may help in the identification of patients with active myocarditis. Cardiac catheterization may be used to measure filling pressures, detect shunts, demonstrate constrictive or restrictive physiology, evaluate left ventricular systolic function, obtain biopsy specimens, measure valve gradients or quantitate regurgitation, and diagnose coronary artery disease. It also may allow therapeutic intervention with coronary angioplasty or valvuloplasty. Peak oxygen consumption, or peak V_{O_2}, is a useful objective index of a patient's functional capacity. Peak V_{O_2} can also be used to help determine prognosis and the appropriate time for referral for heart transplantation. Programmed electrical stimulation (PES) can be used in selected patients to determine the risk of ventricular arrhythmias and can help guide therapy with either antiarrhythmic drugs or implantable defibrillators.

TREATMENT

CHF may present clinically as an acute syndrome with volume overload and/or hemodynamic compromise, or it may develop insidiously with chronic congestion and exercise intolerance. Treatment strategies are reviewed according to presentation and cause (systolic versus diastolic dysfunction).

Acute Congestive Heart Failure

SYSTOLIC DYSFUNCTION
Acute CHF with pulmonary edema is a medical emergency. It is characterized by severe pulmonary congestion that may be

accompanied by evidence of poor tissue perfusion; the risk of
mortality is high without aggressive therapeutic and diagnostic
measures. The most immediate concerns of the clinician
should be to improve respiratory status and restore hemodynamic balance. Emergency treatment should be instituted with
IV furosemide (Lasix) (40 to 80 mg initially; this dose may be
doubled hourly if the response is not adequate), intravenous
morphine sulfate (2 to 10 mg, in increments according to blood
pressure), and 1 to 2 sublingual nitroglycerin tablets followed
by an intravenous nitroglycerin drip (20 to 100 μg/min, dosed
according to blood pressure and clinical status). Intravenous
inotropic agents such as dobutamine (Dobutrex) (2 to 10 μg/kg/
min) may be helpful in improving cardiac output, and arterial
dilators such as sodium nitroprusside (Nipride) (0.05 to 5 μg/
kg/min) may decrease afterload resistance and improve forward flow in vasoconstricted patients. Supplemental oxygen
should be provided by a 100% O_2 nonrebreather face mask (in
the absence of a history of obstructive lung disease, in which
case the lowest O_2 concentration providing an oxygen saturation of 85 to 90% should be used). Mechanical ventilation may
be required if respiratory status does not improve rapidly.
Urine output should be monitored with a Foley catheter in
place. If the patient fails to improve or if the hemodynamic picture makes clinical judgment of intravascular volume status
difficult, a pulmonary artery catheter should be placed for
diagnostic purposes and for evaluation of therapeutic interventions. An immediate evaluation of the cause of pulmonary
edema should be performed concurrently with treatment,
including a history of precipitating factors (medical or dietary
noncompliance, chest pain, bleeding, infection), a complete
physical examination, an ECG for ischemia and arrhythmias,
cardiac enzymes, an echocardiogram for acute valve disruption
or other mechanical disorders, in addition to standard blood
work and chest x-ray. Therapy for the specific cause of pulmonary edema may then be instituted and may include
additional medications, cardioversion for restoration of sinus
rhythm, or surgical intervention.

DIASTOLIC DYSFUNCTION
Patients with acutely decompensated heart failure on a diastolic basis are rarely identified prior to the initiation of
emergent therapy; the acute clinical presentation is indistinguishable from that of a patient with systolic dysfunction.
However, if the decompensation is known to be diastolic, acute
treatment should focus on improving respiratory status with

cautious diuresis as outlined above, prolonging diastolic filling time by slowing tachycardias, and improving ventricular compliance by relief of ischemia. Overly aggressive use of diuretics and venodilating agents can decrease ventricular filling and lead to hypotension in these patients; accordingly, these drugs should be used judiciously. Inotropic support generally is not required in patients with only diastolic dysfunction. The remainder of the treatment and evaluation should be pursued as discussed above. The maintenance of sinus rhythm is of particular importance, since atrial kick contributes significantly to cardiac output in patients with impaired diastolic filling.

Chronic Congestive Heart Failure

SYSTOLIC DYSFUNCTION

The initial approach to the management of the patient with chronic CHF should include an attempt to treat any possible reversible causes of the syndrome. Hypertension should be controlled to reduce excess loading. In patients with ischemic cardiomyopathies, poorly contracting but viable myocardium supplied by diseased vessels may have improved systolic function after percutaneous transluminal coronary angioplasty (PTCA) or bypass surgery. Thus, even though patients with poor LV function are generally considered to be at high risk for revascularization procedures, important benefits often outweigh the risks. Valve repair or replacement in patients with CHF due to primary valve lesions may also reverse progressive ventricular dysfunction. Of note, recent data indicate that immunosuppressive therapy to treat viral or other inflammatory causes of myocarditis is not of benefit.

Once these options have been exhausted, long-term therapeutic measures should be instituted to stabilize the process. Patient education is paramount. All patients with CHF must receive detailed instruction regarding dietary restrictions (2 g/day sodium diet, low fat/cholesterol as indicated), regular exercise, and the proper use of and possible side effects from medications. Both patients and their family members should be informed of warning signs and symptoms that would require hospitalization; family members may consider training in methods of basic life support.

The medical therapy for chronic CHF has advanced dramatically in the past decade. Most effective drugs are now directed at breaking the cycle of peripheral maladaptation and progressive neurohumoral activation. Other drugs still have useful roles in alleviating symptoms, although they may not ultimately prolong survival. The following discussion will focus on the most important agents.

Diuretics Ankle edema and/or pulmonary congestion are common early problems in heart failure. Diuretic agents are useful in relieving the congestive symptoms that result from salt and water retention and in lowering elevated intracardiac pressures. For patients with mild to moderate symptoms, initial therapy with a thiazide diuretic (hydrochlorothiazide 25 to 50 mg daily) is likely to provide symptomatic benefit. For patients with more severe symptoms, therapy is generally undertaken with loop diuretics [e.g., furosemide (Lasix), 40 to 300 mg daily, with dose determined and divided according to clinical status and renal function]. If the resultant diuresis is inadequate, efficacy can be increased with the concomitant use of a thiazide diuretic or metolazone (Zaroxolyn, Diulo, Metenix, 5 to 20 mg daily). Inhibition of aldosterone with spironolactone (Aldactone, 25 to 200 mg in one or two doses daily) may also enhance effectiveness over a longer period. The IV forms should be used in decompensated patients who may not absorb oral drugs well (see also Chap. 42).

Diuretics are part of the standard treatment of patients with CHF, but recognition of their effectiveness must be combined with an awareness of their possible side effects. Overdiuresis may lead to compromised cardiac output, resulting in symptoms of low flow and increased impairment of renal function, as demonstrated by an increase in serum creatinine. In addition, chronic diuretic therapy may create electrolyte imbalances such as hypokalemia, which can provoke or increase lethal arrhythmias, especially in those prone to arrhythmic death or on digitalis glycosides. Close monitoring of potassium and magnesium should accompany diuretic therapy. Once excess fluid retention is controlled, the diuretic dose may be reduced or the drug used intermittently to avoid hypovolemia, hypokalemia, and hypomagnesemia.

Vasodilators The aim of vasodilator therapy is to reduce the load against which the ventricle must contract, increasing ventricular performance for any degree of myocardial contractility. Load is reduced by a decrease in systolic blood pressure as well as a reduction in ventricular size, which lowers wall tension for any developed pressure. Moreover, vasodilators are used to reduce filling pressures, ameliorate pulmonary congestion, and to reverse the inappropriate vasoconstriction that ultimately contributes to the decline of patients with CHF. Nitrate preparations (e.g., isosorbide dinitrate 10 to 40 mg three times daily) have long been part of standard therapy for patients with CHF symptoms. Nitrates predominantly dilate the venous beds, relieving pulmonary congestion, and the intravenous forms are useful in acute pulmonary congestion. The direct arterial dilators prazosin (Minipress, 1 to 5 mg, two or three times daily) and hydralazine (Apresoline, 10 to 50 mg four times daily) may improve low-output symptoms by unloading the LV. The combination of hydralazine and isosorbide dinitrate has been shown to reduce mortality. However, tolerance to the effects of these medications is a problem for long-term therapy. Angiotensin converting enzyme (ACE) inhibitors have proven symptomatic and mortality benefits that exceed those of the direct vasodilators in patients with both mild and severe CHF. They have also been shown to prevent the progression from asymptomatic LV dysfunction to overt heart failure and to prevent numerous other cardiovascular complications in patients with reduced ejection fractions and recent infarctions. Therefore, patients across the spectrum—from asymptomatic LV dysfunction to advanced or decompensated CHF—should all receive therapy with an ACE inhibitor [such as captopril (Capoten) 6.25 to 25 mg three times daily, or enalapril (Vasotec) 2.5–40 mg daily, in one or two divided doses] unless a specific contraindication exists (see Chap. 44).

A new class of drugs, direct angiotensin receptor blockers, has been shown to have similar hemodynamic effects as ACE inhibitors but with fewer side effects. This class of drugs may allow for more complete blockade of the effects of angiotensin II than ACE inhibitors, as non-ACE–dependent pathways that form angiotensin II are not blocked by ACE inhibitor therapy, but angiotensin II formed by any pathway is blocked by direct angiotensin receptor blockers. The benefit of these drugs in the treatment of patients with heart failure is currently under evaluation.

Calcium channel blockers are generally not recommended in the treatment of patients with heart failure; recently, however, amlodipine was shown to be a safe antihypertensive agent in the management of these patients.

Inotropic Agents   The use of these drugs is predicated on the belief that myocardial contractility is reduced and can be increased. Digitalis glycosides, which augment intracellular calcium by inhibiting surface membrane Na^+-K^+-ATPase, increase cardiac output and lower arterial peripheral resistance when ventricular performance is depressed. At this time, these are the only long-term oral agents available, and their beneficial effects add to those of vasodilating agents such as the ACE inhibitors. Digoxin (Lanoxin, 0.125 to 0.500 mg once daily, adjusted for renal function) has proven symptomatic benefit in CHF and may be especially useful in patients with associated supraventricular tachycardias. Withdrawal of digoxin from patients who have been maintained on it chronically may precipitate episodes of decompensation. Digoxin toxicity is potentially lethal and is not always reflected by blood levels. Patients taking digoxin should be monitored closely, especially since toxicity is exacerbated by hypokalemia and most of these patients receive coincident diuretic therapy. Effective therapy for digoxin toxicity exists in the form of an antibody fragment. Other oral inotropic drugs have been associated with excess mortality compared with placebo, and newer experimental agents are under study at this time (see Chap. 43).

For patients with end-stage disease refractory to other forms of therapy, treatment with intravenous inotropic agents may be palliative. These drugs include the beta-agonists dobutamine (Dobutrex, 2 to 10 μg/kg/min) and dopamine (Intropin, 2 to 5 μg/kg/min), and the phosphodiesterase III (PDE III) inhibitor milrinone (Primacor, 50 μg/kg/bolus over 10 min, followed by 0.375 to 0.5 μg/kg/min infusion; adjust for creatinine clearance). All three ultimately work by increasing intracardiac cyclic AMP: the beta-agonists by increasing production, and the phosphodiesterase inhibitor by decreasing degradation. The impact of these drugs on myocardial oxygen consumption depends on the balance between their effect on increasing contractility and lowering filling pressures. Beta-agonists increase heart rate and contractility out of proportion to their effect on filling pressures and thus increase myocardial oxygen demand. PDE III inhibitors produce arterial and venous dilation with only a modest inotropic action, so oxygen consumption is not greatly increased. Milrinone may be used in conjunction with dobutamine and may permit the use of a lower dose of dobutamine. Milrinone's vasodilator effect limits its use in hypotensive patients, and it is contraindicated in patients with significant renal insufficiency (see Chap. 43).

Beta-Blocking Drugs The use of beta-adrenergic blocking agents in the treatment of patients with heart failure is gaining wider acceptance. Once felt to be absolutely contraindicated, beta blockade is now thought to have benefits deriving from the interruption of detrimental chronic endogenous adrenergic stimulation. Many studies have demonstrated a beneficial effect of beta blockade on left ventricular remodeling, as assessed by left ventricular ejection fraction, with an average increase of 5 to 10 percent. Patients treated with beta blockers have fewer hospitalizations as well as improved resting and exercise hemodynamics. The effects of beta blocking agents on maximal exercise capacity, NYHA functional class, and quality of life have been mixed, with more consistent benefits demonstrated with beta$_1$-selective blocking agents rather than nonselective agents. Carvedilol, a nonselective agent with vasodilating and antioxidant properties, has been approved for use by patients with both ischemic and nonischemic cardiomyopathy; however, the effect of this agent on mortality is not conclusive. Importantly, the CIBIII trial using bisoprolol was recently stopped, as the primary endpoint, a mortality reduction of at least 25 percent in treated patients, was reached. In general, these agents are well tolerated provided that the starting doses are extremely low, one-tenth to one-twentieth of standard doses, and the titration is gradual, with increases at most every 2 weeks. The optimal target doses for many beta-blocking agents have not been well established; in general, however, many of the benefits appear to have a dose-response, so it seems reasonable to titrate the patient to the highest tolerated or recommended dose.

Antiarrhythmic Drugs   Arrhythmias are common in heart failure. Atrial dilation may induce atrial flutter or fibrillation, and the resultant tachycardia may provoke ischemia and limit ventricular filling. Ventricular extrasystoles and nonsustained or sustained ventricular tachycardia may degenerate into ventricular fibrillation. Sudden arrhythmic death occurs in 25 percent or more of patients with heart failure. Vigorous treatment of the heart failure is the initial therapeutic approach. Antiarrhythmic drugs should be used with great caution in patients with poor ventricular function. Proarrhythmic side effects may be more pronounced in these patients, and efficacy is limited. The use of antiarrhythmic drugs in the primary prevention of death in patients with heart failure has been disappointing. Class I agents have not been demonstrated to have benefit and may increase the likelihood of arrhythmias. Results from trials with the class III agent amiodarone have been mixed, with some data suggesting no benefit, some suggesting benefit in patients with nonischemic cardiomyopathy, and some suggesting a reduction in sudden death but no improvement on total mortality. The use of D-sotalol has been shown to increase mortality in these patients. Use of these drugs should be restricted to life-threatening arrhythmias under appropriate guidance (see also Chaps. 2 and 45).

Implanted Cardiac Defibrillators   The use of implanted cardiac defibrillators (ICDs) has, in contrast, met with greater success. While conclusive data on the use of these devices in the primary prevention of sudden death are not yet available, they do appear to have benefit in patients who have already experienced cardiac arrest or syncope found to result from ventricular arrhythmias. In general, patients with the poorest functional capacity will have the greatest reduction in sudden death with the use of an ICD, but the death rate from progressive heart failure will remain substantial. PES is likely to help identify asymptomatic patients who will benefit from an ICD as primary prevention for sudden death. Thus, the challenge for the future will be to identify a patient population that will have meaningful benefit from the use of ICDs in primary prevention, thus balancing the risks, benefits, and costs of these devices.

Exercise Training   A growing body of data supports the use of exercise training programs in the treatment of patients with heart failure. Standard exercise training using workloads of 70 to 80 percent of peak aerobic capacity for 12 weeks improves exercise capacity in patients with heart failure. Recently, several trials have established that regular exercise at lower levels—i.e., <50 percent of peak aerobic capacity—confers similar benefits with lower intracardiac pressures than higher-level exercise. Large-scale trials are needed to evaluate the effect of long-term training on morbidity and mortality in these patients.

Cardiac Transplantation   This treatment option is reserved for patients with advanced disease whose prognosis is limited despite available medical and surgical therapy. Candidates are restricted by age and by ability to comply with rigorous medical regimens and follow-up. Other factors are coexisting diseases, renal status, pulmonary vascular resistance, and continued smoking, drinking, or substance abuse. The availability of donor hearts is the limiting factor for most transplant candidates. Survival following transplant is approximately 90 percent at 1 year and 85 percent at 5 years. Long-term outcome is complicated and is limited by immunologic rejection, renal hypertension, and increased disease of the small and large arteries. The ongoing development of implantable mechanical left ventricular assist devices may reduce the current demand for donor hearts and provide treatment for those who are not candidates for transplantation.

DIASTOLIC DYSFUNCTION

Patients with heart failure due to diastolic dysfunction should be evaluated for such reversible causes as valvular and coronary artery disease. Control of hypertension may lead to regression of LV hypertrophy and improved compliance. As in the treatment of systolic dysfunction, control of sodium intake and compliance with medical therapy should be emphasized. The use of diuretics to control pulmonary congestion and peripheral edema should be monitored carefully; overdiuresis may further limit cardiac filling. Vasodilating and inotropic drugs generally do not play a role in the management of these patients, as ventricular volumes are normal to low and systolic function is preserved. The pharmacologic agents shown to be of use in diastolic dysfunction include both beta blockers and calcium-channel blockers (e.g., propranolol 160 to 240 mg daily in divided doses or verapamil 120 to 360 mg daily in divided doses). These agents prolong ventricular filling time, improve compliance, and decrease myocardial oxygen demand (see also Chaps. 47 and 48).

Some patients with hypertrophic obstructive cardiomyopathy (idiopathic hypertrophic subaortic stenosis) improve with surgical excision of a portion of the obstructing septum (myomectomy). Dual-chamber pacing to alter the pattern of ventricular contraction and relieve outflow tract obstruction is currently under investigation (see Chap. 18).

SUGGESTED READINGS

Barry F. Uretsky, Richard G. Sheahan: Primary prevention of sudden cardiac death in heart failure: will the solution be shocking? *J Am Coll Cardiol* 1997; 30: 1589–1597.

Brater DC: Diuretic therapy. *N Engl J Med* 1998; 339: 387–395.

Hjalmarson A, Kneider M, Waagstein F: The role of β-blockers in left ventricular dysfunction and heart failure. *Drugs* 1997; 54:501–510.

LeJemtel TH, Sonnenblick EH (eds): Heart failure—adaptive and maladaptive processes. *Circulation* 1993; 87(7): 1–121.

LeJemtel TH, Sonnenblick EH, Frishman WH: Diagnosis and management of heart failure. In: Alexander RW, Schlant RC, Fuster V, O'Rourke RA, Roberts R, Sonnenblick EH (eds): *Hurst's The Heart*, 9th ed. New York, McGraw-Hill, 1998:745–781.

Packer M, Gheorghiade M, Young JB, et al: Withdrawal of digoxin from patients with chronic heart failure treated with angiotensin-converting enzyme inhibitors. *N Engl J Med* 1993; 329:1–7.

Remme WJ: Prevention of worsening heart failure: future focus. STICARE, Cardiovascular Research Foundation, Rotterdam, The Netherlands. *Eur Heart J* 1998; 19(suppl B):B47–B53.

The SOLVD Investigators: Effect of enalapril on survival in patients with reduced left ventricular ejection fractions and congestive heart failure. *N Engl J Med* 1991; 325: 293–302.

Schlant RC, Sonnenblick EH, Katz AM: Pathophysiology of heart failure. In: Alexander RW, Schlant RC, Fuster V, O'Rourke RA, Roberts R, Sonnenblick EH (eds): *Hurst's The Heart*, 9th ed. New York, McGraw-Hill, 1998: 687–726.

Zelis R, Sinoway LI, Musch TI, et al: Regional blood flow in congestive heart failure: Concept of compensatory mechanisms with short and long time constants. *Am J Cardiol* 1988; 62:2E–8E.

Szabó BM, van Veldhuisen DJ, de Graeff PA, Lie KI: Alterations in the prognosis of chronic heart failure; an overview of the major mortality trials. *Cardiovasc Drugs Ther* 1997; 11:427–434.

CHAPTER

2

ARRHYTHMIAS AND CONDUCTION DISTURBANCES

Simon Chakko
Robert J. Myerburg

Effective management of cardiac arrhythmias and conduction disturbances requires accurate identification of the specific rhythm disturbances, analysis of the clinical settings in which they occur, and identification of a safe and effective endpoint of therapy.[1] Clinical settings may be divided into those that cause acute and transient electrophysiologic abnormalities and those that are chronic and provide the substrate for persistent or recurrent arrhythmias. The former include acute ischemia, acute phase of myocardial infarction, electrolyte abnormalities, and proarrhythmic drugs; the latter include chronic ischemic heart disease, cardiomyopathies, and the anatomic and physiologic substrate for the various supraventricular tachyarrhythmias. Arrhythmias may be aggravated by hemodynamic, electrolyte, metabolic, and respiratory abnormalities. Correction of such abnormalities is essential to the treatment of arrhythmia. The decision to treat an arrhythmia depends upon the presence and type of symptoms and the potential for morbidity and mortality. Some arrhythmias may produce bothersome symptoms but may not affect long-term prognosis, whereas other arrhythmias that cause few or no symptoms may predict a poor outcome.

PRINCIPLES OF RHYTHM ANALYSIS

The standard 12-lead electrocardiogram (ECG) and rhythm strips are the most easily accessible tools for the diagnosis of a cardiac rhythm disturbance. Recognition of the P-wave and QRS morphology and their relative timing may be the only information needed to diagnose the arrhythmia correctly. When the standard ECG does not provide enough information, special lead systems, such as a bipolar esophageal lead (to record left atrial activity) or an intraatrial electrode catheter (to record right atrial activity) may be used simultaneously with the standard ECG to help identify P waves and to provide further information.

Continuous monitoring of cardiac rhythm at the bedside or with ambulatory recording devices with simultaneous two-lead recordings, usually lead II and MCL-1, may improve diagnostic yield. For infrequently occurring arrhythmias, patient- or observer-activated event recorders may be used. These allow device activation when an event occurs and subsequent retrieval of the arrhythmia recording for physician review at a later time.

Exercise stress testing is useful for evaluating exercise-induced arrhythmias, particularly premature ventricular contractions (PVCs) and ventricular tachycardia; to distinguish between autonomic and structural disease-related mechanisms; for sinus node or AV node dysfunction; and to evaluate for rate-dependent proarrhythmic effects of antiarrhythmic drugs. Stress testing also may be used to estimate roughly the refractory period of the accessory pathway in patients with Wolff-Parkinson-White syndrome. Signal-averaged electrocardiography employs amplification of low-amplitude signals occurring after termination of QRS complex using high-amplification techniques. These signals represent delayed activation of parts of ventricular muscle mass. Absence of late potential after healing of myocardial infarction identifies a greater than 97 percent probability of remaining free of ventricular arrhythmias. Positive predictive value is poor.

Heart rate variability estimates the sympathetic and parasympathetic balance by measuring the variations in sinus rate from Holter recordings. Blunted heart rate variability is found in subgroups of myocardial infarction and cardiac arrest survivors who appear to be at increased risk for recurrent events.

Intracardiac electrophysiologic testing is used to diagnose arrhythmias and conduction disturbances when surface electrocardiography is insufficient. Multielectrode catheters positioned in various intracardiac sites allow mapping of the sequence of activation in the atria, AV junction, and ventricle. They also permit the localization of sites of anomalous pathways and the origin of supraventricular tachyarrhythmias as well as identification of the mechanisms of ventricular tachyarrhythmias. In addition, the level of AV block can be identified during electrophysiologic testing.

Intraoperative mapping of the origin of both supraventricular and ventricular tachyarrhythmias may be performed using hand-held probes or specialized multielectrode arrays to identify areas for surgical ablation. Previously, the management of cardiac arrhythmias was limited to pharmacologic or surgical choices. However, surgical mapping for most supraventricular arrhythmias and some forms of ventricular arrhythmias has largely been replaced by catheter mapping and ablation techniques performed in the electrophysiology laboratory. Implantable devices are commonly used for long-term management of both tachyarrhythmic episodes and bradyarrhythmic events.

SUPRAVENTRICULAR ARRHYTHMIAS

Sinus Rhythm and Sinus Tachycardia

Normal sinus rhythm is defined as a rate of 60 to 100 beats per minute, originating in the sinus node; the rhythm is regular. *Sinus arrhythmia* is present when the variation between the longest and the shortest cycle on a resting tracing is above 0.12 s. This is a normal variant seen most commonly in children; it decreases with age. If the cycle lengths shorten with inspiration and lengthen with expiration, it is defined as *phasic sinus arrhythmia*. If unrelated to the respiratory cycle, it is defined as nonphasic.

Sinus tachycardia results from automatic discharge of the sinoatrial pacemaker cells at a rate that exceeds 100 beats per minute. It is characterized by normal sinus P waves at a rapid rate that usually does not exceed 130 to 140 beats per minute under resting conditions but can be as high as 180 to 200 beats per minute, particularly during exercise. Sinus tachycardia is a normal physiologic response to exercise or emotional stress or may be pharmacologically induced by such drugs as epinephrine, ephedrine, or atropine. Exposure to alcohol, caffeine, or nicotine can also cause sinus tachycardia. Persistence of sinus tachycardia usually signals an underlying disorder (such as heart failure, pulmonary embolism, hypovolemia, or hypermetabolic states). Vagotonic maneuvers, such as carotid sinus massage or Valsalva maneuver, may help differentiate sinus tachycardia from other SVTs. Gradual slowing of the rapid rate, followed by gradual return to that rate, is typical for sinus tachycardia. In contrast, vagal maneuvers may abruptly terminate other SVTs or block conduction across the AV node. Sinus tachycardia usually requires nonspecific treatment; management should be directed toward the underlying disorder. When pharmacologic slowing of sinus tachycardia is desired, beta-adrenergic blocking agents often are effective. However, it must first be determined that the tachycardia is not a necessary compensatory response, as in heart failure.

Premature Atrial Impulses

Atrial extrasystoles or premature atrial contractions (PACs) are impulses that arise in an ectopic atrial focus and are premature in relation to the prevailing sinus rate. The early P wave has a different vector than the sinus P wave, and the PR interval of the conducted PAC is usually normal or minimally prolonged, but occasionally may be markedly prolonged. If the coupling interval of the PAC to the previous sinus P wave is short, aberrant intraventricular conduction may occur, making the diagnosis dependent on recognition of the P wave distorting the previous T wave. The *hallmark* of timing of PACs is the less than fully compensatory pause; however, this may not always occur. The significance of atrial extrasystoles depends on the clinical setting in which they occur. Often they are found in completely normal individuals; however, they may be associated with myocardial ischemia, rheumatic heart disease, myopericarditis, congestive heart failure, and a variety of systemic abnormalities including acid-base and electrolyte disturbances and pulmonary diseases. Caffeine, tobacco, or alcohol use, as well as emotional stress, may initiate or exacerbate premature atrial contractions. Asymptomatic patients with no underlying heart disease require no treatment other than removal of the underlying or precipitating factors. If patients are symptomatic, beta-adrenergic blocking agents may provide relief. Both digitalis and verapamil have been tried, but their efficacy has not been proven.

Premature atrial impulses that initiate sustained arrhythmias, such as atrial fibrillation or atrial flutter, reentrant supraventricular arrhythmias, or, rarely, ventricular arrhythmias may require treatment. Conventional membrane active antiarrhythmic agents, particularly the class I-A antiarrhythmic drugs (procainamide, quinidine, or disopyramide), may be effective in suppressing the triggering PACs, but their use is rarely indicated because of a high risk/benefit ratio. The class

I-B and I-C drugs also may be effective in selected individuals, but there are limited data on their use. None are approved for this indication (see also Chap. 45).

Supraventricular Tachyarrhythmias

All tachyarrhythmias that originate above the bifurcation of the bundle of His or incorporate tissue proximal to it in a reentrant loop are classified as supraventricular arrhythmias. The atrial rate must be 100 or more beats per minute for a diagnosis, but the ventricular rate may be less when AV conduction is incomplete. Supraventricular tachycardias usually have narrow QRS configurations, but they may be wide because of aberrant conduction through the intraventricular conduction tissue, preexisting bundle branch block, or conduction via an accessory pathway. Atrial activity may be identified by using a long rhythm strip with multiple leads. Recording the rhythm strip at rapid paper speed (e.g., 50 mm/s) may be helpful. Other diagnostic aids include vagal maneuvers to slow the ventricular response rate or an esophageal lead to identify atrial activity. Intraatrial electrograms are occasionally required.

Supraventricular tachycardias (SVTs) may be classified as *paroxysmal* (lasting seconds to hours), *persistent* (lasting days to weeks), or *chronic* (lasting weeks to years). Consideration of not only the duration of the tachyarrhythmia but also its electrophysiologic mechanism is essential to the appropriate management of supraventricular arrhythmias.

Paroxysmal supraventricular tachycardia (PSVT) may occur in the presence or absence of heart disease and in patients of all ages. Electrophysiologically, it is most often due to reentry, usually within the AV node or involving an accessory pathway; infrequently, sinus node reentry or intraatrial reentry is the mechanism.

PVST DUE TO AV NODAL REENTRY

AV nodal reentry is the most common cause of PSVT and is characterized electrophysiologically by two functionally distinct pathways (slow and fast) within or near the AV node. In the common form of AV nodal reentrant tachycardia, antegrade conduction occurs over the slow pathway and retrograde conduction over the fast pathway, resulting in almost simultaneous activation of the atria and ventricles. Electrocardiographically, retrograde P waves are buried within the QRS complex or appear immediately after it. In the uncommon form, in which antegrade conduction occurs over the fast pathway, the retrograde P wave occurs well after the end of the QRS complex and is characterized by a long R-P interval and a short PR interval with an inverted P wave in II, III, and AVF. In the absence of ischemia or significant valvular heart disease, PSVT due to AV nodal reentry is a benign rhythm and may be treated acutely with rest, sedation, and vagotonic maneuvers. If these physiologic interventions are unsuccessful, intravenous adenosine, intravenous calcium antagonists, digoxin, or beta-adrenergic blockers may be used (Table 2-1). Adenosine, 6 mg intravenously, followed by one or two 12-mg boluses if necessary, has an extremely short half-life (10 s), causes no hemodynamic complications, and is usually the first choice for treatment of PSVT. Its use is contraindicated in patients with hypotension or high-grade AV block. Intravenous verapamil 5-mg bolus, followed by one or two additional 5-mg boluses

TABLE 2-1

DRUGS FOR ACUTE MANAGEMENT
OF SUPRAVENTRICULAR TACHYCARDIA

Drug	Dosage
Adenosine	IV: 6 mg rapidly; if unsuccessful within 1–2 min, 12 mg rapidly
Diltiazem	IV: 0.25 mg/kg body wt over 2 min; if response inadequate, wait 15 min, then 0.35 mg/kg over 2 min; maintenance 10–15 mg/h
Digoxin[a]	IV: 0.5 mg over 10 min; if response inadequate 0.25 mg q 4 h to a maximum of 1.5 mg in 24 h
Esmolol	IV: 500 µg/kg per min × 1 min followed by 50 µg/kg per min × 4 min, repeat with 50-µg increments to maintenance dose of 200 µg/kg/min
Procainamide	IV: 10–15 mg/kg at 25 mg/min as loading dose, then 1–4 mg/min
Propranolol	IV: 0.1 mg/kg in divided 1-mg doses
Verapamil	IV: 5 mg ovaer 1 min; if unsuccessful, 1–2 5-mg boluses 10 min apart

[a]Contraindicated in patients with Wolff-Parkinson-White syndrome.

10 min apart if the initial bolus does not convert the arrhythmia, has been effective in up to 90 percent of patients. Intravenous diltiazem beginning with a 0.25 mg/kg bolus over 2 min also is effective. If the response is not satisfactory, a repeat bolus of 0.35 mg/kg may be given after 15 min. Intravenous digoxin, 0.5 mg over 10 min, followed by an additional 0.25 mg every 4 h to a maximum dose of 1.5 mg in 24 h may be used. Intravenous beta-adrenergic blockers, propranolol or esmolol, may be effective. Class IA antiarrhythmic agents occasionally may be effective if the other drugs fail. Hemodynamic instability dictates immediate cardioversion; low-energy shocks (10 to 50 W/s) are usually sufficient. Rapid atrial pacing is an alternative if other methods fail or cardioversion should be avoided.

Although the QRS complexes usually are narrow in PSVT due to AV nodal reentry, occasionally aberrant intraventricular conduction with resultant wide QRS complexes (either RBBB or LBBB patterns) may occur. However, unless preexisting bundle branch block or aberrant conduction has *clearly* been documented, a wide-QRS complex tachycardia should be assumed to be ventricular tachycardia. The use of verapamil for the treatment of a wide complex tachycardia, on the *assumption* that it is a supraventricular tachycardia with aberration, may lead to severe hemodynamic compromise and even death if the arrhythmia has been incorrectly diagnosed.[2] Low-energy cardioversion or class I antiarrhythmic agents are logical alternatives when the diagnosis of the arrhythmia is uncertain.

Long-term therapy for control of recurrent PSVT due to AV nodal reentry is most frequently achieved today with either pharmacologic methods or catheter ablation techniques. No chronic therapy may be necessary in patients who have infrequent, short-lived, well-tolerated attacks and/or who respond to physiologic maneuvers. Patients who have more frequent attacks, who are intolerant to medications, and/or whose SVTs cause hemodynamic compromise are offered radiofrequency catheter ablation for curative therapy. Ablation of the common form of AV nodal reentry is achieved by selective ablation of the slow pathway to abolish the reentrant loop. Less commonly, ablation of the fast pathway will be performed, but the risk of iatrogenic heart block is higher. In either case, experience has demonstrated that this is a safe and effective technique for the management of AV nodal reentry.

Pharmacologic therapy is an alternative for patients who do not desire radiofrequency ablation or who have few, well-tolerated occurrences. Beta-adrenergic blocking agents, verapamil, or digoxin in standard doses may be used. In patients with no structural heart disease, class IC agents may be used (Table 2-2).

PSVT DUE TO WOLFF-PARKINSON-WHITE (WPW) SYNDROME

This is the second most common form of reentrant SVT. When conduction during an SVT occurs antegrade through the AV node and retrograde via the accessory pathway, it is referred to as an *orthodromic* reciprocating tachycardia. This is the common form of PSVT in WPW syndrome; the ECG pattern is a narrow QRS tachycardia at rates ranging from 160 to 240 per minute. Vagal maneuvers, adenosine, verapamil, diltiazem, propranolol, and the class IA antiarrhythmic agents may be used acutely to convert the arrhythmia. However, digoxin is contraindicated, since it may shorten the refractory period of the bypass tract and cause extremely rapid conduction across the bypass tract during atrial flutter (>250 beats per minute), leading to hemodynamic collapse or ventricular fibrillation, or it may accelerate the rate of orthodromic tachycardia. If the patient has a history of atrial fibrillation or flutter, verapamil should also be avoided, as it may accelerate the ventricular rate. *Antidromic* SVT, referring to retrograde conduction through the normal pathway and antegrade conduction via the accessory pathway, is uncommon; the QRS complexes are wide and are similar to fully preexcited impulses during sinus rhythm or premature atrial contractions. Differentiation from venticular tachycardia may be difficult.

Intracardiac electrophysiologic studies permit characterization of the accessory pathway and its associated tachyarrhythmias. Electrophysiologic testing is recommended for patients who have frequent or poorly tolerated tachyarrhythmias, a history of atrial fibrillation or atrial flutter (particularly with antegrade bypass tract conduction), and a family history of WPW syndrome and sudden death. Radiofrequency catheter ablation of the accessory pathway has revolutionized the treatment of PSVT due to WPW syndrome, and today it is the preferred method. Intracardiac mapping utilizing multielectrode catheters allows for localization of the accessory pathway and subsequent application of radiofrequency energy to abolish the reentrant loop and prevent the recurrence of arrhythmias. Medical therapy with class IC agents may be used temporarily in patients without structural heart disease while the patient awaits ablative therapy, or in patients who do not desire ablation. Class III agents are also effective, but not all are approved for this indication. The threshold for use of amiodarone should be high because of its side-effect profile. Surgical therapy currently is reserved for patients whose arrhythmias have not been amenable to catheter ablation.

Concealed WPW syndrome is an entity in which the accessory pathway is incapable of antegrade conduction. However, the ability to conduct retrograde across the bypass tract permits orthodromic PSVT. There is little if any danger of atrial fibrillation degenerating to ventricular fibrillation in this instance. Class IC agents, calcium-channel antagonists, and beta-adrenergic blocking agents may be utilized in patients with documented SVTs. Catheter ablation is recommended for definitive, curative therapy.

OTHER REENTRANT SVTS

PSVT due to sinus node reentry or intraatrial reentry is distinguished electrocardiographically from PSVT due to AV node reentry or WPW syndrome by the presence of P waves preceding the QRS complexes, with normal or short PR intervals. Nodoventricular pathways (Mahaim's tracts) may cause PSVT with wide QRS complexes having a left bundle branch block pattern. There is no standard effective treatment for these PSVTs; membrane-active antiarrhythmic agents, beta-adrenergic blockers, and calcium-channel antagonists have all been used. Radiofrequency catheter ablation is helpful in some instances. Intracardiac electrophysiologic studies may be required to optimize therapy in some patients.

ECTOPIC ATRIAL TACHYCARDIAS

These arrhythmias are characterized by an abnormal P-wave vector, a tendency to low P-wave amplitude, and rapid atrial rates (range, 160 to 240 beats per minute). Although ectopic atrial rates in excess of 200 beats per minute usually are accompanied by 2:1 AV conduction, an ectopic atrial rhythm associated with a high-grade block and a relatively slow ventricular rate (so-called PAT with block) suggests digitalis intoxication.

Antiarrhythmic agents may provide effective treatment if no reversible cause can be found. Cardioversion is rarely helpful. Since ectopic atrial tachycardias commonly have precipitating factors, however, removal, reversal, or control of the inciting factors [e.g., digitalis intoxication, decompensated chronic obstructive pulmonary disease (COPD), electrolyte imbalance, metabolic abnormalities, hypoxia, thyrotoxicosis]

TABLE 2-2
ANTIARRHYTHMIC DRUGS: DOSAGE AND KINETICS

Drug		Usual Dosing Range[a]	Half-Life, h	Therapeutic Range, μg/mL	Plasma Protein Binding, %	Major Route of Excretion
Class IA:	Quinidine	Oral sulfate: 200–600 mg q 6 h	5–7	2.3–5	80	H
		Oral long acting: 330–660 mg, q 8 h or q 6 h				
	Procainamide	Oral: 250–750 mg, q 4 h or q 6 h	3–5	4–10	15	R[b]
		Oral long acting: 500–1500 mg, q 8 h or q 6 h				
		IV: 10–15 mg/kg at 25 mg/min, then 1–6 mg/min				
	Disopyramide	Oral: 100–200 mg q 8 h or q 6 h	8–9	2–5	35–95	H/R
	Moricizine[c]	Oral: 150–300 mg q 12 h to q 8 h	6–13	—	95	H
Class IB:	Lidocaine	IV: 1–3 mg/kg at 20–50 mg/min, then 1–4 mg/min	1–2	1–5	60	H
	Tocainide	Oral: 400–600 mg 8–12 h	15	4–10	10	H
	Mexiletine	Oral: 200–400 mg q 8 h	10–12	0.5–2.0	55	H
Class IC:	Flecainide	Oral: 100–200 mg q 12 h	20	0.4–1.0	40	H
	Encainide	Oral: 25–50 mg q 8 h	3–4+	0.5–1.0[d]	80	H
	Propafenone[e]	Oral: 150–300 mg q 8 h	2–10	0.5–1.5[d]	95	H
Class II:	Propranolol	Oral: 10–100 mg q 6 h	4–6	0.04–0.10	95	H
		IV: 0.1 mg/kg in divided 1-mg doses				
	Esmolol	IV: 500 μg/kg per min × 1 min followed by 50 μg/kg per min × 4 min, repeat with 50-μg increments to maintenance dose to 200 μg/kg per min	9 min	—	55	H

26

Class	Drug	Dosage[a]				H/R
Class III:	Acebutolol	Oral: 200–600 mg bid	3–4	—	26	H
	Amiodarone	Oral: 600–1600 mg/day × 1–3 weeks, then 200–400 mg/day; IV: 15 mg/min × 10 min, then 1 mg/min × 6 h, then maintenance at 0.5 mg/min	50 days	1–2.5	96	H
	Bretylium	IV: 5–10 mg/kg at 1–2 mg/kg, then 0.5–2.0 mg/min	?	?	—	R
	Sotalol[d]	Oral: 80–320 mg q 12 h	8–14	0.5–1.5	0	R
	Ibutilide	IV: (for >60 kg): 1 mg over 10 min; may repeat × 1 10 min after completion of initial dose[g]	10–15 / 2–12	—	40	H
Class IV:	Verapamil	Oral: 80–320 mg q 6–8 h; IV: 5–10 mg in 1–2 min	3–8	0.1–0.15	90	H
	Diltiazem	IV: 0.25 mg/kg body wt over 2 min; if response inadequate, wait 15 min, then 0.35 mg/kg over 2 min; maintenance 10–15 mg/h	3.5–5.0	0.1–3.0	70–80	H
Other:	Digoxin	Oral: 1.25–1.5 mg in divided doses over 24 h followed by 0.125–0.375 mg/day; IV: Approximately 70% of oral dose	36	0.8–1.4 ng/mL	30	R
	Adenosine	IV: 6 mg rapidly; if unsuccessful within 1–2 min, 12 mg rapidly	10s	—	—	—

[a] All dosing should follow FDA-approved guidelines as outlined in package insert or *Physicians' Desk Reference*. See also Chap. 45. Does not include pediatric use in infants and young children.

[b] Parent compound metabolized to active metabolite (NAPA) in liver; both active metabolite and unmetabolized parent compound excreted by kidneys.

[c] Shares classes IB, IC activities

[d] Active metabolite limits significance of these measurements.

[e] Shares class II activity.

[f] D/C upon arrhythmia conversion or for ventricular or prolongation of QT or QT$_c$.

NOTE: H = hepatic; R = renal.

is the primary therapy. In patients in whom no reversible cause can be identified, intracardiac localization of the arrhythmia focus and subsequent radiofrequency ablation may be attempted.

MULTIFOCAL ATRIAL TACHYCARDIA

This tachycardia is identified electrocardiographically by three or more P-wave morphologies and a chaotic, irregular rhythm. The rate is usually <150 beats per minute. When the average rate is <100 beats per minute, it is not a tachycardia and is referred to as chaotic or multifocal atrial *rhythm*, but the implications are similar. It occurs most commonly in chronic lung disease but is also seen in patients with severe metabolic abnormalities or sepsis. Although calcium-channel antagonists have been tried with some success, the most effective approach to therapy has been to correct the underlying hypoxia or other metabolic disturbance. There is no role for cardioversion, surgery, or catheter ablation.

ATRIAL FLUTTER

Atrial flutter is characterized electrocardiographically by broad, atrial deflections, "F" or flutter waves, which have a sawtooth configuration in leads II, III, and a VF. There are two types of atrial flutter, classic or type I and type II. Type I flutter can be entrained and interrupted with atrial pacing techniques. It has an atrial rate of 280 to 320 per minute. Type II flutter has an atrial rate faster than 340 per minute and cannot be terminated by pacing. Usual atrioventricular conduction ratio is 2:1 or 4:1. However, partial treatment with antiarrhythmic drugs (especially class IC drugs) may slow the flutter rate to as low as 220 beats per minute, facilitating 1:1 conduction. The *paroxysmal* form of atrial flutter may rarely occur in healthy individuals or transiently in acute pulmonary embolism or thyrotoxicosis. Most commonly it is associated with some form of chronic heart disease, such as mitral valve disease, congenital heart disease, or cardiomyopathy.

Carotid sinus massage may slow the ventricular response, but it tends to increase the flutter rate; occasionally it will convert flutter to fibrillation and very rarely to sinus rhythm. The treatment of choice in hemodynamically compromising atrial flutter is low-dose (10 to 50 W/s) electrical cardioversion. Otherwise treatment is directed first to controlling the ventricular rate with digoxin, verapamil, or diltiazem and then adding a class IA antiarrhythmic agent to convert the rhythm to sinus. Class IC antiarrhythmic agents are less predictably effective in converting atrial flutter to sinus rhythm. If antiarrhythmic agents are unsuccessful, elective cardioversion is an alternative. Patients receiving digoxin who do not have signs of digoxin toxicity may safely undergo cardioversion with digoxin levels under 1.9 mg/mL.

Although recent studies using transesophageal echocardiography have demonstrated a significant incidence of thrombus in the left atrial appendage, an increased incidence of thromboembolic events after cardioversion has not been reported. Anticoagulation, therefore, is not recommended. If cardioversion is contraindicated, rapid atrial pacing may convert the atrial flutter to sinus rhythm or atrial fibrillation. Pacing is usually performed from the right atrium with rapid bursts faster than the atrial flutter rate. A special pacemaker generator is required for overdrive pacing, and insertion of a right atrial catheter usually requires fluoroscopy. Alternatively, in the patient who has recently undergone open-heart surgery, atrial pacing wires are already in place, and these may be used. Chronic control of recurrences of paroxysmal and persistent atrial flutter includes class IA antiarrhythmics to prevent the arrhythmia and digitalis to control the ventricular response rate during recurrences. Class III agents, particularly amiodarone, also are effective. A new class III drug, intravenous ibutilide, has been reported to convert 63 percent of patients with atrial flutter, but it must be used under close monitoring, since it may prolong the QT interval acutely with the risk of short-term torsades de pointes. In patients with frequent symptomatic episodes of type I flutter, radiofrequency catheter ablation to interrupt the reentrant loop permanently has been effective, with maintenance of normal AV nodal conduction. Alternatively, digoxin, beta-adrenergic blocking agents, and calcium-channel antagonists are frequently used to control the ventricular rate.

Chronic atrial flutter usually occurs in the setting of advanced heart disease and commonly heralds atrial fibrillation. The goal of therapy for chronic atrial flutter is adequate control of the ventricular rate. Digoxin, beta-adrenergic blocking agents, and calcium-channel blockers may be tried; in some patients, catheter ablation for AV nodal modification may control the rate. As with paroxysmal atrial flutter, catheter ablation may abolish chronic atrial flutter in select patients or direct surgical ablation may be utilized.

ATRIAL FIBRILLATION

Atrial fibrillation is characterized by disorganized atrial deflections and an irregular AV conduction sequence resulting in a grossly irregular pattern of the QRS complexes. Atrial fibrillatory waves are best seen in standard lead V_1 and are usually evident in II, III, and aV_F. They may be large and coarse or fine to imperceptible. In the latter case, the grossly irregular ventricular rhythm suggests atrial fibrillation.

First Episode of Atrial Fibrillation   The first episode of atrial fibrillation requires a thorough clinical investigation to determine whether the arrhythmia is a primary electrical abnormality or secondary to a hemodynamic abnormality. The incidence of atrial fibrillation increases with age and the presence of structural heart disease. Atrial fibrillation occurring in the absence of structural heart disease is called *lone atrial fibrillation.* Significant mitral or aortic valve disease, hypertension, coronary artery disease, cardiomyopathy, atrial septal defect, and myopericarditis are all disease processes frequently associated with the development of atrial fibrillation. Pulmonary emboli and thyrotoxicosis are well-known causes of atrial fibrillation. Consumption of coffee, tobacco, or alcohol and extreme stress or fatigue also predispose to atrial fibrillation.

In the absence of organic heart disease or WPW syndrome, removal of precipitating factors and observation for recurrences are sufficient. In the presence of significant heart disease, however, therapy should be directed toward treatment of that particular cardiac abnormality; otherwise, recurrence is high, even with pharmacologic therapy and/or electrical cardioversion (Table 2-3). Cardioversion is warranted for the first episode of atrial fibrillation if the patient requires the benefit obtained from the hemodynamic contribution of atrial contraction (e.g., aortic stenosis) or slowing of the ventricular rate to prolong the diastolic filling period (e.g., mitral stenosis).

Paroxysmal Atrial Fibrillation   In the absence of underlying heart disease, rest, sedation, and treatment with digitalis is the treatment of choice for short paroxysms. Chronic therapy is based on the need to control the ventricular rate during recurrences and may be accomplished with digitalis, beta blockers, or calcium channel blockers as described for atrial flutter.

In the presence of heart disease, the development of hemodynamic compromise or congestive heart failure (CHF) requires immediate reversion to sinus rhythm. Immediate cardioversion is mandatory to prevent or reverse the development of pulmonary edema when atrial fibrillation occurs in the presence of hemodynamically significant mitral or aortic stenosis. Direct-current countershock synchronized to the QRS complex using energies ranging from 100 W/s for the initial shock to

TABLE 2-3

ENERGIES FOR EXTERNAL CARDIOVERSION/DEFIBRILLATION

Atrial flutter	10–50 W/s (synchronized)
Atrial fibrillation	50–200 W/s (synchronized)
Supraventricular tachycardia	25–50 W/s (synchronized)
Ventricular tachycardia	50–200 W/s (synchronized, except for very rapid VT)
Ventricular fibrillation	200–360 W/s (asynchronized)
Torsades de pointes	Not indicated; use methods to increase HR, e.g., temporary external or transvenous pacing, isoproterenol 1 μg/min and titrate to increased HR (contraindicated in ischemia)

200 W/s for the second and subsequent shocks is the preferred method. If the patient is hemodynamically stable, the ventricular rate can be controlled with intravenous digoxin, beta blocker, or calcium antagonists. Currently, intravenous verapamil or diltiazem is preferred because of their rapid onset of action. Second, unlike digoxin, which loses its vagal effect as sympathetic tone predominates, verapamil maintains its depressant effect on the AV node, although dose adjustment may be necessary over time. The class IA antiarrhythmics—quinidine, procainamide and disopyramide—are effective in chemically converting atrial fibrillation to sinus rhythm and maintaining sinus rhythm. Quinidine has been the most frequently used. Conventional dosing schedules (200 to 600 mg orally every 6 to 8 h) are now used—unlike the highly aggressive, potentially toxic quinidine protocols of the past. During attempted chemical cardioversion, careful monitoring of QT intervals for excessive prolongation (i.e., QTc 25 percent greater than QTc prior to initiation of treatment) should be performed in addition to monitoring of serum drug levels. The class IC agents (flecainide and propafenone) may convert atrial fibrillation to sinus rhythm when class IA agents have failed. They have also been useful for maintaining sinus rhythm (see also Chap. 45). Intravenous formulations of these drugs are also effective but are not available in the United States. A new class III drug, intravenous ibutilide, has also been demon-

strated to restore sinus rhythm in 31 percent of patients, but it must be used under close monitoring, since it may prolong QT interval acutely with the risk of short-term torsades de pointes (see Chap. 45).

Amiodarone, a class III agent, also is effective for preventing recurrences of atrial fibrillation. The major limitation of amiodarone is its adverse side-effect profile and its extraordinarily long half-life, which limits flexibility in changing therapy. However, low doses of amiodarone (Cordarone) (200 to 300 mg daily) have been shown to decrease the adverse side effects significantly. Sotalol (Betapace), another class III agent, also may be useful in preventing recurrence of atrial fibrillation. These class III drugs are currently not approved for this indication in the United States. In patients with atrial fibrillation refractory to all medical therapy, both conventional and experimental, and in whom the arrhythmia is accompanied by severe disabling symptoms, catheter ablation adjacent to the His bundle to modify conduction or to produce complete AV block is an alternative. However, because this often results in pacemaker dependency, the procedure should be used as a last resort for the control of ventricular rate during atrial fibrillation.

Persistent Atrial Fibrillation   If recurrent episodes of persistent atrial fibrillation (lasting days to weeks) are well tolerated hemodynamically, most physicians avoid repeated electrical cardioversions. This pattern of atrial fibrillation leads eventually to chronic atrial fibrillation; therefore, the best approach is control of ventricular rate during recurrences. Membrane-active antiarrhythmic agents may be used in an attempt to decrease the frequency of recurrences, but their efficacy is unpredictable and the risk of side effects is high. Prevention of episodes of atrial fibrillation may be achieved with class IA, IC, or III drugs. Efficacy is uneven and proarrhythmic or toxic adverse effects are of concern. In the absence of structural heart disease, class IC drugs may be used. Amiodarone, a class III agent, is also effective and is preferred in patients with cardiomyopathies. If symptoms during recurrences are disturbing enough, catheter ablation of the AV node should be considered.

Chronic Atrial Fibrillation   Pharmacologic or electrical cardioversion of chronic atrial fibrillation is indicated primarily if the patient will gain some hemodynamic benefit. Usually no more than one attempt at electrical cardioversion is warranted in the presence of adequate levels of a membrane-active antiarrhythmic agent, since the chance of long-term maintenance of sinus rhythm is very low if the patient reverts back to chronic atrial fibrillation after cardioversion. Management of the patient is then directed toward control of the ventricular response rate as outlined above.

Anticoagulation of Patients with Atrial Fibrillation   The purpose of anticoagulation is to limit the morbidity and mortality from systemic and pulmonary embolization.[3] The decision to initiate anticoagulation for atrial fibrillation depends on the balance between the relative risk of an embolic event versus the risk of a major bleeding complication secondary to anticoagulant therapy. Table 2-4 lists indications and relative contraindications for anticoagulation in patients with atrial fibrillation. These same indications apply to elective cardioversion of recent-onset persistent atrial fibrillation, or chronic

---

TABLE 2-4

ANTICOAGULATION IN PATIENTS WITH ATRIAL FIBRILLATION

Indications

In the presence of hypertension, previous transient ischemic attack or stroke, congestive heart failure, dilated cardiomyopathy, clinical coronary artery disease, mitral stenosis, prosthetic heart valves, or thyrotoxicosis.

Prior to (≥ 3 weeks) and after (≥ 4 weeks) after elective cardioversion

Age ≥ 65 years

Controversial or limited data

Coronary or hypertensive heart disease with normal left atrial size, after first episode of paroxysmal atrial fibrillation

Elective cardioversion of atrial fibrillation of short duration (2 days) with normal left atrial size

Not Indicated

Lone atrial fibrillation in patients aged < 65 years

Relative contraindications

Inability to control prothrombin time

Dementia, malignancy

Prior serious bleeding events

Uncontrolled severe hypertension

atrial fibrillation. Anticoagulation with warfarin (Coumadin) is begun 3 weeks before elective cardioversion and maintained for 4 weeks after cardioversion because of the higher risk of embolic phenomena in the early days after reversion to sinus rhythm. When anticoagulation is used, warfarin is given in doses sufficient to prolong the prothrombin time to an International Normalized Ratio (INR) of 2.0 to 3.0 (see also Chap. 46).

AV JUNCTIONAL AND ACCELERATED VENTRICULAR RHYTHMS

AV junctional rhythms originate within or just distal to the immediate vicinity of the AV node. This category includes premature AV junctional impulses, accelerated junctional rhythms, and AV junctional tachycardias that may be automatic or reentrant. Various forms of reentrant tachyarrhythmias that incorporate the AV junction as a part of a larger reentrant pathway are discussed in the section on PSVTs, above. In AV junctional rhythm, the impulse travels antegrade and retrograde at the same time from the AV junction and is characterized by a normal QRS complex (unless coexistent BBB or aberrancy is present) and a retrograde P wave. Depending on the site of origin and the rate of conduction in each direction, the P wave may occur shortly before the QRS complex, may follow the QRS, or may be lost within it. The rates of AV junctional escape rhythms are usually in the range of 40 to 60 beats per minute; therefore, these rhythms become manifest only when the sinus impulse fails to reach the AV node within physiologic ranges of rate. These rhythms are secondary, occurring as a result of sinus depression, sinoatrial block, or AV block, and are a normal physiologic phenomenon. Failure of these escape rhythms can result in significant bradycardia. This is discussed in the section on bradyarrhythmias, below.

Another type of secondary rhythm is an *accelerated ventricular rhythm*. This occurs because the sinus rate is slow enough to permit an ectopic ventricular rhythm to escape. The ectopic pacemaker is accelerated above its normal physiologic rate of 20 to 40 beats per minute and overrides the sinus rate, which may be relatively depressed. The rate of an accelerated ventricular rhythm is usually between 50 and 100 beats per minute, and the QRS complexes are wide. The rhythm commonly begins with one or two fusion beats and then is regular; however, it may show progressive acceleration or deceleration until it terminates spontaneously.

AV Junctional Premature Beats

An AV junctional premature beat or AV nodal extrasystole arises from an AV nodal or junctional focus; these are premature in relation to the prevailing sinus rhythm. They are relatively uncommon and may be difficult to distinguish from atrial or ventricular extrasystoles. In the absence of heart disease, they usually require no treatment. Even if AV extrasystoles occur in heart disease, they usually require no active treatment. Correction of any hemodynamic abnormalities present may abolish the arrhythmias.

Accelerated Junctional and Accelerated Ventricular Rhythms

If the AV junctional rate exceeds 60 beats per minute but is less than 100 beats per minute, it is referred to as an *accelerated junctional rhythm*. These rhythms are seen commonly in patients with acute myocardial infarction (MI) (particularly inferior MI). They have also been associated with digitalis intoxication, electrolyte abnormalities, hypertensive heart disease, cardiomyopathy, and congenital and rheumatic heart disease. The mechanism of the accelerated AV junctional rhythm is enhanced phase 4 depolarization in the AV junction or the intraventricular conduction system, which allows these normally subordinate pacemakers to usurp pacemaker function when the sinus rate falls below the accelerated focus rate. In ischemia, ventricular pacemaker acceleration is associated with sinus node depression. Accelerated ventricular rhythms are discussed here with accelerated junctional rhythms because both are due to enhanced automaticity. AV junctional rhythms, however, are supraventricular in origin, whereas accelerated ventricular rhythms originate in the His-Purkinje network of the ventricles. These rhythms usually require no treatment; in fact, the use of antiarrhythmic agents may suppress a subordinate pacemaker required for maintenance of an adequate rate. If a faster ventricular rate is required to maintain adequate hemodynamics, atropine, 0.6 to 1.2 mg intravenously, may be given to increase the sinus rate, or temporary pacing may be used.

AV Junctional Tachycardia

Occasionally, the rate of accelerated AV junctional rhythm increases abruptly to the tachycardia range (i.e., ≥100 beats per minute). This phenomenon probably represents an autonomic focus firing at the faster rate with 2:1 exit block that abruptly changes to 1:1 conduction. Usually no treatment is needed except in ischemia, when faster heart rates are unacceptable.

Persistent AV junctional tachycardia (sometimes referred to as *nonparoxysmal* junctional tachycardia) occasionally occurs in patients with chronic heart disease. The response to treatment is unpredictable, and the rhythm may be resistant to conventional antiarrhythmics. Catheter ablation has been utilized in some patients.

Permanent junctional reciprocating tachycardia (PJRT) is characterized by a long RP-short PR re-entry pattern and may be due to a very slow conducting retrograde accessory pathway. This rhythm tends to occur in children, and is difficult to treat pharmacologically, although class IC agents have been reported to be somewhat effective. Catheter ablation also has been used successfully.

VENTRICULAR ARRHYTHMIAS

Effective management of ventricular arrhythmias depends on identification of the risk associated with a particular clinical setting, assessment of the risk-benefit ratio of antiarrhythmic drug treatment, awareness of nonpharmacologic methods of treatment, and an ability to set realistic goals of therapy. The equilibrium between the risk implied by an arrhythmia and the proarrhythmic risk of a drug has been dramatically emphasized by the CAST study.[4] The conventional definition of VT—three or more consecutive ventricular ectopic impulses at a rate of ≤120 beats per minute—is no longer applicable to current methods of evaluation and treatment. A distinction between short runs or *salvos* of three to five consecutive impulses, bursts of nonsustained VT lasting 30 s or less (not resulting in hemodynamic compromise), and sustained VT lasting more than 30 s is necessary to evaluate patients properly. In addition, the ECG pattern of the VT, uniform morphology, polymorphic VT, torsades de pointes, right ventricular outflow pattern, and bidirectional tachycardia have clinical implication for risk and treatment. Finally, the presence or absence of heart disease and the left ventricular function (ejection fraction) markedly influence the management approach. Risk increases with the severity of structural heart disease and left ventricular dysfunction.

Premature Ventricular Contractions

A ventricular extrasystole or premature ventricular contraction (PVC) is an impulse that arises in an ectopic ventricular focus and is premature in relation to the prevailing rhythm. PVCs are characterized by a tendency to constant coupling intervals, wide QRS complexes, and secondary ST-segment and T-wave changes.

Occasionally, PVCs will have narrow complexes, and the QRS and/or T vector will change only minimally. The sinus cycle length usually is not interrupted, resulting in a fully compensatory pause.

PVCS IN THE ABSENCE OF SIGNIFICANT STRUCTURAL DISEASE

Routine treatment of PVCs in this setting is not indicated because there is little or no increased risk of lethal arrhythmias in patients with no underlying cardiac disease. If a patient is bothered by symptoms of palpitations, particularly if frequent, the problem must be addressed to improve the quality of life. First, aggravating factors such as tobacco, caffeine, stress, or other stimulants should be removed. If the symptoms persist, therapy with a mild antianxiety medication or a low-dose beta-adrenergic blocking agent, such as 5 to 20 mg propranolol qid or an equivalent dose of another beta blocker, may be helpful. Class I antiarrhythmic agents are rarely indicated because of their potential side effects. In the absence of structural heart disease, even more advanced forms of PVCs, such as salvos, need not be treated because there is no increased risk of sudden death.[5]

Although PVCs are common in mitral valve prolapse (MVP), only a special subgroup of patients with MVP appear to be at higher risk for serious ventricular arrhythmias. Patients with nonspecific ST-T wave changes in leads II, III, and aV_F as well as more advanced forms of arrhythmias and a redundant valve by echocardiography may require more aggressive treatment. Management of MVP patients with sustained VT or VF is similar to that used in other clinical settings (see also Chap. 13).

PREMATURE VENTRICULAR CONTRACTIONS IN ACUTE SETTINGS

PVCs are frequently seen in acute MI. Although the classic teaching is that PVCs are a "warning" for more severe arrhythmias and therefore require aggressive treatment, the predictive value of such warning arrhythmias is not substantiated. Clinical management of arrhythmias in acute MI ranges from routine treatment of all patients to prevent PVCs, as well as VT or VF, to a threshold for treatment of certain frequencies of PVCs. Intravenous lidocaine, 50- to 200-mg bolus, followed by a 2- to 4-mg/min continuous infusion, is the treatment of choice for PVCs associated with an acute MI. Second, intravenous procainamide, 100 mg every 5 min to a total dose of 10 to 20 mg/kg body weight, followed by a 1- to 4-mg/min continuous infusion, may be used if lidocaine cannot be used or has proved ineffective. Both drugs have significant side effects, particularly with improper dosing. Although their "routine" use is supported by practice, these drugs have not been shown to alter hospital mortality in patients for whom prompt medical attention and electrical defibrillation are available.

Other acute syndromes associated with the appearance of PVCs are those characterized by transient myocardial ischemia and coronary reperfusion, such as Prinzmetal's angina, thrombolysis in acute MI, and balloon deflation during percutaneous transluminal coronary angioplasty (PTCA). Reperfusion arrhythmias are usually transient and self-limiting, but they do have the potential to progress to sustained VT or VF. Arrhythmias associated with acute ischemia initially may be treated with intravenous lidocaine or intravenous procainamide, although prevention of recurrences should include control of the ischemia. In the setting of reperfusion arrhythmias, intravenous lidocaine is used in the same dosages as for acute MI, and prophylactic administration may be used during thrombolysis or PTCA.

Frequent and advanced forms of PVCs are common in severe heart failure and acute pulmonary edema as well as in acute and subacute myocarditis and myopericarditis. Antiarrhythmic therapy is given until the hemodynamic abnormality or acute disease process has resolved. In the case of myocarditis and myopericarditis, antiarrhythmic therapy should be continued for at least 2 months after resolution of clinical symptoms. At that time, the patient should be reevaluated by 24-h ECG monitoring while off antiarrhythmatic medications. If no advanced forms reappear, antiarrhythmic therapy is not restarted. If complex forms do reappear, drug therapy is reinstituted for 2 to 3 months, after which a drug-free 24-h ECG monitoring is repeated. Usually no antiarrhythmic therapy is required after 6 months.

PREMATURE VENTRICULAR CONTRACTIONS IN CHRONIC CARDIAC DISEASES

The presence of chronic heart disease heightens the clinical significance of PVCs. Sudden and total death rates are increased in patients having frequent or repetitive PVCs in the presence of chronic ischemic heart disease, hypertensive heart disease, and the cardiomyopathies. A reduced ejection fraction (<30 percent) further increases mortality risk. Conversely, patients with ejection fractions >40 percent and single PVCs have a much smaller increase in mortality risk.

In post-MI patients, the management of frequent and repetitive forms of PVCs moved away from the use of membrane-active agents after the CAST study demonstrated that PVC suppression with the class IC agents, flecainide and encainide, demonstrated a significant excess risk of sudden and total cardiac death among the treatment groups.[4] In addition, met-analyses of data from previous smaller randomized studies testing the effect of antiarrhythmic drugs on mortality in post-MI patients also suggested an adverse effect of most antiarrhythmic agents. Accordingly, the drugs used in CAST are now contraindicated in post-MI patients with asymptomatic or mildly symptomatic PVCs, and there is a trend away from the use of any membrane-active agent for asymptomatic PVCs in this setting. Beta-adrenergic agents, however, have a beneficial effect on long-term outcome and have evolved as the drug of choice for post-MI patients with asymptomatic or mildly symptomatic PVCs (see also Chap. 10). Patients with repetitive forms of PVCs, especially when accompanied by low left ventricular ejection fraction are at a higher risk for sudden death. Regardless of the depressed ejection fraction, beta-adrenergic blocking agents should be tried first. Use of antiarrhythmic therapy guided by programmed electrical stimulation has been reported to be effective. A recently completed trial comparing conventional antiarrhythmic therapy (80 percent amiodarone) with implantable defibrillators in a population of patients with nonsustained VT, ejection fraction ≤35 percent and inducible sustained VT resistant to intravenous procainamide demonstrated a major benefit in favor of automatic implanted defibrillator.[6]

Cardiomyopathies represent the other major category of heart disease associated with PVCs. The risk of sudden cardiac death is high in both dilated and hypertrophic cardiomyopathies; the efficacy of antiarrhythmic medications in the suppression and prevention of ventricular arrhythmias is not certain. In patients with heart failure secondary to idiopathic dilated or ischemic cardiomyopathy, chronic PVCs are very common and nonsustained VT is a marker for sudden death; but in the CHF-STAT study, amiodarone did not improve survival.[7] Subgroup analysis of the CHF-STAT study and the GESICA trial suggest that amiodarone may improve survival in patients with heart failure secondary to idiopathic dilated cardiomyopathy.

The selection of antiarrhythmic medications for suppression of chronic PVCs in higher-risk patients involves several considerations: incidence of proarrhythmia, occurrence of intolerable side effects, and myocardial depression. Hospitalization of patients during initiation of antiarrhythmic therapy is recommended because of the increased risk of proarrhythmia. Class IA agents all have significant risks of proarrhythmia, but most of these responses are not life-threatening. Among high-risk proarrhythmic effects, they may cause torsades de pointes. Procainamide may be associated with a lupus-like reaction, gastrointestinal discomfort, or agranulocytosis. Quinidine most commonly causes diarrhea but is also associated with cinchonism, allergic reactions, or thrombocytopenic purpura. Disopyramide has such side effects as urinary retention, dry mouth, and abdominal discomfort attributable to its anticholinergic properties. Importantly, in the patient with a reduced ejection fraction, disopyramide may induce congestive heart failure (CHF) (see Chap. 45).

The oral class IB agents, tocainide and mexiletine, although they may be effective, are associated with a high incidence of gastrointestinal and neurologic side effects. Flecainide and encainide, class IC agents, are contraindicated for the treatment of PVCs after MI on data from the CAST.

Occasionally, a combination of antiarrhythmic agents such as a classes IA and IB may be effective when a single agent is not effective. Class III agents have been approved only for life-threatening arrhythmias, although amiodarone may be useful in patients with severe LV dysfunction and long runs of nonsustained VT. The class II agents, beta-adrenergic blockers, suppress PVCs. The class IV agents, calcium-channel blockers, have no role in the treatment of chronic PVCs.

The appropriate endpoint of therapy for chronic PVCs appears to be suppression of advanced forms of PVCs (couplets, salvos, nonsustained VT) if these forms were present on baseline ambulatory monitoring. The goal of therapy includes suppression of 70 to 80 percent of total PVCs in a 24-h period and complete suppression of salvos and nonsustained VT.

Nonsustained Ventricular Tachycardia

Nonsustained runs of VT (salvos of 3 to 5 consecutive ventricular impulses) and nonsustained VT of 6 to 30 s are considered indicators of high risk for potentially lethal arrhythmias (sustained VT or VF) in all patients except those with no underlying or limited heart disease. Patients with no underlying or limited heart disease appear to be at low risk for potentially lethal arrhythmias. Patients with cardiomyopathies or advanced coronary artery disease and low ejection fractions are among those at highest risk. Treatment for these conditions is similar to that for other forms of PVCs, although recent data suggest that patients with prior MI who have a low ejection fraction, nonsustained VT, and are inducible into sustained VT during electrophysiologic testing may benefit from the implantation of a defibrillator.[6]

REPETITIVE MONOMORPHIC VENTRICULAR TACHYCARDIA

This unusual form of ventricular tachycardia is characterized by short paroxysms of VT that last a few beats or a few seconds. The paroxysms may be separated by only one sinus beat and are not aggravated by effort. Occasionally the paroxysms become continuous, and then the arrhythmia becomes a sustained VT. This rhythm disturbance is more common in women and usually is benign. The QRS pattern of the tachycardia on a 12-lead ECG suggests a right ventricular outflow tract origin. Treatment usually is not needed unless there is concomitant structural heart disease or the patient is extremely symptomatic. Beta-adrenergic blocking agents or calcium antagonists are effective in some patients. Catheter ablation also may be considered.

Sustained Ventricular Tachycardia

Sustained VT is defined as a succession of ventricular impulses at a rate ≥ 100 per minute and lasting more than 30 s, or resulting in severe hemodynamic compromise in less than 30 s. In the absence of hemodynamic compromise, intravenous antiarrhythmic therapy may be used. In addition, a 12-lead ECG should be recorded to characterize VT morphology, and a blood sample to measure plasma concentrations of known prescribed antiarrhythmic medications should be obtained prior to institution of acute drug therapy. Although plasma drug levels are not usually available for initial acute management, knowledge of plasma drug concentrations may help later to determine whether or not the rhythm disturbance was secondary to inadequate treatment.

The electrocardiographic distinction between sustained ventricular tachycardia and supraventricular tachyarrhythmias with aberrant intraventricular conduction is based on a complex set of electrocardiographic criteria (Table 2-5). Ventricular/atrial dissociation with clearly discernible P waves, independent of a regular QRS rhythm, is strongly suggestive of VT, as is the presence of P waves associated with alternate QRS complexes. In the presence of ventricular/atrial dissociation, a sinus impulse fused with the wide QRS complex due to VT and producing a single cycle having an altered (usually narrowed) QRS complex known as a *fusion beat* helps to distinguish VT from SVT. A QRS duration above 0.14 s favors VT, but also can occur in patients with SVT and a preexisting BBB. Concordantly positive or negative QRS complexes across the precordium from V_1 to V_6 strongly favor VT over SVT with aberrant conduction. In V_1, a RBBB pattern that is monophasic (R) or biphasic (qR) suggests VT, while a triphasic pattern (rSR) suggests SVT with aberrant conduction. R-wave amplitude in V_1 during the tachycardia that exceeds that during sinus rhythm favors VT. Polymorphic tachyarrhythmias, wide QRS tachycardias with a LBBB configuration in the precordial leads and right axis deviation in the frontal leads, and bidirectional tachycardias are almost always ventricular in origin.

TABLE 2-5

ELECTROCARDIOGRAPHIC SIGNS THAT FAVOR THE DIAGNOSIS OF VENTRICULAR TACHYCARDIA

AV dissociation

Fusion beats

QRS duration >0.14 s

Positive or negative concordance of QRS complexes across precordium

RBBB pattern with monophasic (R) or biphasic (qR)

Polymorphic tachyarrhythmias

Wide QRS tachycardias with LBBB in precordial leads, right axis deviation in frontal leads

Bidirectional tachycardias

SUSTAINED UNIFORM MORPHOLOGY VENTRICULAR TACHYCARDIA

Management of this form of ventricular tachycardia depends on the clinical characteristics of the VT and the setting in which it occurs. When it appears within the first 24 h of an acute MI, aggressive treatment is mandatory because of the high risk that the VT will degenerate into VF. If the patient is hemodynamically unstable, immediate DC cardioversion is required, followed by an infusion of lidocaine that is maintained for 24 to 48 h. If the patient is hemodynamically stable, intravenous lidocaine, 75 to 100 mg, followed by a 1 to 4 μg/min infusion is the first line of therapy (Table 2-6). Cardioversion is necessary if the arrhythmia does not revert immediately or the patient develops hemodynamic compromise (Table 2-3). If VT recurs in spite of lidocaine therapy, procainamide is the next drug of choice; 100-mg boluses of procainamide are infused at 5-min intervals to a total loading dose of 10 to

TABLE 2-6

ANTIARRHYTHMIC AGENTS FOR VENTRICULAR ARRHYTHMIAS IN ACUTE MYOCARDIAL ISCHEMIA/INFARCTION

Drug	Dosage
Lidocaine[a]	IV: 50–200 mg bolus, followed by 2–4 mg/min continuous infusion
Procainamide	IV: 100 mg every 5 min to a total dose of 10–20 mg/kg body weight, followed by 1–4 mg/min continuous infusion
Bretylium	IV: 5 mg/kg body weight over 15 min, repeated if necessary, and followed by 0.5–2.0 mg/min infusion; total dose not to exceed 25 mg/kg per 24 h
Amiodarone	IV: 150 mg over 10 min, followed by a continuous infusion of 1 mg/min for 6 h, then maintenance infusion 0.5 mg/min

[a]Treatment of choice.

20 mg/kg body weight, followed by a constant infusion of 2 to 4 mg/min. If neither lidocaine nor procainamide is effective in suppressing the arrhythmias, bretylium tosylate (Bretylol) or intravenous amiodarone (Cordarone) may be used. Bretylium, 5 mg/kg body weight, is infused over 15 min, repeated if necessary, and followed by an infusion of 0.5 to 2.0 mg/min. Total dose should not exceed 25 mg/kg/24 h. Bretylium is usually well tolerated, but it often causes an initial increased systemic arterial pressure followed by mild hypotension. Therefore, caution should be exercised when using this medication. Amiodarone is administered intravenously with a loading dose of 150 mg infused over 10 min, followed by a continuous infusion of 1 mg/min for 6 h and then a maintenance infusion at the rate of 0.5 mg/min. Hypotension is the most common adverse effect. Antiarrhythmic therapy may be discontinued 48 to 72 h later, since the risk of recurrence is small at that point.

VT occurring during the convalescent phase after MI, which is most common in patients with large anterior wall infarctions, has a far more serious long-term implication and higher mortality rate than VT occurring during the acute phase. Patients should undergo both cardiac catheterization and invasive electrophysiologic studies to define the effective therapy. For both convalescent VT (up to 8 weeks after infarction) and VT appearing later (i.e., chronic phase), electrophysiologic testing is used to demonstrate the characteristics of arrhythmias induced at baseline study and to guide therapy. Left main coronary artery disease or unstable angina are relative contraindications to baseline testing. At least 80 percent of patients with chronic ischemic heart disease and recurrent monomorphic VT can have their clinical VT induced at baseline off all antiarrhythmic drugs. Identification of a drug or combination of drugs that will prevent induction of the same VT is associated with a lower risk of recurrent VT at 1 year compared with the risk of VT if the patient remains inducible on therapy (10 to 15 percent if therapy results in noninducibility compared with 30 to 40 percent if not). If electrophysiologic testing is not available, antiarrhythmic therapy may be guided by ambulatory ECG monitoring or exercise testing. This requires identification of at least 10 ectopic impulses per hour and/or couplets, salvos, nonsustained VT at baseline monitoring, or VT induced during exercise testing. The reduction of PVC frequency by 80 percent or more and abolition of complex forms is an acceptable endpoint for therapy when ambulatory ECG monitoring is used to define therapy.

Alternatives to pharmacologic antiarrhythmic therapy include antiarrhythmic surgery, catheter ablation, and defibrillator implantation. Patients with discrete ventricular aneurysms and bypassable coronary artery lesions may undergo coronary bypass, aneurysmectomy, and surgical cryoablation guided by E-P mapping if a hemodynamically stable VT was induced preoperatively. If the arrhythmia is still inducible after antiarrhythmic surgery, the patient may undergo repeat drug trials, since the substrate for the arrhythmias may have changed. If this approach is unsuccessful, an implantable cardioverter-defibrillator (ICD) is indicated. The development of ICDs with antitachycardia pacing (ATP) and programmed tiered therapy has expanded the therapeutic options for patients with inducible ventricular tachycardias that can be pace-terminated. The technique avoids electrical cardioversions in patients with frequent VT. Medication may be used to slow the rate of VT in some patients, thus rendering the VT pace-terminable. ICDs are also indicated for patients who have life-threatening recurrent arrhythmias that cannot be managed surgically and are refractory to medical therapy. The relative proportion of patients managed by these different techniques (i.e., pharmacologic, surgical, catheter ablation and ICD) is changing, with fewer surgical interventions, fewer antiarrhythmic drug trials after one or two drug failures, and broader use of ICD therapy. ICDs with antitachycardia pacing capabilities obviate the need for antiarrhythmic surgery in some cases.

Electrophysiologically guided therapy is applicable to patients with sustained VT and dilated cardiomyopathy, but mortality may not be altered by drug therapy. Patients who have had clinical VF or nonsustained VT do not appear to benefit from electrophysiologic testing. ICD implantation is useful in these patients, although the degree of LV dysfunction is probably more influential in determining long-term outcome.

LESS COMMON CAUSES OF SUSTAINED VENTRICULAR TACHYCARDIA

Sustained VT may be mediated by catecholamines or other neurophysiologic influences. Beta-adrenergic blocking agents are useful for those sustained VTs seen in association with emotion and stress, as they probably are catecholamine-related. Another small group of patients with VT may respond to calcium antagonists. This group includes adenosine-sensitive VT and VT with RBBB/left axis deviation originating in the low septum. Arrhythmogenic right ventricular dysplasia is com-

monly associated with nonsustained or sustained VT. The VT
in this condition has a LBBB morphology, reflecting its origin
from the right ventricle. During sinus rhythm anterior precordial T-wave inversions are commonly present.

Sustained VT or VF that may be seen years after repair of
complex congenital heart defects such as tetralogy of Fallot
and transposition of the great vessels are potentially lethal and
must be treated. Bidirectional VT is very infrequent and usually associated with digitalis toxicity.

POLYMORPHIC VENTRICULAR TACHYCARDIA

VT with a continuously varying QRS morphology is called
polymorphic VT. It is often associated with acute ischemia and
tends to be more electrically unstable than monomorphic VT.
Hemodynamic collapse is common and serial electrophysiologic testing is not possible.

Polymorphic ventricular tachycardia includes a specific
variant referred to as *torsades de pointes*, which is characterized by QRS peaks that "twist" around the baseline occurs in
the presence of repolarization abnormalities and a prolonged
QT interval—although the same ECG patterns may occur in
the presence of a normal QT interval. Most commonly, it
occurs as a proarrhythmic response to a class IA or III antiarrhythmic agent, but it may also be seen rarely in association
with class IC agent. Other causes of torsades de pointes include
chronic bradycardia (complete AV block with a slow ventricular response rate), hypokalemia, hypomagnesemia, phenothiazines, tricyclic antidepressants, and the use of drugs such as
terfenadine or cisapride in combination with cytochrome P450
inhibitors (ketoconazole, miconazole, erythromycin). Treatment is first directed toward the correction of precipitating
factors. Cardioversion usually only interrupts the torsades de
pointes transiently. Intravenous magnesium sulfate, 2 g over
2 min, is often effective. Acceleration of the heart rate by
pacing or isoproterenol will shorten the QT interval, but the
latter may precipitate ischemia.

Torsades de pointes may also be caused by congenital long-
QT syndromes that are present from childhood and
characterized by the presence of long-QT intervals and/or
prominent U waves on the 12-lead ECG. Arrhythmias may
occur at rest, under emotional stress, or with exercise. Recent
progress in molecular genetics has clearly demonstrated that
inherited defects in the membrane ion channel's molecular
structure and function underlie the disease. Congenital QT prolongation with associated symptomatic ventricular arrhythmias

may require beta-adrenergic blockade and/or partial sympa-thectomy. Placement of an ICD should be considered in patients with resistant arrhythmias.

Ventricular Fibrillation

Ventricular fibrillation is a terminal rhythm characterized electrocardiographically by gross disorganization without identifiable repetitive waveforms or intervals. It requires immediate defibrillation with 200 W/s or more. Cardiopulmonary resuscitation (CPR) measures should be performed until defibrillation is successful. Some antiarrhythmic drugs may increase the defibrillation threshold, and thresholds may be decreased by the use of bretylium, lidocaine, or epinephrine. Once the patient's rhythm has returned to normal, prophylactic antiarrhythmic agents (lidocaine or procainamide, or, in resistant recurrent cases, bretylium) should be administered, and all metabolic and electrolyte disturbances corrected. Ventricular fibrillation occurs most commonly in acute ischemia.

Ventricular fibrillation is the underlying arrhythmia in most patients with sudden cardiac arrest. If it is not associated with an acute MI, electrophysiologic testing is warranted to guide therapy. Drug testing may be guided by ambulatory ECG monitoring or exercise testing if electrophysiologic testing is unavailable; however, this requires that the patient have high-risk forms on ambulatory monitoring or during exercise. Patients who are refractory to drug therapy and who do not have cardiac disease amenable to surgery should undergo ICD implantation. Concomitant antiarrhythmic therapy may be necessary to avoid excessive shocks for recurrent arrhythmias.

BRADYARRHYTHMIAS

Bradyarrhythmias may be secondary to abnormalities of cardiac impulse formation or to AV conduction abnormalities. Although they are often asymptomatic, symptoms of hypoperfusion may occur, necessitating immediate treatment. The first step in management is to increase the heart rate. Atropine sulfate, 0.5 to 1.0 mg (0.01 mg/kg), should be administered intravenously and repeated two to three times if necessary. Sympathomimetic amines such as isoproterenol (Isuprel) may be used, but with caution; they should be avoided in patients with ischemic symptoms. Temporary external pacing is a simple, noninvasive way to increase the heart rate and in some instances has supplanted transvenous pacing. It offers a rapid alternative for patients in whom venous access is difficult or is relatively contraindicated (e.g., patients receiving thrombolytic agents). Moreover, it is ideal for patients who have transient bradycardic episodes. Its limitation is the number of patients in whom capture is inconsistent or fails. Temporary transvenous demand pacing provides a stable and reliable increase in ventricular rate when necessary. Dual chamber pacing is indicated for patients who will benefit from synchronized atrial contraction.

In addition to increasing the heart rate, any medications known to cause bradycardia should be discontinued. Beta-adrenergic blocking agents, calcium channel blockers, digitalis, clonidine, and lithium are common offenders; less frequently, quinidine, procainamide, and lidocaine cause bradyarrhythmias. Patients with persistent symptomatic bradyarrhythmias and no identifiable reversible cause require permanent pacing.

Failure of Impulse Formation

SINUS BRADYCARDIA

Sinus bradycardia is a rhythm in which each cardiac impulse arises normally from the sinoatrial node, but the rate is <60 beats per minute. The P-wave morphology is identical to that observed in normal sinus rhythm; occasionally the P-R interval is prolonged. Sinus bradycardia occurs normally in patients with no underlying heart disease (particularly in well-conditioned individuals), in acute MI, and in association with certain medications, autonomic imbalance, hypothermia, hypothyroidism, or hyperkalemia. The rhythm requires no treatment unless the patient is symptomatic. Removal of aggravating factors is the first step in therapy. If this is not successful, or if the patient requires negative chronotropic agents as part of medical management, permanent pacing may be necessary.

SICK SINUS SYNDROME

Sick sinus syndrome is a condition characterized by abnormal cardiac impulse formation, commonly accompanied by disordered intraatrial electrical activity and atrioventricular conduction. The syndrome is associated with a wide spectrum of brady- and tachyarrhythmias. Some patients exhibit fixed or intermittent sinus bradyarrhythmias; others have sinus bradyarrhythmias alternating with normal sinus rhythm and/or supraventricular tachyarrhythmias (the "tachy/brady syndrome").

Therapy should be reserved for patients with electrocardiographic documentation of bradyarrhythmias or tachyarrhythmias and symptoms corresponding to the periods of arrhythmias. Patients with sick sinus syndrome may be particularly susceptible to bradycardias induced by beta-adrenergic blocking agents, calcium-channel blockers, and antiarrhythmic drugs. Since these medications are frequently used to treat the tachyarrhythmias associated with sick sinus syndrome, concomitant permanent pacing may be required.

AV Conduction Abnormalities

FIRST-DEGREE ATRIOVENTRICULAR BLOCK

Isolated first-degree AV block is characterized electrocardiographically by a P-R interval that exceeds 200 ms. It may occur as the result of increased vagal tone, vagotonic drugs, digitalis, beta-adrenergic receptor blockade, hypokalemia, acute carditis, tricuspid stenosis, Chagas' disease, and some forms of congenital heart disease. A prolonged P-R interval may occur in normal individuals, reflecting increased vagotonia. Isolated first-degree AV block is never symptomatic and temporary or permanent pacing is not indicated.

SECOND-DEGREE AV BLOCK

Mobitz type I AV block, or the Wenckebach phenomenon, is characterized electrocardiographically by consecutively conducted impulses with progressively increasing P-R intervals until an impulse is blocked and the P wave is not followed by a QRS complex. This is the most common form of second-degree AV block and is usually not symptomatic. It usually does not progress to high-grade AV block; therefore, prophylactic pacing is not necessary unless the patient is symptomatic and vagolytic therapy is ineffective. The presence of Mobitz type I AV block usually does not adversely affect a patient's prognosis.

In contrast, the less common Mobitz type II block implies more significant distal or infrahisian conduction system disease. It is characterized by consecutively conducted impulses with fixed P-R intervals and a sudden block of impulse conduction. It is almost always associated with organic heart disease, including disease in the AV conducting system distal to the AV node, and may progress to complete AV block. For this reason permanent pacing may be indicated, primarily to protect the patient from symptomatic events.

Paroxysmal AV block refers to runs of consecutive atrial impulses that fail to conduct to the ventricles and may last 10 to 20 s. Unless a reversible cause is clearly identified, permanent pacing is required.

COMPLETE AV BLOCK

Complete heart block, or third-degree AV block, is characterized by a complete interruption of conduction within the AV junctional tissues; supraventricular impulses are unable to propagate to and activate the ventricles. The ventricles are subsequently activated by a subsidiary idionodal or idioventricular pacemaker at a rate of 20 to 50 beats per minute. Two independent pacemakers then control the rhythm of the heart: one for the atria and one for the ventricles. The two rhythms are asynchronous, since each pacemaker discharges at its own rate.

Acute, symptomatic complete heart block requires immediate treatment with either pharmacologic agents (atropine or isoproterenol) or temporary external or transvenous intracardiac pacing. Isoproterenol should be avoided in ischemia. Complete heart block in the setting of an acute inferior MI usually is transient but may take up to 2 weeks to resolve. In contrast, complete AV block in association with acute anterior wall MI may be permanent and require permanent pacing. Even if AV block in this setting is not permanent, it indicates high risk for future events. Some have proposed permanent pacemakers even after transient AV block in anterior infarction.

CONGENITAL AV BLOCK

Complete heart block may occur as an isolated congenital anomaly. The QRS complex usually is normal or near normal, since the site of the block is almost always within the AV node or the bundle of His. The resting heart rate usually is in the range of 45 to 65 beats per minute, and the patients frequently are asymptomatic; syncopal attacks are rare. Diagnostic evaluation should include exercise testing to ascertain whether or not the patient can mount an adequate heart rate in response to stress. If the patient is asymptomatic and the heart rate increases appreciably with exercise, no further therapy may be needed. On the other hand, congenital complete AV block associated with structural congenital cardiac abnormalities implies higher risk.

AV DISSOCIATION

Electrocardiographically, the diagnosis of AV dissociation is made when the P waves of sinus rhythm (or other forms of atrial electrical activity) are dissociated from and bear no fixed relationship to the QRS complexes of the ectopic idionodal or idioventricular rhythm. The presence of AV dissociation suggests an abnormality of normal intrinsic pacemaker activity. AV dissociation may occur as a result of slowing of normal pacemaker acitivy, acceleration of a subordinate focus, or the presence of complete AV block. The diagnosis of AV dissociation, however, is not synonymous with complete AV block.

If a patient with AV dissociation is symptomatic, the underlying rhythm disturbance responsible for the symptoms must be identified and treatment directed toward that rhythm disturbance. For example, a ventricular ectopic focus may become predominant during extreme sinus bradycardia. Suppression of these ventricular escape beats with antiarrhythmic therapy may worsen the underlying bradycardia and the patient's symptoms. Appropriate treatment in this instance would be directed toward the bradycardia; pacing would alleviate the bradycardia, abolish the patient's symptoms, and suppress the ventricular ectopy.

Indications for Pacing

Cardiac pacing is indicated for the treatment of symptomatic bradyarrhythmias that are unresponsive to medical therapy. In the setting of an acute anterior wall MI, temporary pacing is indicated for AV block if it is associated with excessively slow heart rates and/or a reduction in cardiac output. The development of a new LBBB or RBBB, particularly in association with a left hemiblock, frequently heralds the development of complete heart block and traditionally requires prophylactic pacing (see also Chap. 3).

External pacing techniques in some instances have supplanted the need for insertion of temporary transvenous pacemakers. This is particularly applicable in acute inferior wall MI, where AV block is usually transient. New left bundle branch block, or a preexisting right or left bundle branch block, does not require pacing. Permanent pacing is often recommended for patients who have had transient complete heart block during an acute anterior wall MI, although it is unclear whether or not mortality is affected. Permanent pacing is rarely required after complete AV block in inferior wall infarction. In the absence of an acute or recent MI, permanent pacing is the therapy of choice for fixed or intermittent symptomatic bradyarrhythmias that have no identifiable reversible cardiac or noncardiac cause (see also Chap. 3).

REFERENCES

1. Chakko S, Kessler KM: Recognition and management of cardiac arrhythmias. *Curr Probl Cardiol* 1995; 20:53–120.
2. Stewart RB, Bardy GH, Greene LH: Wide complex tachycardia: misdiagnosis and outcome after emergent therapy. *Ann Intern Med* 1986; 104:766–771.
3. Laupacis A, Albers G, Dalen J, et al: Antithrombotic therapy in atrial fibrillation. *Chest* 1995; 108:352S–359S.
4. Echt DS, Liebson RP, Mitchell B, et al: Mortality and morbidity in patients receiving encainide, flecainide, or placebo: The Cardiac Arrhythmia Suppression Trial. *N Engl J Med* 1991; 324:781–788.
5. Kennedy HL, Whitlock JA, Spague MK, et al: Long-term follow-up of asymptomatic healthy subjects with frequent and complex ventricular ectopy. *N Engl J Med* 1985; 313:193–197.
6. Moss AM, Hall WJ, Cannom DS, et al: For the Multicenter Automatic Defibrillator Implantation Trial Investigators: improved survival with an implanted defibrillator in patients with coronary artery disease at high risk for ventricular arrhythmia. *N Engl J Med* 1996; 335:1933–1940.
7. Singh SN, Fletcher RD, Fisher SB, et al: For the CHF STAT Investigators: Amiodarone in patients with congestive heart failure and asymptomatic ventricular arrhythmia. *N Engl J Med* 1995; 333:77–82.

SUGGESTED READING

Akhtar M, Breithardt G, Camm AJ, et al: CAST and beyond: implications of the Cardiac Arrhythmia Suppression Trial. *Circulation* 1990; 81:1123–1127.

Brugada P, Brugada J, Mont L, et al: A new approach to the differential diagnosis of a regular tachycardia with a wide QRS complex. *Circulation* 1991; 83:1649–1659.

Calkins M, Sousa J, el-Atassi R, et al: Diagnosis and cure of the Wolff-Parkinson-White syndrome or paroxysmal supraventricular tachycardias during a single electrophysiological test. *N Engl J Med* 1991; 324:1612–1618.

Chakko CS, Gheorghiade M: Ventricular arrhythmias in severe heart failure: incidence, significance, and effectiveness of anti-arrhythmic therapy. *Am Heart J* 1985; 109:497–504.

Cox JL: The status of surgery for cardiac arrhythmias. *Circulation* 1985; 71:413–417.

Gregoratos G, Cheitlin M, Cinill A, et al: ACC/AHA guidelines for implantation of cardiac pacemakers and antiarrhythmia devices: executive summary. *Circulation* 1998; 97:1325–1335.

Jackman WM, Wang X, Friday KJ, et al: Catheter ablation of accessory atrioventricular pathways (Wolff-Parkinson-White syndrome) by radiofrequency current. *N Engl J Med* 1991; 324:1601–1611.

Lee MA, Morady F, Kadish A, et al: Catheter modification of the atrioventricular junction with radiofrequency energy for control of atrioventricular nodal reentry tachycardia. *Circulation* 1991; 83:827–835.

Mason JW, for the ESVEM Investigators: A comparison of electrophysiologic testing with Holter monitoring to predict antiarrhythmic drug efficacy for ventricular tachyarrhythmias. *N Engl J Med* 1993; 329:445–451.

Myerburg RJ, Kessler KM, Castellanos A: Recognition, clinical assessment, and management of arrhythmias and conduction disturbances. In: Alexander RW, Schlant RC, Fuster V, O'Rourke RA, Roberts R, Sonnenblick EH (eds): *Hurst's The Heart*, 9th ed. New York, McGraw-Hill, 1998:873–941.

Myerburg RJ, Castellanos A: Evolution, evaluation and efficacy of implantable defibrillator technology. *Circulation* 1992; 86:691–693.

Prystowsky EN, Benson W, Fuster V, et al: Management of patients with atrial fibrillation. *Circulation* 1996; 93: 1262–1277.

Reifel JA, Estes NAM, Waldo AL, et al: A consensus report on antiarrhythmic drug use. *Clin Cardiol* 1994; 17:103–116.

Ryan T, Anderson JL, Antman EM, et al: ACC/AHA guidelines for the management of patients with acute myocardial infarction: executive summary. *Circulation* 1996; 94:2341–2350.

Scheinman MM, Laks MM, DiMarco J, et al: Current role of catheter ablative procedures in patients with cardiac arrhythmias. *Circulation* 1991; 83:2146–2153.

Scheinman MM: Treatment of cardiac arrhythmias with catheter-ablative techniques. In: Alexander RW, Schlant RC, Fuster V, O'Rourke RA, Roberts R, Sonnenblick EH (eds): *Hurst's The Heart*, 9th ed. New York, McGraw-Hill, 1998:995–1001.

3

CARDIAC PACEMAKERS

Raul D. Mitrani

Pacemakers are coded by specific abbreviations according to the type of pacemaker and mode of pacing. The first letter refers to the chamber(s) being paced and the second letter to the chamber(s) being sensed. The letters A and V indicate atrial or ventricular pacing and/or sensing. If dual chambers are paced and/or sensed, the designation D is used. The third letter refers to the response to a sensed event, whether it is a normal complex, a premature complex, or a sensing artifact. The pacemaker inhibits (I) pacing output from one or both of its leads, or triggers (T) pacing at a programmable interval after the sensed event. If a pacer can inhibit atrial output and trigger a ventricular paced complex after a sensed atrial complex, then the third letter is designated by D. In common usage rate responsiveness is noted by a fourth letter, R.

In order for an electrical signal from a pacer to cause electrical depolarization, a minimal threshold of current is necessary. This is a function of the pacer lead, longevity, and other clinical factors. The current delivered is a function of the pacemaker voltage and pulse width, which is generally programmed to deliver two to four times the threshold current in order to have adequate safety margin.

A pacemaker senses intrinsic cardiac activity by measuring intracardiac electrograms. The ranges for atrial and ventricular electrograms are 1 to 5 and 5 to 10 mV respectively; therefore, pacemaker sensitivities are programed at 0.25 to 2 mV in the atrial channel and 2 to 4 mV in the ventricular channel in order to have an adequate safety margin for sensing.

TEMPORARY PACING

Temporary pacing is used to treat intermittent or persistent hemodynamically relevant bradyarrhythmias or to provide standby pacing for patients at increased risk for sudden and complete heart block. Occasionally, temporary pacing is used to control sustained atrial or ventricular tachyarrhythmias. The end point for temporary pacing is either resolution of a temporary indication for pacing or implantation of a permanent pacemaker for a continuing indication.

Transcutaneous pacing is a common, rapid method for non-invasively pacing patients who require a prophylactic temporary pacer or emergent pacing. The unit incorporates two large pads placed in an anterior and posterior position. The main drawback is the high energy requirement (50 to 100 mA at 20 to 40 ms), which causes skeletal muscle stimulation and pain.

INDICATIONS FOR PERMANENT PACEMAKERS

Many indications for pacemaker implantation are predicated upon the presence of symptoms. However, many symptoms—such as fatigue or subtle symptoms of congestive heart failure—may be recognized only in retrospect, after placement of a permanent pacemaker.

Pacing for Specific Causes and Patterns of AV Block

In general, complete heart block, permanent or intermittent, at any anatomic level associated with symptoms such as dizziness, lightheadedness, congestive heart failure, or confusion is an indication for the implantation of a permanent pacemaker. In awake, asymptomatic patients, a ventricular rate below 40 beats per minute or pauses greater than 3.0 s are also considered indications for pacemaker implantation.

Asymptomatic third-degree or type II second-degree AV block is generally an indication for a permanent pacemaker. Asymptomatic type I second-degree AV block is generally not considered a precondition for permanent pacing unless there is electrophysiologic evidence that such a block is in the His-Purkinje system. First-degree AV block is also not considered to be an indication for permanent pacing unless there are symptoms associated with marked PR prolongation (usually in the setting of left ventricular dysfunction) and documentation of alleviation of symptoms or hemodynamic improvement with pacing at a shorter AV interval.

In the setting of acute myocardial infarction (MI), a pacemaker is indicated for high-grade or complete block in the His-Purkinje system. In the setting of inferior infarction, AV block typically occurs at the level of the AV node and may be due to reversible injury and/or autonomic tone; therefore, AV block usually subsides in time. Transient advanced infranodal AV block with associated bundle branch block is also an indication for pacing; however, electrophysiologic studies may be required to determine the level of block.

In the presence of bifascicular or trifascicular block, intermittent third-degree or type II second-degree AV block usually indicates the need for a permanent pacemaker. When such a patient presents with syncope, a pacemaker may be required. An electrophysiologic study may be required to rule out other causes of syncope (i.e., ventricular tachycardia), particularly if structural heart disease is present.

Pacing in Congenital AV Block

The site of AV block in congenital heart block is usually at the level of the AV node. Congenital AV block is associated with serious and possibly fatal complications, including syncope and sudden death. Cardiac pacing is indicated in all symptomatic patients. In asymptomatic patients, the coexistence of excessive bradycardia, left ventricular dysfunction, chronotropic incompetence, and a prolonged QT interval also indicates a need for cardiac pacing.

Pacing in Sinus Nodal Dysfunction

Sinus nodal dysfunction has become the most common indication for pacing in the United States, especially in the presence of symptoms correlating with bradyarrhythmias; however, it may be difficult to correlate electrocardiographic (ECG) findings with symptoms. Furthermore, symptoms may be nebulous. For instance, the presence of fatigue and dyspnea may be due to a bradyarrhythmia, to lack of conditioning, or to some other cardiac dysfunction.

Therefore, pacing is indicated for symptomatic sinus nodal dysfunction when there is documented bradycardia or there are pauses that may be secondary to essential long-term drug therapy. Patients with asymptomatic bradyarrhythmias should be evaluated carefully prior to placing a pacemaker. Athletes commonly have physiologic bradycardia, even with heart rates less than 40 beats per minute, due to enhanced vagal tone. Other etiologies, such as sleep apnea, can also cause asymptomatic nocturnal bradyarrhythmias.

Pacing for Carotid Sinus Syndrome

The diagnosis of carotid sinus syndrome is typically made by demonstrating asystolic pauses of more than 3 s with carotid sinus massage or a vasodepressor response of >50 mmHg associated with clear symptoms provoked by carotid sinus stimulation such as wearing a tight shirt or turning the head. Improvement of symptoms and suppression of syncope has been demonstrated by treating patients with cardiac pacing, particularly dual-chamber pacing.

Cardiac Pacing in Neurocardiogenic Syncope

Cardiac pacing plays little role for most patients with neurocardiogenic syncope because they still experience hypotension, vasodilatation, and other associated symptoms despite cardiac pacing. Some studies suggest that even though pacemakers do not prevent these vasovagal episodes, pacing may prevent syncope and convert a vasovagal episode to an episode of dizziness and presyncope.

Pacing in Hypertrophic Cardiomyopathy (HCM)

DDD pacing with a short AV interval reduces left ventricular outflow tract gradients and improves symptoms in some patients with obstructive hypertrophic cardiomyopathy. Recent randomized studies, however, have yielded conflicting results about the long-term hemodynamic benefits of pacing for hypertrophic cardiomyopathy. Therefore, pacing in these patients remains controversial. Patients with nonobstructive symptomatic HCM do not benefit from DDD pacing (Chap. 18).

PACEMAKER HARDWARE

Pacemaker leads can be unipolar or bipolar. Unipolar leads use a distal electrode in the catheter as the cathode and the shell of the pacemaker generator as the anode. Therefore, the myocardium and adjacent tissue complete the circuit. A bipolar lead consist of two separate conductors and electrodes within the lead. Since the electrodes for sensing in a bipolar lead are much closer together, bipolar signals are sharper, with less extraneous noise.

There are several disadvantagaes to unipolar lead systems. Because the unipolar lead uses body tissue to complete the circuit, there is the possibility of causing muscle stimulation. Unipolar sensing is far more likely to detect extracardiac signals, including myopotentials, far-field sensing of remote cardiac potentials, and electromagnetic interference. Finally, unipolar pacing is generally contraindicated in patients who have a concomitant implantable defibrillator.

PACEMAKER FUNCTION AND MODES

Magnet Mode

Magnets cause asynchronous pacing in virtually all pacemakers. The magnet response varies according to manufacturer and pacemaker model. In patients who are pacemaker-dependent and experiencing oversensing, thereby inhibiting pacemaker output, a magnet is a convenient short-term method to ensure pacing. Pacemakers usually have one magnet rate for a battery that is intact and another one for a battery that is at elective replacement index or at the end of battery life, which allows for noninvasive assessment of battery status.

VVI Mode

VVI mode ensures that a minimum ventricular rate is maintained by ventricular pacing at the pacemaker rate unless there is an intrinsic ventricular rate greater than the pacemaker lower rate. This is useful in patients with atrial fibrillation or for those who need backup pacing.

Hysteresis is a programmable function in which the ventricular escape interval is longer after a sensed ventricular event than after a paced ventricular event. This feature is intended to conserve battery life and maintain an intrinsic rhythm, because the effective rate at which a pacer begins to pace is lower than the actual lower rate of the pacemaker.

AAI Pacing

AAI pacing is similar to VVI pacing except that the pacemaker is stimulating the atrium. AAIR is an excellent mode of pacing in a patient with sinus node dysfunction and normal AV nodal and His-Purkinje function.

DDD Pacing

DDD pacing is the most common pacing mode for dual-chamber pacemakers. The timing sequences for DDD pacing are described in Fig. 3-1. This mode is used for patients with AV node and/or sinus node dysfunction.

1. *DDD pacing in patients with sinus node dysfunction.* Patients with sinus node dysfunction may have intermittent or chronic sinus bradycardia requiring intermittent or continuous atrial pacing. If patients have intact AV conduction, the pacemaker functions as an AAI pacer. Because patients with sinus node dysfunction frequently have AV nodal or His-Purkinje disease, patients with DDD pacemakers frequently demonstrate fused ventricular complexes originating from ventricular stimulation and through the natural AV conduction system. If the QRS complex appears normal with a pacing spike in the middle or end of the QRS complex, this is still consistent with normal pacemaker function and is termed *pseudofusion.*

2. *Patients with AV block and normal sinus node function.* In the DDD mode, if the lower rate of the pacer is programmed at a sufficiently low value to permit atrial tracking, the pacemaker stimulates the ventricle synchronously with intrinsic P waves. If a patient does not require atrial pacing, it may be reasonable to implant a dual-chamber pacer with a single tripolar or quadripolar lead that allows atrial sensing and ventricular pacing and sensing. These VDD pacing systems allow for ease of implantation and bipolar atrial sensing.

DDI Pacing

This is a useful mode for pacing in patients with a tachycardia-bradycardia pattern of sick sinus syndrome who have intact AV conduction. During atrial tachyarrhythmias, a pacer in the DDI mode will pace the ventricles at the lower rate. During episodes of bradyarrhythmia, the pacer functions in an atrial or AV pacing mode. DDI pacing is inappropriate for patients with permanent or intermittent AV block.

USE OF PACEMAKERS IN DIFFERENT CLINICAL SITUATIONS

Paroxysmal Atrial Fibrillation, Flutter, and Other Tachyarrhythmias

In order to prevent inappropriate upper tracking behavior during atrial tachyarrhythmias, a pacer can be reprogrammed to DDI or DDIR if the patient has intact AV conduction. Alternatively, a pacer feature, *automatic mode switch*, automatically changes the pacer mode from DDD[R] to VVI[R] or DDI[R] when the atrial rhythm changes from sinus rhythm to atrial fibrillation, flutter, or other types of tachycardia.

Patients with Complete or Intermittent Third-Degree AV Block

Patients with complete or intermittent third-degree AV block generally receive a DDD pacemaker. If patients have intact sinus function, a VDD pacer, utilizing a single lead, may be a reasonable alternative.

Patients with Carotid Sinus Syndrome and Vasovagal Syncope

Patients with one of the neurally mediated syncope syndromes generally require intermittent AV pacing during their episodes. Additionally, they generally benefit from an interventional pacing rate at 75 to 100 beats per minute during these episodes. In order to avoid chronic pacing at such an elevated rate, certain pacemakers offer features that allow for pacing at an intervention rate for a short period of time, being triggered by a precipitous drop in a patient's intrinsic heart rate. These features are known as *scanning hysteresis* or *rate drop response*.

HEMODYNAMICS OF CARDIAC PACING

AV Synchrony

The importance of AV synchrony has been recognized in several clinical situations, particularly in patients with left ventricular hypertrophy or diastolic dysfunction whose cardiac output is dependent on preload. Several acute and long-term hemody-

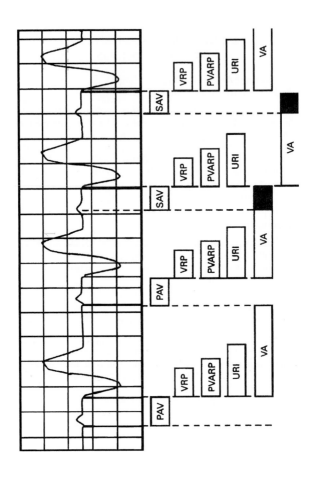

FIGURE 3-1

Schematic diagram of DDD pacing with selected timing cycles and refractory periods. After a paced atrial complex, the paced AV interval (PAV) begins. If there is no ventricular depolarization before this interval expires, the pacemaker response is to output a ventricular impulse. After a paced ventricular output, several refractory periods and timing cycles are initiated. The ventricular refractory period (VRP) is the time during which a ventricular event will not reset the timing intervals. The postventricular refractory period (PVARP) represents the time during which an atrial event will not be sensed or will not reset the timing intervals. The upper rate interval represents the shortest interval (maximum rate) that a pacemaker will ventricular pace corresponding to the programmed upper tracking rate. The ventricular-atrial escape interval (VA) represents the time during which, if there is no sensed atrial electrogram, atrial pacing occurs. The programmed lower rate corresponds to the AV interval and the VA interval. During the first two complexes, there were no sensed atrial complexes; therefore, the VA interval expired and atrial pacing occurred. During the third and fourth complexes, there were sensed atrial electrograms following the PVARP and before the VA interval expired (shaded area in the VA bar). Note that atrial sensing usually occurs after the start of the P wave, representing the atrial conduction time to the atrial electrodes. The programmed AV interval following a sensed atrial complex (SAV) may be programmed at a value lower than the PAV to obtain equivalent PR intervals.

63

namic studies have demonstrated the advantage of atrial-based pacing (AAI or DDD) as compared with VVI pacing.

Pacemaker Syndrome

The pacemaker syndrome is a constellation of signs and symptoms representing adverse reaction to VVI pacing. The basis for the pacemaker syndrome is not only loss of AV synchrony but also the presence of VA conduction. Most of the symptoms relate to loss of AV synchrony and also to retrograde conduction. These can include orthostatic hypotension, near syncope, fatigue, exercise intolerance, malaise, weakness, cough, awareness of heartbeat, chest fullness, neck fullness, headache, chest pain, and other symptoms that may be nonspecific. These patients may have intermittent or persistent cannon *A* waves and possible liver pulsation. The management of pacemaker syndrome in patients with sick sinus syndrome usually requires restoration of AV synchrony.

PACEMAKER COMPLICATIONS

Infections related to pacemaker implantation are rare. Early infections may be caused by *Staphylococcus aureus* and can be aggressive. Late infections are commonly related to *Staphylococcus epidermidis* and may have a more indolent course. Signs of infection include local inflammation and abscess formation, erosion of the pacer, and fever with positive blood culture but without an identifiable focus of infection. Transesophageal echocardiography may help to determine whether vegetations are present on the pacemaker leads. If the pacemaker is infected, removal of the pacemaker leads and generator is usually required.

The insulation on pacer leads may break or leads may fracture, leading to problems with oversensing (due to electrical noise), undersensing, and failure to capture (due to current leak). This problem is often manifest intermittently and may be difficult to detect during a routine pacer check. The patient may complain of pectoral muscle stimulation due to current lead around an insulation break.

Electromagnetic Interference (EMI) of Pacemaker Function

Unipolar pacemakers are usually more susceptible to EMI interference than bipolar pacemakers because the sensing circuit encompasses a larger area than in bipolar sensing. A magnetic resonance imaging scan is contraindicated in patients with pacers. Cellular phones have the potential to adversely affect pacemaker function. Therefore, it is recommended that patients either use analog cellular telephones or keep their digital cellular phones (with power outputs below 3 W) 20 cm away from their pacemakers.

PACEMAKER MALFUNCTION

Pacemaker malfunction can be categorized as loss of capture, abnormal pacing rate, undersensing, oversensing, or other erratic behavior. The approach to diagnosing pacemaker malfunction is to inspect the ECG carefully, interrogate the pacemaker, and to check pacing and sensing thresholds, lead impedances, and battery voltage/magnet rate. A chest x-ray is also in order. Many instances of pacemaker malfunction actually represent normal function of the pacemaker.

Abnormal Pacing Rates

Abnormal pacing rates can be due to normal or abnormal pacing function. Failure of the pacemaker to output is usually due to oversensing (Fig. 3-2). Occasionally, there is pacemaker output that is not visible because bipolar pacing is producing very low amplitude pacing artifacts (artifacts from digital ECG recording are commonly difficult to visualize). Conversely, absence of pacing stimuli may be due to interruption of current flow from a lead fracture, insulation break, or a loose set screw.

Abnormally fast pacing rates are usually due to normal pacing function. They may occur in response to rate-adaptive sensors. In DDD pacing, upper-rate pacing may be due to sinus tachycardia, atrial tachyarrhythmias, or pacemaker-mediated tachycardia.

Loss of Capture

The loss of pacemaker capture occurs when there is a visible pacing stimulus and no atrial or ventricular depolarization (Fig. 3-2). This may be intermittent or persistent. Etiologies include elevation of pacing threshold, lead dislodgment, lead fracture or insulation break, and loose set screws. Battery depletion may also cause pacing failure.

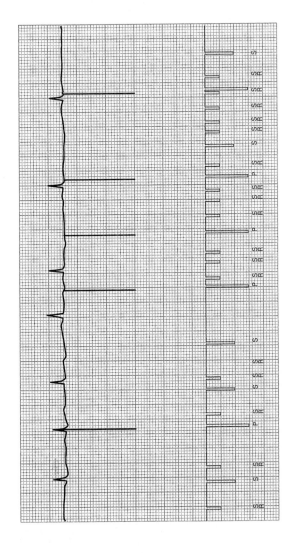

FIGURE 3-2

Electrocardiogram and marker channels are shown for a patient with a ventricular lead impedance break. Note that based on the ECG, there is failure to sense, as manifested by the second and fourth pacing outputs coming very shortly after the QRS complex. There is failure to capture, demonstrated by the second and third pacing outputs, which should capture the ventricle. There is also evidence of oversensing, as demonstrated by the long pause between the fourth and fifth pacing outputs during a diastolic period that exceeds the interval between the previous two pacing outputs. In general, when there is evidence of oversensing, undersensing, and failure to capture, then the likely etiology is either a lead insulation break, lead fracture, or other mechanical problem. The marker channels confirm the above ECG findings. There are sensed ventricular events (S or SR) that do not correspond to surface QRS complexes, consistent with oversensing. Additionally, the erratic pattern of sensed ventricular events is consistent with electrical noise. There are also lack-of-sense markers corresponding to QRS complexes; finally, there are ventricular pace markers (P) that fail to capture the ventricle.

Oversensing

This problem leads to inappropriate pauses. The sources of oversensing can be intracardiac, extracardiac, or due to EMI. Analysis of the ECG, especially with pacemaker interrogation and pacemaker marker channels, may help to determine the cause. If the oversensing is regular, analysis of the pauses may suggest T- or P-wave oversensing. T-wave oversensing can usually be eliminated by reprogramming the pacer.

Oversensing due to lead fracture, insulation break, or other electrode problems will be random and erratic. With early lead problems, the malfunction is intermittent and may be exacerbated by certain body positions or motions. In later stages, the combination of oversensing, undersensing, and failure to capture is almost always diagnostic of a lead-related problem. Programming to an asynchronous mode may temporarily control this problem while awaiting a lead replacement. Replacement of a faulty lead should be carried out promptly.

Myopotential oversensing is usually a problem in unipolar but not bipolar systems. This problem can usually be solved by reprogramming pacer sensitivity.

Undersensing

An inadequate intracardiac signal can lead to undersensing. Etiologies include inflammation or scar formation at the tissue lead interface, drugs, electrolyte abnormalities, infarction, ischemia, lead fracture or insulation breaks, and cardiac defibrillation. Usually, undersensing is a greater problem in the atrium than in the ventricle. The optimal solution is to program an enhanced sensitivity (decrease sensing level). Other etiologies for undersensing occur when intrinsic atrial or ventricular complexes fall within one of the programmed refractory periods. Undersensing can also result from a pacer functioning in an asynchronous mode (as occasionally occurs with battery depletion or resetting of the pacemaker generator).

SUGGESTED READING

Gregoratos G, Cheitlin MD, Conhill A, et al: ACC/AHA Guidelines for Implantation of Cardiac Pacemakers and Antiarrhythmia Devices: a report of the ACC/AHA Task Force on Practice Guidelines (Committee on Pacemaker Implantation). *J Am Coll Cardiol* 1998; 31:1175–1206.

Mitrani RD, Myerburg RJ, Castellanos A: Cardiac pacemakers. In: Alexander RW, Schlant RC, Fuster V, O'Rourke
RA, Roberts R, Sonnenblick EH (eds): *Hurst's The Heart*,
9th ed. New York, McGraw-Hill, 1998:1023–1055.

4

SYNCOPE

Harisios Boudoulas
Steven D. Nelson
Stephen F. Schaal
Richard P. Lewis

CLASSIFICATION

Syncope is a sudden, transient loss of consciousness. Studies have documented the multiple causes and the widely divergent mortality risks associated with an episode of syncope. On the basis of these studies, patients with a transient episode of altered consciousness (presyncope) and those with complete loss of consciousness (syncope) can be classified into three broad categories: (1) syncope unassociated with cardiac disease (*noncardiac syncope*), (2) *syncope associated with cardiac disease*, and (3) *syncope of undetermined cause*.

SYNCOPE UNASSOCIATED WITH CARDIAC DISEASE (NONCARDIAC SYNCOPE)

A classification of noncardiac syncope is given in Table 4-1. By far the most common cause is *neurocardiogenic syncope*, the *common faint*. Neurocardiogenic syncope occurs frequently in early life but can present at any age. Syncope often occurs as a response to emotional stress, real or perceived injury, sudden pain, the sight of blood, or an uncomfortable environment. Fatigue, hunger, fever, blood loss, and bed rest are predisposing factors. Neurocardiogenic syncope occurs typically in the upright posture but occasionally may occur in the sitting position. Neurocardiogenic syncope usually is preceded by pallor, perspiration, nausea, and blurred vision. Characteristic findings are low systolic blood pressure and relative bradycardia.

TABLE 4-1

CLASSIFICATION OF NONCARDIAC SYNCOPE

Neurocardiogenic
Orthostatic
Cerebrovascular
Seizure disorders
Carotid sinus hypersensitivity
Situational
 Cough
 Swallowing
 Valsalva
 Micturition
 Defecation
 Diver's
 Postprandial
Metabolic—drugs
 Hypoxia
 Hypoglycemia
 Hyperventilation, panic attacks
 Ethanol—other drugs
Other forms of syncope or conditions mimicking
syncope
 Vertigo
 Migraine
 Psychiatric

SOURCE: From Boudalas H, Nelson SD, Schaal SF, Lewis RP: Diagnosis and management of syncope. In: Alexander RW, Schlant RC, Fuster V, O'Rourke RA, Roberts R, Sonnenblick EH (eds): *Hurst's The Heart*, 9th ed. New York, McGraw-Hill, 1998:1059–1080, with permission.

Orthostatic hypotension is a syndrome characterized by a fall in arterial pressure, producing inadequate cerebral perfusion upon assumption of the upright posture. The most common causes of orthostatic hypotension are shown in Table 4-2. Hypotension may be due to venous pooling and/or blood-volume depletion, pharmacologically induced vasodilation, disorder of the autonomic nervous system, or circulating endogenous vasodilators. Often, multiple inciting factors coexist. Postural hypotension is common in the elderly as well as in patients with severe systemic arterial hypertension treated with vasodilators, in patients with diabetes mellitus, in patients on

TABLE 4-2

CAUSES OF ORTHOSTATIC SYNCOPE

Venous pooling or volume depletion
    Prolonged bed rest
    Prolonged standing
    Pregnancy
    Venous varicosities
    Blood loss
    Dehydration
Pharmacologic agents
    Antihypertensives
    Sympathetic blocking agents
    Calcium channel blockers
    Converting enzyme inhibitors
    Nitrates
    Diuretics
    Antidepressants, antipsychotics
    Phenothiazines
    Tranquilizers
    Antiparkinsonian
    Central nervous system depressants
Neurogenic
    Diabetes mellitus
    Alcoholic neuropathy
    Spinal cord disease
    Amyloidosis
    Multiple sclerosis
    Multiple cerebral infarcts
    Parkinsonism
    Tabes dorsalis
    Syringomyelia
    Idiopathic orthostatic hypotension
    Shy-Drager syndrome (multiple system atrophy)
Circulating endogenous vasodilators
    Hyperbradykinism
    Mastocytosis
Carcinoid syndrome

SOURCE: From Boudalas H, Nelson SD, Schaal SF, Lewis RP: Diagnosis and management of syncope. In: Alexander RW, Schlant RC, Fuster V, O'Rourke RA, Roberts R, Sonnenblick EH (eds): *Hurst's The Heart*, 9th ed. New York, McGraw-Hill, 1998:1059–1080, with permission.

prolonged bed rest, and in certain patients with mitral valve prolapse syndrome. Antihypertensive agents, anti-ischemic agents, and drugs used for congestive heart failure are commonly associated with postural hypotension.

Syncope in *cerebrovascular disease* is most commonly a complication of hypotension in patients with severe occlusive atherosclerosis of the major extracranial arteries. Rarely, syncope is a presenting symptom in patients with transient ischemic attacks (TIAs). Cerebral embolism is usually manifest by focal neurologic findings but may present as syncope. Nonatherosclerotic occlusive disease (Takayasu's arteritis) and mechanical obstruction from skeletal deformities may cause cerebrovascular syncope. In the *subclavian steal syndrome*, which is caused by occlusive disease of the subclavian artery proximal to the origin of the vertebral artery, syncope occurs during upper extremity exercise. Associated findings are diminished brachial arterial pulse and a supraclavicular bruit on the affected side (see also Chap. 37).

Carotid sinus syncope (CSS) is associated with a hypersensitive carotid sinus reflex. Syncope is due to either profound bradycardia, to a vasodepressor reaction, or both. CSS may be observed with neoplasms (e.g., carotid body tumor) and other masses in the carotid sinus area. Syncope in CSS may be initiated by minor stimulation of the carotid sinus—e.g., turning the head, shaving, or wearing a tight collar. Diagnostic testing for carotid sinus hypersensitivity usually is performed with the patient supine, but occasionally it can be performed with the patient upright; brief (2- to 4-s) massage on the first attempt should be performed. Reproduction of symptoms at the time of carotid sinus stimulation is helpful to confirm the diagnosis of CSS.

Situational syncopal disorders are classified according to their precipitating events. Syncope is due to a combined reflex vagal and/or a vasodepressor reaction in response to mechanical stimuli. This type of syncope may be caused by painful stimulation of the pharyngeal, laryngeal, bronchial, esophageal, or visceral mucosa or of the pleura or peritoneum. Neuropsychiatric causes of syncope include convulsive disorders, syncopal migraine, hysteria, and vertigo. Convulsive disorders usually have a distinctive clinical picture. When syncope is abrupt and associated with prolonged hypotension, transient seizure-like activity may occur, which may make the distinction difficult. When metabolic disorders produce loss of consciousness, the clinical presentation is usually distinguishable from true syncope.

SYNCOPE ASSOCIATED WITH CARDIAC DISEASE (CARDIAC SYNCOPE)

Cardiac syncope occurs when obstruction to cardiac output, a disturbance in cardiac rhythm, or both, produce a marked diminution in cerebral perfusion. The most common causes associated with cardiac syncope are summarized in Table 4-3.

TABLE 4-3

COMMON DISORDERS ASSOCIATED WITH CARDIAC SYNCOPE

Obstruction of cardiac output
Left-sided heart
    Aortic stenosis
    Hypertrophic cardiomyopathy
    Prosthetic valve malfunction
    Mitral stenosis
    Left atrial myxoma (rare)
Right-sided heart
    Eisenmenger syndrome
    Tetralogy of Fallot
    Pulmonary embolism
    Pulmonary stenosis
    Primary pulmonary hypertension
    Cardiac tamponade
Cardiac arrhythmia
    Sinoatrial disease
    Atrioventricular block
    Supraventricular tachycardia
    Ventricular tachycardia/fibrillation
Pacemaker related
    Pacemaker syndrome
    Pacemaker malfunction
    Pacemaker-induced tachycardia

SOURCE: From Boudalas H, Nelson SD, Schaal SF, Lewis RP: Diagnosis and management of syncope. In: Alexander RW, Schlant RC, Fuster V, O'Rourke RA, Roberts R, Sonnenblick EH (eds): *Hurst's The Heart*, 9th ed. New York, McGraw-Hill, 1998:1059–1080, with permission.

Syncope Related to Obstruction to Cardiac Output

Obstruction to cardiac output sufficient to cause syncope is classified as left- or right-sided, according to the predominant site of the hemodynamic impediment. Either a neurocardiogenic response or a transient arrhythmia is often the mechanism for syncope in these patients. Commonly, obstructive syncope is precipitated by exercise (*effort syncope*).

Syncope Related to Cardiac Arrhythmia

Syncope due to cardiac arrhythmia may complicate all forms of heart disease. Either an extremely high or low heart rate can induce syncope on a hemodynamic basis, or, in certain instances, the hemodynamic changes may elicit a neurocardiogenic reaction as the syncopal mechanism.

Syncope in the *sick sinus syndrome* may be associated with episodes of extreme bradycardia or supraventricular tachycardia (the tachycardia-bradycardia syndrome). *Supraventricular tachycardias* may produce syncope either entirely on the basis of excessive heart rate or by producing a neurocardiogenic response. *Ventricular tachycardia*, probably the most common cause of arrhythmic syncope, occurs with all types of cardiac disease. Syncope in the *long-QT syndrome* is almost always due to polymorphic ventricular tachycardia, usually *torsades de pointes*. Congenital long-QT syndromes are often associated with deafness. The *acquired long-QT syndromes* are caused by antiarrhythmic drugs, electrolyte disorders (hypokalemia, hypocalcemia, hypomagnesemia), antidepressants, phenothiazines, and a liquid protein diet. Other drugs commonly implicated in arrhythmic syncope not due to QT prolongation are digitalis, beta-blocking agents, calcium-channel blocking agents, theophylline, beta-agonist derivatives, caffeine, and alcohol. Pacemaker-induced syncope should be considered in patients with syncope who have VVI pacemakers and persistent atrial activity (see also Chap. 3).

SPECIAL PROBLEMS IN SYNCOPE

Syncope in the Elderly

Syncope is particularly common in the elderly because of changes related to aging. Many older patients, who have only marginal cerebral oxygen delivery at rest, frequently have multisystem disease and are likely to be taking medications that may aggravate the tendency to syncope. Arrhythmias also are commonly present in elderly patients, which renders the diagnosis of arrhythmic syncope difficult. Postprandial syncope is seen almost exclusively in the elderly.

Syncope Due to Multiple Causes

In many situations, the occurrence of syncope requires that a constellation of events occur either simultaneously or in sequence. Without the full complex, the patient may note only light-headedness (presyncope) or perhaps no definable symptoms.

Certain predisposing factors such as prolonged bed rest, old age, and fever may determine whether or not syncope occurs. Furthermore, whether or not a neurocardiogenic reaction develops determines if a given stimulus initiates syncope. Thus, it is likely that many cases of syncope of "unknown cause" are in fact due to multiple causes.

Syncope and Sudden Death

Patients with cardiac syncope have a high incidence of sudden death, suggesting that in such patients syncope can be a harbinger of death (see also Chap. 2).

Recurrent Syncope

Syncope is recurrent in up to one-third of patients, most of whom have neurocardiogenic syncope. It is necessary to establish a correct diagnosis and to institute proper therapy so as to minimize recurrence. In refractory cases, restrictions on activity and on and off the job may be necessary.

DIAGNOSTIC EVALUATION

The differential diagnosis of syncope based on history and physical examination is shown in Figs. 4-1 and 4-2. The basic clinical and laboratory examination, along with the electrocardiogram, the chest x-ray, and echo-Doppler study, permit correct diagnosis in most patients and provide a firm differential diagnosis for the remainder (see Fig. 4-3). The extent of evaluation for patients in whom no diagnosis is evident after initial evaluation should be based on the perceived risk of mor-

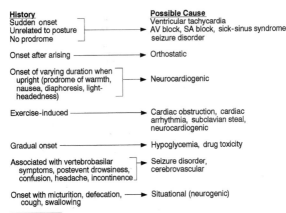

History		Possible Cause
Sudden onset		Ventricular tachycardia
Unrelated to posture	→	AV block, SA block, sick-sinus syndrome
No prodrome		seizure disorder

Onset after arising → Orthostatic

Onset of varying duration when upright (prodrome of warmth, nausea, diaphoresis, light-headedness) → Neurocardiogenic

Exercise-induced → Cardiac obstruction, cardiac arrhythmia, subclavian steal, neurocardiogenic

Gradual onset → Hypoglycemia, drug toxicity

Associated with vertebrobasilar symptoms, postevent drowsiness, confusion, headache, incontinence → Seizure disorder, cerebrovascular

Onset with micturition, defecation, cough, swallowing → Situational (neurogenic)

FIGURE 4-1

Differential diagnosis of syncope based on history. AV = atrioventricular; SA = sinoatrial. (From Boudalas H, Nelson SD, Schaal SF, Lewis RP: Diagnosis and management of syncope. In: Alexander RW, Schlant RC, Fuster V, O'Rourke RA, Roberts R, Sonnenblick EH (eds): *Hurst's The Heart*, 9th ed. New York, McGraw-Hill, 1998:1059–1080, with permission.)

tality and morbidity (high in patients with cardiac syncope, the elderly, and those with abrupt syncope). The head-up tilt test has been widely used to evaluate patients with syncope. Although this test is still being evaluated to determine its sensitivity and specificity, it has proved extremely useful for reproducing the various forms of "neurocardiogenic" syncope. The test is noninvasive, but it is a sophisticated technique that requires a skilled laboratory to achieve useful results. Recently, head-up tilt also has been used to evaluate therapy. Signal-averaged electrocardiography is another new and sophisticated noninvasive technique for identifying patients with an anatomic/electrophysiologic substrate for ventricular tachycardia (Fig. 4-4). The test is most useful in patients with ischemic heart disease.

While the initial workup often established the diagnosis of obstructive cardiac syncope, cardiac catheterization usually is necessary, particularly when corrective cardiac surgery is contemplated. When no arrhythmia can be documented noninvasively and cardiac disease is present, an invasive electrophysiologic study usually is indicated. If these studies are used

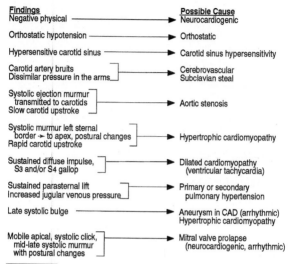

Findings		Possible Cause
Negative physical	→	Neurocardiogenic
Orthostatic hypotension	→	Orthostatic
Hypersensitive carotid sinus	→	Carotid sinus hypersensitivity
Carotid artery bruits Dissimilar pressure in the arms	→	Cerebrovascular Subclavian steal
Systolic ejection murmur transmitted to carotids Slow carotid upstroke	→	Aortic stenosis
Systolic murmur left sternal border → to apex, postural changes Rapid carotid upstroke	→	Hypertrophic cardiomyopathy
Sustained diffuse impulse, S3 and/or S4 gallop	→	Dilated cardiomyopathy (ventricular tachycardia)
Sustained parasternal lift Increased jugular venous pressure	→	Primary or secondary pulmonary hypertension
Late systolic bulge	→	Aneurysm in CAD (arrhythmic) Hypertrophic cardiomyopathy
Mobile apical, systolic click, mid-late systolic murmur with postural changes	→	Mitral valve prolapse (neurocardiogenic, arrhythmic)

FIGURE 4-2

Differential diagnosis of syncope based on physical examination. CAD = coronary artery disease. (From Boudalas H, Nelson SD, Schaal SF, Lewis RP: Diagnosis and management of syncope. In: Alexander RW, Schlant RC, Fuster V, O'Rourke RA, Roberts R, Sonnenblick EH (eds): *Hurst's The Heart*, 9th ed. New York, McGraw-Hill, 1998:1059–1080, with permission.)

properly, the diagnosis of arrhythmic syncope can be made in nearly three-quarters of those suspected of having arrhythmic syncope.

TREATMENT OF NONCARDIAC SYNCOPE

When a neurocardiogenic syncopal effect occurs, the patient should be placed recumbent until he or she recovers. Elevation of the legs to increase venous return may be helpful; in extreme cases, intravenous atropine or vasopressors are required. Chronic therapy for neurocardiogenic syncope is required only when syncope is recurrent. Beta-adrenergic blocking drugs are efficacious in some instances and are often the drugs of choice for initial therapy. In *orthostatic hypotension*, primary emphasis must be placed on the treatment of reversible causes, such

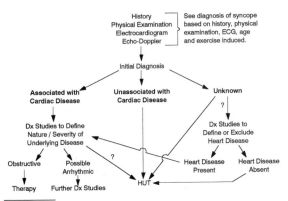

FIGURE 4-3

Basic schema for diagnostic evaluation of syncope. ECG = electrocardiogram; Dx = diagnostic; HUT = head-up tilt. (From Boudalas H, Nelson SD, Schaal SF, Lewis RP: Diagnosis and management of syncope. In: Alexander RW, Schlant RC, Fuster V, O'Rourke RA, Roberts R, Sonnenblick EH (eds): *Hurst's The Heart*, 9th ed. New York, McGraw-Hill, 1998:1059–1080, with permission.)

as replacement of blood volume, prevention of venous pooling, deletion of hypotensive drugs, and use of support stockings. Syncope in *cerebrovascular disease* demands treatment of the primary vascular disorder, including antiplatelet agents (aspirin, ticlopidine) for TIAs. Surgical endarterectomy is indicated when high-grade stenosis is present. In severe cases with CSS, a permanent atrioventricular pacemaker is indicated, because ventricular pacing may accentuate hypotension. *Situational syncope* necessitates intervention directed at the initiating mechanisms. *Neuropsychiatric and metabolic syncope* demand specific attention to the primary cause.

TREATMENT OF CARDIAC SYNCOPE

In syncope due to *obstructive heart disease*, surgical correction of the hemodynamic obstruction, if feasible, is the treatment of choice. Syncope in hypertrophic cardiomyopathy may respond to pharmacotherapy; surgical management must be considered for intractable syncope. In certain cases, implantation of an atrioventricular sequential pacemaker may improve symptoms

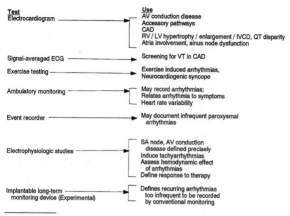

Test	Use
Electrocardiogram	AV conduction disease Accessory pathways CAD RV / LV hypertrophy / enlargement / IVCD / QT disparity Atria involvement, sinus node dysfunction
Signal-averaged ECG	Screening for VT in CAD
Exercise testing	Exercise induced arrhythmias, Neurocardiogenic syncope
Ambulatory monitoring	May record arrhythmias; Relates arrhythmia to symptoms Heart rate variability
Event recorder	May document infrequent paroxysmal arrhythmias
Electrophysiologic studies	SA node, AV conduction disease defined precisely Induce tachyarrhythmias Assess hemodynamic effect of arrhythmias Define response to therapy
Implantable long-term monitoring device (Experimental)	Defines recurring arrhythmias too infrequent to be recorded by conventional monitoring

FIGURE 4-4

Diagnostic tests that can be used for the evaluation of arrhythmic syncope. AV = atrioventricular; CAD = coronary artery disease; RV/LV = right ventricular/left ventricular; IVCD = intraventricular conduction defect; VT = ventricular tachycardia; SA = sinoatrial. (From Boudalas H, Nelson SD, Schaal SF, Lewis RP: Diagnosis and management of syncope. In: Alexander RW, Schlant RC, Fuster V, O'Rourke RA, Roberts R, Sonnenblick EH (eds): *Hurst's The Heart*, 9th ed. New York, McGraw-Hill, 1998:1059–1080, with permission.)

(see Chap. 18). In high-risk patients, the use of an automatic internal cardioventer defibrillator may be indicated. For severe pulmonary hypertension, there are few effective therapeutic options; heart and/or lung transplantation may be indicated in selected patients (see Chap. 24).

Treatment of *arrhythmic syncope* should be directed at the underlying cardiovascular disease and mitigation of precipitating events. Use of arrhythmia-inciting agents (e.g., caffeine, alcohol) should be forbidden. Effective antiarrhythmic drug therapy of supraventricular arrhythmias requires careful identification of the type of tachycardia; an antiarrhythmic drug–induced proarrhythmic effect must be avoided. Catheter ablation for a variety of supraventricular tachycardias affords a clinical cure with avoidance of antiarrhythmic drugs. Combined use of pacemaker and drug therapy may be required for patients with the "tachycardia-bradycardia syndrome." The multicausal nature of syncope in patients with sick-sinus syndrome demands careful evaluation before the institution of permanent pacemaker therapy. Most patients with the sick-sinus syndrome, however, with suspected bradyarrhythmias as a cause of syncope, are dramatically improved with dual-chamber pacemaker therapy. For ventricular arrhythmias, amiodarone and sotalol are often the antiarrhythmic agents of choice, but implantable defibrillators are being used increasingly in conjunction with drug therapy. In the *congenital long-QT syndrome*, beta-blocking drugs often control recurrent syncope; left sympathetic stellectomy may be attempted for intractable syncope. In *acquired long-QT syndromes*, discontinuation of offending drugs and restoration of metabolic balance usually prevents recurrence. A proarrhythmic cause must be considered when syncope occurs after antiarrhythmic therapy has begun. Coronary bypass surgery or coronary angioplasty may prevent syncope related to ventricular tachycardia initiated by myocardial ischemia. When revascularization is not feasible, beta-adrenergic blocking agents and nitrates should be used (see also Chap. 8).

SUGGESTED READING

Abbond F: Neurocardiogenic syncope. *N Engl J Med* 1993; 328:1117–1120.

Boudoulas H, Nelson SD, Schaal SF, Lewis RP. Diagnosis and management of syncope. In: Alexander RW, Schlant RC, Fuster V, O'Rourke RA, Roberts R, Sonnenblick EH (eds): *Hurst's The Heart*, 9th ed. New York, McGraw-Hill, 1998:1059–1080.

Boudoulas H, Weissler AM, Lewis RP, Warren JV: The clinical diagnosis of syncope. *Curr Probl Cardiol* 1982; 7:6–40.

Bousser MG, Dubois B, Castaigne P: Transient loss of consciousness in ischemic cerebral events: a study of 557 ischemic strokes and transient ischemic attacks. *Ann Intern Med* 1980; 132:300–307.

Grech ED, Ramsdale DR: Exertional syncope in aortic stenosis: evidence to support inappropriate left ventricular baroreceptor response. *Am Heart J* 1991; 121:603–606.

Jansen RWMM, Lipsitz LA: Postprandial hypotension: epidemiology, pathophysiology, and clinical management. *Ann Intern Med* 1995; 122:286–295.

Linzer M, Yang EH, Estes NA, et al: Diagnosing syncope: Part 1. Value of history, physical examination, and electrocardiography. *Ann Intern Med* 1997; 126:989–996.

Moss AJ, Schwartz PJ, Crampton RS, et al: The long QT syndrome: prospective longitudinal study of 328 families. *Circulation* 1991; 84:1136–1144.

Moya A, Permanyer-Miralda G, Sagrista-Sauleda J, et al: Limitations of head-up tilt test for evaluating the efficacy of therapeutic interventions in patients with vasovagal syncope: results of a controlled study of etilefrine versus placebo. *J Am Coll Cardiol* 1995; 25:65–69.

Schaal SF, Nelson SD, Boudoulas H, Lewis RP: Syncope. *Curr Probl Cardiol* 1992; 17:211–264.

Tea SH, Mansourati J, L'Heveder G, et al: New insights into the pathophysiology of carotid sinus syndrome. *Circulation* 1996; 93:1411–1418.

CHAPTER

5

SUDDEN CARDIAC DEATH

Erica Diana Engelstein
Douglas P. Zipes

EPIDEMIOLOGY

Incidence

Sudden cardiac death accounts for approximately 300,000 to 400,000 deaths yearly in the United States, depending on the definition used. A definition using a time interval of 1 h or less between onset of symptoms and hemodynamic collapse identifies sudden cardiac death populations having a high proportion (up to 91 percent) of arrhythmic deaths. In autopsy-based studies, a cardiac etiology of sudden death has been reported in 60 to 70 percent of sudden death victims. Sudden cardiac death is the most common and often the first manifestation of coronary heart disease and is responsible for half the mortality from cardiovascular disease, which remains the main cause of death in this country. The National Center for Health Statistics estimated the annual incidence of sudden cardiac death in the United States in 1985 to be 1.9 per 1000 in men and 0.6 per 1000 in women, resulting in 223,864 deaths of a total of 399,324 deaths from ischemic heart disease. In the United States, populations-based studies have documented a decline (15 to 19 percent) in the incidence of sudden cardiac deaths caused by coronary heart disease since the early 1980s.

Influence of Age, Race, and Gender

The epidemiology of sudden cardiac death parallels to a great extent that of coronary heart disease.

AGE

The incidence of sudden cardiac death increases with age, in both men and women, whites and nonwhites, due to the higher prevalence of ischemic heart disease at older ages. Among patients with coronary artery disease, however, the proportion of coronary deaths that are sudden decreases with age.

RACIAL DIFFERENCES

The annual age-adjusted incidence of sudden cardiac arrest is higher in blacks than in whites: 3.4 verus 1.6 percent per 1000 population ($p < 0.001$). Not only is the sudden cardiac death rate higher, but the overall survival is also lower in blacks than in whites (10.2 versus 16.7 percent). The differences in outcome could not be accounted for by differences in emergency medical team response time or administration of advanced cardiac life support. It is more likely that the higher incidence of sudden cardiac death and poorer outcomes in blacks are due to differences in health and socioeconomic status.

GENDER

Sudden cardiac death has a much higher incidence in men than in women, reflecting gender differences in the incidence of coronary heart disease. Between 70 and 89 percent of sudden cardiac deaths occur in men, and their annual incidence of sudden cardiac death in men is overall three to four times higher than in women. Among survivors of cardiac arrest, women are more likely than men to have valvular heart disease, idiopathic dilated cardiomyopathy, or "normal" heart as opposed to coronary heart disease.

Sudden Cardiac Death in the Young

Sudden cardiac death accounts for 19 percent of sudden deaths in children between 1 and 13 years of age and for 30 percent in persons between 14 and 21 years of age. Structural cardiac abnormalities can be identified in over 90 percent of young victims of sudden cardiac death. About 40 percent of sudden cardiac deaths in the pediatric population occur in patients with surgically treated congenital cardiac abnormalities. In the majority of young victims, however, sudden cardiac death is often the first manifestation of the underlying cardiac disease in otherwise healthy-appearing individuals. The most common causes of sudden cardiac death in the first three decades of life are myocarditis, hypertrophic cardiomyopathy, congenital anomalies of the coronary arteries, atherosclerotic coronary heart disease, conduction system abnormalities, mitral valve prolapse, and aortic dissection. Among young people with known cardiac disease, aortic stenosis and primary or secondary pulmonary vascular obstruction were most common in patients without prior cardiac surgery, whereas tetralogy of Fallot and transposition of the great vessels were more common in postoperative patients.

Risk Factors for Sudden Cardiac Death

Since a majority of sudden cardiac deaths in the adult population occur in patients with coronary heart disease, risk factors for sudden cardiac death are similar to those for coronary artery disease, making a high-risk profile for coronary heart disease a high-risk profile for sudden cardiac death. *Left ventricular dysfunction has been identified as the strongest independent predictor of sudden cardiac death.* In the Framingham Study, a multivariate prediction model based on known coronary risk factors such as age, systolic blood pressure, left ventricular hypertrophy, intraventricular block or nonspecific abnormalities on electrocardiography (ECG), serum cholesterol glucose intolerance, vital capacity, smoking, relative weight, and heart rate identified as being in the upper decile of multivariate risk 53 percent of men and 42 percent of women at risk of sudden cardiac death. Despite the fact that numerous population-based epidemiologic studies have shown a strong relationship of cardiovascular risk factors to the incidence of coronary heart disease and that of sudden cardiac death, none of them has identified a single set of risk factors specific for sudden cardiac death. The only coronary risk factors that seem to carry a disproportionately high risk for sudden cardiac death are those related to lifestyle and psychosocial factors.

CIGARETTE SMOKING
In the Framingham Study, the annual incidence of sudden cardiac deaths increased from 13 in 1000 among nonsmokers to 31 in 1000 among those smoking more than 20 cigarettes per day. People who stopped smoking had a prompt reduction in mortality due to coronary heart disease compared with those who continued to smoke, irrespective of the duration of prior smoking habits.

STRESS AND SOCIOECONOMIC STATUS
There are many reports linking stress, particularly emotional stress, to sudden cardiac death. Based on the difference of average and actual daily sudden cardiac death rates following an earthquake in California, it was estimated that as many as 40 percent of sudden cardiac deaths are precipitated by emotional stress.

PHYSICAL ACTIVITY

There is increasing evidence that regular physical activity may help prevent coronary heart disease and its complications. However, triggering of sudden cardiac death and acute myocardial infarction by vigorous exercise has been reported. Emergency medical records show that in adults, 11 to 17 percent of victims of cardiac arrest collapsed during or immediately after exertion, although the amount of exertion is rarely quantified. There is increasing experimental evidence that regular exercise may prevent ischemia-induced ventricular fibrillation and death by altering autonomic function, specifically by increasing vagal reflexes.

Sudden Cardiac Death in Competitive Athletes

Sudden cardiac death in competitive athletes is an extremely rare event. Between 10 and 25 sports-related sudden deaths from cardiac causes occur annually in the United States. Although, unfortunately, sudden cardiac death is often the first manifestation of their disease, a majority of sudden cardiac deaths in athletes occur in persons with underlying cardiac pathology. In athletes below 35 years of age, a vast majority of sudden cardiac deaths arise from a variety of congenital cardiovascular diseases, most commonly hypertrophic cardiomyopathy, arrhythmogenic right ventricular cardiomyopathy, congenital coronary artery anomalies, and aortic rupture associated with Marfan's syndrome. Atherosclerotic coronary artery disease is found in only about 10 percent of athletes in this younger age group, compared with 80 percent in athletes older than 35 years.

MECHANISM OF SUDDEN CARDIAC DEATH

The Relationship between Structure and Function in Sudden Cardiac Death

A vast majority of patients who are victims of sudden cardiac death have cardiac structural abnormalities. In the adult population, these consist predominantly of coronary artery disease, cardiomyopathies, valvular heart disease, and abnormalities of the conduction system. These structural changes provide the substrate for ventricular tachyarrhythmias, which represent the cause of sudden cardiac death in most cases. It is important to recognize the role of triggering factors—such as fluctuations in the autonomic nervous system, electrolyte abnormalities, and proarrhythmic effects of drugs—in the initiation of ventricular arrhythmias resulting in sudden cardiac death (Fig. 5-1).

Tachyarrhythmias versus Bradyarrhythmias in Sudden Cardiac Death

Ventricular fibrillation is the first recorded rhythm in approximately 70 percent of patients who have cardiac arrest. Sustained ventricular tachycardia is only rarely (<2 percent) documented as the initial rhythm, but it is unknown how often it precedes and precipitates ventricular fibrillation. In a series of 157 patients who were wearing an ambulatory ECG (Holter) recorder at the time of their cardiac arrest, primary ventricular fibrillation was documented in 8 percent, ventricular tachycardia degenerating into ventricular fibrillation in 62 percent, and *torsades de pointes* in 13 percent. Electromechanical dissociation and asystole are found in about 30 percent of cardiac arrest patients, and this finding is usually related to the time interval from collapse to first monitoring of the rhythm, suggesting that it is a later manifestation of cardiac arrest. Ambulatory ECG recordings demonstrated that even in patients with atrioventricular or intraventricular conduction defects, ventricular tachyarrhythmias were most often the mode of recurrent cardic arrest.

Role of Ischemia

Ventricular arrhythmias during experimental acute ischemia occur in two peaks, one between 2 and 10 min following coronary occlusion and the second at 15 to 20 min. Rapid polymorphic ventricular tachycardias and ventricular fibrillation are the characteristic arrhythmias during the early stages of ischemia and are the cause of sudden cardiac death. The second peak of ventricular arrhythmias coincides with a peak in catecholamine release; mechanisms such as abnormal automaticity or triggered activity have been invoked. Ventricular tachyarrhythmias also occur frequently during reperfusion of the infarct zone, as seen after administration of thrombolytic

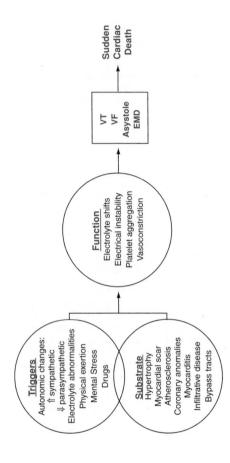

FIGURE 5-1

Interaction between structural cardiac abnormalities, functional changes, and triggering factors in the pathophysiology of sudden cardiac death. The role of triggering factors, such as changes in autonomic tone or reflexes, is being increasingly recognized. EMD, electromechanical dissociation; VF, ventricular fibrillation; VT, ventricular tachycardia.

agents. In the subacute phase of myocardial infarction (within the first 3 days), sudden cardiac death may occur due to ventricular fibrillation in the setting of frequent premature ventricular contractions (PVCs). These ventricular arrhythmias have been shown to be predominantly due to abnormal impulse initiation, consistent with abnormal automaticity, and are clinically manifest as accelerated idioventricular rhythm or idioventricular tachycardia. These arrhythmias appear to arise for the most part from surviving Purkinje fibers in the subendocardial border zone of a transmural infarction. They usually subside after 2 to 3 days, in parallel with the normalization of the resting membrane potential and action potential duration of Purkinje fibers, and have no prognostic significance for the development of late arrhythmias.

Role of the Autonomic Nervous System in the Genesis of Arrhythmias

There is increasing evidence that cardiac abnormalities associated with a high risk of sudden cardiac death are accompanied by changes in the autonomic innervation of the heart. Myocardial infarction, for instance, has been shown to cause regional cardiac sympathetic and parasympathetic denervation, not only in the infarcted area but also in the region apical to the infarct due to the interruption of afferent and efferent nerve fibers transversing the infarct zone. The denervated areas show supersensitivity to catecholamine infusion, with disproportionate shortening of action potential duration and refractoriness. This autonomic heterogeneity may predispose to the development of arrhythmia by creating dispersion of refractoriness and/or conduction.

Whereas sympathetic activation favors the onset of life-threatening cardiac arrhythmias, vagal activation has been suggested to have a protective effect in the presence of tonic sympathetic stimulation due, at least in part, to antiadrenergic effects at the prejunctional (via reduction in norepinephrine release) and cellular levels (inhibition of adenylate cyclase via inhibitory G proteins). Because it is difficult to study directly the effects of vagal activity on ventricular electrophysiological properties, the behavior of the sinus node has been used as a surrogate by measuring indices of heart rate variability (reflecting primarily tonic vagal activity) and evaluating baroreflex sensitivity (a measure of reflex vagal activity).

CARDIAC DISEASES ASSOCIATED WITH
SUDDEN CARDIAC DEATH (TABLE 5-1)

Ischemic Heart Disease

CORONARY ATHEROSCLEROSIS

In survivors of cardiac arrest, coronary heart disease is found in 40 to 86 percent of patients, depending on the age and gender of the population studied. Although the majority of patients who suffer sudden cardiac death have severe multivessel coronary artery disease, fewer than half of the patients resuscitated from ventricular fibrillation evolve evidence of myocardial infarction by elevated cardiac enzymes and less than one-quarter have Q-wave myocardial infarction. Healed infarctions are present in 44 to 82 percent of hearts of sudden cardiac death victims at autopsy and in 38 to 91 percent of survivors of cardiac arrest. Since coronary artery disease is the major substrate of sudden cardiac death, risk stratification following myocardial infarction is an important step in the prevention of sudden cardiac death. Few variables—mainly reduced left ventricular ejection fraction (<40 percent), frequent PVCs (>10/h), and use of digitalis—are independent risk factors for sudden versus nonsudden cardiac death following myocardial infarction. The incidence of sudden cardiac death in the first 2 years after myocardial infarction ranged from 11 to 18 percent. The variables identified to predict sudden cardiac death following myocardial infarction are better in selecting a low-risk population for sudden cardiac death rather than predicting who will die suddenly. In the absence of frequent PVCs and with a normal left ventricular ejection fraction following myocardial infarction, the risk of sudden cardiac death is low (less than 2 percent in the first year). Heart rate variability, baroreflex sensitivity, nonlinear dynamics, T-wave alternans, and imaging of the cardiac autonomic innervation are currently under investigation as noninvasive methods for risk/stratification. Electrophysiologic studies can help to identify high-risk patients for sudden death in select subgroups with ischemic heart disease.

NONATHEROSCLEROTIC DISEASE
OF THE CORONARY ARTERIES

Several nonatherosclerotic diseases of the coronary arteries are associated with increased risk of sudden cardiac death precipitated by cardiac ischemia. *Congenital coronary artery anomalies*, found in approximately 1 percent of all patients undergoing angiography and in 0.3 percent of patients undergoing autopsy, have been complicated by sudden cardiac death in up to about 30 percent of patients and are often exercise-related. Life-threatening ventricular arrhythmias and sudden cardiac death have been described in 13 percent of patients with documented *coronary artery spasm*. Calcium-channel blockers are effective in many patients in preventing coronary spasm and appear also to protect from malignant ventricular arrhythmias if the attacks can be completely abolished.

Cardiomyopathies

IDIOPATHIC DILATED CARDIOMYOPATHY

Idiopathic dilated cardiomyopathy is the substrate for approximately 10 percent of sudden cardiac deaths in the adult population. In an overview of 14 studies including 1432 patients with idiopathic dilated cardiomyopathy, the mean mortality rate after a follow-up of 4 years was 42 percent, with 28 percent of deaths classified as sudden. Sudden cardiac death in idiopathic dilated cardiomyopathy is usually attributed to ventricular tachyarrhythmias based on the high frequency of complex ventricular ectopy found in these patients. The terminal event can, however, also be asystole or electromechanical dissociation, especially in patients with advanced left ventricular dysfunction. Factors potentially contributing to the generation of arrhythmias in idiopathic dilated cardiomyopathy are mechanoelectrical feedback, electrolyte depletion due to chronic diuretic therapy, excessive activation of the sympathetic nervous and renin-angiotensin systems, and proarrhythmic effects.

Risk stratification of patients with idiopathic dilated cardiomyopathy is difficult because there are few clinical predictors specific for sudden cardiac death. Patients with idiopathic dilated cardiomyopathy have a very high incidence of ventricular ectopy, with simple PVCs, complex PVCs, and nonsustained ventricular tachycardia present in 94, 76, and 40 percent, respectively, limiting their prognostic value by a low specificity. The only clinical variable that identifies patients with a higher risk of sudden cardiac death in this population is syncope. The value of electrophysiologic testing in patients with idiopathic dilated cardiomyopathy is limited since it frequently results in the initiation of polymorphic ventricular tachycardia or ventricular fibrillation, which appear to

TABLE 5-1

CARDIAC ABNORMALITIES ASSOCIATED WITH SUDDEN CARDIAC DEATH

Ischemic heart disease
  Coronary atherosclerosis
  Acute myocardial infarction
  Chronic ischemic cardiomyopathy
  Anomalous origin of coronary arteries
  Hypoplastic coronary artery
  Coronary artery spasm
  Coronary artery dissection
  Coronary arteritis
  Small vessel disease
Nonischemic heart disease
  Cardiomyopathies
  Idiopathic dilated cardiomyopathy
  Hypertrophic cardiomyopathy
  Hypertensive cardiomyopathy
  Right ventricular cardiomyopathy

  Pulmonary vascular obstructive disease
  Congenital aortic stenosis
  Primarily electrical abnormalities
  Long-QT syndrome
  Wolff-Parkinson-White syndrome
  Congenital heart block
  Idiopathic ventricular fibrillation
  Syndrome of right bundle branch block,
    ST-segment elevation and sudden death
    (Brugada syndrome)
  Nocturnal sudden death in Southeast
    Asian men (Pokkuri syndrome)
  Drug-induced and other toxic agents
  Antiarrhythmic drugs (class Ia, Ic, III)
  Erythromycin
  Terfenadine

94

Infiltrative and inflammatory heart disease
Sarcoidosis
Amyloidosis
Hemochromatosis
Myocarditis
Valvular heart disease
Aortic stenosis
Aortic regurgitation
Mitral valve prolapse
Infective endocarditis
Congenital heart disease
Tetralogy of Fallot
Transposition of the great vessels
    (post-Mustang/Senning)
Ebstein's anomaly

Pentamidine
Psychotropic drugs (tricyclic antidepres-
sants, chlorpromazine)
Cocaine
Alcohol
Phosphodiesterase inhibitors
Electrolyte abnormalities
Hypokalemia
Hypomagnesiemia
Hypercalcemia
Anorexia nervosa and bulimia
Liquid protein dieting
Diuretics

be nonspecific findings; and the absence of inducible ventricular tachyarrhythmias in this population does not accurately predict a low risk for sudden cardiac death (Chap. 17).

HYPERTROPHIC CARDIOMYOPATHY

The incidence of sudden cardiac death in patients with hypertrophic cardiomyopathy is 2 to 4 percent per year in adults, and 4 to 6 percent per year in children and adolescents. A review of 78 patients with hypertrophic cardiomyopathy who died suddenly or survived a cardiac arrest episode showed that 71 percent were younger than 30 years of age, 54 percent were without functional limitation, and 61 percent were performing sedentary or minimal physical activity at the time of cardiac arrest. The mechanism of sudden cardiac death in hypertrophic cardiomyopathy is not clear. Primary arrhythmias, hemodynamic events with diminished stroke volume, and/or ischemia have been implicated (see also Chap. 18).

There are few predictors of sudden cardiac death in patients with hypertrophic cardiomyopathy. A clinical history of aborted sudden cardiac death or sudden death in family members indicates a worse prognosis, as does onset of symptoms in childhood. Hemodynamic and echocardiographic variables such as left ventricular wall thickness or the presence of outflow tract obstruction are not useful in identifying patients at high risk for sudden cardiac death. Ambulatory ECG (Holter) monitoring has been reported to be of some value in identifying patients with hypertrophic cardiomyopathy who are at risk for sudden cardiac death. Sustained ventricular tachyarrhythmias, predominantly rapid polymorphic ventricular tachycardia, have been induced in 27 to 43 percent of patients with hypertrophic cardiomyopathy at electrophysiologic study, but the prognostic significance of these data is controversial. In approximately 50 percent of families with hypertrophic cardiomyopathy, different missense mutations in the beta-cardiac myosin heavy-chain gene have been identified, and the location of the mutation appeared to influence survival.

HYPERTENSIVE CARDIOMYOPATHY

Left ventricular hypertrophy as detected by electrocardiography is an independent risk factor for cardiovascular deaths and, in particular, sudden cardiac death in patients who also have a history of hypertension. In the Framingham Study, ECG evidence of left ventricular hypertrophy doubled the risk of sudden cardiac death.

ARRHYTHMOGENIC RIGHT VENTRICULAR DYSPLASIA (ARVD)

ARVD, a predominantly right ventricular cardiomyopathy characterized by fatty or fibromatous replacement of myocardium and recurrent ventricular tachycardia with multiple left bundle branch block morphologies, is a rare cause of sudden cardiac death except in a few endemic regions. In patients with ARVD, ventricular tachycardia is often precipitated by exercise, and its induction is usually catecholamine-sensitive at electrophysiologic study. The course and prognosis of ARVD is highly variable and difficult to predict. The annual incidence of sudden cardiac death in ARVD has been estimated to be about 2 percent despite various treatments.

Valvular Heart Disease

The risk of sudden cardiac death in asymptomatic patients with aortic stenosis or regurgitation appears to be low. In contrast, in the presurgical era, sudden cardiac death was one of the three most common types of death in symptomatic patients with aortic stenosis, the other two being bacterial endocarditis and congestive heart failure. In 831 patients receiving a Bjork-Shiley prosthesis in the aortic, mitral, or double-valve position, the incidence of sudden cardiac death in the subgroups was 1.8, 3.5, and 4 percent, respectively, over a follow-up of 7 years.

MITRAL VALVE PROLAPSE

Whether or not mitral valve prolapse is a cause of sudden cardiac death is controversial. The prevalence of mitral valve prolapse is so high (4 to 5 percent of the general population and up to 17 percent of young women) that its presence may be just a coincidental finding in victims of sudden cardiac death and not causally related. The overall 8-year probability of survival in a group of 237 asymptomatic or minimally symptomatic patients with echocardiographically documented mitral valve prolapse was not significantly different from that for a matched control population. On the other hand, mitral valve prolapse may not always be benign, since, in a significant number of sudden cardiac death victims, mitral valve prolapse is the only structural cardiac disease found. Patients with mitral valve prolapse associated with mitral regurgitation and left ventricular dysfunction are clearly at higher risk for complications such as infective endocarditis, cerebroembolic events, and sudden

cardiac death. Several risk factors for sudden cardiac death
have been identified in asymptomatic or mildly symptomatic
patients with mitral valve prolapse without significant mitral
regurgitation, including mitral valve annular circumference,
thickness of the anterior and posterior mitral valve leaflets, the
presence and extent of endocardial plaque, and the presence or
absence of redundant mitral valve leaflets on M-mode echocardiography.

Inflammatory and Infiltrative Myocardial Disease

Any inflammatory myocardial disease can cause sudden cardiac
death due to either ventricular tachyarrhythmias or complete
heart block. Histologic findings suggestive of *myocarditis* have
been reported in 10 to 44 percent of young victims of sudden
cardiac death. In adults, the diagnosis of myocarditis is made
much less frequently, perhaps because of concurrent structural
heart disease or because the late manifestations of the disease
are indistinguishable from idiopathic dilated cardiomyopathy.
In South America, however, myocarditis due to specific
pathogens such as trypanosomes (causing Chagas' disease) is
the most frequent cause of cardiomyopathy and related sudden
cardiac death. Patients with infective endocarditis may also be
at risk for sudden cardiac death due to acute coronary emboli
from valvular vegetations. More often, sudden cardiac death
during or following infective endocarditis is caused by acute
hemodynamic deterioration due to valvular failure. An intramyocardial abscess can also be a cause of ventricular tachycardia
leading to sudden cardiac death.

Infiltrative cardiomyopathies, such as primary or secondary *amyloidosis*, *hemochromatosis*, or *sarcoidosis*, have
been associated not only with predominantly cardiac conduction defects but also with ventricular tachyarrhythmias and
sudden cardiac death. Ventricular tachycardia is sometimes
the mode of presentation of sarcoidosis; it can usually be
reproduced by programmed electrical stimulation and is associated with a high rate of arrhythmia recurrence and sudden
cardiac death.

Congenital Heart Disease

An increased risk for sudden cardiac death due to an arrhythmia has been found predominantly in four congenital condi-

tions: tetralogy of Fallot, transposition of the great vessels, aortic stenosis, and pulmonary vascular obstruction. Patients who have undergone reparative surgery for *tetralogy of Fallot* have a reported risk of sudden cardiac death of 6 percent before age 20. A QRS duration ≥ 180 ms was found to be the most sensitive predictor of sudden cardiac death and ventricular tachyarrhythmias in 178 adult survivors of tetralogy of Fallot repair, and to correlate with other parameters of right ventricular volume overload. *Transposition of the great vessels* (*post–Mustard/Senning*) is associated with a 2 to 8 percent rate of late sudden cardiac death, which is due in some cases to sinus node dysfunction and in others to ventricular tachyarrhythmias. Sudden cardiac death is often (45 to 60 percent) the mode of death in patients with *primary* or *secondary pulmonary hypertension*. Death is often precipitated by general anesthesia, dehydration, exertion, or pregnancy. The sudden cardiac death risk in *congenital aortic stenosis* is estimated to be 1 percent pear year and occurs predominantly in symptomatic patients with severe left ventricular hypertrophy. *Ebstein's anomaly* is frequently (up to 25 percent) associated with the presence of accessory pathways and Wolff-Parkinson-White (WPW) syndrome, which carries a small risk of sudden cardiac death. *Congenital heart block* without associated structural heart disease occurs in 1 of 20,000 infants, and a moderate decrease in heart rate is usually well tolerated. A maternal risk factor is systemic lupus erythematosus. Patients with severe bradycardia, however, have a tendency to develop ventricular arrhythmias. Pacemaker therapy has virtually eliminated the risk of sudden cardiac death in this population.

Primary Electrical Abnormalities

LONG-QT SYNDROME

Sudden cardiac death is one of the hallmarks of the idiopathic long-QT syndrome (LQTS). The prolonged QT interval reflects abnormal prolongation of repolarization. Other characteristics of this disorder include abnormal T-wave contours, relative sinus bradycardia a family history of early sudden death, and a propensity for recurrent syncope and sudden cardiac death due to polymorphic ventricular tachycardia (*torsades de pointes*) and ventricular fibrillation. Over 90 percent of the congenital forms of LQTS have been linked to four specific chromosomal defects, resulting in a genetically based classification (LQT 1–4) with important functional and prognostic implications. Carriers of the LQT gene have been reported to have a 5 percent incidence of aborted sudden cardiac death and a 63 percent incidence of recurrent syncope. Multivariate analysis in a registry population identified female gender, congenital deafness, history of syncope, and a documented episode of *torsades de pointes* or ventricular fibrillation as independent risk factors for syncope or sudden cardiac death. Genetic typing in the future may facilitate risk stratification, providing valuable information not only about the underlying abnormality but also about the expected severity of the disease and preferred therapy.

WOLFF-PARKINSON-WHITE (WPW) SYNDROME

Ths risk of sudden cardiac death in patients with WPW syndrome is less than 1 per 1000 patient-years of follow-up. Although a rare event, WPW syndrome is important to consider since it usually occurs in otherwise healthy individuals and, in the era of catheter ablation of accessory pathways, is a curable cause of sudden cardiac death. The mechanism of sudden cardiac death in most patients with WPW syndrome is the development of atrial fibrillation with rapid ventricular rates due to conduction over an accessory pathway and subsequent degeneration of the ventricular rhythm into ventricular fibrillation. The best predictor for the development of ventricular fibrillation is a rapid ventricular response over the accessory pathway during atrial fibrillation, with the shortest interval between preexcited ventricular beats (i.e., those conducted over the accessory pathway) ≤ 250 ms. An electrophysiologic study offers the opportunity to assess the conduction properties of the accessory pathways, to gauge the propensity to develop tachyarrhythmias, and to cure the patient with catheter ablation at minimal risk.

IDIOPATHIC VENTRICULAR FIBRILLATION

A definite cause of sudden cardiac death cannot be established in 1 to 10 percent of patients dying suddenly or after successful resuscitation from cardiac arrest. The incidence of idiopathic ventricular fibrillation is higher in selected populations such as younger patients (up to 14 percent in patients less than 40 years of age who experienced sudden cardiac death) or female survivors of sudden cardiac death unrelated to myocardial infarction (10 percent). The risk of recurrent ventricular fibrillation in this young and otherwise healthy patient population ranges between 22 and 37 percent at 2 to 4 years. In survivors of cardiac arrest due to idiopathic ventricular fibrillation, the diagnosis is made by exclusion if extensive cardiac workup (including physical examination, laboratory tests for acute myocardial infarction and electrolyte abnormalities, ECG, exercise test, echocardiography, cardiac catheterization, and electrophysiologic study) reveals no abnormality that is thought to account for the ventricular fibrillation episode.

Several distinct clinical or electrophysiologic patterns in patients with idiopathic ventricular fibrillation have been described. They include a syndrome of "nocturnal sudden cardiac death in Southeast Asian men" among previously healthy refugees from Southeast Asia, a similar entity in young Japanese men (Pokkuri disease), and a "syndrome of right bundle branch block, persistent ST-segment elevation, and sudden cardiac death due to ventricular fibrillation" (or Brugada syndrome).

Drugs and Other Toxic Agents

PROARRHYTHMIA

The apparent paradox that antiarrhythmic agents can cause arrhythmias has been recognized since introduction of quinidine in 1918. The results of the Cardiac Arrhythmia Suppression Trial (CAST) showed an increased mortality in postinfarction patients treated with encainide, flecainide, and moricizine as compared with placebo despite effective antiarrhythmic efficacy as documented by the suppression of PVCs. Besides *antiarrhythmic drugs*, many other agents with diverse actions have been implicated in the induction of tachyarrhythmias. Among commonly used drugs associated with the risk of producing ventricular arrhythmias leading to sudden cardiac death are *erythromycin, terfenadine, hismanal, pentamidine, cisapride*, and certain psychotropic drugs such as *tricyclic antidepressants* and *chlorpromazine*, which generally affect repolarization. *Phosphodiesterase inhibitors* and other positive inotropic agents that increase intracellular calcium loading have also been shown to be proarrhythmic and to increase the risk for sudden cardiac death despite their beneficial effects on hemodynamic parameters.

COCAINE AND ALCOHOL

The increasingly widespread use of cocaine in the United States has led to the realization that this drug can precipitate life-threatening cardiac events, including sudden cardiac death. In

a series of 41 survivors of cardiac arrest due to ventricular fibrillation 18 to 35 years of age, one-third had ingested alcohol
or drugs (cocaine, heroin, or tricyclic agents). The combination
of alcohol and cocaine is especially dangerous due to the generation of a unique metabolite, cocaethylene, that has enhanced
cardiotoxicity. Cocaine causes coronary vasoconstriction, increases cardiac sympathetic effects, and precipitates cardiac
arrhythmias irrespective of the amount ingested, prior use, or
whether or not there is an underlying cardiac abnormality.

Electrolyte Abnormalities

Hypokalemia is often found in patients during and following resuscitation from a cardiac arrest. Although it is often a
secondary phenomenon due to catecholamine-induced potassium shift into the cells, primary hypokalemia can also be arrhythmogenic. Many of the electrophysiologic effects of
hypokalemia are similar to those caused by digitalis and catecholamine stimulation, explaining the high risk of ventricular
arrhythmias when a combination of these factors is present. An
association between *magnesium deficiency* and sudden cardiac
death has been reported in humans, especially as a cofactor in
drug-induced *torsades de pointes*. Hypomagnesemia in humans
is generally associated with congestive heart failure, digitalis
use, chronic diuretic use, hypokalemia, and hypocalcemia,
making it difficult to establish whether the hypomagnesemia
alone caused the sudden cardiac death. Acute administration of
magnesium has been successfully used in the treatment of
drug-induced *torsades de pointes*, although hypomagnesemia is not usually documented in this situation. *Increases
in intracellular calcium* are believed to play a significant
role in arrhythmias associated with digitalis glycosides,
catecholamine-induced ventricular tachycardia, reperfusion
arrhythmias, and the proarrhythmic effect seen with phosphodiesterase inhibitors and other positive inotropic agents.

Several studies in patients with hypertension who received
treatment with diuretics suggested an increased risk of sudden
cardiac death due to therapy with *non-potassium-sparing
diuretics*. Drug-induced potassium or magnesium depletion
leading to cardiac arrhythmias has been suggested as the
underlying mechanism. Electrolyte abnormalities are thought
to be the cause of sudden cardiac death in patients with eating
disorders, such as *anorexia nervosa*, *bulimia*, or *liquid protein
dieting*.

CHAPTER 5

CLINICAL PRESENTATION
AND MANAGEMENT OF THE PATIENT
WITH CARDIAC ARREST

Out-of-Hospital Cardiac Arrest

Cardiac arrest is characterized by abrupt loss of consciousness, which would uniformly lead to death in the absence of an acute intervention. Prodromal symptoms such as chest pain, dyspnea, fatigue, and palpitations may be present in up to half of patients presenting with cardiac arrest, but these are generally nonspecific and lead to medical evaluation in only a minority of patients. About 75 percent of cardiac arrests occur at home and about two-thirds are witnessed. Individuals who live alone and women appear more likely to have unwitnessed deaths. The average age of cardiac arrest victims is around 65 years, and 70 to 80 percent are men.

As discussed above, the most common mechanisms of cardiac arrest are ventricular tachyarrhythmias, followed by bradyarrhythmia, or asystole. The most important determinant of successful resuscitation is the time interval from cardiovascular collapse to initial intervention. Since most patients are found in ventricular fibrillation, time to successful defibrillation is a key element in the acute management of the cardiac arrest victim. The importance of early intervention is reflected in the "chain of survival" concept of emergency cardiac care systems: early access, early cardiopulmonary resuscitation (CPR), early defibrillation, early advanced cardiac life support. In order to improve the time to initial defibrillation, early defibrillation by nonmedical personnel, such as firefighters, has been advocated. The use of an automatic external defibrillator has the potential to significantly improve the availability of early defibrillation.

Survival and Prognosis
after Cardiac Arrest

Marked differences in survival following out-of-hospital cardiac arrest have been reported in different communities, being lowest in large cities such as New York (1.4 percent) and Chicago (4 percent) and highest (28 percent) in Seattle, a midsize urban community where many of the early intervention concepts have been pioneered. The in-hospital mortality following successful resuscitation outside the hospital remains high, in the range of 30 to 50 percent. The most important factors associated with increased in-hospital mortality after out-of-hospital cardiac arrest are cardiogenic shock after defibrillation, age ≥ 60 years, requirement of four or more shocks for defibrillation, absence of an acute myocardial infarction, and coma on admission to the hospital.

Survival depends largely on the initial recorded rhythm. Some 40 to 60 percent of patients who are found in ventricular fibrillation are successfully resuscitated, but only about one-fourth of patients survive to be discharged form the hospital. The outcome is much better in the small (<7 percent) group of patients in whom ventricular tachycardia is the initial documented rhythm: 88 percent survive to the hospital and 76 percent are discharged alive. Bradycardia and electromechanical dissociation as the presenting rhythms are associated with the worst prognosis, and few (<1 percent) of these patients survive to discharge form the hospital. Other factors associated with improved survival are a low "comorbidity index," reflecting chronic conditions such as history of heart failure, diabetes, hypertension, gastrointestinal disorders, etc., as well as recent symptoms prior to the event.

MANAGEMENT OF SURVIVORS OF CARDIAC ARREST AND RISK STRATIFICATION FOR SUDDEN CARDIAC DEATH

Establishing the Underlying Cardiac Pathology

The initial management following successful resuscitation from cardiac arrest consists of allowing a period of hemodynamic and respiratory stabilization, after which every effort should be made to establish the cause of the cardiac arrest. For this, the underlying cardiac disease should first be determined. History and physical examination may provide the first clues. Myocardial infarction has to be excluded by serial enzymes and ECG changes. Echocardiography can help in evaluating left ventricular function, regional wall motion abnormalities, valvular heart disease, or other cardiomyopathies. Stress-imaging studies can demonstrate inducible ischemia. Cardiac catheterization is often recommended to evaluate the coronary anatomy and right and left ventricular hemodynamic parameters. Other tests such as radionuclide studies, magnetic resonance imaging (MRI), or cardiac biopsy may be necessary in selected patients. As discussed above, an underlying cardiac disease can be found in nearly all patients.

Primary versus Secondary Cardiac Arrest

One of the important questions following cardiac arrest is whether it was primarily due to acute circulatory or respiratory failure or to an arrhythmia. Although all these events are usually present during the arrest, it is important to distinguish whether the arrhythmia preceded or followed the hemodynamic collapse. While several clinical and historical clues help to answer this question, the distinction sometimes cannot be made with certainty. Separating primary from secondary cardiac arrest has important prognostic and therapeutic consequences. In 142 survivors of cardiac arrest who had coronary artery disease, 1-year survival was 89, 80, and 71 percent in the patients classified as having had cardiac arrest secondary to acute myocardial infarction, secondary to an ischemic event, and due to a primary arrhythmia, respectively. Patients who present with cardiac arrest secondary (and within 48 h) to an acute transmural myocardial infarction have a similar prognosis as those who have an acute myocardial infarction without an arrhythmia. Specific antiarrhythmic therapy is therefore usually not recommended if cardiac arrest occurs during or within 2 days of an acute Q-wave myocardial infarction. In contrast, if the arrhythmia is the primary event and myocardial infarction developed secondary to the acute hemodynamic deterioration during the arrhythmia, antiarrhythmic therapy with a drug or device is recommended unless a transient or reversible cause is identified.

Every effort should be made to exclude potentially reversible causes for sudden cardiac death (Table 5-2), including transient ischemic episodes in patients who are candidates for complete revascularization and in whom the onset of the arrhythmia is clearly preceded by ischemic ECG changes or symptoms. Other reversible etiologies for cardiac arrest can be transient severe electrolyte disturbances and proarrhythmic effects of antiarrhythmic drugs and other pharmacologic agents.

TABLE 5-2

**POTENTIALLY REVERSIBLE CAUSES OF CARDIAC
ARREST DUE TO VENTRICULAR FIBRILLATION**

Myocardial ischemia
Prinzmetal's angina
Proarrhythmia
   Antiarrhythmic agents
   Other drugs
Electrolyte abnormalities
Hypoxia
Acute congestive heart failure

PREVENTION AND THERAPY OF SUDDEN CARDIAC DEATH

Pharmacologic treatment options for patients at risk for sudden cardiac death include specific antiarrhythmic drugs such as amiodarone and sotalol as well as other cardiovascular agents that treat the underlying cardiac abnormality (beta blockers, angiotensin converting enzyme inhibitors). Nonpharmacologic treatment options for patients at risk for sudden cardiac death include automatic implantable cardioverter defibrillators (ICDs), catheter or surgical ablation of the arrhythmia substrate, and pacemaker. The value of the various therapeutic strategies in preventing sudden cardiac death has been evaluated in primary and secondary prevention trials (Table 5-3).

Primary prevention trials are aimed at populations thought to be at high risk for sudden cardiac death to prevent the first episode. Most trials have enrolled patients following myocardial infarction or with low left ventricular ejection fraction with and without ventricular ectopy. In summary, the primary prevention trials have shown that:

- Beta blockers reduce mortality, including that due to sudden cardiac death, after myocardial infarction.
- Encainide, flecainide, moricizine, and D-sotalol increase mortality after myocardial infarction.
- Amiodarone either prolongs survival or, at worst, does not adversely affect survival in patients after myocardial infarction or those with congestive heart failure.

- ICDs are associated with better survival than drug therapy in a select subgroup of patients with coronary artery disease and nonsustained ventricular tachycardia who are inducible at electrophysiologic study.
- Prophylactic ICD implantation at the time of bypass surgery does not prolong survival.

The major problem with primary prevention of sudden cardiac death is identifying the patients at high risk. Several clinical variables have been identified (Table 5-4), but they all lack a high positive predictive value. The risk-benefit ratio between any intervention and the risk of sudden death has to be carefully evaluated. For example, in patients at low risk of sudden cardiac death, proarrhythmia or surgical mortality might outweigh the benefits achieved with an antiarrhythmic intervention. On the other hand, in patients at high risk for recurrent cardiac arrest, the risk-benefit profile of antiarrhythmic treatment strategies may be more favorable. In an era of limited health care resources, the cost-effectiveness of different treatment strategies is another element to be considered in choosing therapy. Many patients may need to be treated to benefit a few. Last but not least, quality of life is an important aspect in the selection of the most appropriate therapy for a patient.

Secondary prevention trials are aimed at patients with sustained ventricular tachyarrhythmias or survivors of cardiac arrest in order to prevent a recurrent event. In summary, these trials have shown that

- Neither ambulatory ECG (Holter) nor electrophysiologic studies are highly reliable in predicting drug efficacy.
- Sotalol is better than several class I drugs (but there are no placebo data).
- Propafenone is inferior to ICD, amiodarone, or beta blocker.
- Empiric amiodarone is superior to electrophysiologically guided conventional drug therapy.
- There is a survival benefit of ICD therapy over amiodarone therapy (Fig. 5-2).

The limitation of secondary prevention trials is that, given the poor survival after the initial cardiac arrest event, patients with near-fatal ventricular tachyarrhythmias represent only the tip of the iceberg of the population at risk for sudden death. Natural history or placebo information is also often lacking in these trials.

TABLE 5-3

SELECTED PRIMARY AND SECONDARY PREVENTION TRIALS FOR SUDDEN CARDIAC DEATH

Study	N	Patient Population	Protocol	Main Results
Primary prevention trials				
Beta blocker trials (metaanalysis)	19,000	Post-MI	Beta blocker vs. placebo	30% RR in sudden cardiac death, particularly in pts with CHF
BASIS	312	Post-MI, PVCs	Amiodarone vs. placebo	Reduction in total mortality from 13 to 5% in the amiodarone group
EMIAT	1,486	Post-MI (5–21 days), LVEF ≤49%	Amiodarone vs. placebo	No change in overall mortality, 35% RR in arrhythmic mortality
MADIT	196	Prior MI (>3 weeks), LVEF ≤35%, NS-VT, inducible, non-suppressible VT	ICD vs. conventional antiarrhythmic therapy (80% got amiodarone)	Survival in ICD group at 2 years was 84%, vs. 61% in the conventional arm
GESICA	516	CHF (NYHA II–IV), LVEF ≤35%, NS-VT	Amiodarone vs. placebo	Amiodarone reduced 2-year mortality from 41 to 34%, largely attributed to reduction in arrhythmic death

	pts	Characteristics	Comparison	Results
CHF-STAT	674	CHF (NYHA II–IV), LVEF ≤40%, PVCs (>10/h)	Amiodarone vs. placebo	Amiodarone did not improve survival but suppressed PVCs and improved LV function
Secondary prevention trials				
CASCADE	228	Cardiac arrest (not associated with Q-wave MI)	Amiodarone vs. conventional Holter- or EP-guided therapy	Amiodarone reduced 2-year cardiac and arrhythmic mortality from 44 to 24%
AVID	1,016	VF or hemodynamically unstable VT	Amiodarone vs. ICD	27% RR in mortality with ICD
CASH	400[†]	Cardiac arrest	Amiodarone vs. ICD vs. metoprolol vs. propafenone	ICDs improved survival vs. amiodarone or beta blocker* Mortality at 1 year was 25% higher in the propafenone group vs. others No difference between amiodarone and metoprolol*

Key: CHF, congestive heart failure; EP, electrophysiologic study; LVEF, left ventricular ejection fraction; MI, myocardial infarction; NS-VT, nonsustained ventricular tachycardia; PVCs, premature ventricular contractions; pts, patients; RR, risk reduction; SAECG, signal-averaged ECG; VT, ventricular tachycardia; VF, ventricular fibrillation.

*Preliminary results presented at the American College of Cardiology meeting, Atlanta, March 1997.

[†]Projected numbers.

TABLE 5-4

RISK FACTORS FOR SUDDEN CARDIAC DEATH

Left ventricular ejection fraction ≤35%
Congestive heart failure
More than one previous myocardial infarction
Active or provocable ischemia
Inducible ventricular tachycardia
Autonomic dysfunction
    Reduced heart rate variability
    Decreased baroreceptor sensitivity
Complex ventricular ectopy
Positive family history of sudden cardiac death
Syncope in the presence of heart disease
Left ventricular hypertrophy
Cigarette smoking

SUMMARY

Sudden cardiac death affects more than 300,000 individuals in the United States and accounts for half the mortality from coronary heart disease. A vast majority of people who experience sudden cardiac death have underlying structural heart disease, which, in the adult population, is most frequently coronary heart disease. Ventricular tachycardia/fibrillation and less often bradycardia/asystole are responsible for sudden cardiac death. Autonomic changes such as increased sympathetic and decreased parasympathetic reflexes appear to be important triggers of sudden cardiac death. Long-term survival following a cardiac arrest episode is still poor (<10 percent). The time delay to defibrillation and/or bystander administration of cardiopulmonary resuscitation (CPR) directly influences survival. Automatic implantable cardioverter/defibrillator and amiodarone are effective therapeutic options to treat survivors of cardiac arrest. Primary prevention of sudden cardiac death remains a problem, since we cannot identify with acceptable sensitivity and specificity a large percentage of the individuals who are at risk for sudden cardiac death.

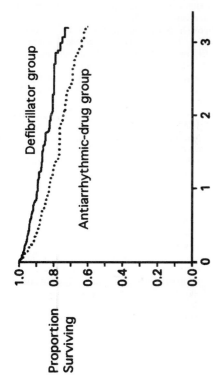

FIGURE 5-2

Overall survival in the trial of antiarrhythmics versus implantable defibrillators (AVID). Survival was better among patients treated with the implantable cardioverter-defibrillator (<0.02). (Reprinted with permission from The Antiarrhythmics Versus Implantable Defibrillators (AVID) Investigators: A comparison of antiarrhythmic drug therapy with implantable defibrillators in patients resuscitated from near-fatal ventricular arrhythmias. *N Eng Med* 1997; 337:1576–1583.)

SUGGESTED READING

Cummins RO, Ornato JP, Thies WH, Pepe PE: Improving survival from sudden cardiac arrest: the "chain of survival" concept. *Circulation* 1991; 83:1832–1847.

Engelstein ED, Zipes DP: Sudden cardiac death. In: Alexander RW, Schlant RC, Fuster V, O'Rourke RA, Roberts R, Sonnenblick EH (eds): *Hurst's The Heart*, 9th ed. New York, McGraw-Hill, 1998:1081–1112.

Green HI: Sudden arrhythmic cardiac death: mechanisms, resuscitation and classification: the Seattle Perspective. *Am J Cardiol* 1990; 65:4B–12B.

Kannel WB, Thomas HE Jr: Sudden coronary death: The Framingham Study. *Ann NY Acad Sci* 1982; 382:3–20.

Liberthson RR: Sudden death from cardiac causes in children and young adults. *N Engl J Med* 1996; 334:1039–1044.

Maron BJ: Heart disease and other causes of sudden death in athletes. *Curr Probl Cardiol* 1998; 23(9):477–532.

Muharji J, Rude R, Poole K, et al: Risk factors for sudden death after acute myocardial infarction: two year follow-up. *Am J Cardiol* 1994; 54:31–56.

Myerburg R, Kessler KM, Bassett AL, Castellanos A: A biological approach to sudden cardiac death: structure, function and cause. *Am J Cardiol* 1989; 63:1512–1516.

The Antiarrhythmics Versus Implantable Defibrillators (AVID) Investigators: A comparison of antiarrhythmic drug therapy with implantable defibrillators in patients resuscitated from near-fatal ventricular arrhythmias. *N Engl J Med* 1997; 337:1576–1583.

Zipes DP: Sudden cardiac death: future approaches. *Circulation* 1992; 85(suppl I):I-160–I-166.

6

CARDIOPULMONARY RESUSCITATION

Nisha Chibber Chandra

DIAGNOSIS AND IDENTIFICATION OF CARDIAC ARREST

Cardiac arrest is defined as the sudden cessation of effective cardiac pumping function as a result of either ventricular asystole (electrical or mechanical) or ventricular fibrillation. Rapid diagnosis and treatment are essential because (1) more than a few minutes of total cardiac arrest results in permanent cerebral anoxic damage and (2) the success of resuscitative measures is related to the rapidity with which they are instituted following arrest. Based on these and other observations, the concept of early activation of Emergency Medical Systems (EMS) has evolved for victims of out-of-hospital cardiac arrest.

Preliminary Patient Evaluation and Triage

Cardiac arrest should be considered in the differential diagnosis of sudden collapse in any patient. It can be clinically confirmed by pulseless major vessels and absent heart sounds. Although respirations (agonal respirations) may continue for a minute or two, the patient with cardiac arrest becomes rapidly cyanotic and unconscious.

Once the diagnosis of cardiac arrest is made and if no trauma is suspected, the unconscious patient should be positioned supine on a firm surface and the airway opened using the head tilt–chin lift technique or alternative strategies, as described below under "Ventilation during CPR." The patient should immediately receive rescue breathing either with a bag–valve mask device or with mouth-to-mouth breathing. Simple airway barrier devices that are easily deployed can be used to minimize direct patient contact and are perceived as being more "hygienic" during mouth-to-mouth resuscitation. Following airway opening and rescue breathing, chest compressions should be promptly initiated at approximately 80 to 100 per minute.

If available, an electrocardiogram (ECG) can confirm the diagnosis and identify asystole, ventricular fibrillation, or electromechanical dissociation as the mechanism of arrest. Cardiopulmonary resuscitation (CPR), however, should be initiated immediately once the clinical diagnosis is made without delaying to obtain this information. If a defibrillator but not an ECG is immediately available, a 200-J countershock should be administered without delay. Prehospital CPR studies confirm in several patients that the mechanism of cardiac arrest is usually ventricular fibrillation and that survival is critically dependent on the time to defibrillation. Most hospitals and paramedics are now equipped with defibrillators with "quick look" paddles that simultaneously allow the ECG rhythm to be analyzed. Guided by the rhythm, one can then explore an etiology for the arrest in a more focused way and initiate appropriate therapy.

AUTOMATIC EXTERNAL DEFIBRILLATORS (AEDs)

Based on these observations, automatic external defibrillators (AED) were developed for use by first (minimally trained) responders and were shown to dramatically improve survival from prehospital arrest. AEDs have been successfully used by nontraditional health care professionals (airline crews and police) with dramatic improvement in patient survival. All AED programs have had strict physician-guided training and supervision.

RESPIRATORY ARREST

Respiratory arrest is the cessation of effective respiratory effort. It can result from airway obstruction (due to a foreign body or other causes), drowning, smoke inhalation, drug overdose, head trauma, cerebrovascular accident, or suffocation. When respiratory arrest occurs suddenly (as with foreign-body obstruction), the patient rapidly becomes cyanotic, though a palpable pulse with blood pressure, consciousness, and ineffective respiratory efforts may be maintained for several minutes. Opening the airway and/or rescue breathing may be all that is necessary to resuscitate such a patient.

The Heimlich maneuver is recommended for relieving foreign-body airway obstruction. It is implemented by standing behind the victim and delivering a series of sharp thrusts to the upper abdomen with a closed fist. Abdominal thrusts can also be used directly in the unconscious supine patient to help dislodge a foreign body mechanically. The Heimlich maneuver can also be self-administered by placing the fist between the navel and xiphoid process and delivering a series of quick upward thrusts. If incorrectly administered, this maneuver can lead to visceral damage. When properly used, however, the technique is both safe and effective. Manual removal of a foreign body should be used only in the unconscious victim. This can be achieved by opening the victim's mouth and manually attempting to dislodge any obvious foreign body with a finger. As a single method, back blows may not be as effective as the Heimlich maneuver in adults. For this reason, the Heimlich maneuver is considered the technique of choice.

VENTILATION DURING CPR

Clearing the airway is of the utmost importance. Foreign bodies, loose dentures, or any other oral obstruction should be removed. Next, the head tilt–chin lift technique, which causes the tongue to move anteriorly, is used to open the airway. The chin is lifted forward, with the fingers of one hand supporting the jaw, and the head is tilted back by the other hand, which has been placed on the patient's forehead. The head tilt–neck lift method of opening the airway is also commonly employed and is an acceptable technique for use by the skilled rescuer. Here, the head is tilted back with one hand on the forehead and the other hand is placed behind the neck, lifting it upward to open the airway. If no spontaneous respirations are present, mouth-to-mouth (or mouth-to-nose) ventilation is immediately initiated, with adequacy being judged by the rise and fall of the patient's chest with each breath. To minimize gastric distention, it is necessary to deliver slow (1- to 2-s) ventilatory breaths.

Several invasive airway adjuncts have also been developed for use by nonphysician health care providers in prehospital situations. The esophageal obturator airway (EOA), esophageal gastric tube airway (EGTA), the Combitube, and the pharyngotracheal lumen airway are among those that have been used in the prehospital setting. Considerable training is required, as well as skill, in placing and using these devices properly. Serious, life-threatening complications have been reported following the use of the EOA or EGTA; after successful resuscitation, balloon deflation frequently results in the regurgitation of gastric contents if an EOA is used. As a consequence, the recent trend in most EMS operations has been to train paramedics in endotracheal intubation, which can be successfully implemented in the field. Following intubation, whenever possible, a nasogastric tube should be inserted to drain the stomach and thus decrease the chances of aspiration.

The optimal requirements for ventilation during CPR in human beings remain unknown. No study has clearly identified the optimal timing, sequence in relation to chest compression, or tidal volume needed during CPR. American Heart Association (AHA) recommendations advise 10 to 12 slow ventilatory breaths per minute with a tidal volume of 800 to 1200 mL per breath.

CHEST COMPRESSION DURING CARDIOPULMONARY RESUSCITATION

The American Heart Association (AHA) has published "Standards for Cardiopulmonary Resuscitation and Emergency Cardiac Care". In reference to external chest compression, they advise (1) 80 to 100 sternal compressions per minute, (2) 50 percent of each compression-relaxation cycle to be compression, and (3) one slow (1- to 2-s) ventilation for every five compressions if two trained rescuers are performing CPR and two slow (1- to 2-s) ventilatory breaths every 15 chest compressions if lay rescuers or one trained person is performing CPR. In addition to these recommendations, it is critical when performing chest compression to use sufficient force to depress the sternum by approximately 2 in. (5 to 6 cm). As this is usually difficult to gauge, sufficient chest compression force should be used to generate a palpable femoral or carotid arterial pulse.

DEFINITIVE THERAPY

The 1992 AHA Standards Guidelines for Emergency Cardiac Care have adopted a new classification for therapeutic recommendations, which allows a relative therapeutic value to be assigned to a given strategy of treatment:

Class I: Definitely helpful
Class IIA: Acceptable, probably helpful
Class IIB: Acceptable, possibly helpful, probably not harmful
Class III: Not indicated, may be harmful

In the text that follows, these specific therapeutic classifications will be mentioned when appropriate.

During cardiac arrest, the ECG will usually show rapid ventricular tachycardia or fibrillation, asystole, or heart block or it may be near normal.

Ventricular Tachycardia or Fibrillation

With ventricular fibrillation, an attempt at electrical defibrillation should be made as quickly as possible. Successful defibrillation is accomplished by the passage of adequate electrical current (amperes) through the heart. Current flow is dependent on the energy chosen (joules) and the transthoracic impedance (ohms), or resistance to current flow. Factors that affect transthoracic impedance include the energy selected, electrode size, skin-paddle coupling material, the number and time interval of previous shocks, the distance between the electrodes (size of the chest), phase of ventilation, and paddle electrode pressure. Human transthoracic impedance ranges from 15 to 150 ohms, with the average adult impedance being 70 to 80 ohms. If transthoracic impedance is high, low-energy shocks are ineffective in generating enough current to achieve successful defibrillation. Transthoracic impedance can be reduced by firm pressure on hand-held electrode paddles and a gel/cream, or saline-soaked gauze pads, between the electrode and the skin. In addition, proper electrode/paddle placement is essential; one electrode should be placed to the right of the upper sternum, below the clavicle, and the other to the left of the nipple, with the center of the electrode in the midaxillary line. An acceptable alternative is one electrode anteriorly over the left precordium and the other posteriorly behind the heart in the right infrascapular location. The latter positioning is best achieved by using preadhesive rather than hand-held electrodes. In female patients with large breasts, the electrodes are best placed right of the upper sternum and either under or lateral to the left breast. Direct current is employed during defibrillation. The paddles, coated with low-resistance gel, are applied firmly to the chest and then discharged with 200 J, which is repeated at 200 to 300 J if the first shock is unsuccessful. The current AHA standards suggest that a third 360-J shock may be delivered if ventricular fibrillation persists.

When ECG shows "fine" fibrillation waves, defibrillation efforts are often unsuccessful. The administration of epinephrine (5 to 10 mL of 1:10,000) intravenously results in a more vigorous and coarse fibrillation that is more responsive to defibrillation. This effect is likely due to improved coronary flow following epinephrine administration and perhaps direct myocardial effects on the electrical properties for defibrillation. If defibrillation fails, it is likely that marked acidosis or hypoxemia is present. Emphasis should be on hyperventilation with supplemental oxygen to correct both hypoxemia and metabolic acidosis. Sodium bicarbonate might then be administered (1 meq/kg) to aid in the management of acidosis, and defibrillation should be repeated with 300 to 360 J.

For recurrent ventricular fibrillation, the administration of 75 to 100 mg of lidocaine IV followed by repeat defibrillation may increase the likelihood of returning to a stable rhythm. Lidocine is an effective antiarrhythmic agent for recurrent ventricular fibrillation. Amiodarone (a 150- to 300-mg bolus over 10 min, 1.0 to 2.0 mg/min for 6 h, and then 0.5 to 1.0 mg/min for 6 to 24 h) intravenously has recently been shown to be of modest benefit for recurrent ventricular fibrillation in patients failing treatment with lidocaine alone. Procainamide can additionally be used in patients failing lidocaine, but it can cause considerable hypotension. For recurrent ventricular fibrillation in the setting of ischemia, intravenous propranolol or other intravenous beta blockers are remarkably effective. They seem particularly helpful in the setting of primary ventricular fibrillation complicating acute myocardial infarction.

Hyperkalemia is a readily treated condition that can cause AV block, impaired intraatrial and intraventricular conduction, and occasionally ventricular fibrillation or, less commonly, asystole. It can be recognized by the development of tall, peaked T waves with a normal QT interval and sine wave–like ventricular tachycardia. Life-threatening hyperkalemia responds most readily to calcium infusion; 10 to 30 mL of 10% calcium gluconate is infused intravenously over 1 to 5 min under constant electrocardiographic monitoring. Calcium counteracts the adverse effects of potassium on the neuromuscular membranes but does not alter plasma potassium. Its effect, though immediate, is transient. Hyperkalemia should subsequently be treated by glucose-insulin infusion, or ion-exchange resins. Sodium bicarbonate is also used as an agent to lower potassium.

With ventricular tachycardia, cough may revert the arrhythmia without defibrillation; repeated cough can also maintain the conscious state as a result of the rise in intrathoracic pressure. The efficacy of the precordial thump (precordial chest blows) has been variably reported in patients with ventricular tachycardia. A thump is generally ineffective for terminating prehospital ventricular fibrillation and may be deleterious for ventricular tachycardia, converting it to ventricular fibrillation or asystole. Hence, it should never be used in the patient with ventricular tachycardia and a pulse unless a defibrillator is available immediately.

Asystole or Heart Block

For patients with prehospital cardiac arrest, asystole has been shown to be an ominous rhythm, with a very low likelihood of successful resuscitation. On the other hand, asystole due to vagal stimulation is the commonest cause of cardiac arrest associated with anesthesia induction and surgical procedures. Asystole also occurs as a result of heart block or sinus node disease. Atropine (0.5 mg) given intravenously and repeated in 5 min can be used acutely to prevent or reverse severe bradycardia in many of these settings.

If asystole is witnessed or of short duration, vigorous blows to the precordium may sometimes restart the heart. Rhythmic chest blows may maintain limited perfusion and can be continued, if needed, while palpating the femoral or carotid pulse until other treatment is available. If the chest blows fail, cardiopulmonary resuscitation should be initiated and intravenous epinephrine (5 to 10 mL of 1:10,000) administered. Possible treatable causes of asystole such as acidosis, hypoxemia, hyper- or hypokalemia, and hypothermia should be considered and appropriately treated if suspected. If overdose of a calcium channel blocker is suspected, calcium chloride, 1 g given as an intravenous bolus, may be very effective (class IIA recommendation). Resuscitation measures may result in the return of a slow ventricular rhythm, which can subsequently be supported with atropine (1 to 2 mg IV) until a temporary pacemaker is placed. Temporary pacing is the optimal treatment for true asystole or profound bradycardia. Obviously, considerable skill and training are required for temporary transvenous pacemaker placement. Transcutaneous pacing has been developed as a noninvasive and simple pacing technique that can be rapidly implemented. However, prehospital studies of transcutaneous pacing for asystole have not confirmed an improvement in survival. It may be of some benefit for patients early in asystole (class IIB intervention). Clinical evidence does not support its routine use in all patients with asystole.

In rare instances, very fine ventricular fibrillation may result in a nearly straight line on a single-lead ECG and thus be mistaken for "asystole." In such cases, where the diagnosis of asystole is in question, it is suggested that a perpendicular ECG lead be viewed. Rotation of "quick-look" ECG paddles by 90 degrees easily achieves this. If ventricular fibrillation is present, the perpendicular ECG lead will demonstrate a typical fibrillation pattern, whereas in true asystole, a straight line will be seen in all ECG leads. If ventricular fibrillation is diagnosed, the initial treatment should be according to the outline above; i.e., three successive countershocks. There is little value in defibrillating true asystole.

Electromechanical Dissociation

In electromechanical dissociation (EMD), there is evidence of organized electrical activity on the ECG at a reasonable rate but failure of effective perfusion (no pulse or blood pressure). This condition is also referred to as pulseless electrical activity (PEA) in the literature. The most treatable causes of this condition are hypovolemia due to severe hemorrhage, pericardial tamponade, tension pneumothorax, hypoxia, hypothermia, acidosis, hyperkalemia, and massive pulmonary embolism. Signs of these problems should be sought and definitive therapy undertaken with fluids and/or blood replacement, pericardiocentesis, placement of a pleural needle or tube, endotracheal intubation, and other maneuvers as deemed necessary. These conditions should also be strongly considered if CPR results in no palpable pulse or evidence of perfusion. Unfortunately, many patients with electromechanical dissociation have primary myocardial failure. Following the diagnosis of EMD, one should optimize ventilation and administer epinephrine. Calcium chloride has been used to treat EMD, but prospective studies have not shown it to improve survival. In acute myocardial infarction, sudden EMD is a sign of myocardial rupture. In such cases, pericardiocentesis and surgical repair can rarely result in survival.

Establishment of an Intravenous Route

Drug administration during CPR should be accomplished only from a source above the diaphragm, since there is little cephalad flow from veins below the diaphragm. If a peripheral vein cannot be cannulated, a cutdown should be attempted or a central venous line placed by a percutaneous route. If cardiopulmonary resuscitation is properly performed, drugs administered through a peripheral line will often reach the arterial circulation within 15 to 30 s. Recent data suggest that a 20-mL fluid bolus significantly improves peripheral drug delivery to the central compartment. Intracardiac injections are unnecessary except when there is no intravenous access. If an intravenous route is unavailable, epinephrine (1 to 2 mg in 10 mL of sterile distilled water) and lidocaine (50 to 100 mg in 10 mL of sterile distilled water) can be administered by way of the endotracheal tube into the bronchial tree. The drug should be injected through a long catheter passed beyond the tip of the endotracheal tube. Cardiac compression should be withheld, and several insufflations with an Ambu bag should immediately follow to aid drug absorption through aerosolization.

MAJOR DRUGS USED DURING CARDIOPULMONARY RESUSCITATION

Drugs used for the treatment of various arrhythmias are mentioned above. Catecholamines are used in cardiac arrest to (1) increase arterial and coronary perfusion during and following cardiopulmonary resuscitation, (2) stimulate spontaneous contraction during asystole, (3) make fine ventricular fibrillation more responsive to defibrillation, and (4) act as inotropic agents.

Epinephrine is effective in achieving all these goals. Once the diagnosis of cardiac arrest is established and CPR initiated, epinephrine should be administered as soon as possible. The recommended dose is 0.5 to 1 mg IV, and this dose should be repeated at approximately 3- to 5-min intervals unless effective cardiac activity is restored. Although it seemed promising in the animal model, large clinical trials failed to demonstrate improved survival with high-dose epinephrine. Hence, most experts would use 1 mg IV uniformly. Higher doses may be used for the second and third doses (3 and 5 mg) if the initial dose, CPR, and defibrillation fail (class IIB recommendation). If an intravenous route is not available, epinephrine can be administered down the endotracheal tube; 10 mL of a 1:10,000 solution should be used, and this can also be repeated every 3 to 5 min.

Norepinephrine is a potent vasoconstrictor and generally produces a rise in blood pressure; it is also an inotropic agent. Its disadvantage is renal and mesenteric vasoconstriction, and it should not be used in the initial phase of resuscitation. This agent is most useful where severe hypotension is present but where the chronotropic effects of epinephrine are not desirable (as in acute myocardial infarction or severe ischemia). This agent should be administered cautiously, since severe tissue injury results from extravasation around an intravenous site. A large prehospital trial failed to identify any differences in survival following treatment with norepinephrine, high-dose epinephrine, or standard epinephrine.

Similarly, dopamine (a chemical precursor of norepinephrine) and dobutamine (a synthetic catecholamine) are preferred for use as inotropic agents because of their lesser chronotropic effect. Isoproterenol (a synthetic catecholamine) is a pure adrenergic agonist and effective vasodilator. Therefore, its use during CPR is contraindicated since it can significantly decrease perfusion pressures in vital organs. In patients with a palpable pulse, however, it is useful for treatment of bradycardia due to heart block or asystole until a temporary pacemaker is placed.

Sodium Bicarbonate

As with other types of metabolic acidosis, if adequate alveolar ventilation is achieved, the metabolic acidosis of arrest is partially corrected through P_{CO_2} excretion. Recent clinical trials failed to demonstrate improved outcome from cardiac arrest with buffer therapy. Rather, several deleterious effects of bicarbonate administration—including metabolic acidosis, hypernatremia, and hyperosmolality—have been reported. Ideally, sodium bicarbonate should be given according to the results of measurement of arterial blood pH, P_{CO_2} determination, and calculation of the base deficit. Bicarbonate should be used, if at all, only after more established interventions such as defibrillation, ventilation with endotracheal intubation, and pharmacologic therapies (epinephrine and antiarrhythmic drugs) have been tried. If needed, 1 meq/kg of sodium bicarbonate should be administered; then no more than half this dose may be repeated every 15 min. On the other hand, bicarbonate may be most useful during the immediate postresuscitation period when a profound metabolic acidosis occurs. In most instances during CPR, its use should be considered as a class IIB recommendation.

Calcium Chloride

Calcium chloride (5 to 7 mg/kg) enhances the contractile state of the heart and is indicated in treating severe hypotension due to overdose of a calcium channel blocker or hyperkalemia. It is no longer recommended for use in asystole or electromechanical dissociation.

TERMINATION OF CARDIOPULMONARY RESUSCITATION

Despite resuscitative efforts, the patient in cardiac arrest may not regain spontaneous circulation. The decision to end (or even initiate) cardiopulmonary resuscitation should be based on a physician's assessment of the patient's prior advance directives (if known) and the cerebral, cardiovascular, and general status of the patient. Failure is likely if there is absence of organized ventricular electrocardiographic activity and/or peripheral perfusion after 10 to 15 min of adequate CPR and appropriate therapy. Recent studies have demonstrated that continued in-hospital CPR efforts (in patients failing prehospital advanced cardiac life support) are not only expensive but also unsuccessful.

Other potential life-threatening problems in the postarrest period include acute renal failure, bowel infarction, infection, adult respiratory distress syndrome, and sepsis. Patients regaining consciousness may have postarrest amnesia or may develop psychotic behavior.

CHAIN OF SURVIVAL

The concept of a "chain of survival" has been adopted by several agencies and underscores the importance of an integrated public education and health care system if outcome from prehospital cardiac arrest is to be optimized. "Early access (to EMS systems), early CPR (to include bystander CPR), early defibrillation (to include the use of AEDs), early ACLS care" are the major links in the chain, and any one weak link weakens the whole chain of survival.

If mortality from out-of-hospital arrest is to be reduced, public education programs to increase awareness of the warning signs of a heart attack and to teach CPR are critical. Individuals teaching CPR should reassure the public learning CPR that the likelihood of disease transmission is minimal: 70 percent of arrest victims collapse at home, and if any person is still unwilling to do mouth-to-mouth CPR, he or she should be taught to at least activate EMS ("call") and start chest compressions ("pump"). Ventilation ("blow") could then be started by suitably equipped, trained (EMS) rescuers. The outcome from prehospital arrest can only be improved if each community strives to optimize its own chain of survival.

SUGGESTED READING

Callaham M, Madsen CD, Barton CW, et al: A randomized clinical trial of high dose epinephrine and norepinephrine vs standard-dose epinephrine in prehospital cardiac arrest. *JAMA* 1992; 268:2667–2672.

Chandra NC, Weisfeldt ML: Cardiopulmonary resuscitation and the subsequent management of the patient. In: Alexander RW, Schlant RC, Fuster V, O'Rourke RA, Roberts R, Sonnenblick EH (eds): *Hurst's The Heart*, 9th ed. New York, McGraw-Hill, 1998:1113–1124.

Criley JM, Blaufuss AN, Kissel GL: Cough-induced cardiac compression. *JAMA* 1976; 236:1246–1250.

Eisenberg MS, Horwood BT, Cummins RO, et al: Cardiac arrest after resuscitation: a tale of 29 cities. *An Emerg Med* 1990; 19:179–186.

Emergency Cardiac Care Committee and Subcommittees, American Heart Association: Guidelines for cardiopulmonary resuscitation and emergency cardiac care. *JAMA* 1992; 268:2171–2302.

Kouwenhoven WB, Jude JR, Knickerbocker GG: Closed chest cardiac massage. *JAMA* 1960; 173:1064–1067.

Lombardi G, Gallagher J, Gennis P: Outcome of out-of-hospital cardiac arrest in New York City: the pre-hospital arrest survival evaluation (PHASE) study. *JAMA* 1994; 271: 678–683.

Niemann JT: Cardiopulmonary resuscitation. *N Engl J Med* 1992; 327:1075–1080.

White RD, Asplin BR, Bugliosi TF, et al: High release survival from out-of-hospital ventricular fibrillation with rapid defibrillation by both police and paramedics. *Acad Emerg Med* 1996; 3:422.

CHAPTER

7

DIAGNOSIS AND MANAGEMENT OF HYPERLIPIDEMIA

Scott M. Grundy

DEFINITIONS

Hypercholesterolemia

SERUM CHOLESTEROL
- Desirable serum cholesterol: less than 200 mg/dL (5.2 mmol/L)
- Borderline high serum cholesterol: 200 to 239 mg/dL (5.2 to 6.2 mmol/L)
- High serum cholesterol: 240 mg/dL or higher (6.2 mmol/L or higher)

LOW-DENSITY-LIPOPROTEIN (LDL) CHOLESTEROL
- Desirable LDL cholesterol: less than 130 mg/dL (3.4 mmol/L)
- Borderline high-risk LDL cholesterol: 130 to 159 mg/dL (3.4 to 4.1 mmol/L)
- High-risk LDL cholesterol: 160 mg/dL or higher (4.1 mmol/L)

Hypertriglyceridemia (High Serum Triglycerides)

- Normal serum triglyceride: less than 200 mg/dL (2.26 mmol/L)
- Borderline high triglycerides: 200 to 400 mg/dL (2.26 to 4.5 mmol/L)
- High triglycerides: 400 to 1000 mg/dL (4.5 to 11.3 mmol/L)
- Very high triglycerides: over 1000 mg/dL (11.3 mmol/L)

Combined Hyperlipidemia

- Serum cholesterol over 240 mg/dL and serum triglyceride between 200 and 400 mg/dL

Hypoalphalipoproteinemia [Low High-Density Lipoprotein (HDL) Cholesterol]

- Serum HDL, cholesterol less than 35 mg/dL (0.9 mmol/L)

CAUSES OF DYSLIPIDEMIA

Causes of Hypercholesterolemia (High LDL Cholesterol)

DIETARY CAUSES
- Excess dietary saturated fatty acids
- Excess dietary cholesterol
- Excess caloric intake (obesity)

GENETIC CAUSES
- Familial hypercholesterolemia (deficiency of LDL receptors)
- Familial combined hyperlipidemia
- Polygenic hypercholesterolemia (multiple genetic defects occurring alone or in combination)

SECONDARY CAUSES
- Nephrotic syndrome
- Hypothyroidism
- Diabetes mellitus
- Obstructive liver disease
- Dysproteinemias (multiple myeloma, macroglobulinemia)

Causes of Hypertriglyceridemia (High Triglycerides)

- Genetic deficiency of lipoprotein lipase or apolipoprotein C-II
- Familial hypertriglyceridemia
- Obesity
- Excessive alcohol intake
- Diabetes mellitus
- Beta-adrenergic blocking agents

Causes of Hypoalphalipoproteinemia (Low HDL Cholesterol)

- Cigarette smoking
- Obesity
- Lack of exercise
- Non–insulin-dependent diabetes mellitus
- Hypertriglyceridemia
- Beta-adrenergic blocking agents
- Genetic disorders of HDL metabolism

MANAGEMENT OF HYPERCHOLESTEROLEMIA

Coronary Heart Disease (CHD) Risk Status as a Guide to Intensity of Therapy

The intensity of treatment of individuals depends on an evaluation of risk status. Those at highest risk should receive the most aggressive cholesterol-lowering therapy. Three risk categories of patients with high serum cholesterol are identified: (a) *very-high-risk* (includes patients with prior CHD or other atherosclerotic disease—peripheral vascular disease or symptomatic carotid disease); (b) *high risk* (includes patients without evident CHD but who have high serum cholesterol together with other CHD risk factors); (c) *moderate risk* (includes middle-aged and elderly patients who have high serum cholesterol but minimal other CHD risk factors; and (d) *otherwise low risk* (includes patients with high blood cholesterol but who are otherwise at low risk: this group comprises young adult men (<35 years) or premenopausal women).

Secondary Prevention vs. Primary Prevention

For patients with established CHD, clinical trials have shown that serum cholesterol lowering reduces both morbidity and mortality from CHD as well as reducing total mortality. Treatment of elevated LDL cholesterol in patients with previous CHD and/or other atherosclerotic disease is called *secondary prevention*. Prevention of new-onset CHD is called *primary prevention*. Patients with clinically manifest atherosclerotic disease are at very high risk for future events and deserve aggressive cholesterol lowering as part of a secondary prevention regimen. High-risk patients without CHD also deserve cholesterol management in the clinical setting.

Detection of High Serum Cholesterol

For the purpose of primary prevention, total serum cholesterol should be measured in all adults 20 years of age and over every 5 years; HDL cholesterol should be measured at the same time. For patients who have a desirable total cholesterol and HDL cholesterol over 35 mg/dL, general educational materials should be provided about dietary modification, physical activity, and other risk-reduction activities, and these patients should be advised to have repeat total cholesterol and HDL cholesterol analysis in 5 years. Those with HDL cholesterol below 35 mg/dL should proceed to lipoprotein analysis. The latter includes measurement of fasting levels of total cholesterol, total triglyceride, and HDL cholesterol. From these values, LDL cholesterol is calculated as follows:

LDL cholesterol
   = total cholesterol − HDL cholesterol − (triglyceride/5)

All patients with total cholesterol over 200 mg/dL should have lipoprotein analysis. For individuals with total cholesterol levels of 200 to 239 mg/dL, the level of HDL cholesterol and the presence (or absence) of other risk factors should determine follow-up. The CHD risk factors that should be used to decide the intensity of therapy include the following:

Positive CHD Risk Factors

- Age: Male ≥ 45 years
- Female ≥ 55 years or premature menopause without estrogen-replacement therapy
- Family history of premature CHD
- Smoking
- Hypertension
- HDL cholesterol <35 mg/dL
- Diabetes mellitus

Negative CHD Risk Factor

- HDL cholesterol ≥ 60 mg/dL

Patients whose LDL-cholesterol levels are 130 to 159 mg/dL and whose HDL cholesterol is 35 mg/dL or higher and have fewer than two other risk factors should be given information on dietary modification, physical activity, and other risk-reduction activity. They are advised to repeat total cholesterol and HDL in 1 to 2 years.

Primary Prevention: Determining Risk Status and Goals of Therapy

Patients with LDL cholesterol over 160 mg/dL or those with LDL cholesterol of 130 to 159 mg/dL *and* two or more risk factors should be elevated clinically and should begin active cholesterol-lowering therapy. The goals of therapy are to lower LDL cholesterol to (1) below 160 mg/dL if fewer than two other risk factors are present (LDL cholesterol <130 mg/dL is a desirable level); and (2) below 130 mg/dL if two (or more) CHD risk factors are present. Dietary therapy is the first line of treatment. If an elevated LDL cholesterol persists after a trial of dietary therapy, drug therapy should be considered for patients with: (1) multiple CHD risk factors (and LDL cholesterol $\geq$ 160 mg/dL) or (2) severe forms of hypercholesterolemia (LDL cholesterol $\geq$ 190 mg/dL), with or without other risk factors. Often drug therapy can be delayed in young adult men (<35 years) and premenopausal women who have LDL cholesterol <220 mg/dL. Some investigators, however, would consider drug therapy in these latter patients when LDL cholesterol is in the range of 190 to 220 mg/dL. Most patients should receive drug therapy when LDL cholesterol exceeds 220 mg/dL. In very high-risk patients without CHD (e.g., diabetic patients), consideration can be given to using cholesterol-lowering drugs when LDL cholesterol levels are in the range of 130 to 159 mg/dL.

Secondary Prevention

For secondary prevention in patients with established CHD (or other atherosclerotic disease), cholesterol-lowering therapy should be initiated if the LDL cholesterol level is over 100 mg/dL. This is regardless of other CHD risk factors. The goal is to reduce LDL cholesterol to 100 mg/dL or below. Maximal dietary therapy should be employed in all patients. If LDL cholesterol is >130 mg/dL at baseline, drug treatment should be considered. However, if the LDL cholesterol is 100 to 129 mg/dL with maximal dietary therapy, clinical judgment is needed to decide whether to use drugs. Also, if one drug lowers LDL to this range, judgment is needed as to whether to add a second drug.

Dietary Therapy

The essential aim of a serum cholesterol-lowering diet is to decrease the intake of saturated fatty acids, cholesterol, and total calories (if the patient is obese). Increased physical activity is also recommended. The first step of dietary therapy is to reduce saturated fatty acids to less than 10 percent of total calories, total fat to 30 percent or less of total calories, and cholesterol to less than 300 mg per day. This can be achieved by reducing the intake of dairy fats (whole milk, ice cream, cheese, and cream), fatty meats, baked goods, eggs, and organ meats.

For primary prevention, after starting the step 1 diet, the patient should be tested for lipoproteins at 6 weeks and at 3 months to determine whether or not the goals of therapy have been obtained. If not, the patient should proceed to the step 2 diet, which further reduces saturated fatty acids to less than 7 percent and cholesterol intake to less than 200 mg per day. This second step of dietary therapy should be continued for another 3 months. If the goals of therapy are not achieved, consideration can be given to the use of cholesterol-lowering drugs. If the goals are obtained, the patient should be continued on the step 2 diet indefinitely.

For secondary prevention, maximal dietary therapy (step 2 diet) should be started immediately. If it becomes apparent that the goals of therapy will not be reached by diet alone, consideration can be given to starting drugs. A 3- to 6-month trial of dietary therapy is not required for secondary prevention.

Drug Therapy

Cholesterol-lowering drugs are classified into major drugs [bile acid sequestrants, nicotinic acid, and HMG-CoA reductase inhibitors (statins)] and other drugs (fibric acids and probucol). Estrogen replacement therapy in postmenopausal women can be considered an alternative (or adjunct) to drug therapy in those with elevated LDL cholesterol.

Bile acid sequestrants include cholestyramine (Questran) and colestipol (Colestid). These drugs bind bile acids in the intestinal tract and prevent their reabsorption by the intestine. This increases conversion of cholesterol into bile acids in the liver, which reduces hepatic content of cholesterol and thereby increases the activity of LDL receptors. Cholestyramine, 8 g twice daily, or colestipol, 10 g twice daily, reduces LDL cholesterol by 20 to 25 percent. At half these doses, reductions of

15 to 20 percent are common; these lower doses are usually better tolerated. The major side effects are constipation and various other gastrointestinal complaints. Sequestrants can interfere with absorption of acidic drugs, and they can raise triglyceride levels. Their use as a single drug cannot be recommended in patients with definite hypertriglyceridemia.

Nicotinic acid inhibits hepatic production of lipoproteins, reduces VLDL and LDL cholesterol, and raises HDL cholesterol. Therapeutic doses are 500 to 1500 mg three times daily. Side effects include flushing and itching of the skin, gastric distress, abnormalities in liver function, hyperuricemia, and glucose intolerance. Flushing and gastric distress can be minimized by starting with low doses and gradually increasing them. (Aspirin may also reduce flushing.) This drug is especially valuable to hypercholesterolemic patients who have borderline hypertriglyceridemia or low HDL cholesterol.

HMG-CoA reductase inhibitors (statins) partially block the synthesis of cholesterol in the liver and thereby increase LDL receptor activity. They lower both LDL and VLDL cholesterol levels. The starting dose of lovastatin (Mevacor), pravastatin (Pravacol), and fluvastatin (Lescol) for most patients is 40 mg at bedtime per day; the starting dose for simvastatin (Zocor) is 20 mg at bedtime. The starting dose for atorvastatin (Lipitor) is 10 mg at bedtime. These doses can be doubled for severe hypercholesterolemia. LDL cholesterol levels are reduced by 30 to 40 percent. The major side effect is myopathy, which typically occurs in one of three forms: muscle pain and weakness; moderately elevated creatine levels; or both. Rarely, a patient develops severe myopathy with myoglobinuria and even acute renal failure. (This is most likely to occur with concomitant use of cyclosporine, gemfibrozil, or nicotinic acid; these combinations should be used with caution and careful monitoring of patients.) The use of these drugs with bile acid sequestrants, which can be highly effective in severe hypercholesterolemia, does not increase risk for myopathy.

Gemfibrozil (Lopid) is a fibric acid and resembles other drugs in this class—clofibrate (Atromid-5), fenofibrate (Tricor), and bezafibrate (Bezalip). The dose of gemfibrozil is 600 mg twice daily. Although these drugs potentially lower triglyceride levels, they may cause a 10 to 20 percent reduction in LDL-cholesterol levels in patients with hypercholesterolemia. The major side effects are gastric distress, myopathy, and cholesterol gallstones. These drugs are used primarily to treat hypertriglyceridemia, especially severe hypertriglyceridemia.

MANAGEMENT OF HYPERTRIGLYCERIDEMIA (HIGH TRIGLYCERIDES)

For all patients with hypertriglyceridemia, the primary aim of lipid-lowering therapy is to achieve the target value for LDL cholesterol. For patients with borderline high triglycerides, dietary therapy should be tried first to achieve triglyceride reduction. This includes weight reduction and decreased intakes of alcohol and carbohydrates. For many patients drug therapy will not be required; but if it is deemed necessary, nicotinic acid and fibric acids are effective triglyceride-lowering drugs. When serum triglycerides exceed 500 mg/dL, patients are at risk for acute pancreatitis; if the triglyceride level cannot be driven below 500 mg/dL by use of diet, drug therapy is indicated for prevention of pancreatitis.

MANAGEMENT OF COMBINED HYPERLIPIDEMIA

Combined elevations of cholesterol and triglycerides call first for reduction of LDL-cholesterol levels to achieve the target value for LDL cholesterol. Secondarily, triglycerides should be reduced to near normal if possible. Weight reduction and exercise are first line of therapy. Nicotinic acid or fibric acids can also be used to lower triglycerides, often in combination with a statin, which is used for LDL lowering. When these combinations are used, the patient must be carefully monitored for the development of myopathy.

MANAGEMENT OF HYPOALPHALIPOPROTEINEMIA (LOW HDL CHOLESTEROL)

First-line therapy for low HDL cholesterol is to remove causes (smoking, obesity, lack of exercise, uncontrolled diabetes, hypertriglyceridemia, and beta-adrenergic blocking agents). If the patient has an elevated LDL cholesterol level, LDL-lowering drugs (e.g., statins) can be considered. Treatment of low HDL cholesterol in the absence of known causes and other risk factors may not be justified; but in high-risk patients, nicotinic acid, a statin, or gemfibrozil can be considered.

SUGGESTED READING

Expert Panel on Detection, Evaluation, and Treatment of High Blood Cholesterol in Adults. Second Report of the National Cholesterol Education Program (NCEP) Expert Panel on Detection, Evaluation, and Treatment of High Blood Cholesterol in Adults (Adult Treatment Panel II). Summary of report published in *JAMA* 1993; 269:3015–3023. Full report published in *Circulation* 1994; 89:1333–1345.

Downs JR, Clearfield M. Whitney E, et al: Primary prevention of acute coronary events with lovastatin in men and women with average cholesterol levels: results of AFCAPS/TexCAPS. *JAMA* 1998; 279:1615–1622.

Grundy SM: HMG-CoA reductase inhibitors for treatment of hypercholesterolemia. *N Engl J Med* 1988; 319:24–33.

Grundy SM, Balady GJ, Criqui MH, et al: When to start cholesterol-lowering therapy in patients with coronary heart disease. A statement for healthcare professionals from the American Heart Association Task Force on risk reduction. *Circulation* 1997; 95:1683–1685.

Grundy SM, Vega GL: Fibric acids: effects on lipids and lipoprotein metabolism. *Am J Med* 1987; 83(suppl 5B): 9–19.

Jones PH, Grundy SM, Gotto AM Jr: Assessment and management of lipid abnormalities. In: Alexander RW, Schlant RC, Fuster V, O'Rourke RA, Roberts R, Sonnenblick EH (eds): *Hurst's The Heart*, 9th ed. New York, McGraw-Hill, 1998:1553–1581.

Mostaza JM, Schulz I, Vega GL, Grundy SM: Comparison of pravastatin with crystalline nicotinic acid monotherapy in treatment of combined hyperlipidemia. *Am J Cardiol* 1997; 79:1298–1301.

Sacks FM, Pfeffer MA, Moye LA, et al: The effect of pravastatin on coronary events after myocardial infarction in patients with average cholesterol levels. *N Engl J Med* 1996; 335:1001–1009.

Scandinavian Simvastatin Survival Study Group: Randomised trial of cholesterol lowering in 4444 patients with coronary heart disease: the Scandinavian Simvastatin Survival Study (4S). *Lancet* 1994; 344:1383–1389.

Shepherd J, Cobbe SM, Ford I, et al: Prevention of coronary heart disease with pravastatin in men with hypercholesterolemia. *N Engl J Med* 1995; 333:1301–1307.

Smith SC Jr, Blair SN, Criqui MH, et al: Preventing heart attack and death in patients with coronary disease. *Circulation* 1995; 92:2–4.

CHAPTER

8

STABLE ANGINA PECTORIS

Robert C. Schlant
R. Wayne Alexander

DEFINITION

Angina pectoris (literally, "strangling" in the chest) is a syndrome of recurrent discomfort in the chest or related areas associated with myocardial ischemia or dysfunction but without myocardial necrosis. Some of the characteristic terms used to describe the discomfort include *pressure, tightness, constricting, dull, fullness, burning, swelling, ache, like a vise*, and *like a toothache*. It usually lasts 1 to 15 min and is located in the retrosternal area but may radiate to, or be confined to, the jaw, throat, shoulder, back, or the left (or less frequently the right) arm or wrist.

When myocardial ischemia occurs, it is first associated with diastolic dysfunction, then systolic dysfunction, then electrocardiographic (ST segment) changes, and, last, chest discomfort. Similarly, many patients have numerous episodes of asymptomatic or "silent" myocardial ischemia for each episode associated with discomfort. In some patients, myocardial ischemia produces a significant transient decrease in the diastolic or systolic function of the ventricle that is responsible for exertional dyspnea or fatigue. In this situation, these symptoms produced by myocardial ischemia are *angina equivalents*.

Angina pectoris is said to be *stable* when there has been no change in the frequency, duration, precipitating factors, or ease of relief during the preceding 60 days.

ETIOLOGY

Coronary artery disease (CAD) due to atherosclerosis is the most common cause of angina pectoris. Other causes include congenital coronary artery abnormalities, coronary artery spasm, coronary thromboembolism, coronary vasculitis, aortic stenosis, mitral stenosis with pulmonary hypertension, pulmonic stenosis, pulmonary hypertension, hypertrophic cardiomyopathy, and systemic arterial hypertension. Other conditions that are less frequently associated with angina pectoris include aortic regurgitation, idiopathic dilated cardiomyopathy, other specific forms of heart muscle disease, and syphilitic heart disease. The common feature in those conditions associated with angina pectoris but without coronary artery disease is myocardial oxygen demand that is in excess of what can be delivered by angiographically normal coronary arteries.

CLASSIFICATION

The Canadian Cardiovascular Society grading scale (Table 8-1) is widely used to classify the severity of angina pectoris.

TABLE 8-1

CANADIAN CARDIOVASCULAR SOCIETY FUNCTIONAL CLASSIFICATION OF ANGINA PECTORIS

I. Ordinary physical activity, such as walking and climbing stairs, does not cause angina. Angina results from strenuous or rapid or prolonged exertion at work or recreation.

II. Slight limitation of ordinary activity. Walking or climbing stairs rapidly, walking uphill, walking or stair climbing after meals, in cold, in wind, or when under emotional stress, or only during the few hours after awakening. Walking more than two blocks on the level and climbing more than one flight of ordinary stairs at a normal pace and under normal conditions.

III. Marked limitations of ordinary physical activity. Walking one to two blocks on the level and climbing more than one flight under normal conditions.

IV. Inability to carry on any physical activity without discomfort—anginal syndrome *may be* present at rest.

Source: Modified from Campeau L: Letter to the editor. *Circulation* 1976; 54:522. Reproduced with permission from the American Heart Association Inc. and the author.

DIAGNOSIS OF ANGINA PECTORIS

Symptoms

Angina pectoris should be assessed in terms of discomfort location, duration, precipitating events, associated symptoms, and mechanisms of relief. Typically, the discomfort of angina pectoris is produced by exertion or emotions, particularly anger, and is relieved in a few minutes by rest or sublingual nitroglycerin. In many patients it is reproduced by the same amount of exertion, whereas in others the amount of exertion necessary to precipitate angina varies from day to day. It is often worsened by severe anemia, tachycardia, systemic hypertension, fever, hypothyroidism, hyperthyroidism, or other causes of high cardiac output. Cold weather, cigarette smoking, increased heat or humidity, and eating may precipitate or lower the exertion threshold for angina pectoris. Emotions—particularly anger, excitation, and frustration—may also precipitate angina. Table 8-2 lists descriptions of discomfort that is unlikely to be produced by myocardial ischemia.

In most patients the discomfort has a characteristic crescendo nature and develops and increases to a plateau over 10 to 30 s and disappears several minutes after exertion is discontinued. The discomfort usually lasts a few minutes, occasionally 10 to 15 min. Very rarely, it may last up to 30 min. In addition to exertion, drugs that increase heart rate or blood pressure can precipitate angina, as can cocaine, which can also increase heart rate and blood pressure and produce coronary artery vasoconstriction or thrombosis.

Nocturnal angina is the occurrence of angina at night. Possible mechanisms include dreaming, unrecognized tachycardia, marked decrease in blood pressure, or an increase in blood

TABLE 8-2

DESCRIPTORS OF DISCOMFORT NOT LIKELY
TO BE ANGINA PECTORIS

Knife-like	Shooting	Pricking
Sticking	Jabbing	Cutting
Stabbing	Twitching	Like an ice pick
Needle-like	Itching	
Stinging	Tingling	

volume and cardiac output (see also Chap. 9, "Unstable Angina").

ANGINA EQUIVALENTS

As noted above, some patients have episodes of myocardial ischemia that are asymptomatic, or "silent." In these patients continuous ambulatory electrocardiographic (Holter) recordings or continuous records of left ventricular function may document the asymptomatic ischemic episodes. In general, the frequency and duration of episodes of "asymptomatic" myocardial episodes are proportional to the number of symptomatic episodes. It is rare for patients to have only episodes of asymptomatic angina pectoris due to CAD.

Patients with angina pectoris due to coronary atherosclerosis have an increased frequency of the usual risk factors for atherosclerosis, including hypercholesterolemia, cigarette smoking, systemic arterial hypertension, diabetes mellitus, and a family history of CAD in a parent or close relative before the age of 55. In women, the use of birth control pills and the postmenopausal state without estrogen replacement may contribute.

PHYSICAL EXAMINATION

Patients with hyperlipidemia may have arcus senilis, xanthelasma, eruptive xanthomata, tendon xanthomata, or tuberous xanthomata.

During an attack of angina, many patients appear pale and quiet. Sweating is frequent. The heart rate and blood pressure are frequently slightly elevated and there may be palpable systolic impulse at the apex or medial to the apical area. A new gallop sound, usually an S4 and occasionally an S3, may be heard, at times in association with a new or louder apical systolic murmur. The discomfort of angina is often relieved by performing the Valsalva maneuver or, very cautiously, carotid sinus massage.

Patients who have had a previously myocardial infarction may have signs of chronic congestive heart failure, including evidence of jugular venous distention, left ventricular dilatation, mitral regurgitation, pulmonary congestion, and peripheral edema.

FACTORS INFLUENCING PROGNOSIS

The prognosis of patients with stable angina pectoris has been related to many historical features and findings on special tests. Among the historical features are the patient's age, gender, progression of symptoms of angina pectoris, history of diabetes mellitus, history of prior myocardial infarction, and history of heart failure or continued use of digoxin. Among the many objective findings that have been related to prognosis are the presence of signs of heart failure or cardiac enlargement on physical examination, left ventricular function (ejection fraction and contraction pattern) at rest and during exercise, ST segment depression on the resting electrocardiogram (ECG), the magnitude of ST segment depression during exercise and the level of exercise at which it occurs, exercise capacity, maximal oxygen consumption during exercise, chronotropic incompetence during exercise, exertional hypotension, extent of myocardial reperfusion on radionuclide scintigraphy, the accumulation of radioactivity in the lungs during exercise radionuclide scintigraphy, and the severity and extent of coronary artery disease on coronary arteriography.

SPECIAL TESTS

In patients with suspected stable angina pectoris, special tests may be performed either to establish the diagnosis or to obtain prognostic information. In general, men with a history of "classic," or "typical," angina pectoris have a greater likelihood of having significant CAD on coronary arteriography than do women with similar symptoms. The same is true for men and women with "atypical" angina. In patients with a very low or very high pretest likelihood of CAD, many noninvasive tests are not cost-effective from a diagnostic standpoint, although they may provide valuable prognostic information. It is important to know the sensitivity, specificity, and reproducibility of various tests in one's own facility.

Exercise-Stress Electrocardiography

This is the most widely used test to obtain objective evidence of myocardial ischemia as well as prognostic information in patients with known CAD. The mean sensitivity is about 68 percent and the mean specificity about 77 percent. The diagnostic value of the test is significantly decreased by the presence of abnormalities on the resting electrocardiogram (ECG), such as bundle branch block, ST-T wave changes, or left ventricular hypertrophy. Digitalis can produce false-positive ST-segment depression. The duration of exercise and maximal oxygen consumption correlate reasonably well with prognosis. Exercise is usually performed using either a treadmill or a bicycle ergometer.

Radionuclide Scintigraphy

This is more expensive than exercise ECG testing, but it provides higher sensitivity and specificity. It is usually performed with the injection of thallium 201 at peak exercise. In patients unable to exercise, pharmacologic stress testing may be employed, usually using intravenous dipyridamole (Persantine), dobutamine (Dobutrex), or adenosine (Adenocard). Some scintigraphic techniques employ technetium-99m sestamibi, at times in combination with thallium 201.

Radionuclide Angiography

The determination of left ventricular ejection fraction and left ventricular wall motion abnormalities, at rest and during exercise, may be useful in the diagnosis of CAD, although numerous other conditions can produce similar abnormalities.

Echocardiography

Echocardiography, including Doppler echocardiography, can be performed at rest, during exercise, or immediately afterwards. Pharmacologic stress echocardiography is performed following the injection of dobutamine, dipyridamole, or adenosine. In general, the diagnostic value is comparable to radionuclide scintigraphy if it is performed by skilled and experienced operators. It is less expensive than radionuclide scintigraphy.

Positron Emission Tomography

Positron emission tomography (PET) is more accurate than standard thallium-201 computed tomography, but it is very expensive. PET can also provide information regarding coronary blood flow and myocardial viability.

Magnetic Resonance Imaging

This is moderately expensive. It provides information regarding coronary blood flow in different regions of the heart as well as overall and localized left ventricular function, and it can also provide useful information regarding myocardial viability.

Coronary Arteriography

This is the "gold standard" for the diagnosis of CAD, although it is invasive and moderately expensive. In general, it is performed only if the results might have a significant influence upon patient management—i.e., coronary revascularization by coronary angioplasty or coronary artery bypass surgery. Occasionally, it is appropriate for diagnostic purposes in patients with atypical symptoms or those unable to undergo a noninvasive test.

DIFFERENTIAL DIAGNOSIS

Some of the more common conditions mimicking angina pectoris due to coronary atherosclerosis are listed in Table 8-3 (on pages 00–00).

PATHOPHYSIOLOGY

The basic cause of angina pectoris is an imbalance between myocardial oxygen supply and oxygen need. Decreased supply may be the result of obstruction produced by atherosclerosis of the epicardial coronary vessels or by vasoconstriction or failure of the epicardial coronary arteries or the smaller intramyocardial arterioles to dilate. Patients with unstable angina (Chap. 9) or myocardial infarction (Chap. 10) may have coronary artery obstruction from thrombosis secondary to rupture of an atheromatous plaque. Patients with coronary atherosclerosis have endothelial dysfunction that may be manifest by a failure of the coronary vasculate to dilate in response to normal vasodilatory stimuli such as increased flow, exercise, tachycardia, acetylcholine, or cold pressor testing. Most patients with angina pectoris due to coronary atherosclerosis have myocardial ischemia caused by both epicardial coronary obstruction and endothelial dysfunction of both large and small vessels. Coronary blood flow, especially to the subendocardium, can

TABLE 8-3
DIFFERENTIAL DIAGNOSIS OF ANGINA PECTORIS

Cardiovascular
Myocardial ischemia
Coronary atherosclerosis
Coronary vasospasm
Congenital coronary artery disease
Anomalous origin
Aberrant coronary artery
Coronary arteriovenous fistula
Kawasaki's disease
Small vessel disease
Microvascular angina (syndrome X)
Systemic arterial hypertension
Hypertrophic cardiomyopathy
Idiopathic dilated cardiomyopathy

Severe anemia, hypoxia
High-dose x-irradiation
Withdrawal from chronic nitroglycerin exposure
Nonmyocardial ischemia
Aortic dissection
Discrete thoracic aortic aneurysm
Mitral valve prolapse
Tachycardia, bradycardia
Palpitations
Pericarditis
Thoracic-respiratory
Pulmonary embolism, infarction
Pneumothorax
Pneumomediastinum (mediastinal emphysema)

Splenic flexure syndrome
Neuromuscular/skeletal
Chest wall pain
Costochondritis (Tietze's syndrome)
Cervical or thoracic degenerative arthritis, nerve compression, radiculopathy
Cervical vertebral disk
Intercostal neuralgia
Thoracic outlet (scalenus anticus) syndrome
Shoulder arthropathies
Shoulder hand syndrome
Fibromyalgia (myofascial pain syndrome; fibromyositis)
Pectoral, intercostal, serratus anterior

Aortic valve disease
Coronary artery dissection (Marfan's syndrome)
Pulmonary hypertension
Right ventricular hypertension
Chronic obstructive pulmonary disease
Syphilitic aortitis, coronary ostial disease
Connective tissue disease
Pseudoxanthoma elasticum
Cystic medial necrosis
Homocystinuria
Gargoylism
Periarteritis nodosa
Systemic lupus erythematosus
Rheumatoid arthritis
Cardiac amyloid
Cardiac tumors

Pleuritis
Epidemic pleurodynia (Bornholm's disease)
Mediastinitis
Intrathoracic malignancy
Café coronary
Gastrointestinal
Gastroesophageal reflux, esophagitis
Esophageal spasm
Esophageal rupture (Mallory-Weiss syndrome; Boerhaave's syndrome)
Esophageal impaction
Hiatal hernia
Cholecystitis, gallstones
Gastritis
Peptic uler disease
Pancreatitis
Splenic infarction

Precordial catch syndrome
Cardiac causalgia
Bursitis
Superficial thrombophlebitis of thoracic veins (Mondor's syndrome)
Xiphoidalgia
Diaphragmatic flutter
Neurocutaneous
Herpes zoster
Breast
Pendulous breast syndrome
Brassiere syndrome
Psychologic
Anxiety
Hyperventilation
Panic attacks
Depression
Self-gain
Munchhausen syndrome

also be impaired by an elevation of left ventricular diastolic pressure (see below). The major determinants of myocardial oxygen demand are heart rate, myocardial contractility, and systolic wall tension, which is related to left ventricular systolic pressure and volume.

Microvascular Angina

Some patients without other cause for myocardial ischemia may have exertional angina with normal coronary arteries. These patients should have documented objective evidence of abnormal coronary vasodilator reserve before the diagnosis is accepted. This is sometimes referred to as *syndrome X* (see below).

Circadian Rhythm of Coronary Ischemia

Angina pectoris, including "silent" or asymptomatic ischemia, is more frequent during the first few hours after awakening. A similar pattern is found in patients with acute myocardial infarction. Factors that might contribute to the circadian variation include increased heart rate, blood pressure, plasma concentrations of catecholamines and cortisol, and increased platelet aggregability.

MEDICAL THERAPY OF ANGINA PECTORIS

General

Patients with angina pectoris due to coronary atherosclerosis should be evaluated for risk factors for coronary disease and, whenever possible, the risk factors should be corrected. Tobacco should be avoided in all forms, and hypertension and diabetes mellitus should be controlled. Ideal body weight should be achieved. A low-fat, low-cholesterol diet should be instituted and a lipid profile determined.

In most patients with angina due to CAD, the low-density lipoprotein (LDL) cholesterol should be lowered to or below 100 mg/dL (2.6 mmol/L). This will often require the use of drugs in addition to diet (see Chap. 7). Patient participation in a rehabilitation or health-enhancement program will often assist in achieving ideal body weight and controlling risk factors.

Aspirin

In the absence of contraindications, patients with angina pectoris should take a daily enteric-coated aspirin (160 to 325 mg). Lower doses (80 mg/day) may also be satisfactory, particularly in patients with a history of gastritis or peptic ulcer disease.

Nitrates

The standard first-line therapy of angina remains sublingual nitroglycerin (TNG), which usually relieves the symptoms within 1 to 5 min. TNG may be taken acutely either as a sublingual tablet (0.3 to 0.6 mg) or as an oral spray, each puff of which is calculated to deliver 0.4 mg. Monotherapy with sublingual or oral spray TNG may be satisfactory if the episodes occur no more often than once a week.

Long-acting formulations of nitrates are used prophylactically in patients who have more frequent episodes. The many forms of nitrates include a slowly absorbed buccal capsule, a transdermal ointment or patch, and sublingual or oral forms that are more slowly absorbed or metabolized. Table 44-1 lists the commonly used nitrates and their dosages. In general, it is preferable to start with a low dosage and to increase the dose progressively. The most common side effects are headache, dizziness, and postural hypotension.

The development of tolerance is a major disadvantage to chronic nitrate therapy. In general, tolerance can be avoided by providing a 10- to 12-h nitrate-free period. For many patients this can be from about 9 p.m. to about 7 a.m. Nitroglycerin ointment or patches should usually be removed about 8 or 9 p.m. Isosorbide dinitrate (ISDN) (10 to 60 mg) can be given orally in doses of 30 mg twice a day at 8 a.m. and 5 p.m. or three times a day at 8 a.m., 1 p.m., and 5 p.m. Isorbide-5-mononitrate (ISMO), a metabolite of ISDN, is administered orally in 20-mg doses at 7 a.m. and 2 p.m. An extended release form of isosorbide-5-mononitrate (IMDUR) can be taken orally as a single 60-mg dose at 7 or 8 a.m. Patients who have angina at night may need either to have the nitrate-free period during another time or to concurrently use a beta blocker (Chap. 47) or a calcium channel blocker (Chap. 48).

Patients should be instructed that if they have an episode of angina that persists more than 10 min despite taking three TNG tablets, they should report promptly to the nearest medical facility for further evaluation and management.

Beta-Adrenergic Blocking Agents

Beta-adrenergic blocking agents, which reduce heart rate and myocardial contractility both at rest and during exertion, are very effective in the management of patients with angina pectoris. There are many beta blockers available (see Table 47-1). In general, cardioselective agents are preferred in patients who have a history of bronochospastic disease, diabetes mellitus, or peripheral vascular disease. IT should be noted, however, that even cardioselective beta blockers can produce bronchospasm in some patients. All beta blockers can worsen heart block or depress left ventricular function and worsen heart failure.

Beta blockers are usually administered to lower the resting heart rate to 50 to 60 beats per minute. When beta blockers are discontinued, the dosage should be tapered over 3 to 10 days when possible to avoid a rebound worsening of angina pectoris and possible myocardial infarction (see also Chap. 47).

Calcium Channel Blockers

Calcium channel blockers (or calcium antagonists), which impede the entry of calcium into myocytes and smooth muscle cells, decrease myocardial oxygen requirements by producing arterial dilation, reducing arterial blood pressure, and afterload, and reducing myocardial contractility. Many calcium channel blockers also produce coronary vasodilatation and prevent coronary artery spasm. Some calcium channel blockers, such as verapamil and diltiazem, also tend to reduce heart rate. They can also increase heart block, particularly when used in combination with a beta blocker or, occasionally, digoxin.

Most calcium channel blockers are as effective as beta blockers for the treatment of angina pectoris, and they are frequently used concurrently. Calcium channel blockers are of particular value in patients with conditions that may be made worse by beta blockers, such as asthma, chronic obstructive pulmonary disease, insulin-dependent diabetes mellitus, or peripheral vascular disease. The calcium channel blockers are discussed in more detail in Chap. 48.

Combined Drug Therapy

Patients with only occasional episodes of angina pectoris are treated with either sublingual TNG or oral-spray TNG. Patients with more frequent episodes should first be treated with the additional use of a longer-acting nitrate regimen, a beta blocker, or a calcium channel blocker. If symptoms persist, one of the other medications is added. In most patients, therapy with a long-acting nitrate regimen and either a beta blocker or a calcium channel blocker is effective. Some patients appear to benefit from triple therapy with a long-acting nitrate, beta blocker, and calcium channel blocker. The combination of a beta blocker and verapamil is particularly likely to worsen heart failure or to produce heart block, particularly in the elderly.

The commonly used beta blockers and their dosages are described in Chap. 47 and the calcium channel blockers in Chap. 48. The more frequently used nitrates and their dosages are listed in Table 44-1.

Revascularization

Some patients with stable angina pectoris are candidates for revascularization either by coronary artery bypass surgery (CABS) or percutaneous transluminal coronary angioplasty (PTCA), with or without stenting. The two general indications for revascularization are (1) the presence of symptoms that are not acceptable to the patient because of restriction of physical activity and lifestyle or side effects from medications or (2) the presence of coronary arteriographic findings indicating clearly that the patient would have a significantly better prognosis with revascularization than with medical therapy. In general, patients with stable angina should have objective evidence of myocardial ischemia prior to revascularization. Additional major considerations include the age of the patient, presence of other comorbid conditions, grade or class of angina experienced by the patient on maximal therapy, extent and severity of the evidence of myocardial ischemia on noninvasive testing, degree of left ventricular dysfunction, and distribution and severity of coronary artery disease.

CABS provides good relief to most patients who have suitable vessels. It is particularly indicated in patients with severe CAB [left main (equal to or greater than 50 percent diameter stenosis) or three-vessel disease with impaired left ventricular function and some patients with two-vessel disease that includes the proximal left anterior descending coronary artery] and in patients in whom revascularization is indicated but who have lesions that are not amenable to PTCA. Vein grafts have a significant incidence of failure after 10 years. In contrast, internal mammary artery grafts have superior patency at

10 years. Whenever possible, one or both internal mammary arteries should be used during CABS.

PTCA can be successfully performed on many patients with anatomically suitable disease of native vessels or bypass grafts. In general, patients should be refractory to medical therapy or unable to tolerate medical therapy and should have objective evidence of myocardial ischemia. Since emergency CABS may become necessary, patients should generally be sufficiently symptomatic to warrant CABS. Initial relief of angina occurs in about 80 to 90 percent of patients; however, within 6 months, restenosis occurs in about 25 to 40 percent of patients. In this situation, repeat TCA can be repeated with similar or better results. Complications of PTCA include dissection or acute thrombosis of the coronary artery; death in about 1 percent; emergency CABS in about 3 to 5 percent; and myocardial infarction in about 4 percent.

Patients with one-vessel disease who have indications for revascularization and a suitable vessel are usually treated by PTCA, frequently with the use of a stent, whereas patients with left main disease or three-vessel disease with decreased left ventricular function are generally treated with CABS. Patients with two-vessel disease including the proximal left anterior descending are usually treated with CABS. Other patients with two- or three-vessel disease who have indications for revascularization are treated by either PTCA, with or without the use of stents, or CABS, depending upon the individual characteristics of the CAD and the local experience with CABS and PTCA. Many patients with two- or three-vessel disease should be treated medically if their angina is class I or II, their exercise performance is good, and their left ventricular function is good.

HEART FAILURE IN PATIENTS WITH STABLE ANGINA PECTORIS

Patients with angina pectoris associated with chronic heart failure have a poor prognosis; most such patients have had extensive infarction of the left ventricle. In a small percentage of patients, however, the significant depression of left ventricular function is not irreversible but rather is "hibernating," presumably due to chronic ischemia if the myocardium were to contract more vigorously. The left ventricular function of such patients can occasionally be significantly improved by revascularization. Accordingly, it may be important in such patients to attempt to estimate the viability of the myocardium by special techniques, including observing left ventricular function following a premature ventricular contraction, radionuclide scintigraphic techniques, positron emission tomography (PET), or magnetic resonance imaging techniques.

In general, therapy for heart failure in patients with angina pectoris is the same as for other patients with systolic left ventricular dysfunction (see Chap. 1). Special efforts should be made to avoid hypokalemia, hypomagnesemia, and digitalis toxicity.

CHEST PAIN WITH ANGIOGRAPHICALLY NORMAL CORONARY ARTERIES

It has been known for many years that some patients with both stable and apparently unstable angina pectoris can have angiographically normal coronary arteries. In these patients, it is thought that coronary vasospasm is sufficient to produce angina pectoris and, rarely, myocardial infarction or serious arrhythmias. While most patients who have angina due to vasospasm of large, epicardial blood vessels also have associated significant obstructive coronary atherosclerosis, some patients can have this syndrome apparently due only to coronary vasospasm with angiographically normal coronary arteries. Such patients may be more common in Japan than in the United States. These patients, who have variant, or Prinzmetal's, angina are discussed in Chap. 9.

A second and larger group of patients have chest pain with angiographically normal coronary arteries and no evidence of vasospasm of the large coronary arteries. This syndrome of angina with normal coronary arteriography has sometimes been referred to as *syndrome X*. The patients reported with this syndrome are heterogeneous, and some authors have included patients with systemic arterial hypertension, hypertrophic cardiomyopathy, or idiopathic dilated cardiomyopathy.

Clinically, the majority of patients are women with atypical chest pain that often lasts longer than typical angina and may be less responsive to nitroglycerin. Psychiatric disorders are sometimes present.

Most patients with the syndrome have ischemic ST-segment depression and myocardial lactate production on exercise testing and many have evidence of exercise- or stress-induced myocardial ischemia or regional wall-motion abnormalities on radionuclide scintigraphy, PET, or magnetic resonance imaging. A unifying abnormality in most patients appears to be a defective endothelial-dependent vasodilation of the small myocardial arterioles, which are not visible on routine coronary arteriography. This defect results in a decreased coronary flow reserve that is responsible for the myocardial ischemia. These patients may also have enhanced pain perception, cardiomyopathic changes on endocardial biopsy, or impaired vasodilator reserve in forearm vessels. Patients should not be diagnosed as having the syndrome in the absence of objective evidence of myocardial ischemia. Treatment consists of blood pressure control (Chap. 11) in patients who have borderline or definite systemic hypertension, reassurance, and the use of nitrates and calcium channel blockers. Occasionally, aminophylline may be of benefit. The general prognosis is good, although some patients have persistent symptoms.

SUGGESTED READING

Alderman EL, Botas J: Selection of revascularization for patients with stable angina pectoris. *Coronary Artery Disease* 1993; 4:1061–1067.

Alexander RW, Griendling KK: The coronary ischemic syndromes: Relationship to the biology of atherosclerosis, In: Alexander RW, Schlant RC, Fuster V, O'Rourke RA, Roberts R, Sonnenblick EH (eds): *Hurst's The Heart*, 9th ed. New York, McGraw-Hill, 1998, Chap 44, pp 1263–1274.

American College of Cardiology/American Heart Association Task Force on Assessment of Diagnostic and Therapeutic Cardiovascular Procedures: Guidelines and indications for coronary artery bypass graft surgery. *J Am Coll Cardiol* 1991; 17:543–584.

Cannon RO III, Camici PG, Epstein SE: Pathophysiological dilemma of syndrome X. *Circulation* 1992; 85:883–892.

Detrano R, Gianrossi R, Froelicher V: The diagnostic accuracy of the exercise electrocardiogram: a meta-analysis of 22 years of research. *Prog Cardiovasc Dis* 1989; 32:173–206.

Factor SM, Bache RJ: Pathophysiology of myocardial ischemia, In: Alexander RW, Schlant RC, Fuster V, O'Rourke RA, Roberts R, Sonnenblick EH (eds): *Hurst's The Heart*, 9th ed. New York, McGraw-Hill, 1998, Chap 43, pp 1241–1262.

Gersh BJ, Califf RM, Loop FD, et al: Coronary bypass surgery
    in chronic stable angina. *Circulation* 1989; 79 (suppl 1):
    1-46–1-59.

Gibbons RJ, Balady GJ, Beasley JW, et al: ACC/AHA Guide-
    lines for Exercise Testing: a report of the American
    College of Cardiology/American Heart Association Task
    Force on Practice Guidelines (Committee on Exercise
    Testing). *J Am Coll Cardiol* 1997; 30:260–315.

Hammermeister KE, Morrison DA: Coronary bypass surgery
    for stable and unstable angina pectoris. *Cardiol Clin* 1991;
    9:135–155.

Martin TW, Seaworth JF, Johns SP, Pupa LE, Condos WR:
    Comparison of adenosine, dipyridamole, and dobutamine
    in stress echocardiography. *Ann Intern Med* 1992;
    116:190–196.

Maseri A, Crea F, Kaski JC, Davies G: Mechanisms and sig-
    nificance of cardiac ischemic pain. *Prog Cardiovasc Dis*
    1992; 35:1–18.

Nwasokwa ON, Koss JH, Friedman GH, et al: Bypass surgery
    for chronic stable angina: Predictors of survival benefit
    and strategy for patient selection. *Ann Intern Med* 1991;
    114:1035–1049.

Parker JD, Parker JO: Nitrate therapy for stable angina pec-
    toris. *N Engl J Med* 1998; 338:520–531.

Pryor DB, Bruce RA, Chaitman BR, et al: Determination of
    prognosis in patients with ischemic heart disease. *J Am
    Coll Cardiol* 1989; 14:1016–1042.

Ridker PM, Manson JE, Gaziano JM, et al: Low dose aspirin
    therapy for chronic stable angina: a randomized, placebo-
    controlled clinical trial. *Ann Intern Med* 1991; 114:
    835–839.

Ryan TJ, Bauman WB, Kennedy JW, et al: Guidelines for per-
    cutaneous transluminal coronary angioplasty: a report of
    the American College of Cardiology/American Heart
    Association Task Force on Assessment of Diagnostic and
    Therapeutic Cardiovascular Procedures (Subcommittee on
    Percutaneous Transluminal Angioplasty). *J Am Coll Car-
    diol* 1993; 22:2033–2054.

Schlant RC, Alexander RW: Diagnosis and management of
    patients with chronic ischemic heart disease, In: Alexander
    RW, Schlant RC, Fuster V, O'Rourke RA, Roberts R,
    Sonnenblick EH (eds): *Hurst's The Heart*, 9th ed. New
    York, McGraw-Hill, 1998, Chap 45, pp 1275–1305.

Summary of the Second Report of National Cholesterol Edu-
    cation Program (NCEP) Expert Panel on Detection,

Evaluation, and Treatment of High Blood Cholesterol in Adults (Adult Treatment Panel II), *JAMA* 1993; 269: 3015–3023.

The VA Coronary Artery Bypass Surgery Cooperative Study Group: Eighteen-year follow-up in the Veterans Affairs Cooperative Study of Coronary Artery Bypass Surgery for Stable Angina. *Circulation* 1992; 86:121–130.

Waller BF: Nonatherosclerotic coronary heart disease, In: Alexander RW, Schlant RC, Fuster V, O'Rourke RA, Roberts R, Sonnenblick EH (eds): *Hurst's The Heart*, 9th ed. New York, McGraw-Hill, 1998, Chap 42, pp 1197–1240.

Willard JE, Lange RA, Hillis LD: The use of aspirin in ischemic heart disease. *N Engl J Med* 1992; 327:175–181.

CHAPTER

9

UNSTABLE ANGINA

Pierre Théroux

DEFINITION, ETIOLOGY, AND PREVALENCE

Unstable angina marks the transition from a chronic to an acute phase of coronary artery disease and from a stable to an unstable clinical status. It is generally considered as an intermediary phase between stable angina and myocardial infarction. Although unstable angina can occur in patients with multivessel disease as well as in those with single-vessel disease, rapid progression in the severity of one atherosclerotic plaque—the culprit lesion—is generally the cause. Typically the lesion is of only moderate severity, 40 to 60 percent reduction of lumen diameter, but is the site of a rupture or fissure with an overlying thrombus protruding into the lumen of the artery, which further obstructs coronary blood flow. The plaque is highly inflammatory, with degradation of its matrix and thinning of its cap, explaining its vulnerability to rupture under the influence of hemodynamic stress and other local and systemic factors. The reasons for the inflammatory process are suspected to be toxic, infectious, or autoimmune. Circulating platelets adhere to von Willebrand factor and to collagen exposed from the subendothelium; it becomes activated with release of active products and a configurational change in membrane receptor glycoprotein IIb/IIIa, which makes it competent to bind fibrinogen and to mediate platelet aggregation. Tissue factor abundant within the macrophages in the diseased plaque binds factor VIIa in the circulation to trigger locally the coagulation cascade and thrombin generation.

The clinical manifestations of this pathophysiologic sequence of events can be unstable angina, non-Q-wave myocardial infarction, myocardial infarction with ST-segment elevation, or sudden death, depending on the severity and duration of the coronary obstruction and the presence or absence of a collateral circulation. Unstable angina and non-Q-wave

153

myocardial infarction now account for the majority of admissions in coronary care units.

CLINICAL DIAGNOSIS

Unstable angina is a clinical diagnosis based on recognition of symptoms departing from the usual pattern of chest pain in a given patient. The pain may be more severe and more prolonged, triggered by unusually low levels of physical activity, and it may occur at rest or at night. It is less consistently relieved by nitroglycerin. Unstable angina may possess many clinical features that would lead to the diagnosis of atypical chest pain in other clinical settings. Skillful clinical judgment is needed to integrate into the medical diagnosis the likelihood of coronary artery disease based on risk factors and patient's past previous history and the characteristics and momentum of pain and, sometimes, of chest pain equivalents.

DIAGNOSTIC AIDS

The most useful diagnostic help during the acute phase is a 12-lead electrocardiogram (ECG), recognizing that a normal ECG does not rule out unstable angina and that an abnormal ECG does not confirm diagnosis. The ECG is more informative when it is obtained during an episode of chest pain and when the changes are evolving. Changes suggesting the diagnosis are horizontal or downsloping ST-segment depression, transient ST-segment elevation, and deep T-wave inversion. Serial plasma CK and MB-CK and, if available, troponin T or troponin I are obtained during the first 24 h to determine if some degree of myocardial necrosis has occurred; the troponin levels are often elevated with normal CK levels, indicating higher-risk unstable angina. Other diagnostic measures at times useful are two-dimensional echocardiography and myocardial scintigraphy with thallium-201 or technetium-99m sestamibi. The sensitivity of these tests is high, especially when they are obtained during or shortly after an episode of chest pain.

The treadmill exercise test and other provocative tests are generally not indicated during the acute phase of unstable angina except as a means to rule out an unlikely diagnosis. It must be realized that a negative provocative test does not exclude transient thrombotic occlusion superimposed on a plaque of only borderline hemodynamic significance. The provocative tests are more useful past the acute phase to evaluate the presence and severity of the underlying atherosclerotic disease for risk stratification.

New tests not currently used routinely are under investigation and may provide important diagnostic and prognostic information. Thus, many studies have now shown that elevated plasma levels of C-reactive protein could be an independent prognostic marker.

Coronary angiography is indicated in a large proportion of patients, sometimes for diagnostic purpose but more frequently to evaluate the best therapeutic options to offer the patient. Indeed, left main coronary artery disease is present in approximately 10 percent of patients with unstable angina and three-vessel disease is seen in another 25 percent. Indications for coronary angiography are recurrence of chest pain on medical therapy, left ventricular dysfunction, and a provocative test showing significant ischemia. Patients with previously invalidating angina may also profit from an intervention procedure. The availability of percutaneous intervention has widened the indications for coronary angiography.

DIFFERENTIAL DIAGNOSIS

Precipitating causes of unstable angina are ruled out as part of the initial evaluation of accelerating angina. These may be of cardiac or extracardiac origin and are related to problems with oxygen delivery or increased cardiac work. Frequently encountered conditions are stressful environmental situations, anemia, fever, tachyarrhythmias, thyrotoxicosis, aortic valve stenosis, hypertrophic cardiomyopathy, congestive heart failure, and cocaine abuse. Prinzmetal's variant angina is diagnosed when transient ST-segment elevation is present during a chest pain episode. Specific clinical features of the syndrome are chest pain at rest, often in the early morning hours, and repetitive symptoms of short duration promptly relieved with nitroglycerin; syncope during pain is infrequent but highly suggestive. The syndrome may be associated with manifestations of a systemic spastic disease such as Raynaud's phenomenon or migraine headache. Testing with ergonovine maleate (Ergotrate), methylergonovine maleate (Methergine, Ergometrine), or intracoronary acetylcholine can sometimes be used for the diagnosis when the coronary artery lesions are not angiographically critically severe.

CLINICAL CLASSIFICATION

There is no unique classification of unstable angina. The various classifications may refer to the clinical manifestation, clinical context, pathophysiologic mechanisms, and prognosis. Classically, unstable angina is classified as *de novo angina* when the symptoms of angina are of new onset and progressive in severity, *crescendo angina* when they become more severe in a patient with previously stable symptoms, and *prolonged chest pain* (acute coronary insufficiency) when pain lasts 20 min or more. A non-Q-wave myocardial infarction is diagnosed a posteriori when an elevation in the creatine kinase and its MB fraction is documented. The ST-segment depression on the ECG is usually more persistent in non-Q-wave myocardial infarction. Subclasses include early postinfarction angina occurring between 24 h to 1 month after an infarction, recurrence of angina within 6 months after balloon angioplasty usually caused by a restenosis, and unstable angina occurring late after venous bypass graft surgery, often associated with graft disease. The classification of Braunwald is now often used, representing an attempt to include clinical severity, background, and elements of prognosis. Risk evaluation may be more important for patient management than semantics in the classification.

NATURAL HISTORY AND RISK STRATIFICATION

Without medical intervention, the risk of fatal or nonfatal myocardial infarction at 1 week in patients hospitalized for unstable angina is 10 percent; at 3 months, it is 15 percent, with mortality rates of 4 and 10 percent, respectively. Additional events include recurrence of severe refractory angina in hospital in 20 to 25 percent of patients and residual ischemia with a positive treadmill exercise test in another 20 to 30 percent. Although the prognosis often is unpredictable, markers of a higher risk can be identified. These are an older age, ST-segment changes at admission or during chest pain, hemodynamic instability, elevated CK and MB-CK levels, elevated troponin T or troponin I levels and, importantly, the recurrence of chest pain on medical treatment. Predictors of a worse prognosis, should an event occur, include depressed left ventricular function, more extensive coronary artery disease, and a larger area of jeopardized myocardium. The latter can be appreciated by the hemodynamic condition of the patient at the time of chest pain and by the ECG leads changes showing ischemic changes. Other useful tests are thallium-201 lung uptake, left ventricular dysfunction on the echo, one or more large areas of ischemia on radionuclide scintigrams, and prolonged and frequent periods of silent ischemia on Holter monitoring.

MANAGEMENT

Higher-risk patients are hospitalized in a coronary care unit and moderate-risk patients assigned to the coronary care unit or a general ward, depending on the local facilities. Low-risk patients with a more doubtful diagnosis can be investigated in a chest pain unit or promptly as outpatients. The initial treatment is pharmacologic, with the aim of controlling the disease process and relieving symptoms. The etiologic treatment addresses the acute thrombotic process. For this purpose, *aspirin* is administered to all patients unless a contraindication exists or bypass surgery is planned in the short term. The initial dose is 180 to 324 mg, followed by 80 to 160 mg daily. The addition of intravenous unfractionated *heparin* titrated to an aPTT two times the control value further reduces risk. A low-molecular-weight heparin administered subcutaneously is a valid alternative to unfractionated heparin. Heparin is administered for 48 to 72 h and as long as the patient remains unstable. Hirudin, a direct thrombin inhibitor now approved for clinical use in patients with heparin-induced thrombocytopenia and investigated in patients with unstable angina, may show benefit over heparin as an anticoagulant. The most effective therapy to date to prevent death, myocardial infarction, and recurrent severe ischemia in patients with unstable angina and non-Q-wave myocardial infarction is triple antithrombotic therapy combining a GPIIb/IIIa antagonist to aspirin and heparin. These drugs are administered for 48 to 96 h and should be continued during and for 12 h after percutaneous intervention procedures when performed. *Tirofiban* and *eptifibatide* are approved for this purpose. *Abciximab* is also approved for use during coronary intervention in these patients. The dose of tirofiban is 0.4 μg/kg/min for 30 min followed by 0.1 μg/kg/min for the total duration of the infusion. The dose of eptifibatide is a 180 μg/kg bolus and 2 μg/kg/min infusion during the phase of medical therapy, with a decrease of the infusion rate to 0.5 μg/kg/min during angioplasty. *Thrombolysis* is not indicated if pain is not associated with ST-segment elevation or left bundle branch block. Coronary angiography usually is performed following 24 to 72 h of medical therapy, or earlier on an urgent basis if the clinical status deteriorates. Coronary angioplasty can be attempted during the procedure when a suitable stenosis is found. Bypass surgery is recommended for high-risk lesions not amenable to angioplasty or in the presence of left main disease or left main equivalent and in diabetic patients with multivessel disease. Patients with depressed left ventricular function also benefit from an intervention procedure. Appropriate risk stratification with treadmill exercise testing is an alternative to routine intervention procedure in patients with well-controlled disease and no recurrent symptoms. Other possible methods for risk stratification are stress or pharmacologic imaging with echocardiography or nuclear medicine and 24-h electrocardiographic monitoring to detect silent ischemia.

The symptomatic treatment includes administration of a beta blocker, nitrate, or calcium antagonist, and often combination therapy with two and sometimes three classes of drugs. An intravenous infusion of nitroglycerin is recommended in the more unstable patients. The use of a dihydropyridine without a beta blocker is contraindicated. The patient with recurrent angina should be treated more aggressively medically as well as with interventions. Intraaortic balloon counterpulsation also is useful in patients with refractory ischemia on optimal medical treatment and in patients with hemodynamic instability as a bridge to more thorough investigation and correction of the underlying coronary artery lesions.

Patients who have presented with an episode of unstable angina have identified themselves as being at higher risk for recurrence and for rapid progression of atherosclerosis. They should receive aspirin long term, independently of other treatment modalities used, and appropriate antianginal therapy. Most important, an aggressive program of control of risk factors should be instituted, including smoking cessation, lowering of LDL cholesterol levels, and physical fitness. The field of clinical investigation on measures stabilizing the diseased endothelium and preventing thrombus formation is presently very rich; physicians should be aware of the progress made and apply them in selected patients.

SUGGESTED READING

Boden WA, O'Rourke RA, Crawford MH, et al: Outcomes in patients with acute non-Q-wave myocardial infarction randomly assigned to an invasive as compared with a conservative management strategy. *N Engl J Med* 1998; 338: 1785–1792.

Braunwald E, Jones RH, Mark DB, et al: Diagnosing and managing unstable angina. *Circulation* 1994; 90:613–622.

Cairns J, Théroux P, Armstrong P, et al: Unstable angina—report from a Canadian expert roundtable. *Can J Cardiol* 1996; 12:1279–1292.

Clinical practice guidelines: Unstable angina: diagnosis and management. *AHCPR publication No. 04-0602*. Washington, DC: U.S. Department of Health and Human Services, 1994.

Cohen M. Demers C, Gurfinkel EP, et al: Low molecular weight heparin versus unfractionated heparin for unstable angina and non-Q-wave myocardial infarction. *N Engl J Med* 1997; 337:447–452.

Fuster V, Badimon L, Badimon JJ, Chesebro JH: Mechanisms of disease: the pathogenesis of coronary artery disease and the acute coronary syndromes. *N Engl J Med* 1992; 326: 310–318.

Oler A, Whooley MA, Oler J, Grady D: Adding heparin to aspirin reduces the incidence of myocardial infarction and death in patients with unstable angina. *JAMA* 1996; 276: 811–815.

The Platelet Receptor Inhibition in Ischemic Syndrome Management in Patients Limited by Unstable Signs and Symptoms (PRISM-PLUS) Study Investigators: Inhibition of the platelet glycoprotein IIb/IIIa receptor with tirofiban in unstable angina and non-Q-wave myocardial infarction. *N Engl J Med* 1998; 338:1488–1497.

Théroux P, Fuster V: Acute coronary syndromes: unstable angina and non-Q-wave myocardial infarction. *Circulation* 1998; 97:1195–1206.

Théroux P, Waters D: Diagnosis and management of patients with unstable angina. In: Alexander W, Schlant RC, Fuster V, O'Rourke RA, Roberts R, Sonnenblick EH (eds): *Hurst's The Heart*, 9th ed. New York: McGraw-Hill, 1998:1307–1343.

TIMI IIIB Investigators: Effects of tissue plasminogen activator and a comparison of early invasive and conservative strategies in unstable angina and non-Q-wave myocardial infarction. *Circulation* 1994; 89:1545–1556.

CHAPTER

10

DIAGNOSIS AND MANAGEMENT OF PATIENTS WITH ACUTE MYOCARDIAL INFARCTION

R. Wayne Alexander

BACKGROUND AND GENERAL PRINCIPLES

The demonstration in the late 1970s of the role of thrombus formation in the pathogenesis of acute myocardial infarction (AMI) quickly led to the systematic testing of thrombolytic strategies to abort the event. Major multicenter clinical trials on the treatment of AMI demonstrated the efficacy of streptokinase and recombinant tissue plasminogen activator in reducing mortality. Thus, large, adequately powered, randomized studies in this area have helped set a new standard and approach to the goal of treating AMI. The availability of data from clinical trials has permitted the development of practice guidelines for the treatment of AMI. The available data related to each diagnostic and therapeutic alternative have been considered by a panel of experts and graded on the basis of the supporting evidence. The results of these deliberations were expressed in standard format as follows:

Class I: Conditions for which there is evidence and/or general agreement that a given procedure or treatment is beneficial, useful, and effective.

Class II: Conditions for which there is conflicting evidence and/or a divergence of opinion about the usefulness/efficacy of a procedure or treatment.

Class IIa: Weight of evidence/opinion is in favor of usefulness/efficacy.

Class IIb: Usefulness/efficacy is less well established by evidence/opinion.

Class III: Conditions for which there is evidence and/or general agreement that a procedure/treatment is not useful/effective and in some cases may be harmful.

In general, recommendations in this chapter are associated with a class I, II, or III designation to guide the reader in weighing diagnostic and therapeutic options.

The progress that has been made in treatment of AMI has resulted in substantial improvement in outcomes. The increased use of standard coronary care unit (CCU) procedures—including electrocardiographic and hemodynamic monitoring, defibrillation, and beta blockers and, more recently, thrombolytics, coronary interventions, and aspirin—has decreased the mortality, under the best of circumstances, to 5 percent or less.

CLINICAL ASPECTS

Predisposing Characteristics and Circumstances

The risk factors for the development of coronary artery disease are dyslipidemia, family history, relative age, male gender, cigarette smoking, diabetes mellitus, and hypertension. *Careful consideration of the probabilities of the presence of coronary artery disease is centrally important in the initial assessment and evaluation of testing results of any patient with chest pain.* AMI occurs as a result of the disruption of a plaque at a site of a high density of inflammatory cells. Thus, AMI results from the acute exacerbation of a chronic inflammatory response.

Precipitating Events

There is intriguing indirect evidence that external factors might exacerbate the arterial inflammatory response. An association has been noted between acute myocardial infarction and antecedent mild respiratory syndromes. A more specific relationship between acute myocardial infarction and an infectious agent has been posited in the case of *Chlamydia pneumoniae.*

AMI has been associated with emotional or environmental stresses that activate the sympathetic nervous system and with increases in catecholamines. Increased sympathetic drive increases cardiac oxygen consumption by increasing contractility and rate, and it increases stress on vascular lesions by augmenting contraction, torque, and blood pressure. These forces can lead to plaque rupture in an area weakened by inflammation. High catecholamine levels can increase thrombus formation by activating platelets.

Any stressful event can precipitate acute myocardial infarction in a patient with "active," susceptible coronary atherosclerotic lesions. Anesthesia and surgery are well known to enhance the risk of myocardial infarction, and cardiac events are the leading cause of perioperative morbidity.

Circadian Variation

There is a marked circadian periodicity in the occurrence of myocardian infarction, with a peak prevalence between 6:00 A.M. and noon. There was a threefold increase in the frequency of infarction at peak (9:00 A.M.) periods as compared with trough (11:00 P.M.) periods. The rhythms both for the occurrence of myocardial infarction and for deaths from ischemic events are actually bimodal, with a secondary, less pronounced, late-afternoon or early-evening peak (6:00 to 8:00 P.M.). The blunting of the morning peak of AMI by both aspirin and beta-adrenergic blockers emphasizes the contributions of both the sympathetic nervous system and the coagulation pathways to the circadian rhythm of cardiovascular events.

Symptoms

Prodromal symptoms antedating acute myocardial infarction are common and occur in at least 60 percent of patients. Most of these symptoms are anginal or angina-like, especially when assessed retrospectively in the context of the character of the pain of the acute infarct. Considering the general feeling of malaise and fatigue that many patients have prior to AMI, it is obviously relatively unusual for the event to be totally unheralded.

The *classic symptoms* of AMI involve chest discomfort that is commonly retrosternal or precordial in *location* and is described as pressure, aching, burning, crushing, squeezing, heavy, swelling, or bursting in *quality*. The location of chest pain is usually of little help in differentiating ischemia/infarction from other causes, but severe chest pain (as opposed to vague discomfort) and the presence of *associated symptoms* (dyspnea, nausea, diaphoresis, and vomiting) are more commonly associated with MI. The discomfort often *radiates* over the anterior chest and frequently into the left arm or both arms (particularly the medial aspect) and/or into the neck or jaw. In unusual instances, the pain may be in the back, particularly between the scapulae. There may be skip areas with retrosternal pain, associated with jaw, antecubital fossa, or wrist pain, or there may be no pain in between the two sites. Moreover, the pain may appear only in the referral area. The *duration* of the pain is prolonged, lasting by definition longer than 15 min. While the intensity of the pain is usually steady following an initial crescendo, there is occasionally some waxing and waning. Sudden relief of pain may accompany reperfusion. Marked apprehension is common. Occasionally, presenting symptoms include syncope, acute confusion, agitation, stroke, or palpitations.

Approximately 23 percent of myocardial infarctions go unrecognized by patients because of the absence of symptoms, or the lack of recognition of the significance of symptoms. The common symptoms in this latter instance are nonclassic or atypical pain, dyspnea, nausea, vomiting, and/or epigastric pain. A myocardial infarction may also masquerade as the development or worsening of congestive heart failure, the appearance of an arrhythmia, an overwhelming sense of apprehension, profound weakness, acute indigestion, pericarditis, embolic stroke, or peripheral embolus. Presentation with painless myocardial infarction is more common in the elderly than in the nonelderly, and this subgroup has an increased frequency of congestive heart failure as the initial presenting symptom.

Physical Findings

GENERAL EXAMINATION

The patient is frequently sitting up because of a sense of suffocation or shortness of breath. Most patients have some sense of impending doom that is reflected in their facial expression. They may have a grayish appearance, or one of panic or exhaustion. Diaphoresis is frequent. In severe cases, patients may be quite anxious, with an ashen or pale face beaded with perspiration.

It is important to rapidly ascertain the vital signs and the nature, character, and rhythm of the arterial pulse; to observe the jugular venous pulse; to check the peripheral pulses; to palpate the precordium; and to auscultate the chest and precordium. Examination of the extremities should include subjective assessment of the temperature and color of the feet. The presence of very cool feet, especially with acrocyanosis in the setting of tachycardia, suggests low cardiac output.

The heart rate and rhythm in AMI are very important indicators of cardiac function. *A normal rate usually indicates absence of significant hemodynamic compromise.* In patients with inferior myocardial infarction, heart rates in the fifties and sixties are very common, especially in the first few hours. The bradycardia may be associated with secondary hypotension resulting from vagal stimulation. *Persistent sinus tachycardia beyond the initial 12 to 24 h is predictive of a high mortality rate.* The pulse may be low in volume, reflecting decreased stroke volume. The blood pressure is usually normal but may be increased secondary to anxiety, or it may be decreased from cardiac failure. All peripheral pulses should be examined to exclude current occlusion and to provide a baseline in case of future embolic events. The carotid pulse is most useful in assessing systolic upstroke time and stroke volume, which are decreased in the patient with a low output state. The rhythm of the pulse is very important because of the frequency of ectopic atrial and, in particular, ventricular beats in AMI.

The respiratory rate is usually within the normal range. However, patients who are extremely anxious often exhibit hyperventilation, and those with pulmonary edema and cardiac failure have an increased respiratory rate associated with shallow inspirations.

Examination of the jugular venous pulse is important with AMI, especially in patients with an inferior infarction, because insights can be gained into possible involvement of the right ventricle, which is common. It may be manifest by elevated jugular venous pressure or by a prominent *a* wave because of the decreased compliance of the right ventricle (RV). Kussmaul's sign, or an increase in the venous pressure on inspiration, may also be seen in right ventricular infarction because of decreased RV compliance.

EXAMINATION OF THE LUNGS
Basilar rales are frequently detected in AMI. Cardiac failure diagnosed on the basis of mild signs of pulmonary congestion occurs in 30 to 40 percent of otherwise uncomplicated patients.

CARDIAC EXAMINATION
Palpation of the precordium may reveal evidence of regional wall motion abnormalities and should be performed with the patient initially lying supine; this is often adequate to ascertain whether there is a localized normal apical impulse or a dyskinetic impulse. In the left lateral decubitus position, one may palpate a diffuse rather than a localized apical impulse, akine-

sis, or a paradoxical bulging during late systole; and in some patients, a palpable atrial contraction corresponding to an audible S4 gallop may be present.

The first and second heart sounds are often soft because of decreased contractility. The second heart sound is usually normal; however, with extensive damage, it may be single. Paradoxical splitting may reflect severe left ventricular dysfunction. A fourth heart sound is often audible. A third heart sound is heard in probably only about 15 to 20 percent of patients. A pericardial friction rub is heard in only about 10 percent of patients, usually 48 to 72 h after onset. The crescendo-decrescendo, midsystolic murmur of papillary muscle dysfunction is relatively common early and reflects ischemia of the papillary muscle or the myocardial attachment rather than irreversible injury. This murmur usually disappears after the first 12 to 24 h if soft, but if it is moderate to loud in intensity, it may persist much longer. Mitral regurgitation is most commonly due to ischemia of the posteromedial papillary muscle.

Diagnosis of Acute Myocardial Infarction

DIFFERENTIAL DIAGNOSIS
MI has typically been diagnosed on the basis of the triad of chest pain, electrocardiographic (ECG) changes, and elevated plasma enzyme activity. Although AMI occurs without chest pain (20 to 25 percent of cases), chest pain is usually responsible for the patient's presentation. The differential diagnosis of prolonged chest pain is presented in Table 10-1. It is often impossible on the basis of history alone to distinguish ischemia or infarction from other causes of chest pain. Most patients at risk for MI will be admitted for evaluation unless definite noncardiac causes of chest pain—such as chest wall pain, hyperventilation, pleurisy, gastrointestinal pain, and so on—that are not imminently dangerous can be identified. Only about 20 percent of patients admitted with chest pain have AMI.

ELECTROCARDIOGRAPHIC DIAGNOSIS
The ECG is very sensitive for detecting ischemia and infarction but is frequently not powerful enough for differentiating ischemia from necrosis. Serial ECGs during AMI will show some evolutionary changes in the majority of patients. An ECG obtained during cardiac ischemic pain frequently exhibits

TABLE 10-1

DIFFERENTIAL DIAGNOSIS OF PROLONGED CHEST PAIN

AMI
Aortic dissection
Pericarditis
Atypical anginal pain associated with hypertrophic
   cardiomyopathy
Esophageal, other upper gastrointestinal, or biliary
   tract disease
Pulmonary disease
   Pleurisy: infectious, malignant, or immune disease-
     related
   Embolus with or without infarction
   Pneumothorax
Hyperventilation syndrome
Chest wall
   Skeletal
   Neuropathic
Psychogenic

changes in repolarization. The absence of ECG changes during pain provides evidence but not proof that the pain is not ischemic in nature. The early ECG changes of T-wave inversion or ST-segment depression may reflect ischemia or infarction. ST-segment elevation is more specific for AMI and reflects the epicardial injury-associated total occlusion of an epicardial coronary artery. The hallmark of AMI is the development of abnormal Q waves, which appear, on the average, 8 to 12 h from the onset of symptoms but may not develop for 24 to 48 h. Abnormal Q waves usually reflect tissue death and the development of an electrical dead zone but are not useful in initial diagnostic management and triage except to indicate presence of prior MI. The diagnostic serial ECG changes consist of ST-segment elevation with the development of T-wave inversion and the evolution of abnormal Q waves. The appearance of abnormal Q waves is very specific to AMI; however, they are present in less than 50 percent of patients with documented AMI. Most of the other patients who have AMI will have ECG changes restricted to T-wave inversion or ST-segment depression or no change at all and represent the group with non-Q-wave infarction.

Based on autopsy studies, the misnomers *transmural* and *nontransmural infarction* have been replaced by the terms *Q-wave infarction* and *non-Q-wave infarction*, respectively. The evolution of a non-Q-wave infarction is characterized by the appearance of reversible ST-T-wave changes with ST depression that usually return to normal over a few days but are occasionally permanent. There are major differences in the pathogenesis, clinical manifestations, treatment, and prognosis of Q-wave and non-Q-wave MI. The initiating events are identical—namely, thrombus superimposed upon vasoconstriction. In non-Q-wave infarction, early spontaneous reperfusion occurs. In contrast, in Q-wave infarction, the coronary occlusion is sustained for a period that is at least long enough to result in extensive necrosis.

The ECG criteria for the diagnosis of AMI are the presence, in the setting of chest pain, of any one of the following (1) new or presumably new Q waves (at least 30 ms wide and 0.20 mV deep) in at least two leads from any of the following: (a) leads II, III, or aV_F; (b) leads V_1 through V_6; or (c) leads I and aV_L; (2) new or presumably new ST-T segment elevation or depression (≥ 0.10 mV measured 0.02 s after the J point in two contiguous leads of the above-mentioned lead combinations); or (3) new, complete left bundle branch block in the appropriate clinical setting.

The ECG diagnosis of RV infarction offers special challenges. RV infarction occurs in the presence of inferior left ventricular infarction; the resulting ST-segment elevation is usually overwhelmed in the conventional precordial leads overlying the RV (V_2 and V_3) by the ST elevation in the opposing left ventricular myocardium on the inferior surface. ST elevation must be sought in the right chest leads, V_1, and V_3R to V_6R; when found, this provides reasonably strong evidence for the presence of RV infarction.

In view of a lack of sensitivity and specificity of the chest pain history or of the ECG, confirmation of the diagnosis of AMI is based on elevated plasma levels of cardiac-specific isoenzymes.

DIAGNOSTIC MARKERS IN PLASMA

Myocardial necrosis is associated with the release of a variety of proteins that have been evaluated as diagnostic markers for acute myocardial infarction. The use of CK and MB-CK has become routine; these are highly sensitive, specific, and cost-effective for diagnosing myocardial infarction. In the adult human heart, 15 percent of total CK activity is MB-CK and the remainder is MM-CK. Myoglobin is ubiquitously distributed throughout cardiac and skeletal muscles. The cardiac troponin I radioimmunoassay is very specific for myocardial injury. It is also very sensitive. The assay for plasma cardiac troponin T is also very specific but may be less sensitive.

Temporal Profiles of Plasma MB-CK, Myoglobin, Troponin I, and Troponin T   Plasma MB-CK activity following MI is significantly elevated, and reliable diagnostic sensitivity (>90 percent) is reached within 12 to 16 h of the onset of symptoms. Maximal levels are reached between 14 and 36 h and return to normal after 48 to 72 h. Reliable diagnostic sensitivity (≥90 percent) is reached with plasma troponins I and T by 12 to 16 h and maximal activity is reached by 24 to 36 h. Plasma myoglobin is increased within 2 h of the onset of symptoms and remains increased for at least 7 to 12 h.

Early Diagnosis (6 to 10 h of Onset): MB-CK Subforms and Myoglobin   Early, rapid diagnosis is required to triage patients with chest pain in order to reduce costs and to select appropriate therapy, because of the difficulty in distinguishing cardiac ischemia from infarction based on clinical criteria and of the frequent absence of diagnostic ECG changes. The only two plausible candidates as early (<6 h) diagnostic markers are MB-CK subforms and myoglobin.

When MB-CK is released into plasma after myocardial injury, the parent (MB-2) form is converted into MB-1 by proteolytic activity and the subforms can be rapidly separated and detected by electrophoresis. The total assay can be performed in less than 1/2 h. Normally, MB-1 and MB-2 are in equilibrium with a ratio of 1 to 1. *When infarction occurs, MB-2 initially is released into the circulation in minute amounts, so that total MB-CK remains within the normal range; but the ratio of MB-2 to MB-1 changes markedly and provides the basis for an early diagnosis of myocardial infarction.*

MB-CK subforms afford a sensitivity and specificity of about 90 percent for the diagnosis of AMI within 6 h of the onset of symptoms. Myoglobin has a sensitivity of 83 percent. Thus, if a patient has a negative MB-CK subform test at 6 h after the onset of symptoms, one can reliably conclude that the patient does not have infarction. The total MB-CK (activity or mass assay) and troponins T and I afforded a sensitivity of only 65 percent within the first 6 h (Table 10-2). For the same time intervals, myoglobin had a sensitivity of 83 percent. Total MB-CK and troponins I and T have high sensitivity and specificity for the diagnosis of myocardial infarction from 10 to 14 h from the onset of symptoms. The enzymatic criteria for diagnosis of acute MI are summarized in Table 10-3.

TABLE 10-2

DIAGNOSTIC SENSITIVITY OF MYOGLOBIN AND MB-CK SUBFORMS ON ADMISSION AND 1 H LATER

Markers	Sample on Admission	Sample 1 h Later
MB-CK subform	67%	91%
Myoglobin	63%	78%

DIAGNOSIS OF ACUTE MYOCARDIAL INFARCTION 48 H OR MORE FROM THE ONSET OF SYMPTOMS

In patients admitted 48 to 72 h after the onset of symptoms, particularly when associated with minimal myocardial damage, the preferred diagnostic marker has become troponin I or T. Both remain elevated for 10 to 14 days.

Noninvasive Imaging in Acute Myocardial Infarction

CHEST ROENTGENOGRAM

The chest roentgenogram (x-ray) is important and may help to exclude cases of chest pain such as pneumothorax, pulmonary infarction with effusion, aortic dissection, skeletal fractures,

TABLE 10-3

ENZYMATIC CRITERIA FOR DIAGNOSIS OF MYOCARDIAL INFARCTION

Serial increase, then decrease of plasma MB-CK or subform ratio, with a change >25% between any two values

MB-CK >10–13 U/L or >5% total CK activity

Increase in MB-CK activity >50% between any two samples, separated by at least 4 h

If only a single sample available, MB-CK elevation > twofold

Beyond 72 h, an elevation of troponin T or I or LDH-1 >LDH-2 and so on. In the patient with AMI, the chest film can be useful in establishing the presence of pulmonary edema, in assessing heart size to assist in determining whether or not cardiomegaly is present, and in deciding whether heart failure or myocardial or valvular disease is acute or chronic.

ECHOCARDIOGRAPHY

Two-dimensional and Doppler echocardiography are very useful tools in the assessment of the patient with AMI and especially of the patient with a nondiagnostic ECG. The presence of a regional wall motion abnormality provides strong supportive evidence of acute coronary ischemia and is generally present in Q-wave MI. Wall motion abnormalities are less common but still frequently present in non-Q-wave infarction. Echocardiography also provides an assessment of ventricular function; it is useful in predicting the prognosis and in diagnosing RV infarction. It can also provide information concerning alternative diagnoses such as aortic dissection and, coupled with Doppler, can provide information on such complications as ruptured chordae tendineae with mitral regurgitation and ventricular septal defect. It is useful in detecting ventricular thrombus and pericardial fluid.

Measurement of Myocardial Infarct Size

The major determinant of both acute and long-term prognoses following MI is the extent of myocardial damage.

ELECTROCARDIOGRAPHIC ESTIMATES OF INFARCT SIZE

The ECG has long been used to obtain a semiquantitative assessment of the extent of MI. In general, it has been found, for example, that patients with anterior infarcts who develop Q waves in leads V_1 to V_6 usually have extensive damage and an unfavorable prognosis. In general, there is a direct relationship between the number of leads showing ST-segment elevation and mortality.

INFARCT SIZE ASSESSMENT BY IMAGING

For practical reasons, echocardiography is the most commonly used imaging modality in the acute evaluation.

Prehospital Care

Modern in-hospital care of the AMI patient has resulted in a substantial reduction in mortality. Some 40 to 65 percent of deaths from AMI, however, occur within an hour of the onset of symptoms and prior to arrival at a hospital. Most of these deaths are attributable to ventricular fibrillation. Substantial further reductions in acute mortality rates are likely to require marked improvements in prehospital care and in patient responsiveness to symptoms. The goals for community Emergency Medical Systems (EMS) have been discussed extensively and include the availability of 911 telephone access and of personnel trained in defibrillation and, potentially, in initiation of out-of-hospital thrombolysis.

The major contribution that the physician can make to minimize delay from the time that patients first appreciate subjective manifestations of AMI to the time they present at the emergency department (ED) is in educating them beforehand as to the proper responses to ischemic coronary symptoms. The guiding principles are *recognition and response*. Thus, the patient should be taught to recognize and appreciate chest pain as potentially representing coronary ischemia, and, if sustained, threatened AMI. Patients should be warned specifically of the dangers of rationalizing the pain as having a noncardiac origin or of trying extensive "diagnostic trials" of home remedies. They should be instructed in the standard protocol for using nitroglycerin. That is, at the onset of pain, the patient should immediately use nitroglycerin in a form that is absorbed rapidly from the oral muscosa; if the pain is not relieved within 5 min, the dosing should be repeated. If the discomfort persists for another 5 min, a third dose should be administered. If at this point no relief is obtained, the patient should proceed immediately to the nearest ED. The potential risk of fatal arrhythmia in the early course of AMI should be explained. *Educating patients about this protocol is one of the most important functions of the physician caring for patients with coronary artery disease.*

EVALUATION AND MANAGEMENT OF THE CHEST PAIN PATIENT IN THE EMERGENCY DEPARTMENT

Background

The goals of the emergency department (ED) with respect to chest pain patients are as follows: to rapidly identify those patients with AMI with both typical and atypical presentations, so that appropriate therapy can be initiated; to recognize those patients with acute coronary syndromes (unstable angina) but without MI who are thus at high risk; and to assess accurately those patients at low risk who are candidates for noninvasive evaluation and early discharge. The earlier reperfusion therapy is initiated in the subset of patients with diagnostic ST-segment elevation, the more favorable the clinical results.

An important objective, obviously, should be a triage system that minimizes the number of patients at high risk (AMI or unstable angina) who are inadvertently discharged from the ED while also minimizing the admission to high-intensity CCUs of low-risk patients without MI.

Misdiagnosis of AMI is commonly associated with misinterpretation of the ECG. A major problem contributing to the difficulty of diagnosing AMI is that even experienced clinicians appear to be able to achieve a sensitivity and specificity of only about 80 percent in the diagnosis of AMI on clinical grounds alone. The problem extends also to the diagnosis of unstable angina in the absence of infarction. The patients with coronary instability are also at high intermediate term risk. *The clinical focus in the ED should not be simply to "rule out" AMI but, taking a proactive approach, to "rule in" either acute infarction or unstable angina in an expeditious manner. Once these urgent conditions have been excluded or ascertained to be of low probability, the next level of concern is determining the presence of other acute cardiovascular or cardiopulmonary conditions, such as aortic dissection, pulmonary embolus, pericarditis, and so on.*

Initial Approach, Detection, and Assessment of Risk

A major goal in dealing with chest pain patients is establishment of a routine approach that leads to a rapid (10-min) preliminary evaluation, acquisition of a 12-lead ECG, and establishment of intravenous access, continuous electrocardiographic monitoring, and supplemental oxygen (class I) (Fig. 10-1). The initial assessment is guided by the differential diagnosis of chest pain, with the goal of establishing whether or not myocardial ischemia is a likely or possible diagnosis. Blood is drawn for baseline cardiac marker levels; if coronary ischemia is suspected and there are no contraindications, the patient is given aspirin of 160 to 325 mg to chew and swallow. Also, the patient with suspected coronary ischemia is given sublingual nitroglycerin unless the systolic blood pressure is

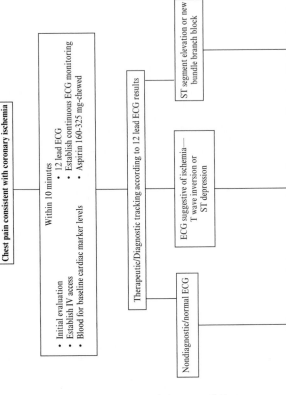

FIGURE 10-1

Algorithm for the initial assessment and evaluation of the patient with acute chest pain in the emergency department. The emergency department should be organized to facilitate the rapid triage of chest pain patients so that the initial evaluation, obtaining a 12-lead ECG, and establishing intravenous access and continuous monitoring are accomplished within 10 min. The path in the decision tree is determined by the results of the 12-lead ECG. The presence of ST-segment elevation diagnostic of AMI or of presumptively new BBB suggestive of this diagnosis should lead to the immediate consideration of the suitability of the patient for reperfusion therapy, which, if indicated, should be initiated within 30 min of the patient's arrival. The primary PTCA option is applicable only in those settings in which it is immediately available and can be performed by highly qualified interventional cardiologists. In general, patients should not be transferred for angioplasty if thrombolysis is an option. Thrombolysis is not indicated in patients with only ST-segment depression.

174

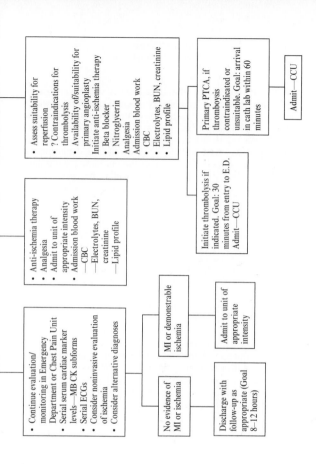

- Assess suitability for reperfusion
- ? Contraindications for thrombolysis
- Availability of/suitability for primary angioplasty
Initiate anti-ischemia therapy
- Beta blocker
- Nitroglycerin
Analgesia
Admission blood work
- CBC
- Electrolytes, BUN, creatinine
- Lipid profile

Primary PTCA, if thromboysis contraindicated or unsuitable. Goal: arrival in cath lab within 60 minutes

Admit—CCU

- Anti-ischemia therapy
- Analgesia
- Admit to unit of appropriate intensity
- Admission blood work
  —CBC
  —Electrolytes, BUN, creatinine
  —Lipid profile

Initiate thrombolysis if indicated. Goal: 30 minutes from entry to E.D.
Admit—CCU

- Continue evaluation/ monitoring in Emergency Department or Chest Pain Unit
- Serial serum cardiac marker levels—MB CK subforms
- Serial ECGs
- Consider noninvasive evaluation of ischemia
- Consider alternative diagnoses

MI or demonstrable ischemia

Admit to unit of appropriate intensity

No evidence of MI or ischemia

Discharge with follow-up as appropriate (Goal 8–12 hours)

175

less than 90 mmHg. This should be avoided with severe bradycardia or tachycardia. The history of chest pain alone usually
dictates entry into the system for evaluation. In general, the
only chest pain patients not systematically evaluated for myocardial ischemia would be those in whom a clear noncardiac
cause, such as chest wall tenderness, can be demonstrated
unequivocally to be the etiology of the presenting symptoms.
Continuous ECG monitoring is essential because of the
propensity of any patient with an acute coronary ischemic syndrome to develop sudden and potentially lethal ventricular
arrhythmias. Intravenous access is essential for therapeutic
interventions under such circumstances as well as for more general purposes. Additionally, paroxysmal changes in the ST
segment may be recognizable on the monitor. The differential
diagnosis of chest pain and the clinical recognition of acute
myocardial infarction were discussed previously. The causes
of chest pain that are not the result of acute pathological
changes compromising the structural integrity of the large
coronary arteries are listed in Table 10-4.

*As a general rule one should begin the evaluation of the
chest pain patient with the assumption that one is dealing with
myocardial ischemia until proven otherwise.* The three most
serious and urgent alternative diagnoses that need to be considered specifically during the initial evaluation are *aortic
dissection, acute pulmonary embolus,* and *acute pneumothorax. Acute pericarditis* and *myopericarditis* need to be
considered as well.

Aortic dissection must be considered and ruled in or out
during the initial evaluation, since specific intervention can
decrease its high mortality. Furthermore, administration of
thrombolytic agents in the presence of aortic dissection is associated with high mortality. Suspicion of dissection should be
heightened, especially in hypertensive patients or in those with
marfanoid habitus.

Pulmonary embolus can be life-threatening and should be
suspected in anyone with a sudden onset of shortness of breath
and chest pressure or pain, especially if there is a history of
being sedentary or immobilized and/or a history of deep
venous thrombosis.

Acute pericarditis may mimic AMI in that the pain can be
substernal and persistent. Frequently, however, there will be a
positional component as well as characteristics of pleurisy,
with accentuation by deep breathing. Furthermore, the diffuse
ST-segment elevation may lead to a misdiagnosis of MI. The
key differentiating features in pericarditis include PR depres-

TABLE 10-4

CAUSES OF CHEST PAIN OTHER THAN ACUTE CORONARY ARTERY SYNDROMES

Cardiovascular
    Aortic dissection
    Aortic stenosis
    Pericarditis
    Mitral valve prolapse
    Microvascular angina
        Hypertrophic cardiomyopathy
        Syndrome X
    Pulmonary embolus
    Arrhythmia/palpitations
Noncardiovascular
    Pleurisy
    Pneumonia
    Pneumothorax
    Costochondritis
    Gastrointestinal
        Esophageal spasm/reflux
        Acid peptic disease
        Cholecystitis
        Gastritis
Psychiatric
    Panic attack
    Cardiac neurosis
    Depression
Malingering

sion, the diffuse nature of ST elevation in most leads, and the absence of reciprocal changes.

It should be kept in mind that esophageal disorders are very common in patients presenting with chest pain in whom cardiac ischemia is ruled out. Because of the high frequency of gastrointestinal disease in chest pain patients, "GI cocktails" or antacids have been used as a diagnostic tool to guide triage and disposition. Only 25 percent of patients with esophageal pain, however, have been reported to obtain pain relief with antacids. Furthermore, coincidental, spontaneous relief of ischemic chest pain at the time of administration of the GI cocktail could be misleading. Similarly, administration of nitroglycerin as a diagnostic strategy for ischemic disease could be misleading because it can relieve esophageal spasm. Moreover, it has been found that pain relief after nitroglycerin did not predict unstable angina or acute myocardial infarction in the chest pain patient. The use of these "response-to-treatment" strategies as major decision points in the evaluation of chest pain has been discouraged.

DETECTION

The 12-Lead Electrocardiogram as a Guide to Management Strategy The results of the 12-lead ECG guide the next level of decision making for the patient with chest pain thought to be compatible with a diagnosis of myocardial ischemia (Fig. 10-1). The ECG interpretation is assigned to one of three categories: (1) ST-segment elevation in two or more leads or a presumptively new bundle branch block implicating acute coronary occlusion, usually thrombotic; (2) ST-segment depression and/or T-wave inversion implying subtotal occlusion and nontransmural ischemia; and (3) normal or nondiagnostic. The group with ST-segment elevation or a left bundle branch block is particularly important to define, as it is this group that has been shown to benefit from thrombolytic therapy. There is no indication as yet of the benefit of thrombolytic therapy or primary angioplasty in those patients without ST-segment elevation or bundle branch block.

The ECG also serves as a basis for initial risk assessment. ST-segment elevation or a new left bundle branch block in the patient with chest pain defines a high-risk group, and in those with elevated ST segments, the mortality correlates positively with the number of leads with the ST changes. The presence of ST-segment depression or T-wave inversion also defines a high-risk group.

The measurement of serum markers of myocardial damage plays a major role in diagnosis, as discussed. Measurement of CK-MB is the benchmark laboratory test, and the specificity and sensitivity of samples taken 2 h apart during serial sampling have been reported to be 91 to 94 percent, respectively.

Two-dimensional echocardiography can be useful as an adjunctive modality. It may be especially useful in detecting wall motion abnormalities in the presence of conduction abnormalities on the ECG.

RISK STRATIFICATION

Stratifying risk in the patient with acute myocardial infarction is an essential part of the management strategy during all phases of care.

Predictors of an increased risk include the following: ECG evidence of ST-segment elevation or Q waves in two or more leads that are not known to have been present previously; ST-segment depression or T-wave inversions consistent with myocardial ischemia and not known to be present previously; pain worse than prior angina or the same as that experienced with prior myocardial infarction; systolic blood pressure of less than 100 mmHg; or rales bilaterally.

Blood levels of cardiac markers are prognostically important. *In particular, the levels of troponins (I and T) at presentation appear to be strong predictors of risk in patients with acute ischemic syndromes.*

INITIAL MANAGEMENT

As discussed, one frequently does not have a definitive diagnosis of AMI in the chest pain patient in the emergency department. Nevertheless, the initial general treatment of the acute coronary syndromes is the same.

Routine General Measures

Oxygen Administration Hypoxemia is not uncommon in patients with acute myocardial infarction, even with an uncomplicated course. Nasal oxygen should be administered to all AMI patients with pulmonary congestion and Sa_{O_2} of less than 90 percent (class I). O_2 should be administered to all uncomplicated AMI patients for the first 2 to 3 h (class IIa). There appears to be little justification for extending use of oxygen administration in uncomplicated myocardial infarction with an Sa_{O_2} of greater than 90 percent beyond 2 to 3 h (class IIb). Oxygen administration should be continued in patients with pulmonary congestion and desaturation.

Analgesia. The alleviation of pain and anxiety remains an essential element in the care of the patient with AMI. The pain and accompanying anxiety contribute to excessive activity of the autonomic nervous system and to restlessness. These factors, in turn, increase the metabolic demands of the myocardium.

The approach to pain consists of relieving ischemia and attacking the pain directly. Anti-ischemic therapy consists of reperfusion, beta blockers (if appropriate), nitrates, and oxygen administration. Morphine, in most instances, is the drug of choice, since it is well tolerated and offers analgesia without significant cardiac depression. It also relieves anxiety and the feeling of doom that is commonly described. Morphine sulfate can be given intravenously at doses of 2 to 4 mg every 15 min until adequate relief has been obtained, which, in some patients may require 25 to 30 mg. If the patient's anxiety is not controlled by the administration of narcotics, mild sedation with a benzodiazepine is appropriate. Diazepam in doses of 5 mg orally every 8 to 12 h or alprazolam in doses of 0.25 mg every 8 h are most often used.

Nitroglycerin. This should be administered for the first 24 to 48 h to AMI patients with CHF; large anterior MI; persistent recurrent angina; or persistent pulmonary congestion (class I). Nitroglycerin should be considered beyond 48 h in an oral or topical form in patients with large or complicated AMI (class IIb).

Nitroglycerin has become very widely used in the treatment of AMI. It is an anti-ischemic agent not only by virtue of its actions to decrease preload and afterload and thus to decrease oxygen demand, but also because of its vasodilator actions on epicardial coronary arteries and on coronary collaterals. The early administration of nitroglycerin limits the extent of myocardial damage and favorably affects survival.

Complications and limitations The most serious complication of nitroglycerin is hypotension. Thus nitroglycerin should be avoided with a systolic pressure of less than 90 mmHg or in the case of severe bradycardia (class III). Caution should be exercised in the case of inferior wall infarction because of the possibility of right ventricular involvement. Nitroglycerin should be used only with extreme caution, if at all, in right ventricular infarction, because the right ventricle in this circumstance becomes extremely dependent upon preload, which can be diminished by the venodilating properties of the drug.

Dosage of nitroglycerin Long-acting nitrates should generally not be used as initial therapy in AMI. Intravenous nitroglycerin is preferable. Dose titration can be assessed by frequent determinations of blood pressure and heart rate. Invasive monitoring is not essential but is probably prudent if high doses are required or if there is hemodynamic instability or uncertainty about the adequacy of ventricular preload.

Treatment should be initiated with a bolus injection of 12.5 to 25 μg and should be followed by infusion by pump of 10 to 20 μg/min with increases of 5 to 10 μg every 5 to 10 min while assessing hemodynamic and clinical responses. Control of symptoms is a major endpoint; in the case of high left ventricular filling pressure, the objective is a decrease of 10 to 30 percent in pulmonary artery wedge pressure. Limitations of nitroglycerin dosing are as follows: (1) a decrease in mean arterial pressure of 10 percent in normotensive patients; or (2) a decrease of 30 percent in hypertensive patients but not below a systolic pressure of 90 mmHg; or (3) an increase in heart rate of 10 beats per min not to exceed 110 beats per minute.

Doses of nitroglycerin of greater than 200 μg are associated with an increased risk of hypotension. Requirements this high may indicate tolerance, and alternative drugs such as angiotensin converting enzyme (ACE) inhibitors or nitroprusside should be considered. If tolerance is the issue, responsiveness should return after a 12- to 18-h period off nitroglycerin.

Aspirin. Aspirin reduces the incidence of vascular events in patients with AMI at 1 month. The patient suspected of having a coronary ischemic syndrome and without contraindication should receive, early in the course, 160 to 325 mg of non-enteric-coated aspirin, which is chewed. If coronary artery disease is confirmed, aspirin treatment should be continued indefinitely (class I). In case of true aspirin allergy, other antiplatelet agents such as dipyridamole or ticlopidine may be substituted (class IIb).

Management after Triage into ECG Subgroups

The initial ECG, as a first approximation, permits assigning of patients with chest pain into subgroups distinguishable by therapeutic responsiveness and risk, as discussed previously. *It must be kept in mind that these initial categorizations do not necessarily define ultimate outcome. Thus, patients with no ST-segment elevation at presentation may, in fact, have unstable angina and ultimately have no infarction or may progress to have either a Q-wave or a non-Q-wave infarction. Similarly, those presenting with ST-segment elevation may have a non-Q-wave infarction, although the majority of these will develop Q waves.* This potential for variable outcomes provides the underlying rationale for close monitoring and continuous reassessment of clinical course, risk, and therapeutic strategies during the period of observation and for monitoring both in the ED and subsequently in other hospital units.

APPROACH TO THE PATIENT WITH ST-SEGMENT ELEVATION

The approach to the patient with chest pain and ST-segment elevation is guided heavily by evidence that members of this subgroup have a high frequency of epicardial coronary artery occlusion by a thrombus and that, in them, thrombolytic therapy has shown clinical benefit. Evaluation and management of the patient with ischemic chest pain and ST-segment elevation is focused on the rapid assessment of suitability for and delivery of reperfusion therapy (Fig. 10-2).

Initial evaluation and management have been discussed. The appropriate next steps are to administer a beta-adrenergic blocker if not contraindicated and to initiate evaluation for reperfusion therapy. The 12-h point defines the time frame in which the risk/benefit ratio is clearly favorable for administering thrombolytic therapy—although, obviously, the earlier the better.

Beta-Adrenergic Receptor Blockers   Beta-adrenergic receptor blockers interfere with the positive inotropic and chronotropic effects of catecholamines and therefore reduce myocardial oxygen consumption.

The available data strongly support the use of beta blockers early in the course of acute Q-wave MI in the absence of contraindications irrespective of concomitant thrombolytic therapy (class I). Patients with continuing or recurrent ischemic pain or with tachyarrhythmias, such as atrial fibrillation with rapid ventricular response, should also be considered for beta-blocker therapy (class I). The evidence is less compelling for beta-blocker therapy in non-Q-wave AMI (class IIb), and patients with moderate or severe congestive heart failure or other contraindications should not receive beta blockers (class III). While metoprolol and atenolol are the only FDA-approved beta blockers for use in the United States for AMI, *therapeutic efficacy is a class effect of beta blockers lacking intrinsic sympathomimetic activity.*

The relative contraindications to beta-blocker therapy are as follows: (1) heart rate less than 60 beats per minute; (2) systolic blood pressure less than 100 mmHg; (3) moderate or severe left ventricular failure; (4) signs of peripheral hypoperfusion; (5) PR interval greater than 240 ms; (6) second- or third-degree AV block; (7) severe chronic pulmonary disease; (8) history of asthma; (9) severe peripheral vascular disease; and (10) insulin-dependent diabetes mellitus. Since these contraindications are relative and not absolute, the clinician has the option of assessing the effects of beta blockade with the short-acting, intravenous beta blocker esmolol, which has an onset of action within 5 to 10 min and a half-life of about 30 min.

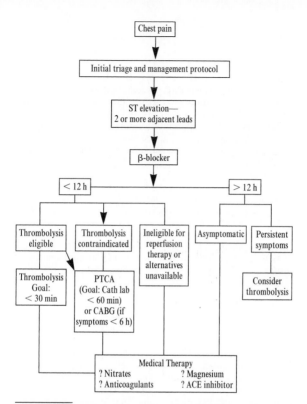

FIGURE 10-2

Evaluation of patients with ST-segment elevation. Algorithm for initial decision making in regard to reperfusion therapy in patients with suspected acute myocardial infarction and ST-segment elevation. Whether or not to administer thrombolytics or to perform primary PTCA is determined by the time from onset of symptoms. When more than 12 h have elapsed since the onset of symptoms, reperfusion should be considered only if there are persistent or recurrent symptoms associated with ST-segment elevation. For patients with ST-segment elevation where the duration of symptoms is between 7 and 12 h, the decision to proceed with a reperfusion strategy requires careful clinical judgment in weighing the risk/benefit issues, as discussed in the text. [From Ryan TJ, Anderson JL, Antman EM, et al: ACC/AHA guidelines for the management of patients with acute myocardial infarction: a report of the American College of Cardiology/American Heart Association Task Force on Practice Guidelines (Committee on Management of Acute Myocardial Infarction). *J Am Coll Cardiol* 1996; 28:1328–1428, with permission.]

Thrombolysis

Indications for Thrombolytic Therapy.   Reperfusion therapy should be given immediate consideration in all patients presenting with AMI. The primary indication for attempts at reperfusion, given an appropriate history, is the findings on the ECG, as discussed above. Patients with ST-segment elevation (>0.1 mV) in two or more contiguous leads or a bundle branch block (BBB) masking ST-segment changes occurring within 12 h of onset of symptoms are candidates for thrombolytic therapy (<75 years of age, class I; >75 years of age, class IIa). *Patients with ongoing symptoms should be repeatedly evaluated by 12-lead ECGs as frequently as every 10 to 15 min in order to identify ST-segment elevation as soon as possible.* Conversely, ST-segment elevation in the absence of suggestive symptoms should suggest such possibilities as early repolarization, pericarditis, and previous infarction with aneurysm formation. Elderly patients should not be excluded from thrombolytic therapy primarily because of their age or because of the increased risk of bleeding.

Large, placebo-controlled clinical trials have consistently demonstrated reduced mortality in patients receiving thrombolytic therapy within 6 h of the onset of an AMI. Because of suggestion of benefit between 6 and 12 h, it has been recommended that the time limit for therapy be up to 12 h from the onset of symptoms. The benefit of thrombolytics given between 6 and 12 h postinfarction is greater in patients classified as having high-risk infarctions, such as those with severe heart failure and large infarctions. Thrombolysis can be considered in the case of ongoing pain and marked ST-segment elevation at times between 12 to 24 h from onset, although there is only a trend for benefit under these circumstances in clinical trials (class IIb). Patients with ST-segment depression, T-wave inversion, or no ECG changes have not been shown to benefit from thrombolytic therapy (class III).

The potential for therapeutic benefit of thrombolysis in the setting of high risk of MI when the blood pressure is markedly elevated (>180 mmHg systolic and/or >110 mmHg diastolic) has to be carefully considered against the increased risk of intracranial hemorrhage under these circumstances. Lowering the blood pressure pharmacologically before administering thrombolytics has been recommended but is of unproven benefit (class IIb). If available, coronary artery bypass grafting or primary percutaneous transluminal coronary angioplasty (PTCA) should be considered (class IIb).

Contraindications to Thrombolytic Therapy. The absolute and relative contraindications to thrombolytic therapy are summarized in Table 10-5.

Choice of Thrombolytic Agent. Four thrombolytic agents have been approved in the United States for routine use: streptokinase (SK), recombinant tissue plasminogen activator (rt-PA), anistreplase (APSAC), and reteplase (r-PA). Each has been shown to limit infarct size, preserve ventricular function, and improve survival rates.

TABLE 10-5

ABSOLUTE AND RELATIVE CONTRAINDICATIONS TO THROMBOLYTIC THERAPY

Absolute Contraindications	Relative Contraindications
Active internal bleeding	History of nonhemorrhagic cerebrovascular accident in distant past with complete recovery
Intracranial neoplasm or recent head trauma	
Prolonged, traumatic CPR	
Suspected aortic dissection	Recent trauma or surgery >2 weeks previously
Pregnancy	Active peptic ulcer disease
History of hemorrhagic cerebrovascular accident or recent nonhemorrhagic cerebrovascular accident	Hemorrhagic retinopathy
	History of severe hypertension with diastolic blood pressure >100
Recorded blood pressure >200/120	Bleeding diathesis or concurrent use of anticoagulants
Trauma or surgery that is a potential bleeding source within previous 2 weeks	
Allergy to SK or APSAC if being considered	Previous treatment with SK or APSAC if being considered (does not apply to rt-PA)

KEY: CPR, cardiopulmonary resuscitation; SK, streptokinase; APSAC, anistreplase; rt-PA, recombinant tissue plasminogen activator.

Dosage and Administration of Thrombolytic Agents. Streptokinase is given in a dose of 1.5 million units intravenously over 30 to 60 min. Since antibodies develop and may persist for several years, a subsequent need for thrombolytic therapy, as for early or late reocclusion, would require the use of rt-PA or r-PA. APSAC is identical to SK as a thrombolytic agent but can be given as a rapid infusion of 30 U over 5 to 10 min. The FDA-approved dose of rt-PA is an initial bolus of 15 mg, followed by an infusion of 50 mg or 0.75 mg/kg over the next 30 min and an infusion of 35 mg or 0.50 mg/kg over the subsequent 60 min, for a total of up to 100 mg given over 90 min. Reteplase is given as an initial bolus of 15 megaunits (MU), followed by a second bolus of 15 MU in 30 min. Thrombolytic therapy is rapidly evolving, and both the specific agent and various combinations as well as the specific doses and regimens of administration are changing rapidly.

Overall Strategy for Reperfusion of Patients with Acute Myocardial Infarction The criteria for initiating thrombolytic therapy are as follows (Table 10-6):

1. Patients presenting with chest pain suggestive of myocardial ischemia, having ST-T-segment elevation greater than 1 mm in two contiguous limb leads or greater than 2 mm in two contiguous precordial leads or new left bundle branch block and who are within 6 h of the onset of symptoms should receive thrombolytic therapy if there are no contraindications. In patients presenting between 6 and 12 h, the bias should be toward thrombolytic therapy the higher the risk of the AMI. Patients presenting after 12 h are not routinely considered for thrombolytic therapy.
2. Contraindications for thrombolytic therapy are absolute or relative, as discussed earlier (Table 10-5).

TABLE 10-6

CRITERIA FOR INITIATING THROMBOLYTIC THERAPY

Chest pain consistent with angina
ECG changes
    ST ↑ ≥ 1 mm, ≥2 contiguous limb leads
    ST ↑ ≥ 2 mm, ≥2 contiguous precordial leads
    New left bundle branch block
Absence of contraindications

3. In patients receiving rt-PA or r-PA, it is recommended that heparin be given as a bolus infusion of 5000 U, followed by a continuous infusion of 1000 U/h, adjusted to keep the PTT at one-half to two times the normal control for 24 to 48 h. Heparin should be given to patients who have received SK or APSAC who are at high risk for systemic embolization. Aspirin (160 to 325 mg) should be administered as soon as possible and continued indefinitely. Beta blockers, nitrates, and occasionally calcium-channel blockers may be given as indicated with or without thrombolytic therapy.

4. Patients allergic to SK or APSAC who require thrombolytic therapy should receive rt-PA or r-PA. Patients who received SK or APSAC and who again require thrombolytic therapy should receive rt-PA or r-PA.

5. Patients presenting with ST-T-segment depression and chest pain are not candidates for thrombolytic therapy and need to be triaged, as indicated in Fig. 10-1.

6. PTCA, as a primary procedure, is an alternative to thrombolytic therapy only if performed in a timely fashion by individuals skilled in the procedure and supported by experienced personnel in high-volume centers (class I). PTCA is indicated in patients with a contraindication to thrombolytic therapy or in those in cardiogenic shock (class IIa).

7. Elective angioplasty should be reserved for patients who develop ischemia or reinfarction or in whom thrombolytic therapy appears ineffective. In patients in whom angioplasty cannot be performed and who develop recurrent ischemia with possible infarction, the possibility of readministering a thrombolytic agent should be considered.

Heparin as Conjunctive or Adjunctive Therapy

It is recommended that heparin not be started immediately but that an activated partial thromboplastin time (aPTT) be drawn at 4 h and that heparin be started when the aPTT returns to less than twice control (about 70 s). Lysis of a thrombus by any thrombolytic agent induces a highly thrombogenic surface. Furthermore, lysis with either rt-PA or SK is associated with marked elevation of plasma levels of thrombin. The use of heparin during the initial 24 to 48 h is critical to prevent rethrombosis and reocclusion.

Heparin is not necessary to achieve reperfusion but is essential in the first 24 h to maintain patency rates with rt-PA. While heparin may be beneficial when SK is being used, subcutaneous administration of heparin appears adequate in this circumstance. At present, heparin is recommended in a bolus of 5000 U intravenously followed by an infusion of 1000 to 1200 U/h to keep the PTT at 1.5 to 2.0 times normal. It is recommended that the PTT not be measured until 4 h after initiating heparin therapy because it has not yet reached a steady state. If the PTT has increased more than twofold over normal, the same dose of heparin should be continued; if PTT exhibits less than a twofold increase, the infusion rate of heparin should be increased. Initiation of heparin is recommended either during or following completion of thrombolytic therapy and should be maintained in the uncomplicated patient for 24 to 48 h.

The use of heparin has also been recommended conjunctively in patients with acute myocardial infarction who are not being treated with the drug for other reasons, i.e., postthrombolysis or post–primary PTCA. Currently, guidelines recommend heparin 7500 U twice daily subcutaneously as prophylaxis against deep venous thrombosis. Given the enhanced risk of stroke after acute myocardial infarction in patients with atrial arrhythmias, those with large and especially anterior and apical infarction, and those with history of previous stroke, guidelines have incorporated this recommendation for broader prophylaxis against systemic embolization. In high-risk patients, the intravenous route is probably preferable. Heparin therapy should be continued for 48 h and judgment should be made at that point about continuation based on individual patient characteristics.

EARLY CORONARY ANGIOGRAPHY IN PATIENTS WITH ST-SEGMENT ELEVATION NOT UNDERGOING PRIMARY PTCA

Routine immediate or delayed angioplasty is not recommended as a standard mode of therapy following thrombolysis. *At present, the most widely accepted recommendation is to perform cardiac catheterization for possible angioplasty or bypass surgery in patients who develop angina or manifest evidence of myocardial ischemia during submaximal exercise testing or who develop hemodynamic or ischemic instability.* Thus, if intervening with PTCA generally offers no demonstrable benefit after thrombolysis, there is little apparent reason to perform early coronary angiography routinely.

Patients with cardiogenic shock have a very high ($>$70 percent) mortality with or without thrombolysis, and some rather small series have provided evidence that outcomes are improved with an aggressive reperfusion strategy. Successful PTCA in conventionally treated patients who had cardiogenic shock reduced mortality from greater than 80 to about 30 percent. *Thus, an aggressive interventional strategy including PTCA seems reasonable, based on available data, in appropriate patients with cardiogenic shock who have failed thrombolytic therapy.*

EMERGENCY OR URGENT CORONARY ARTERY BYPASS SURGERY

In the presence of cardiogenic shock, coronary artery bypass grafting in patients in whom other strategies have either failed or not been indicated has been associated with mortality rates from about 10 to 40 percent. These results are generally better than those associated with PTCA. *Thus, acute myocardial infarction patients with multivessel coronary artery disease or cardiogenic shock who have had unsuccessful thrombolysis and/or PTCA and are within 4 to 6 hours of the onset of symptoms should be considered for emergency coronary artery bypass grafting.*

ARRHYTHMIAS EARLY IN THE COURSE OF ACUTE MYOCARDIAL INFARCTION

Bradycardia   Bradyarrhythmias are relatively common (30 to 40 percent) early in the course of acute myocardial infarction, especially in inferior infarction or after reperfusion of the right coronary artery. Atropine, because of its anticholinergic effects, can be very useful in this situation, since it enhances the discharge rate of the sinus node and facilitates atrioventricular (AV) conduction as well as reversing the peripheral effects of excessive cholinergic activity, such as vasodilation with associated hypotension. Parasympathomimetic effects with bradycardia, hypotension, and nausea and vomiting are also produced by morphine and can be reversed by atropine.

SINUS BRADYCARDIA, ATRIOVENTRICULAR BLOCK, OR VENTRICULAR ASYSTOLE

Atropine is indicated for the treatment of type I second-degree AV block, especially when it is complicating inferior myocardial infarction, and is useful at times in third-degree AV block at the AV node in restoring AV conduction or for increasing the junctional response rate. Treatment of sinus bradycardia or first- or second-degree AV block is generally not indicated in the absence of hemodynamic compromise, and atropine should seldom be used in the treatment of type II AV block (location of block below the AV node). *Symptomatic bradycardia that is unresponsive to atropine should be treated with pacing.* Atropine should be administered intravenously at a dosage of 0.5 to 1.0 mg and titrated carefully as necessary to achieve an adequate heart rate (50 to 60 beats per minute), with doses as needed every 3 to 5 min, up to a total maximum dose of 2.5 mg, which gives complete vagal blockade.

HEART BLOCK

Heart block develops in about 10 percent of patients with AMI and is associated with an increased mortality during hospitalization, but it does not predict long-term mortality in those who survive to be discharged. Intraventricular conduction delay or bundle branch block is also associated with increased in-hospital mortality. The increase in mortality associated with heart block reflects the extent of myocardial damage. Thus, heart block in the setting of anterior MI reflects extensive infarction and concomitant destruction of the conduction system and is associated with relatively high mortality. In contrast, heart block with inferior MI may primarily reflect ischemia of the AV node rather than extensive tissue damage and is associated with a more favorable prognosis. Because of the overwhelming effect of the extent of myocardial damage on prognosis, pacing has not been shown to lessen mortality associated with AV block or bundle branch block. *In AMI, the risk of developing heart block is augmented by the presence of any evidence of conduction system abnormality, including first-degree AV block, Mobitz type I or II AV block, left anterior or posterior hemiblock, or a left or right bundle branch block.*

Temporary Pacing Early in the Course of Acute Myocardial Infarction. Recent guidelines place increased emphasis on transcutaneous pacing in view of the availability of new systems that provide standby status for pacing in AMI in patients who do not require immediate pacing and are at intermediate risk for developing heart block. These systems use a single pair of multifunctional electrodes, permitting ECG monitoring, transcutaneous pacing, and defibrillation. Transcutaneous pacing does not entail the risk and complications of transvenous pacing and, because invasive procedures may be avoided or delayed, is well suited for use in the patient who has undergone thrombolysis. Percutaneous pacing is painful; if prolonged pacing is required, the patient should be switched to transvenous systems.

The following conditions have class I indications for placement of patches or activation (demand) of transcutaneous pacing: (1) sinus bradycardia (rate less than 50 beats per minute) with symptoms of hypotension unresponsive to drug therapy; (2) Mobitz type II second-degree AV block; (3) third-degree heart block; (4) bilateral bundle branch block (alternating left and right bundle branch block or right bundle branch block with alternating left anterior and posterior fascicular block; (5) newly acquired or age-indeterminant left bundle branch block, right bundle branch block, and anterior or posterior fascicular block; and (6) right or left bundle branch block and first-degree AV block.

As noted, transcutaneous pacing is intended to be temporary; if prolonged pacing is required, transvenous pacing should be instituted (discussed subsequently). In addition, patients with a high probability of requiring pacing should have it instituted early on (see also Chap. 3).

VENTRICULAR ECTOPY, TACHYCARDIA, AND FIBRILLATION

Ventricular rhythm abnormalities are common during the early phases of AMI, with an incidence of ventricular fibrillation (VF) within the first 4 h, so-called *primary VF*, of 3 to 5 percent, which declines rapidly thereafter. Primary VF is associated with increased in-hospital mortality but not with increased long-term mortality for patients who survive and are discharged.

Post-AMI ventricular tachycardia (VT) occurs in about 15 percent of patients and is also most commonly manifest during the relatively early period. VT is classified according to its electrocardiographic morphology (*monomorphic or polymorphic*) and by its duration and consequences: *sustained (lasting more than 30 s and/or causing hemodynamic compromise earlier, which requires intervention)* and *nonsustained (not resulting in hemodynamic compromise and lasting less than 30 s)*. Short runs (5 beats or less) of nonsustained VT are very common in the early post-MI period and do not require specific treatment.

Accelerated idioventricular rhythm normally occurs frequently during the first hours of AMI and occurs after thrombolysis as a reperfusion arrhythmia. In neither case is it a premonitory rhythm for VT/VF. *Accelerated idioventricular rhythm should ordinarily be observed and not treated specifically.*

192 HURST'S THE HEART
COMPANION HANDBOOK

Formerly, it was common practice to treat prophylactically
with lidocaine in order to prevent VT/VF. *Routine use of prophylactic lidocaine in AMI in the presence or absence of
thrombolysis is not recommended.* Two prophylactic approaches to the prevention of VT/VF, however, are recommended. Routine administration of beta blockers, as described
previously, has been shown to reduce the incidence of VT/VF.
Also, since evidence suggests that hypokalemia is a risk factor
for VT/VF, it is recommended that serum potassium levels be
kept above 4.0 mEq/L by supplementation, as necessary.

Treatment of Ventricular Tachycardia/Fibrillation. Electrical
cardioversion of VT that is hemodynamically compromising
should be performed immediately. Rapid polymorphic VT
should be considered the equivalent of VF and cardioverted
with an unsynchronized shock of 200 J; monomorphic VT at a
rate of greater than 150 beats/min can be treated initially with
a synchronized discharge of 100 J. Urgent cardioversion for
VT with rates of under 150 beats per minute is usually not
needed. VT that is tolerated hemodynamically can be
approached initially with trials of lidocaine, procainamide, or
amiodarone.

Ventricular fibrillation should initially be treated with an
unsynchronized shock of 200 J, then incrementally at 200 to
300 J, and finally at 360 J as needed. The advanced cardiac life
support (ACLS) protocol recommends the following hierarchi-
cal approach, as needed, to adjunctive therapy of resistant
VF: (1) epinephrine (1 mg IV); (2) lidocaine (1.5 mg/kg IV);
and (3) bretylium (5 to 10 mg/kg IV). Intravenous amiodarone
(150 mg IV bolus) may also be used. In the case of resistant or
recurrent VT/VF, electrolyte imbalances should be sought and
corrected, and ongoing ischemia should be suspected. Beta-
adrenergic blockers should be used in recurrent VT or primary
VF to decrease sympathetic input to the heart and to decrease
ischemia. *If ongoing ischemia is involved, intraaortic balloon
pumping or emergency revascularization should be considered.*

APPROACH TO THE PATIENT WITH ISCHEMIC-TYPE
CHEST PAIN AND WITHOUT ST-SEGMENT ELEVATION

As discussed, the initial criterion differentiating patients with
symptoms compatible with AMI for therapeutic purposes is the
presence or absence of ST-segment elevation; this is a distinction of importance because, in the absence of ST-segment

elevation, there is no therapeutic benefit of thrombolysis. AMI in which Q waves do not develop is categorized as *non-Q-wave MI (NQWMI)*, and most such patients (90 percent) present with ST-segment depression.

NQWMI is precipitated by plaque disruption. Total coronary occlusion demonstrated angiographically is much less common than in Q-wave MI. Because of the residual noninfarcted myocardium at risk distal to a disrupted plaque, moreover, patients with NQWMI have a high propensity for recurrent ischemia, infarction, and death and present an opportunity for secondary prevention. Two important conclusions can be derived from the available data: *(1) thrombolysis cannot be recommended in AMI patients without ST-segment elevation and (2) in the NQWMI group and based on the admission ECG, there is a graded, decremental spectrum of risk ranging from ST-segment depression to T-wave inversion to normal.*

Management of Non-Q-Wave Myocardial Infarction Initially, the NQWMI patient, who by definition does not have diagnostic ST-segment elevation, cannot be distinguished from the patient with unstable angina and no myocardial necrosis. Thus, patients are admitted to the coronary care unit and *other than avoiding thrombolytic therapy, the initial pharmacologic approach is identical* (Fig. 10-1). Serial ECGs and cardiac marker measurements should be performed and, in the case of recurrent pain with the development of ST-segment elevation, thrombolysis or primary PTCA should be performed. If the patient has recurrent stuttering symptoms, angiography should be performed.

The calcium entry blocker diltiazem (immediate-release form) has been shown to be effective in reducing reinfarction in NQWMI in patients with preserved left ventricular function and with no evidence of congestive heart failure (class IIa indication when diltiezem is given to control ongoing ischemia or to control rapid ventricular response in case of atrial fibrillation when beta blockers are ineffective or contraindicated) (class IIb for general prophylaxis and may be added to standard therapy after 24 h and continued for 1 year). Similar results have been seen for verapamil. Both drugs are contraindicated in the presence of significant left ventricular dysfunction or congestive heart failure. Dihydropyridine calcium-channel blockers have not shown benefit and may be detrimental (class III, not indicated). In a recent trial comparing PTCA with conventional therapy there was a decrease in mortality associated with dilti-azem in the arm comparing conventional medical management with or without diltiazem. Parenthetically, there were more deaths in the invasive than in the conservative arm.

It seems prudent to recommend aspirin (160 to 325 mg/day) for NQWMI, and—in patients without evidence of congestive heart failure, pulmonary congestion, or left ventricular dys-function—to add diltiazem to standard therapy after the first 24 h and to continue it for 1 year. Generally, beta blockers have shown no effect on the reinfarction rate in patients recovering from NQWMI. Beta blockers may be given to relieve pain or arrhythmia, as discussed previously for Q-wave MI, and have a class IIb recommendation for use in NQWMI.

MANAGEMENT AFTER HOSPITAL ADMISSION

General Approach

The general issues involved in the management of the patient with suspected or manifest AMI in the intensive or moderate care unit are to provide for adequate monitoring for the detec-tion of arrhythmia, ischemia, and hemodynamic instability; to provide the patient with a calm, supportive, and reassuring environment; to control the level of activity; to begin the edu-cation process for a lifetime of living with coronary heart disease; to control pain and inappropriate anxiety; and to treat adverse events promptly. It is assumed, as previously dis-cussed, that oxygen therapy, beta-adrenergic blockers, aspirin, thrombolytics, heparin, and nitroglycerin have been begun or given, as appropriate, in the ED.

ACTIVITY

Minimizing physical exertion is an important approach, in addition to minimizing sympathetic nervous system drive by administering beta-adrenergic blockers—also, by controlling pain and excessive anxiety, to decrease myocardial oxygen demand and thus myocardial ischemia and necrosis. *Prolonged bed rest and a severe limitation of activities such as self-feed-ing are no longer recommended except in the case of continuing ischemic pain and/or hemodynamic instability.* Constipation should be avoided and stool softeners routinely prescribed. A bedside commode is preferable to a bedpan in all but the most unstable patients.

ANALGESICS AND ANXIOLYTICS

The importance of controlling chest pain and excessive anxiety and the use of morphine and diazepam were previously discussed. Anxiolytics may be useful in treating symptoms of nicotine withdrawal in smokers. Intravenous haloperidol can be useful and safe in treating intensive care unit (ICU) psychosis, particularly in the elderly.

EDUCATION

Education of the AMI patient, by both the coronary care unit (CCU) staff and the physician, with information about the management of symptoms and the prevention of the recurrence, gives a sense of empowerment that is associated with changes in behavior and decreases in anxiety. Because of the substantial risk of cardiac arrest in the 18 months after AMI, family members should be taught cardiopulmonary resuscitation (CPR).

Adjunctive Therapy during the Early in-Hospital Period

ANGIOTENSIN CONVERTING ENZYME (ACE) INHIBITORS

ACE inhibitors reduce left ventricular dysfunction and dilatation and they slow the progression to congestive heart failure in patients with left ventricular dysfunction after AMI. Efficacy may be greatest in those at highest risk—i.e., patients with the worst left ventricular function. Oral therapy with low doses should begin within the first 24 h after hemodynamic stabilization whether or not thrombolytic therapy has been administered. ACE inhibitors should not be given if systolic blood pressure is below 100 mmHg. If there is no evidence of left ventricular dysfunction at 4 to 6 weeks, therapy can be stopped. With significant left ventricular dysfunction, therapy should probably be continued indefinitely.

Management of the Low-Risk Patient

The patient with AMI who has an uncomplicated initial course and is at low risk for development of complications is a candidate for transfer out of the CCU within 24 to 36 h. These patients, including those who have had thrombolysis, may be candidates for early discharge at 3 to 4 days. Excessive diagnostic testing in all post-AMI patients, especially those at low risk, should be discouraged.

If AMI is effectively ruled out using serum markers in the low risk patient (i.e., normal ECG and absence of the characteristics noted above, especially the absence of prolonged initial pain or the recurrence of pain), noninvasive testing can establish the safety of early discharge (3 to 12 h) from the emergency department, or CCU, for further evaluation as an outpatient. In general, such patients do not necessarily need to be admitted to the CCU unless noninvasive testing is positive for ischemic heart disease. Patients with ischemic-type chest discomfort and intermediate probabilities of AMI (i.e., duration of chest pain greater than 20 to 30 min and nondiagnostic ECG changes without significant ST-segment elevation or depression, T-wave inversion, or bundle branch block) and without known coronary artery disease should be admitted to an observation unit or to the CCU. They should be placed on a fast track to rule in AMI or unstable angina. If the clinical course is unrevealing and serum marker enzymes are negative, stress testing and further evaluation can be planned. Clinical decisions can usually be made within 12 h in this setting.

Management of the High-Risk Patient with Acute Myocardial Infarction

The high-risk AMI patient is defined by the presence of one or more of the following: recurrent chest pain, congestive heart failure and low cardiac output, arrhythmias—in particular recurrent or sustained ventricualr tachycardia or fibrillation, mechanical cardiac complications of AMI such as ruptured papillary muscle or intraventricular septum, and/or inducible ischemia and extensive coronary artery disease.

Recurrent Chest Pain

The most common causes of recurrent chest pain after AMI are coronary ischemia and pericarditis.

RECURRENT ISCHEMIA

Recurrence of chest pain in the AMI patient is a serious development and requires immediate diagnosis and treatment, especially if the pain represents recurrent ischemia. Postinfarction angina is chest pain occurring at rest or with limited activity during hospitalization 24 h or more after onset of the AMI. Three categories of patients are at high risk: (1) patients with NQWMI; (2) patients who have received thrombolysis; and (3) patients with multiple risk factors.

The approach to recurrent ischemia is similar to that for the original episode. Coronary arteriography generally should be performed and, if a high-grade stenosis is found, PTCA should be performed (if lesion is suitable) or, if there is associated ST-segment elevation, additional thrombolysis should be administered if mechanical reperfusion is not feasible or available. If multiple high-grade stenoses are found, coronary artery bypass grafting should be considered.

PERICARDITIS

Early Postinfarction Pericarditis. Early postinfarction pericarditis, as reflected by pain and a friction rub, occurs in about 10 percent of patients, usually between days 2 and 4. *The treatment of choice is aspirin (160 to 325 mg daily), although higher doses (650 mg every 4 to 6 h) may be required*; however, other nonsteroidal anti-inflammatory drugs (NSAIDs) may be used. The use of anticoagulants is relatively contraindicated in AMI complicated by pericarditis.

Heart Failure in Acute Myocardial Infarction

PATHOPHYSIOLOGY AND HEMODYNAMICS

Cardiac failure develops when left ventricular function is reduced to 30 percent or more of normal and usually occurs within minutes or hours of onset of a large infarction. Some compromise of cardiac function is associated with perhaps more than two-thirds of cases of AMI and is usually transient. The severity of the failure, its duration, and whether or not it is reversible are predominantly dependent on infarct size.

RIGHT VENTRICULAR INFARCTION

Inferior MI associated with right ventricular infarction defines a high-risk subset with a mortality rate of 25 to 30 percent. This group should be approached aggressively with consideration for reperfusion therapy. *Right ventricular involvement should always be considered and should be specifically sought out in inferior MI with clinical evidence of low cardiac output, because the therapeutic approaches are quite different in the presence of right ventricular involvement from those for predominantly left ventricular failure.*

ST-segment elevation in lead V_{4R}, as noted, is the single most powerful predictor of right ventricular involvement in inferior infarction and identifies a patient subset with a markedly increased in-hospital mortality. All patients with inferior infarction should be screened by recording ECG lead V_{4R}. Echocardiography can also be useful as an adjunctive diagnostic approach.

Treatment of Right Ventricular Ischemia/Infarction   The major objectives in treating right ventricular infarction are to maintain right ventricular preload, to provide inotropic support, to reduce afterload of the right ventricle, and to achieve early reperfusion The recommendations are summarized in Table 10-7.

MANAGEMENT OF CONGESTIVE HEART FAILURE IN ACUTE MYOCARDIAL INFARCTION— GENERAL ISSUES

Hemodynamic Monitoring   The balloon flotation (Swan-Ganz) catheter fundamentally permits one, in the setting of low cardiac output, to distinguish between inadequate ventricular filling pressures and inadequate systolic function. The former is treated with volume expansion and the latter with inotropic support and frequently, afterload reduction. The catheter, even when used correctly, is not totally benign. Class I indications for insertion are severe or progressive congestive heart failure or shock and suspected mechanical complications (ventricular septal defect, papillary muscle rupture, tamponade).

Arterial monitoring in AMI is useful (class I) in all hypotensive patients, but especially in those in shock. Arterial monitoring is also recommended in patients receiving vasopressor agents. The radial artery is the preferred site, although the brachial and femoral arteries can be used.

Intraaortic Balloon Counterpulsation   The intraaortic balloon pump reduces afterload during ventricular systole and increases coronary perfusion during diastole. The decrease in afterload and increased coronary perfusion account for its efficacy in cardiogenic shock and ischemia. It is particularly useful as a stabilizing bridge to facilitate diagnostic angiography as well as revascularization and repair of mechanical complications of AMI (all class I indications).

TABLE 10-7

TREATMENT STRATEGY FOR RIGHT VENTRICULAR ISCHEMIA/INFARCTION

Maintain right ventricular preload
  Volume loading (IV normal saline)
  Avoid use of nitrates and diuretics
  Maintain AV synchrony
    AV sequential pacing for symptomatic high-
      degree heart block unresponsive to atropine
  Prompt cardioversion for hemodynamically
    significant SVT
Inotropic support
  Dobutamine (if cardiac output fails to increase after
    volume loading)
Reduce right ventricular afterload with left ventricular
  dysfunction
  Intra-aortic balloon pump
  Arterial vasodilators (sodium nitroprusside,
    hydralazine)
  ACE inhibitors
Reperfusion
  Thrombolytic agents
  Primary PTCA
  CABG (in selected patients with multivessel
    disease)

KEY: IV, intravenous; AV, atrioventricular; SVT, supraventricular tachycardia; ACE, angiotensin converting enzyme; PTCA, percutaneous transluminal coronary angioplasty; CABG, coronary artery bypass graft.
SOURCE: Ryan TJ, Anderson JL, Antman EM, et al: ACC/AHA guidelines for the management of patients with acute myocardial infarction: a report of the American College of Cardiology/American Heart Association Task Force on Practice Guidelines (Committee on Management of Acute Myocardial Infarction). *J Am Coll Cardiol* 1996; 28:1328–1428, with permission.

Diuretics and Positive Inotropic Agents

Diuretics and Cardiac Failure in Acute Myocardial Infarction.
Diuretics generally should not be the drugs used initially in the
treatment of pulmonary congestion in AMI because intravascular volume is initially normal (unless there was preexisting
congestive heart failure). Their use early in the course should
usually be guided by hemodynamic measurements from a
Swan-Ganz catheter. Diuretic therapy may become appropriate
later if salt and water retention occur and left ventricular filling
pressures become excessively high.

Inotropic Agents in Congestive Heart Failure.   Digoxin is not
the drug of choice in acute heart failure in MI. The primary use
of digoxin in AMI is to control heart rate in atrial fibrillation.
Dobutamine has favorable pharmacologic properties for use in
heart failure in MI. Dopamine has a tendency to increase heart
rate more than dobutamine. With higher doses, it may increase
peripheral resistance and filling pressures, thus offsetting some
of the positive inotropic effects (see also Chap 1).

Management of Uncomplicated Cardiac Failure after AMI   *In*
patients with uncomplicated AMI, there is no need to perform
invasive monitoring if careful clinical observations are made.
If cardiac failure is not complicated by mechanical factors—such as mitral valve rupture, ventricular septal rupture,
pulmonary embolus, or tamponade—the failure in most
patients is transient and of mild to moderate severity. If the cardiac output is normal, aggressive treatment is often not
recommended. In patients with rales at the base of the lungs
with only minimal increase in heart rate and no other signs of
hypoxemia, conventional therapy with morphine, nasal oxygen, nitrates (intravenous, oral, or transdermal), and bed rest is
adequate. In patients with extensive pulmonary edema who are
normotensive and exhibit hypoxia and dyspnea, the treatment
of choice is nitroglycerin given intravenously at 0.1 mg per
kilogram of body weight per minute and increased in increments of 5 to 10 mg/min, stopping at a dose that does not
decrease the systolic blood pressure below 100 mmHg. It is
preferable that hemodynamics be monitored invasively (Swan-
Ganz catheter) when one gives a vasodilator to reduce the
ventricular filling pressure to 15 to 17 mmHg. An intravenous
inotropic agent may be needed. The inotropic agents are generally dobutamine or dopamine. Dobutamine is the preferred
agent. The infusion should be initiated at 2 to 5 mg per kilogram of body weight per minute and should be increased such

that adequate systemic pressure is maintained and the heart rate does not increase by more than 10 to 15 percent. The ventricular filling pressure should be decreased to a range of 14 to 18 mmHg while maintaining adequate cardiac output and blood pressure.

In patients with borderline blood pressure and evidence of peripheral hypoperfusion, therapy should be initiated with an inotropic agent and not a vasodilator. Low doses of dopamine (2 to 7 mg per kilogram of body weight per minute) are usually given and are associated with an increase in stroke volume, cardiac output, renal blood flow, and peripheral resistance to a modest degree. Higher doses of dopamine induce significant vasoconstriction and increased left ventricular filling pressure. Diuretics should be used with caution. If high filling pressure (>18 to 20 mmHg) persists after achieving adequate output with positive inotropic agents and/or vasodilators, diuretics may be added.

Complicated Heart Failure after Myocardial Infarction In fulminating heart failure, administration of high concentrations (60 to 100%) of oxygen via a face mask is essential and endotracheal intubation may be needed. Invasive hemodynamic monitoring is particularly useful in these patients. The therapy for severe pulmonary edema should include intravenous morphine. If systolic blood pressure is adequate (≥100 mmHg), nitroglycerin is administered intravenously. In severe pulmonary edema, nitroprusside may be essential to reduce afterload. If the systolic blood pressure is <100 mmHg, treatment should be initiated with a positive inotropic agent, with the subsequent addition of a vasodilator if adequate blood pressure is achieved.

Hypotension and Cardiogenic Shock. Due to massive ischemia and necrosis, cardiogenic shock usually occurs within hours of the onset of infarction. Reversible causes must be excluded. These include mitral valve rupture, ventricular septal rupture, right ventricular infarction, pulmonary embolus, and cardiac tamponade.

The approaches to pulmonary congestion include the use of morphine and the maintenance of adequate oxygenation together with endotracheal intubation and mechanical ventilation if necessary. Pulmonary artery and arterial catheters should be placed. Urinary output is monitored using an indwelling catheter. The cornerstones of therapy are inotropic and vasopressor agents. If the systemic arterial vasopressure is below 80 to 90 mmHg, a pressor agent such as dopamine should be infused as described. If high doses of dopamine are necessary to maintain adequate perfusion (and pressure of 90 to 100 mmHg), a change to norepinephrine infusion should be considered. On occasion, the severity of cardiac pump dysfunction will require the combined use of nitroprusside and dopamine. Stabilization in cardiogenic shock may be achieved by the intraaortic balloon. Aortic counterpulsation is usually reserved for patients with a potentially reversible condition or in whom cardiac transplantation is being considered.

Restoration of coronary blood flow will probably be the most effective therapy in salvaging patients with cardiogenic shock. Mechanical revascularization appears to improve survival in cardiogenic shock complicating AMI.

Mechanical Dysfunction Contributing to Cardiac Failure
Papillary muscle rupture. This is manifest by the sudden appearance of pulmonary edema, usually 2 to 7 days after the infarction. A mid- or holosystolic murmur with wide radiation is usually audible. The diagnosis can be established by Doppler echocardiographic studies.

Immediate recognition and treatment are essential. Intra-aortic counterpulsation alone, or with vasodilator and inotropic therapy, is frequently required for temporary stabilization. The patient should undergo cardiac catheterization to define coronary anatomy, and surgery for mitral valve replacement or repair and coronary artery bypass grafting should be performed.

Papillary Muscle Dysfunction. The sudden development of an apical systolic murmur after a myocardial infarction is much more often secondary to papillary muscle dysfunction than to rupture. Papillary muscle dysfunction is frequently compatible with long-term survival.

Echocardiography coupled with Doppler flow studies will confirm the presence of mitral regurgitation, grade its severity, and permit assessment of left ventricular function. Ordinarily papillary muscle dysfunction will require no specific therapy, while the unusual patient with severe regurgitation should be treated like those with papillary muscle rupture. In intermediate cases, afterload reduction with ACE inhibitors should be considered.

Ventricular Septal Rupture. There is a higher prevalence of ventricular septal rupture in first infarctions and the majority occur within the first week. Ventricular septal rupture is usually manifest by the appearance of a new harsh, holosystolic murmur along the left sternal border (often associated with a thrill) and sudden clinical deterioration with hypotension and pulmonary congestion. Often the event is heralded by a recurrence of chest pain. The diagnosis can be established by two-dimensional and Doppler echocardiographic studies. Results of these studies and/or the oxygen setup on right heart catheterization would confirm the presence of septal rupture.

Medical therapy can be expected to be ineffective. Prompt but temporary stabilization can be achieved with intraaortic balloon counterpulsation alone or in conjunction with vasodilator and inotropic drug therapy. Cardiac catheterization should be performed in an expeditious manner. An aggressive approach of immediate operative repair of these patients results in a short-term survival rate of 42 to 75 percent.

Arrhythmias and Conduction Disturbances Complicating Acute Myocardial Infarction

VENTRICULAR ECTOPY, VENTRICULAR TACHYCARDIA, AND VENTRICULAR FIBRILLATION

The management of ventricular tachycardia and fibrillation after the first 24 h of hospitalization for AMI is similar to that for the early phase. The occurrence of symptomatic, sustained ventricular tachycardia or of ventricular fibrillation in the later phases of the hospital course, however, suggests that a chronic arrhythmogenic focus may be developing in the damaged ventricle. These ventricular arrhythmias are classified as *secondary* and indicate an increased risk of subsequent sudden cardiac death.

SINUS TACHYCARDIA OR ATRIAL PREMATURE BEATS

Sinus tachycardia following AMI is common and is frequently an unfavorable prognostic sign. Patients with a large area of infarcted myocardium may have sinus tachycardia on the basis of left ventricular dysfunction, which causes reflex sympathetic nervous system activation. Frequent atrial premature complexes are relatively common in AMI and no specific therapy is indicated.

PAROXYSMAL SUPRAVENTRICULAR TACHYCARDIA

Episodes of paroxysmal supraventricular tachycardia occur rather commonly in AMI and are usually transient. Rate control is essential, and the therapeutic approaches may include

carotid sinus massage, adenosine, digoxin, verapamil, or diltiazem.

ATRIAL FLUTTER AND ATRIAL FIBRILLATION

Atrial flutter is relatively uncommon in AMI, whereas atrial
fibrillation has an incidence of 10 to 15 percent. Atrial fibrillation is associated with an increased in-hospital mortality rate.
A rapid ventricular response can worsen ischemia and infarction. Risk of systemic embolization is increased in AMI in the
presence of atrial fibrillation. Thus, heparin therapy is indicated in patients not already receiving it. If the patient
experiences new or worsening pain, ischemic ST changes, or
hemodynamic instability during atrial fibrillation with a rapid
ventricular response rate, immediate electrical cardioversion is
indicated.

HEART BLOCK

First-degree block is seen frequently in AMI, especially in
inferior MI. This is attributable to ischemia or enhanced vagal
activity. Treatment is seldom required.

Second-degree AV block is also relatively common, especially Mobitz type I or Wenckebach block. It is associated with
a narrow QRS and is frequently the result of AV node ischemia
in inferior MI. It is usually transient, and its presence does not
affect the prognosis. Mobitz type II block is uncommon but is
associated with more serious complications and a worse prognosis. It usually occurs with anterior MI and reflects trifascicular block. It is characterized by a wide QRS and a non-
varying PR interval before a nonconducted atrial beat. Heart
block may develop suddenly and is an ominous sign, with a
mortality of about 80 percent. It is usually permanent.

Third-degree AV block, or complete heart block, occurs in
about 5 percent of patients with AMI and is most commonly
seen with inferior infarction, usually with block at the AV
node. There is some increase in in-hospital mortality rates in
this setting, but complete heart block in inferior MI is not an
independent predictor of poor long-term prognosis. In contrast,
patients with anterior infarction who develop third-degree AV
block have a high mortality rate of 80 percent.

INDICATIONS FOR TEMPORARY TRANSVENOUS PACING

The indications generally agreed on for temporary pacemaker
insertion in AMI include asystole, complete heart block in the
setting of anterior MI, new onset of right or left bundle branch

block with persistent Mobitz II second-degree AV block in the setting of anterior MI, or other symptomatic bradycardias unresponsive to atropine. Bundle branch block in the setting of AMI identifies a population at risk for both electrical and mechanical complications. Such patients must be monitored for evidence of transient high-degree heart block (see also Chap. 3).

PERMANENT PACING

The use of permanent pacemakers is reviewed extensively in the American College of Cardiology/American Heart Association guidelines for pacemaker implantation and is summarized in Chap. 34 of *Hurst's The Heart*, 9th ed. The fact that temporary pacing may have been required in the course of AMI does not necessarily indicate the need for permanent pacing. Patients who have had permanent pacemakers inserted after AMI usually have a relatively unfavorable prognosis that is primarily related to the extensiveness of the underlying disease and myocardial damage (see also Chap. 3).

Discharge from the Coronary Care Unit

The length of stay in the CCU should be based on the risk of developing ventricular tachycardia and ventricular fibrillation. The risk of developing primary ventricular fibrillation after AMI decreases exponentially, with the majority of arrhythmic deaths occurring within the first 24 h. A patient with an uncomplicated infarction can be transferred from the CCU on the third day, although some patients need more prolonged cardiac monitoring. Those patients who are prime candidates for late-hospital sudden death manifest, while in the coronary care unit, one or more of the following: (1) the arrhythmias of pump failure (sinus tachycardia, atrial flutter, or atrial fibrillation); (2) the arrhythmias of electrical instability (ventricular tachycardia or ventricular fibrillation); (3) acute interventricular conduction disturbances; (4) evidence of circulatory failure (congestive heart failure, pulmonary edema, or significant hypotension); or (5) large anterior infarction. The effectiveness of prolonged monitoring of this select group of patients in an intermediate care unit following CCU discharge is evident in a doubling of the rate of successful resuscitations.

In an uncomplicated MI, the patient does not need to be confined to bed for longer than 24 h. Upon transfer from a

CCU, the patient should be started on a progressive ambulation program. The speed with which the patient progresses from one stage to the next depends on the severity of the infarction, the presence or absence of complications, the patient's age, and the presence of comorbid conditions. The length of hospitalization following an AMI should likewise depend on these same factors. If the patient has not experienced complications during the first 4 days of hospitalization, he or she is very unlikely to do so at any later time. This patient could probably be discharged after 7 or fewer days in the hospital. The last 2 to 3 days of the hospitalization are generally necessary to resolve questions pertaining to residual ventricular function, the presence or absence of ventricular ectopy, and the adequacy of the remainder of the coronary circulation. In addition, time is needed for instruction in risk-factor modification.

Noninvasive Risk Stratification in Patients Surviving Acute Myocardial Infarction

Survivors of acute myocardial infarction have a substantial risk of facing subsequent cardiovascular events. Noninvasive risk assessment provides useful information to individualize the extent of further workup and therapy: (1) targeting specific long-term therapies; (2) identifying high-risk patients requiring aggressive diagnostic testing; (3) counseling the patient on prognosis; (4) developing an exercise program; and (5) planning modifications of lifestyle.

Three interrelated prognostic factors are the focus of predischarge assessment: (1) assessment of left ventricular function; (2) detection of residual myocardial ischemia (jeopardized myocardium); and (3) assessment of the risk of arrhythmic (sudden cardiac) death. High-risk patients can be identified clinically because of the presence of one or more of the following: decompensated congestive heart failure, angina associated with electrocardiographic changes, in-hospital cardiac arrest, spontaneous sustained ventricular tachycardia, or the development of a high-degree heart block. In the majority of patients who have a relatively benign hospital course, noninvasive testing can accurately identify a group at very low risk whose annual mortality is 1 to 3 percent. The practical consequences of identifying a low-risk group is that emphasis is focused on early discharge and lifestyle modification and targeted prophylactic medical therapy rather than expensive, invasive diagnostic testing.

Early coronary angiography and aggressive interventional therapy is indicated for patients with recurrent episodes of spontaneous or induced (with low level exercise testing) angina or ischemia or with evidence of persistent pulmonary congestion, clinical left ventricular dysfunction, or cardiogenic shock. *There is general agreement that there is not enough evidence in asymptomatic patients to support routine coronary angiography as the initial assessment; therefore, the guidelines dissuade its use as the primary tool for diagnostic evaluation.*

ASSESSMENT OF LEFT VENTRICULAR FUNCTION AND LEFT VENTRICULAR EJECTION FRACTION

Ventricular function is an important determinant of long-term survival after AMI regardless of reperfusion status and, in general, should be assessed by noninvasive or, if otherwise indicated, angiographic techniques.

ASSESSMENT OF MYOCARDIAL ISCHEMIA

Exercise Testing in Uncomplicated Patients *Clinical Significance of Predischarge Submaximal Exercise Testing.* Predischarge exercise testing consistently identifies a group at high risk for recurrent cardiac events (myocardial infarction, unstable angina, etc.) or mortality in the first year after the AMI. Exercise testing also identifies a group at very low risk (1 to 3 percent mortality rate for the first year). A negative test should promote early discharge as well as discourage an aggressive diagnostic approach. Submaximal exercise testing in *uncomplicated* patients, including those who have had thrombolysis, before discharge has a class I indication.

The presence of ischemia generally mandates cardiac catheterization to define the coronary anatomy and to consider revascularization. A recommended approach to post-AMI risk stratification is summarized in Fig. 10-3.

ASSESSMENT OF THE RISK OF ARRHYTHMIC (SUDDEN CARDIAC) DEATH

The identification of patients who are asymptomatic but at high risk for arrhythmic death after AMI has not been associated with the delineation of a successful treatment strategy. Although the presence of asymptomatic spontaneous ventricular arrhythmias as detected on ambulatory monitoring is predictive of increased arrhythmic (sudden) death, the positive predictive value is poor. The dearth of safe, effective antiar-

Clinical Indications of High Risk at Predischarge

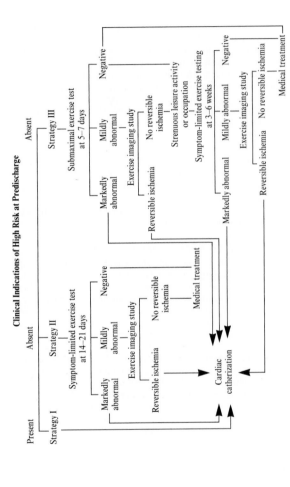

FIGURE 10-3

Strategies for exercise test evaluations soon after myocardial infarction (MI). If patients are at high risk for ischemic events based on clinical criteria, they should undergo invasive evaluation to determine if they are candidates for coronary revascularization procedures (strategy I). For patients initially deemed to be at low risk at time of discharge after myocardial infarction, two strategies for performing exercise testing can be used. One is a symptom-limited test at 14 to 21 days (strategy II). If the patient is on digoxin or if the baseline ECG precludes accurate interpretation of ST-segment changes (e.g., baseline left BBB or left ventricular hypertrophy), then an initial exercise imaging study can be performed. Results of exercise testing should be stratified to determine need for additional invasive or exercise perfusion studies. A third strategy is to perform a submaximal exercise test at 5 to 7 days after myocardial infarction or just before hospital discharge. The exercise test results could be stratified using the guidelines in strategy I. If exercise test studies are negative, a second symptom-limited exercise test could be repeated at 3 to 6 weeks for patients undergoing vigorous activity during leisure or at work. [From Ryan TJ, Anderson JL, Antman EM, et al: ACC/AHA guidelines for the management of patients with acute myocardial infarction: a report of the American College of Cardiology/American Heart Association Task Force on Practice Guidelines (Committee on Management of Acute Myocardial Infarction). *J Am Coll Cardiol* 1996; 28:1328–1428, with permission.]

rhythmic drugs has limited the usefulness of routinely evaluating the asymptomatic post-AMI patient in this regard.

Assessing Arrhythmic Death: Conclusions   None of the noninvasive (ambulatory monitoring) or invasive (electrophysiologic testing) techniques is generally agreed upon to be beneficial, useful, and effective in assessment of arrhythmic death. No clinical trial has demonstrated that the use of any one or a combination of these modalities of testing identifies a high-risk population in whom intervention can result in clinical benefit.

Coronary Angiography and Percutaneous Transluminal Coronary Angioplasty

The selection of patients for cardiac catheterization and coronary angiographic studies prior to hospital discharge should be based on identifying those at risk for ischemic events and on whether or not the information provided by cardiac catheterization and coronary angiography will change patient management. *In general, studies that have compared acute or early cardiac catheterization to a more conservative approach, of performing cardiac catheterization and coronary angiographic studies only for patients with spontaneous recurrent angina or exercise-induced ischemia, have demonstrated no benefit to the strategy of routine catheterization.* Patients who have a complicated clinical course characterized by refractory cardiac failure, unstable angina, an episode of sustained ventricular tachycardia, or cardiac arrest should be studied.

The recommended algorithm for selecting asymptomatic, uncomplicated post-AMI patients for cardiac catheterization is presented in Fig. 10-3. Decision making focuses on the presence or absence of myocardial ischemia. When patients have received thrombolytic therapy, it seems reasonable that those who have evidence of residual ischemia are still at increased risk of future ischemic events and should undergo coronary angiography. Consideration of PTCA following coronary angiographic studies should be based on established clinical and anatomic guidelines. Coronary artery bypass surgery should be considered in those groups in whom it has been shown to be of proven benefit: patients with triple-vessel disease, those with ischemia, and those with significant left ventricular dysfunction.

SECONDARY PREVENTION
AND CARDIAC REHABILITATION

Risk-Factor Reduction

There is now abundant evidence that risk-factor reduction will reduce coronary events in susceptible patients. Thus, since patients who have had AMI are among those at highest risk for recurrence, management strategies to mitigate this risk are very important.

SMOKING

Smoking cessation is an essential goal after AMI, since the recurrence and death rates after AMI are doubled by the continuation of smoking and the risk associated with smoking declines rapidly, within 3 years, to that of the nonsmoking cohort survivors. The role of the physician in motivating the patient to quit smoking is extremely important, and the likelihood of success appears to be directly related to the extent of his or her involvement.

DYSLIPIDEMIA

Recent large secondary prevention trials have provided compelling evidence that after AMI, therapy with HmG-CoA reductase inhibitors to lower serum cholesterol levels that were either initially elevated as in the Scandinavian Simvastatin Survival Study (4S) or within "average" range as in the Cholesterol and Recurrent Events trial (CARE) was effective in reducing both cardiovascular and total mortality as well as cardiovascular events. The guidelines of the expert panel of the National Cholesterol Education Program provide target goals for patients with manifest coronary artery disease. These goals are LDL cholesterol <100 mg/dL (2.59 mmol/L) and HDL cholesterol greater than 35 mg/dL (0.91 mmol/L). *All AMI patients should have serum lipids evaluated and treated intensively in order to achieve target goals.* Treatment should start in the hospital with initiation of the AHA step II diet (see also Chap. 7).

INACTIVITY

A sedentary lifestyle is a risk factor for coronary artery disease. Metanalysis of cardiac rehabilitation studies has shown a reduction in mortality in the exercise group as opposed to a control group. *The greatest benefits of exercise are those observed with moderate, regular exercise as contrasted with no exercise.*

Regular aerobic exercise should be prescribed for post-AMI patients in stable condition at an intensity, duration, and frequency as determined by formal testing and clinical judgment. Optimum benefit is achieved in a supervised program, although asymptomatic, stable patients can exercise without direct supervision. They should, however, receive regular monitoring by a physician.

Drug Therapy

BETA-ADRENERGIC BLOCKERS
The benefits of beta-blocker therapy given early in the course of AMI were discussed. Multiple clinical trials have also demonstrated the benefits of long-term treatment of post-AMI patients with beta blockers. Mortality has been shown to be reduced by about 25 to 35 percent. The beneficial effect is highest in high-risk patients with large (usually anterior) MI and compensated left ventricular dysfunction. The beneficial effects in low-risk patients are less clear but the consensus is that these patients probably should be treated because of the relatively favorable side-effect profile. Beta blockers with intrinsic sympathomimetic activity should not be used in this context.

ASPIRIN
The role of aspirin during the early phases of AMI was discussed. Aspirin use over the long term after AMI is also associated with a reduction in mortality. Aspirin at relatively low doses (81 to 325 mg/day) is recommended for all patients with AMI in the absence of contraindications (see also Chap. 46).

ANTICOAGULATION
Anticoagulation can reduce mortality, recurrent MI, and stroke after AMI. The role of warfarin is rather limited to those at increased risk for developing mural thrombi. In addition, those post-AMI patients with demonstrable left ventricular thrombus and atrial fibrillation should be anticoagulated. The duration of anticoagulation should be limited to 3 months in the case of left ventricular thrombus.

ANGIOTENSIN CONVERTING ENZYME INHIBITORS

The ACE inhibitors and recommendations for their use early in the course of AMI were discussed. Recent studies have documented their efficacy in secondary prevention. Thus, ACE inhibitors are recommended for chronic use after AMI in those patients with significant left ventricular dysfunction, and their use should be considered in those with only mild to moderate left ventricular dysfunction (ejection fraction less than 45 percent).

Modification of Lifestyle and Cardiac Rehabilitation After AMI

Because of the relatively high risk of recurrence and the need for lifelong modification of lifestyles and risk factors, most post-AMI patients should be enrolled in a cardiac rehabilitation program that emphasizes dietary modification, risk-factor reduction and exercise. The low-risk patient does not require prolonged supervised exercise. All patients, however, can benefit from a structured environment to launch a lifetime of healthy living. Cardiac rehabilitation is discussed in detail in Chaps. 47 and 55 of *Hurst's The Heart*, 9th ed.

SUGGESTED READING

Alexander RW, Pratt CM, Roberts R: Diagnosis and management of patients with acute myocardial infarction. In: Alexander RW, Schlant RC, Fuster V, O'Rourke RA, Roberts R, Sonnenblick EH (eds): *Hurst's The Heart*, 9th ed. New York, McGraw-Hill, 1998:1345–1433.

Antman EM, Tanasijevic MJ, Thompson B, et al: Cardiac-specific troponin I levels to predict the risk of mortality in patients with acute coronary syndromes. *N Engl J Med* 1996; 335:1342–1349.

Fibrinolytic Therapy Trialists' (FTT) Collaborative Group: Indications for fibrinolytic therapy in suspected acute myocardial infarction: collaborative overview of early mortality and major morbidity results from all randomized trials of more than one-thousand patients. *Lancet* 1994; 343:311–322.

Fuster V, Badimon L, Badimon JJ, Chesebro JH: The pathogenesis of coronary artery disease and the acute coronary
syndromes (2). *N Engl J Med* 1992; 326:310–318.

GUSTO IIA Investigators, Ohman EM, Armstrong PW, Chris-
tenson RH, et al: Cardiac troponin T levels for risk
stratification in acute myocardial ischemia. *N Engl J Med*
1996; 335:1333–1341.

Hochman JS, Boland J, Sleeper LA, et al: Current spectrum of
cardiogenic shock and effect of early revascularization on
mortality: results of an International Registry. SHOCK
Registry Investigators. *Circulation* 1995; 91:873–881.

Kinch JW, Ryan TJ: Right ventricular infarction. *N Engl J Med*
1994; 330:1211–1217.

Pfeffer MA, Braunwald E, Moye LA, et al: Effect of captopril
on morbidity and mortality in patients with left ventricular
dysfunction after myocardial infarction: results of the Survival and Ventricular Enlargement Trial. The SAVE
Investigators. *N Engl J Med* 1992; 327:669–677.

Puleo PR, Meyer D, Wathen C, et al: Use of rapid assay of subforms of creatine kinase MB to diagnose or rule out acute
myocardial infarction. *N Engl J Med* 1994; 331:561–566.

Ryan TJ, Anderson JL, Antman EM, et al: ACC/AHA guidelines for the management of patients with acute
myocardial infarction: a report of the American College of
Cardiology/American Heart Association Task Force on
Practice Guidelines (Committee on Management of Acute
Myocardial Infarction). *J Am Coll Cardiol* 1996; 28:
1328–1428.

Sacks FM, Pfeffer MA, Moye LA, et al: The effect of pravastatin on coronary events after myocardial infarction in
patients with average cholesterol levels. Cholesterol and
Recurrent Events (CARE) Trial Investigators. *N Engl J
Med* 1996; 335:1001–1009.

Scandinavian Simvastatin Survival Study Group: Randomised
trial of cholesterol lowering in 4444 patients with coronary
heart disease: the Scandinavian Simvastatin Survival
Study (4S). *Lancet* 1994; 344:1383–1389.

Wenger NK: Rehabilitation of the patient with coronary heart
disease. In: Alexander RW, Schlant RC, Fuster V,
O'Rourke RA, Roberts R, Sonnenblick EH (eds): *Hurst's
The Heart*. New York, McGraw-Hill, 1998: 1619–1631.

CHAPTER

11

TREATMENT OF SYSTEMIC ARTERIAL HYPERTENSION

W. Dallas Hall

Initial pharmacologic therapy for primary (essential) hypertension usually begins with low doses of either a diuretic, a beta blocker, an angiotensin-converting enzyme (ACE) inhibitor, or a calcium channel blocker. The goal of therapy recommended by the Joint National Committee (JNC VI) is reduction of systolic blood pressure to below 140 mmHg and diastolic blood pressure to below 90 mmHg.

Disregarding patients' clinical profiles, one finds that less than 50 percent of compliant patients with stage 1 or 2 hypertension respond to monotherapy with any of the above-listed four classes of antihypertensive drugs. This response rate can be improved considerably by the tailoring of therapy to individual patients. For example, hypertensive African Americans generally respond better to initial therapy with diuretics or calcium channel blockers than to initial therapy with ACE inhibitors or beta blockers. Table 11-1 outlines the minimum and maximum dosage range for antihypertensive drugs in common use.

DIURETICS AS INITIAL THERAPY

Diuretic therapy should begin with low doses, such as the equivalent of 12.5 mg daily of chlorthalidone or hydrochlorothiazide. If the serum creatinine level is normal, there usually is no advantage to beginning therapy with a loop diuretic such as furosemide, torsemide, or bumetanide. If renal function is impaired by 50 to 75 percent or more, however, then the loop diuretics usually are necessary to obtain adequate natriuresis. The usual starting dose of furosemide is 40 mg twice daily; of torsemide, 5 mg daily; of bumetanide, 0.5 mg daily.

TABLE 11-1

DOSAGE RANGE FOR ANTIHYPERTENSIVE DRUGS

Drugs	Trade Name in USA	Minimum initial dose (mg/day)	Maximum dose (mg/day)
Diuretics			
Thiazide-type			
Bendroflumethiazide	Naturetin	2.5	5
Chlorthalidone	Hygroton	12.5	50
Hydrochlorothiazide	HydroDiuril, Esidrix, Microzide	12.5	50
Indapamide	Lozol	1.25	5
Metolazone	Zaroxolyn	2.5	10
Metolazone	Mykrox	0.5	1
Trichlormethiazide	Naqua, Metahydrin	2	4
Loop			
Bumetanide	Bumex	0.5	10
Furosemide	Lasix	40	240
Torsemide	Demadex	5	20
Potassium-sparing			
Amiloride	Midamor	5	10
Spironolactone	Aldactone	25	100

		25	100
Triamterene	Dyrenium		

Adrenergic inhibitors

Beta-adrenergic blockers

		25	100
Acebutolol	Sectral	200	800
Atenolol	Tenormin	25	100
Betaxolol	Kerlone	5	20
Bisoprolol	Zebeta	2.5	10
Carteolol	Cartrol	2.5	10
Metoprolol	Lopressor	50	300
Nadolol	Corgard	40	320
Penbutolol	Levatol	10	20
Pindolol	Visken	10	60
Propranolol	Inderal	40	480
Timolol	Blocadren	20	80

Central-acting adrenergic inhibitors

Clonidine	Catapres	0.2	1
Guanabenz	Wytensin	8	32
Guanfacine	Tenex	1	3
Methyldopa	Aldomet	500	2000

Peripheral adrenergic inhibitors

Guanadrel	Hylorel	10	75
Guanethidine	Ismelin	10	75
Reserpine	Serpasil	0.05	0.25

(continued)

TABLE 11-1
(Continued)

Drugs	Trade Name in USA	Minimum initial dose (mg/day)	Maximum dose (mg/day)
Alpha-adrenergic blockers			
Doxazosin	Cardura	1	16
Prazosin	Minipress	1	20
Terazosin	Hytrin	1	20
Combined alpha- and beta-adrenergic blockers			
Labetalol	Normodyne, Trandate	200	1200
Vasodilators			
Hydralazine	Apresoline	50	300
Minoxidil	Loniten	5	60
Angiotensin-converting enzyme inhibitors			
Benazapril	Lotensin	10	40
Captopril	Capoten	25	150
Enalapril	Vasotec	5	40
Fosinopril	Monopril	10	40

218

Lisinopril	Zestril, Prinivil	10	40
Moexipril	Univasc	7.5	15
Quinapril	Accupril	5	80
Ramipril	Altace	1.25	20
Trandolapril	Mavik	1	4
Calcium channel blockers			
Amlodipine*	Norvasc	5	10
Diltiazem	Cardizem, Dilacor, Tiazac	120	360
Felodipine*	Plendil	5	20
Isradipine*	DynaCirc	5	20
Nisoldipine*	Sular	20	60
Nicardipine*	Cardene	60	90
Nifedipine*	Adalat, Procardia	10	90
Verapamil	Isoptin, Calan, Verelan, Covera HS	120	480
Angiotensin receptor blockers			
Irbesartan	Avapro	150	300
Losartan	Cozaar	50	100
Valsartan	Diovan	80	160

*A dihydropyridine.

When diuretic therapy is begun in outpatients with uncomplicated essential hypertension, the serum potassium level should be measured prior to treatment, again within 4 weeks or less, and once or twice a year thereafter. The development of diuretic-induced hypokalemia relates more closely to the magnitude of the natriuretic response and the increase in aldosterone levels than to the particular agent used. In general, either potassium supplements or a potassium-sparing diuretic (e.g., Maxzide, Dyazide, Aldactazide) should be prescribed when a serum potassium level below 3.5 mmol/L is confirmed in a patient with uncomplicated hypertension. In patients with ventricular ectopy and normal renal function, potassium replacement may be desirable with less severe hypokalemia, particularly if there has been a greater than usual postdiuretic decrease in the serum potassium level (e.g., from 4.8 to 3.8 mmol/L). When hypokalemia is treated with potassium supplements, doses of at least 40 mmol daily are typically required. In general, potassium supplements should not be used concomitantly with potassium-sparing diuretics because of the risk of hyperkalemia.

Approximately 40 percent of hypertensive patients are dyslipidemic. High dosages of diuretics often cause a transient worsening of serum lipid levels, especially total cholesterol and triglycerides. Hence, high-risk patients with borderline or elevated lipid levels should receive counseling on a low-cholesterol, low-fat diet and lipid-lowering therapy as appropriate (see Chap. 7).

BETA BLOCKERS AS INITIAL THERAPY

There are several clinical settings in which beta blockers are excellent choices as first-line therapy for hypertension. These include young patients with high-renin essential hypertension or the hyperdynamic beta-adrenergic circulatory state, patients with alcohol-withdrawal hypertension, and, possibly, tachycardia- or anxiety-prone patients with evidence of increased adrenergic tone. In addition, the beta blockers are established as cardioprotective in patients with previous myocardial infarction (see Chap. 10).

Acebutolol, atenolol, betaxolol, bisoprolol, and metoprolol are more cardioselective than carteolol, nadolol, penbutolol, pindolol, propranolol, or timolol. Acebutolol, carteolol, pindolol, and penbutolol possess intrinsic sympathomimetic activity (ISA).

If one must prescribe a beta blocker for a diabetic patient receiving insulin, the cardioselective agents may be better choices. In contrast, the nonselective beta blockers (i.e., those that possess beta$_2$ receptor–blocking properties) are often more effective if beta-blocker therapy is used for the treatment of migraine headaches or intention tremors. Once-a-day therapy is advantageous in some patients and is appropriate for most of the beta blockers. Like diuretics, many beta blockers can also sometimes be associated with worsening of serum lipid levels, especially lowering of the HDL cholesterol.

ANGIOTENSIN-CONVERTING ENZYME (ACE) INHIBITORS AS INITIAL THERAPY

The ACE inhibitors are effective first-line therapy for Caucasian patients with hypertension. They are also useful in the treatment of diabetic patients, because they do not impair glucose tolerance and can reduce proteinuria and slow the progression of diabetic nephropathy. ACE inhibitors are the antihypertensive class of choice in patients with left ventricular systolic dysfunction.

The starting dose of captopril should be no more than 12.5 mg twice daily. An even lower initial dose (i.e., a 6.25-mg test dose) may be warranted in patients with high plasma renin activity—such as most of the patients with congestive heart failure and hyponatremia, accelerated malignant hypertension, or after acute diuresis. Drug interactions are common, especially with nonsteroidal anti-inflammatory agents, which can blunt the antihypertensive efficacy of ACE inhibitors.

A persistent dry, hacking cough is the most frequent effect of ACE inhibitor therapy. Angioneurotic edema is rare but can occur within hours after the initial dose.

CALCIUM CHANNEL BLOCKERS AS INITIAL THERAPY

The calcium channel blockers (e.g., amlodipine, diltiazem, felodipine, isradipine, nicardipine, nifedipine, nisoldipine, and verapamil) are also effective first-line drugs in the treatment of hypertension. They are equally efficacious in all races. The dihydropyridines (see Table 11-1) have a predominant effect on peripheral vascular calcium channels, with a lesser effect than diltiazem or verapamil on atrioventricular conduction and cardiac contractility (see Chap. 47). Diuretics are the preferred class for patients with isolated systolic hypertension, but the long-acting dihydropyridines are acceptable alternatives.

The most common adverse effect of verapamil is constipation, which can be managed with a high-fiber diet and mild laxatives. Caution must be used to avoid precipitation of heart failure or heart block whenever diltiazem or verapamil is used in conjunction with beta-blocker therapy in patients who may have borderline ejection fractions or underlying conduction disturbances. Amlodipine and felodipine have been shown to be safe when used in addition to ACE inhibitors, diuretics, or digoxin for the treatment of angina or hypertension in patients with advanced left ventricular dysfunction. Edema, flushing, and headaches are the most frequent adverse effects of the dihydropyridines. Peripheral edema is due to a redistribution of intravascular volume, not to retention of salt and water. Any calcium channel blocker can occasionally produce a severe rash.

One must be aware of certain clinically important food or drug interactions with the calcium channel blockers. For example, grapefruit juice can exaggerate the hypotensive effect of some calcium channel blockers, including felodipine, nifedipine, and nisoldipine. Verapamil and other nondihydropyridines can lead to an increase in the serum level of digoxin. Diltiazem, nicardipine, and verapamil can lead to significant increases in the blood levels of cyclosporine.

MULTIDRUG THERAPY FOR OUTPATIENTS WITH SEVERE OR RESISTANT HYPERTENSION (SEE TABLE 11-2)

Severe hypertension is defined as a systolic blood pressure of 180 mmHg or more, or a diastolic blood pressure of 110 mmHg or more; *resistant hypertension* exists when the systolic or diastolic blood pressure remains at 160 or 100 mmHg or more, respectively, despite reasonable dosages of three or more antihypertensive drugs, one of which is a diuretic.

TABLE 11-2

FACTORS TO EVALUATE IN OUTPATIENTS WITH SEVERE OR RESISTANT HYPERTENSION

Adherence to therapy
Volume expansion
Secondary causes
   Renovascular disease
   Aldosteronism
   Pheochromocytoma

The most common cause of resistant hypertension is non-compliance with a medication program. Does the patient know all of his or her medications? When was the last dose taken? Has the patient brought all medications to the office visit? Does the number of remaining pills approximate the number that should be present after your last prescription? Does the patient often miss appointments?

The second most common cause of resistant hypertension is volume expansion due either to medications or to excessive intake of dietary salt. As emphasized in JNC VI, if a diuretic is not chosen as the first drug, it is usually indicated as the second-step agent. Adding low doses of a diuretic will often control blood pressure in a patient with an incomplete response to one- or two-drug therapy. Increasing the dosage of diuretic is often necessary to control blood pressure in patients receiving therapy with multiple drugs. This is particularly true for obese patients, those with moderate renal insufficiency (i.e., serum creatinine level of 1.7 mg/dL or more), and patients receiving nonsteroidal anti-inflammatory drugs, corticosteroids, diphenylhydantoin, methyldopa, or minoxidil.

Outpatients with resistant hypertension are good candidates for elective hospital admission. Secondary causes—such as renovascular disease, aldosteronism, and pheochromocytoma—can be excluded by the ACE scintiscan, renal arteriography, saline suppression test, 24-h urinary metanephrine-to-creatinine excretion ratio, and abdominal CT scan. In addition, the patient can be observed for blood pressure response to his or her usual medication, thus differentiating noncompliance from true resistance to therapy.

ACUTE THERAPY FOR PATIENTS ADMITTED WITH URGENT, ACCELERATED, OR MALIGNANT HYPERTENSION

Urgent hypertension generally refers to upper levels of stage 3 hypertension (for example, a systolic or diastolic blood pressure of 240 or 140 mmHg, respectively) when it is desirable to reduce blood pressure within a few hours despite the absence of symptoms or acute retinopathy. *Accelerated hypertension* means that retinal hemorrhages or exudates are present; *malignant hypertension* implies papilledema.

Urgent and accelerated hypertension are sometimes treated with oral medication, whereas malignant hypertension and hypertensive emergencies (i.e., encephalopathy, intracranial hemorrhage, dissecting aneurysm, acute pulmonary edema, severe chest pain) are generally treated with parenteral therapy (see Table 11-3).

Oral Therapy

Reinstitution of previous antihypertensive therapy (that the patient has not been taking) is often efficacious in reducing severely elevated blood pressure within 4 to 6 h. One must be careful, however, not to "overshoot" and induce hypotension in

---

TABLE 11-3

THERAPEUTIC OPTIONS FOR THE ACUTE
MANAGEMENT OF PATIENTS WITH URGENT,
ACCELERATED, OR MALIGNANT HYPERTENSION

Oral therapy
    Reinstitution of previous medications
    Clonidine loading
    Labetalol
Parenteral therapy
    Sodium nitroprusside or nitroglycerin
    Fenoldopam
    Labetalol
    Nicardipine
    Enalaprilat

a patient who has never taken full doses of the prescribed medication. For example, it is wise not to restart initial single doses above 60 mg for long-acting nifedipine, 0.3 mg for clonidine, 5 mg for prazosin, or 10 mg for minoxidil. In addition, one must take care to ascertain that a previously prescribed medication is not contraindicated in the new clinical setting.

SHORT-ACTING NIFEDIPINE
Although short-acting nifedipine capsules reduce blood pressure rapidly, several serious adverse effects have been reported from the sometimes precipitous decreases in blood pressure. Short-acting nifedipine capsules should no longer be used in the treatment of urgent hypertension or hypertensive emergencies.

CLONIDINE LOADING
In patients with urgent hypertension, it is sometimes appropriate to use the clonidine loading regimen. An oral dose of 0.2 mg is followed by 0.1 mg every hour for up to 5 h, until the severely elevated blood pressure is gradually reduced or a total dose of 0.7 mg is reached. This therapy is successful in up to 90 percent of patients with urgent hypertension, but it may be accompanied by sedation, dry mouth, bradycardia, and occasionally orthostatic hypotension.

Parenteral Therapy

The most commonly used intravenous therapies for malignant hypertension and hypertensive emergencies are sodium nitroprusside, nitroglycerin, fenoldopam, labetalol, nicardipine, and enalaprilat.

NITROPRUSSIDE AND NITROGLYCERIN
The starting dose of intravenous sodium nitroprusside is 0.5 µg/kg/min of a solution of 50 mg nitroprusside in 250 mL dextrose and water (i.e., 200 µg/mL). The mixture must be protected from light and the solution changed every 12 h. The onset of action is immediate; the effects dissipate within 2 to 10 min after discontinuation. The rate of administration can be uptitrated to 8 µg/kg/min, but high dosages over several days require monitoring of the serum thiocyanate level, especially in patients with renal impairment. Toxicity often presents as projectile vomiting or acidosis, and occasionally as methemoglobinemia if thiocyanate levels exceed 15 to 20 mg/dL.

When myocardial ischemia is associated with relatively mild degrees of hypertension, nitroglycerin infusion may be preferable to nitroprusside. In patients with more severe elevations of blood pressure, however, nitroprusside or labetalol is more effective for the lowering of blood pressure. The starting dose of nitroglycerin is the same as that for nitroprusside: 0.5 μg/kg/min of a solution of 50 mg mixed into 250 mL dextrose and water. The solution does not need to be shielded from light, but a special nonabsorbing infusion set is required to avoid adherence of nitroglycerin to the plastic or polyvinyl chloride contained in most intravenous lines.

FENOLDOPAM

Fenoldopam is a selective dopamine-1 receptor agonist that reduces blood pressure while increasing renal blood flow and natriuresis. It is especially useful for patients with significantly impaired renal function where the use of sodium nitroprusside is limited. The starting dose is continuous intravenous infusion of 0.1 to 0.3 μg/kg/min, diluted into a solution of dextrose and water. The infusion rate can be increased by 0.05 to 0.10 μg/kg/min at 15-min intervals to a maximum rate of 1.5 μg/kg/min. Adverse effects include tachycardia, headache, nausea, flushing, and occasionally precipitation of acute glaucoma.

LABETALOL

The initial dose of labetalol is 20 mg (4 mL of a 20-mL ampule solution containing 5 mg/mL) given intravenously over 2 min. Fifteen minutes after this test dose, a bolus of 40 mg can be given, followed by repeated bolus doses of 40 to 80 mg as needed. The onset of action is usually within 10 to 20 min and the duration is 3 to 6 h. Once a response is obtained (often after a total dose of 60 to 100 mg), the labetalol can be administered by intravenous infusion of 0.5 to 2.0 mg/min of a mixture containing 100 mg labetalol in 250 mL dextrose and water. Conversion from parenteral to oral therapy is usually accomplished by starting with 100 to 200 mg orally, twice daily.

NICARDIPINE

The starting dose of nicardipine, a calcium channel blocker, is continuous intravenous infusion of 5 mg/h, diluted into a solution of dextrose and water. The infusion rate can be increased by 2.5 mg/h at 15- to 30-min intervals to a maximum rate of

15 mg/h. To minimize the risk of phlebitis, a peripheral vein infusion site should be changed every 12 h.

Nicardipine should be used with caution in patients with ischemia and not at all in patients with advanced aortic stenosis or acute heart failure. The most prominent adverse effect is headache; nausea, vomiting, tachycardia, and flushing are less frequent.

ENALAPRILAT

Enalaprilat is the active metabolite of enalapril. A test dose of 0.625 mg (0.5 mL of a 2-mL vial containing 2.5 mg) is infused over 5 min and used initially to avoid exaggerated hypotension. A second dose of 0.625 mg can be repeated after 1 h if there is an inadequate clinical response. The onset of action is usually within 15 min, but the peak effect can be delayed for up to 4 h. In responders, the maintenance dose is usually 1.25 mg intravenously every 6 h for adult patients with serum creatinine levels below approximately 3 mg/dL, and 0.625 mg every 6 h for those with serum creatinine levels of 3 mg/dL or more. The response to enalaprilat is quite variable.

SUGGESTED READING

Adult Treatment Panel II: Second Report of the National Cholesterol Education Program Expert Panel on Detection, Evaluation, and Treatment of High Blood Cholesterol in Adults. *JAMA* 1993; 269:3015–3023.

Frohlich ED, Re RN: Pathophysiology of systemic hypertension. In: Alexander RW, Schlant RC, Fuster V, O'Rourke RA, Roberts R, Sonnenblick EH (eds): *Hurst's The Heart*, 9th ed. New York, McGraw-Hill, 1998:1635–1650.

Gifford RW Jr: Treatment of patients with systemic arterial hypertension. In: Alexander RW, Schlant RC, Fuster V, O'Rourke RA, Roberts R, Sonnenblick EH (eds): *The Heart*, 9th ed. New York, McGraw-Hill, 1998:1673–1696.

Hall WD: Cardiovascular therapy in black patients. In: Messerli FH (ed): *Cardiovascular Drug Therapy*, 2d ed. Philadelphia, WB Saunders, 1996:280–291.

Hall WD: Hypertensive crises. In: Kassirer J, Greene HL (eds): *Current Therapy in Adult Medicine*, 4th ed. St. Louis, Mosby–Year Book, 1997:587–590.

Joint National Committee on Prevention, Detection, Evaluation and Treatment of High Blood Pressure: The Sixth report of the Joint National Committee on Prevention, Detection, Evaluation, and Treatment of High Blood Pressure. *Arch Intern Med* 1997; 157: 2413–2446.

Kaplan NM (ed): *Clinical Hypertension*, 7th ed. Baltimore, Williams & Wilkins, 1998.

Kasiske BL, Ma JZ, Kalil RSN, et al: Effects of antihypertensive therapy on serum lipids. *Ann Intern Med* 1995; 122: 131–141.

12

AORTIC VALVE
DISEASE

Robert A. O'Rourke

AORTIC STENOSIS

Definition, Etiology, and Pathology

Aortic stenosis (AS) is obstruction to outflow of blood from the left ventricle to the aorta. The obstruction may be valvular, supravalvular, or subvalvular in location.

The most common causes of AS are congenital, rheumatic, and calcific (degenerative). Calcific AS is seen in patients 35 years of age or older and results from calcification of a congenital or rheumatic valve or a normal valve that has undergone "degenerative" changes. An autoimmune reaction to antigens in the valve may be present in some patients. Rare causes of AS include obstructive infective vegetations, homozygous type II hyperlipidproteinemia, Paget's disease of bone, systemic lupus erythematosus, rheumatoid disease, ochronosis, and irradiation.

At the present time, calcific AS in middle-aged and elderly patients is the most common valve lesion coming to valve replacement. Among patients under the age of 70, congenital bicuspid valves account for one-half of the surgical cases; degenerative changes are the cause in 18 percent. In contrast, in those at or above 70 years of age, degenerative changes account for almost one-half of the surgical cases, and the congenital bicuspid valve for approximately one-quarter of the cases.

Supravalvular and subvalvular aortic stenosis are usually congenital. The exception may be hypertrophic cardiomyopathy (see Chap. 18).

A congenital bicuspid valve can produce severe obstruction to left ventricular (LV) outflow after the first few years of life. Immobilization of the cusps by calcium appears to be the most

229

important factor in converting the bicuspid valve into a stenotic lesion. The process of collagen generation, lipid deposition, and calcification in the bicuspid valve is similar to that noted much later in life in some tricuspid valves. The similarity in the pathologic findings suggest that stenosis in the bicuspid valve is a degenerative change accelerated by the abnormal mechanical stress imposed on the valve tissue by its bicuspid configuration.

Rheumatic AS results from adhesions and fusions of the commissures and cusps. The leaflets and valve ring become vascularized, which leads to retraction and stiffening of the cusp. Calcification occurs, and the aortic valve orifice is reduced to a small triangular or round opening that is frequently regurgitant as well as stenotic. Rheumatoid AS is extremely rare and results from nodular thickening of the valve leaflets. In severe forms of hypercholesterolemia, lipid deposits occur not only in the aortic wall but also in the aortic valve, occasionally producing AS.

The LV is concentrically hypertrophied. The hypertrophied cardiac muscle cells are increased in size. There is an increase of connective tissue and variable amounts of fibrous tissue in the interstitial tissue. Myocardial ultrastructural changes may account for the LV systolic dysfunction that occurs late in the disease. Subclinical calcific emboli are commonly found in calcific AS if diligently sought at autopsy.

The aortic valve must be reduced to one-fourth of its natural size before significant changes in the circulation occur. Since the orifice area of the normal adult valve is approximately 3.0 to 4.0 cm^2, an area exceeding 0.75 to 1.0 cm^2 is not usually considered to be severe AS. In large patients, however, a valve area of 1.0 cm^2 may be seriously stenotic; by contrast, a valve area of 0.7 cm^2 may be adequate for small patients.

Based on the variety of hemodynamic and natural history data, the degree of AS is usually graded as mild (area < 1.5 cm^2), moderate (area 1.1 to 1.5 cm^2), or severe (area ≤ 1.0 cm^2). When the stenosis is severe and the cardiac output is normal, the mean transvalvular pressure gradient generally exceeds 50 mmHg. Some patients with severe AS remain asymptomatic while others develop symptoms with only moderate stenosis.

Pathophysiology

With reduction in the aortic valve area (AVA), energy is dissipated during the transport of blood from the LV to the aorta.

The AVA has to be reduced by about 50 percent of normal before a measurable gradient can be demonstrated in humans. The outflow obstruction imposes a pressure overload on the LV, which compensates by an increase in ventricular wall thickness and mass. The concentric hypertrophy normalizes systolic wall stress and preserves ventricular function. Even though systolic function may be preserved, however, abnormal diastolic compliance accompanies the development of concentric hypertrophy. A sustained pressure overload on the LV eventually will result in impaired contractility and mild to modest dilatation. Left atrial contraction is a considerable benefit to patients with AS. Loss of effective atrial contraction, because of either atrial fibrillation or an inappropriately timed atrial contraction, results in an elevation of mean left atrial pressure, reduction of cardiac output, or both; ineffective atrial contraction often precipitates clinical heart failure with pulmonary congestion.

In most patients with AS, cardiac output is in the normal range and initially increases appropriately with exercise. Later, as the severity of AS increases progressively, the cardiac output remains within the normal range at rest; during exercise, however, it no longer increases in proportion to the amount of exercise undertaken or does not increase at all. In severe AS, myocardial oxygen demands are increased because of an increased LV muscle mass, elevations in LV pressure, and prolongation of the systolic ejection time. As a result, patients may have classic angina pectoris in the absence of coronary artery disease.

Clinical Manifestations

HISTORY

Patients with congenital valve stenosis may give a history of a murmur since childhood or infancy; those with rheumatic stenosis may have a history of rheumatic fever. Most patients with valvular AS, including some with severe AS, are asymptomatic.

The characteristic triad of clinical symptoms of AS are angina pectoris, syncope, and heart failure. Sudden death may occur. Although symptoms tend to occur late in the course of this condition, 3 to 5 percent of patients may die suddenly during an asymptomatic period; however, most of these have had some cardiac symptoms before the fatal episode.

Angina pectoris usually is the initial clinical manifestation and most common, occurring in 50 to 70 percent of symptomatic patients. Syncope may result from reduced cerebral perfusion. Syncope occurring on effort is caused by systemic vasodilatation in the presence of a fixed or inadequate cardiac output, an arrhythmia, or both. This manifestation occurs much less commonly than angina, with a reported frequency of 15 to 30 percent in symptomatic patients. Dyspnea on exertion, orthopnea, paroxysmal nocturnal dyspnea, and pulmonary edema often result from varying degrees of pulmonary venous hypertension.

There is an increased incidence of gastrointestinal arteriovenous malformations in AS. As a result, these patients are susceptible to gastrointestinal hemorrhage and anemia. Calcific systemic embolism may occur.

PHYSICAL FINDINGS

There is a spectrum of physical findings in patients with AS, depending upon the severity of the stenosis, stroke volume, LV function, and the rigidity and calcification of the valve.

Typical physical findings of significant aortic stenosis include a delayed upstroke of the carotid arterial pulse (*parvus et tardus*), a diamond-shaped, harsh midsystolic murmur, and evidence of LV hypertrophy. The arterial pulse pressure may be narrowed and a systolic thrill may be felt in the carotid arteries.

The LV cardiac impulse is heaving and sustained in character and there may be a palpable fourth heart sound (S_4). An aortic systolic thrill is often present at the base of the heart. In 80 to 90 percent of adult patients with severe AS, there is an S_4 gallop sound, a midsystolic ejection murmur that peaks late in systole, and a single second heart sound (S_2) because A_2 and P_2 are superimposed or A_2 is absent or soft. The murmur commonly is loudest in the second right intercostal space, with radiation to the carotids and to the apex. Occasionally the murmur is loudest in the apical area, where it characteristically has a higher pitch. This is particularly common in elderly patients, where the murmur of AS frequently masquerades as a pure-frequency musical cooing murmur that is heard best at the apex.

A faint early diastolic murmur of minimal aortic regurgitation is often present. In the young patient with valvular aortic stenosis, a systolic ejection sound initiates the systolic murmur but later often disappears as AS becomes severe. The S_2 may be paradoxically split due to a late A_2.

In patients 60 years of age or older, the clinical features of aortic stenosis tend to be somewhat different from those typical of younger patients. Systemic hypertension is common, being present in about 20 percent of the patients, half of whom have moderate or severe systolic and diastolic hypertension. As may as 20 percent of these patients present in congestive heart failure. The male:female ratio is 2:1. Because of the thickening of the arterial wall and its lack of distensibility, the arterial pulse rises normally or even rapidly, and the pulse pressure can be normal or wide.

ELECTROCARDIOGRAM

Sinus rhythm is usual in valvular AS, regardless of severity. Most patients with significant AS have electrocardiographic (ECG) evidence of LV hypertrophy. Except in elderly patients, atrial fibrillation suggests coexisting mitral valve disease or coronary artery disease. The typical ECG of AS shows increased QRS voltage associated with depressed ST segments and T-wave inversion in the leads with the most prominent R waves. Conduction defects are common and range from first-degree heart block to left bundle branch block.

CHEST ROENTGENOGRAM

The characteristic finding is a normal-sized heart with a dilated proximal ascending aorta (poststenotic dilatation). Calcium in the aortic valve may be seen on the lateral film but is better appreciated by fluoroscopy. In the current era, calcification is most commonly recognized on two-dimensional (2D) echocardiography. Calcium in the aortic valve is the hallmark of AS in adults 40 to 45 years of age. Above age 45, the diagnosis of severe AS is doubtful if there is no calcium in the aortic valve. On the other hand, the presence of calcium does not necessarily mean that the valve is stenotic or that AS is severe. The cardiac silhouette on chest x-ray shows only mild to moderate enlargement if any. In patients with heart failure, the cardiac size becomes increased because of dilatation of the left ventricle and left atrium; the lung fields show pulmonary edema and pulmonary venous congestion, and the right ventricle and the right atrium may be dilated.

Laboratory Studies

ECHOCARDIOGRAPHY/DOPPLER

Echocradiography is an extremely important and useful noninvasive test. On the echocardiogram, the aortic valve leaflets normally are barely visible in systole, and the normal range of aortic valve opening is 1.6 to 2.6 cm. In the presence of a bicuspid valve, eccentric valve leaflets may be seen. The aortic valve leaflets may appear to be thickened as a result of calcification or fibrosis. The aortic valve may have a reduced opening; however, this also occurs in other conditions with a low cardiac output. The LV hypertrophy often results in thickening of both the interventricular septum and the posterior LV wall. The LV cavity size is usually normal. When LV systolic function is impaired, the LV and LA are dilated and the percent of LV dimensional shortening is reduced.

When properly applied, Doppler echocardiography is extremely useful for estimating the valve gradient and AVA noninvasively. The cornerstone of the ultrasound evaluation of AS is continuous-waveform Doppler interrogation through the aortic valve. The calculated gradient using the peak Doppler velocity [4 (AS velocity)^2] correlates closely with the peak instantaneous gradient measured at catheterization. Values for aortic valve area can be calculated using the continuity equation by measuring the velocity of the jet across the aortic valve with continuous-waveform Doppler; the velocity in the LV outflow tract just proximal to the valve with pulse-wave Doppler; and by deriving the aortic valve of the outflow tract from the diameter of the aortic annulus. Results from the continuity equation usually correlate well with the aortic valve calculated based on catheterization data and the Gorlin formula.

Transesophageal echocardiography is very useful in defining the aortic valve abnormality and assessing the severity of AS when adequate examination cannot be obtained with the transthoracic technique.

CARDIAC CATHETERIZATION AND ANGIOGRAPHY

Cardiac catheterization remains the usual technique for accurately assessing the severity of AS. When simultaneous LV and ascending aortic pressures are measured as well as the cardiac output by either the Fick principle or the indicator-dilution technique, the AVA can be calculated. The state of LV systolic pump function can also be quantitated by measuring LV enddiastolic and end-systolic volumes and ejection fraction. It must be recognized that ejection fraction may underestimate myocardial function in the presence of the increased afterload of severe AS.

The presence of coronary artery disease and its site and severity can be estimated only by selective coronary angiography, which should be performed in all patients 35 years of age or older who are being considered for valve surgery and in those below age 35 if there is a history of angina or two or more risk factors for premature coronary artery disease are present.

OTHER LABORATORY STUDIES

Gated blood pool radionuclide scans may provide information on ventricular ejection fraction (LVEF), similar to that provided by 2D echocardiography and LV cineangiography.

Exercise tests of any kind are usually not recommended for patients with severe AS unless there is a specific reason for such studies. Ambulatory ECG recordings may occasionally be informative in patients suspected of having arrhythmias or painless ischemia.

Natural History and Prognosis

The natural history of AS in the adult consists of a prolonged latent period during which morbidity and mortality are very low. The rate of progression to stenotic lesions has been estimated in a variety of hemodynamic studies performed largely in patients with moderate AS. Catheterization studies indicate that some patients exhibit a decrease in valve area that ranges from 0.1 to 0.3 cm^2 per year; the systolic pressure gradient across the valve may increase by as much as 10 to 15 mmHg each year. On the other hand, more than half of the reported patients had little or no progression over a 3- to 9-year period.

While it appears that the progression of AS can be more rapid in patients with degenerative calcific disease than in those with congenital or rheumatic fever, it is not possible to predict the rate of progression in an individual patient. Eventually, after a long latent period, symptoms of angina, syncope, or heart failure develop and the outlook changes dramatically. After the onset of symptoms, the average survival is less than 3 years. Thus, the development of symptoms identifies a critical point in the natural history of AS.

Medical Therapy

All patients with AS need antibiotic prophylaxis against infective endocarditis (Chap. 21). When the valve lesion is of rheumatic origin, additional prophylaxis against recurrence of rheumatic fever is indicated. Mechanical obstruction to LV outflow cannot be altered by medical therapy.

Patients with moderate AS should avoid moderate to severe physical exertion in competitive sports. If atrial fibrillation should occur in patients with mild or moderate AS, it should be reverted rapidly to sinus rhythm. In severe AS, reversion to sinus rhythm often becomes a matter of urgency.

Surgical Therapy

Operation should be advised for the symptomatic patient who has severe AS. In young patients, *if the valve is pliable and mobile*, simple commissurotomy or valve repair may be feasible; the operative mortality is less than 1 percent. It will relieve outflow obstruction to a major degree. In this minority of patients, catheter balloon valvuloplasty is the procedure of choice in centers with experienced and skilled staff and may postpone valve replacement for many years. Older patients and even young patients with calcified, rigid valves need valve replacement. Thus, catheter balloon valvuloplasty is a temporary palliative procedure for most adults and is reserved for high-risk elderly patients with advanced symptoms, particularly those undergoing emergency noncardiac surgical procedures who later become candidates for appropriate valve replacement.

The natural history of symptomatic patients with severe AS is dismal, with a 10-year mortality rate of 80 to 90 percent; however, there is good outcome after surgery, particularly in patients without any comorbid cardiac and noncardiac conditions. Given the unknown natural history of the asymptomatic patient with severe AS, it may be reasonable to recommend surgery even for certain asymptomatic patients. Clearly if the patient has LV dysfunction, then valve replacement should be performed. Some recommend valve replacement for most asymptomatic patients with severe AS, while others would recommend it for those with an AVA ≤ 0.75 cm^2 and in selected patients with an AVA of 0.76 to 1.0 cm^2.

The operative mortality of valve replacement is 5 percent or less. In patients without associated coronary artery disease, heart failure, or other comorbid factors, it may be equal to or less than 1 to 2 percent in centers with experienced and skilled staff. Patients with associated coronary artery disease should have coronary bypass surgery at the same time as valve surgery because it results in a lower operative and late mortality.

In severe AS, valve replacement results in an improvement of survival, even in those with normal preoperative LV function. Aortic valve replacement is not recommended for asymptomatic patients with severe aortic stenosis but normal LV function simply to prevent the occurrence of sudden death.

Two general types of valve replacement are available: mechanical and tissue prostheses (see Chap. 15). The mechanical prosthesis has the advantage of greater durability but the disadvantage of a propensity for thromboembolism. The tissue prosthesis lessens the risk of thromboembolism but tends to degenerate sooner. This degenerative process appears to occur more rapidly in younger patients. There is an increase in utilization of cryopreserved allografts to replace the aortic valve. Present data suggests that tissue viability of the allograft is maintained and that the degenerative changes are much less than with porcine prostheses. Patients with mechanical prostheses should be maintained on lifetime warfarin therapy (INR 2.5 to 3.5) (see Chap. 15).

AORTIC REGURGITATION

Definition, Etiology, and Pathology

Aortic regurgitation is a flow of blood in diastole from the aorta into the left ventricle (LV) due to incompetence of the aortic valve. The two most common causes of acute aortic regurgitation (AR) are infective endocarditis and prosthetic valve dysfunction. Common causes include dissection of the aorta, systemic hypertension, and trauma.

In North America, the most common cause of chronic, isolated, severe AR is aortic root/annular dilatation that is presumably the result of medial disease. Common causes include a congenital bicuspid valve, previous infective endocarditis, and rheumatic disease. Syphilis used to be a common cause, but it is now a rare one in North America. Chronic aortic regurgitation occurs in association with a variety of other diseases, particularly those that result in dilatation of the aortic root. Between 40 and 60 percent of the surgically removed valves from patients with isolated severe AR are classified as idiopathic. Half of these show histologic evidence of myxomatous changes. Incompetence of the aortic valve can result from alteration of the leaflets, including contracture scarring, perforation, redundance, loss of intrinsic rigidity, tissue destruction, or dilatation or dissection of the aortic wall.

In AR, volume overload of the LV is the basic hemodynamic abnormality. The extent of this volume overload depends on the volume of the regurgitant blood flow, which is determined by the area of the regurgitant orifice, the diastolic pressure gradient between the aorta and the LV, and the duration of diastole.

The LV diastolic pressure-volume relationship plays a very important role in the pathophysiology of *acute aortic valve regurgitation*. The ability of the left ventricle to dilate acutely is limited; as a result, the volume overload of severe AR produces a rapid increase in LV diastolic pressure. If the LV is somewhat dilated from a previous lesion, the LV pressure will initially rise more gradually with acute AR but may subsequently rise to the same high levels as those seen with a normal or stiff LV.

Acute AR that is mild produces little or no hemodynamic abnormality, while acute AR that is severe results in a large volume of regurgitant blood. In the latter case, the increased LV diastolic pressure results in increases in mean left atrial and pulmonary venous pressures and produces various degrees of pulmonary edema. Two compensatory mechanisms are utilized by the LV: an increase in myocardial contractility and a compensatory tachycardia to maintain an adequate forward cardiac output. As a result, the forward cardiac output may be appropriate initially. Subsequently LV systolic pump function may become abnormal, leading to more severe manifestations of clinical heart failure.

In *chronic aortic regurgitation*, the regurgitation becomes severe over a longer period of time and the LV diastolic pressure-volume relationships are different from those seen in acute AR. If AR is mild to moderate, the LV end-diastolic volume is increased moderately, the LV diastolic pressure-volume curve is moved to the right and the LV diastolic pressure is usually normal. In severe chronic AR, the LV end-diastolic volume can be quite large without significant elevations of LV end-diastolic pressure. If the LV diastolic volume increases further, however, the LV diastolic pressure will be increased.

Severe chronic AR results in a large regurgitant volume (a large percentage of LV stroke volume). The LV responds by dilating, with the dilatation proportional to the amount of the regurgitant volume. The subsequent large LV stroke volume produces LV systolic hypertension. Both of these increase LV wall stress, which can result in an impairment of LV function. In time, the hemodynamic burden imposed by volume overloading will result in depressed myocardial contractility and decreased LV compliance.

In severe AR, myocardial oxygen needs are increased because of the increase in LV diastolic and systolic volume, LV muscle mass (hypertrophy), and LV pressures as well as by prolongation of the systolic ejection time. Total coronary blood flow is increased. Coronary reserve is significantly reduced, however, and some patients with severe AR may complain of angina pectoris on effort even in the absence of coronary artery disease.

Clinical Manifestations

HISTORY

Patients with mild to moderate AR usually do not have symptoms that can be attributed to the heart. Even patients with severe chronic AR may be asymptomatic for many years. The onset of disability is usually insidious. The earliest symptoms may be an awareness of the increased force of cardiac contraction and of the beating of a dilated left ventricle that undergoes a large volume change in systole. The main symptoms of severe AR result from elevated pulmonary venous pressure and include dyspnea on exertion, orthopnea, and paroxysmal nocturnal dyspnea. When heart failure occurs, patients complain of fatigue and weakness. Angina pectoris occurs in 20 percent of patients and may be present even in the absence of coronary artery disease.

PHYSICAL FINDINGS

The physical findings in severe chronic AR are related to the wide arterial pulse pressure, a large LV diastolic volume, and the regurgitant blood flow. The absence of a wide pulse pressure (pulse pressure greater than 50 percent of peak systolic pressure) or the presence of a diastolic pressure greater than 70 mmHg in a patient without congestive heart failure makes *severe chronic aortic regurgitation* unlikely. The LV dilatation with severe AR displaces the apical impulse inferiorly and laterally. The regurgitant flow is characterized by a high-pitched diastolic blowing murmur along the left sternal border. A third heart sound (S_3) gallop and a low-pitched diastolic (Austin-Flint) rumble may be heard at the apex secondary to the regurgitant flow. The arterial pulse is very characteristic and consists of an abrupt distention with a rapid rise and a quick collapse. The arterial pulse may be bisferious, a double impulse during systole. The systolic arterial pressure is increased (usually above 145 mmHg) and the diastolic pressure is reduced (usually below 60 mmHg). Korotkoff's sounds often persists down to 0 mmHg during indirect blood pressure recordings. The wide pulse pressure and left ventricular dilatation of chronic AR are not features of *acute* AR.

ELECTROCARDIOGRAM

The ECG shows LV hypertrophy with or without associated ST-T wave changes. In some patients, ECG evidence of LV hypertrophy is absent in spite of severe AR. Conduction abnormalities such as atrioventricular (AV) block or left or right bundle branch block with or without QRS axis deviation may be present. The PR interval may be prolonged, particularly in patients with ankylosing spondylitis. The rhythm is usually sinus. The presence of atrial fibrillation should suggest associated mitral valve disease or heart failure.

CHEST ROENTGENOGRAM

The LV is increased in size, as indicated by an increase in the cardiothoracic ratio, and is often associated with a dilated ascending aorta. A better noninvasive quantitation of LV size can be obtained by echocardiography. There may be calcium in the aortic valve. With increased LV filling pressures in the later stages, there is often evidence of a large left atrium and increased left atrial and pulmonary venous pressure. This is manifest by redistribution of pulmonary blood flow, pulmonary congestion, or pulmonary edema. In the presence of heart failure, enlargement of the right atrium and superior vena cava may also be present. Calcification limited to the ascending aorta suggests luetic aortitis as the etiology of aortic regurgitation.

Laboratory Studies

ECHOCARDIOGRAPHY/DOPPLER

The echocardiogram can provide information concerning the etiology of the AR as well as regarding LV size and function (2D echo) and the severity of the regurgitation (Doppler). Diastolic fluttering of the anterior leaflet of the mitral valve is often present on M-mode and 2D echo. 2D echo is much superior to the M-mode technique for assessing LV volume and systolic function. A dilated, enlarged aorta can be detected by echocardiography, as can an enlarged left atrium. Aortic valve vegetations suggest infective endocarditis. There may be prolapse of the aortic leaflet into the LV during diastole. Doppler echocardiography is useful for diagnosing and assessing the severity of AR. There is an overlap between the various grades of AR severity by Doppler as compared with contrast aortic angiography. Transesophageal echocardiography is a useful technique when transthoracic echo is unsatisfactory and in certain instances for identifying the anatomy of the valve leaflets and aortic root/annulus. It is essential to determine if the valve is suitable for repair and is also important for assessing the presence or absence of dysfunction of other valves.

CARDIAC CATHETERIZATION AND ANGIOGRAPHY

Cardiac catheterization permits the measurement of intracardiac and intravascular pressures and of cardiac output both at rest and during exercise. In addition, other valvular disease—such as mitral stenosis, aortic stenosis, and mitral regurgitation—can be excluded. LV angiography demonstrates large LV volumes and allows the calculation of LV volumes and LV ejection fraction. Angiography performed with the ejection of contrast media in the ascending aorta demonstrates AR and allows a semiquantitative assessment of its severity. Also, the aortogram demonstrates the dimension of the aortic root and the state of the ascending aorta. The indications for selective coronary arteriography are the same as those for aortic stenosis.

OTHER LABORATORY STUDIES

Gated blood pool radionuclide scans allow the noninvasive measurement of LV volumes and ejection fraction. In addition, using this technique, it is possible to quantify the amount of AR. The radionuclide ventriculogram is particularly useful for evaluating the LV ejection fraction serially in patients who have AR, are initially asymptomatic, and have a normal LV ejection fraction and a mild to moderately dilated left ventricle.

A treadmill exercise test often provides an objective assessment of the degree of functional impairment and may document arrhythmias related to exertion. In some patients, however, the exercise test may remain normal despite deterioration of LV function. Ambulatory ECG recordings may be useful in an occasional patient suspected of having an arrhythmia. Magnetic resonance imaging can determine the severity of AR noninvasively but is rarely used clinically.

Natural History and Prognosis

Patients with mild AR that does not progress should have a normal life expectancy. The major risk is the development of infective endocarditis and further valve destruction. Patients with moderate AR, if their disease does not progress, will be expected to have a life expectancy that is reasonably close to the normal range. The disease does progress, however, and the mortality at the end of 10 years appears to be 15 percent.

Patients with severe AR are known to have a long asymptomatic period before the condition is discovered. In asymptomatic patients with normal LV function, symptoms and/or LV dysfunction develop at the rate of about 3 to 6 percent per year. LV systolic dysfunction at rest is the predictor of the subsequent development of symptoms. In patients with normal LV systolic function at rest, the predictors of the future development of LV systolic dysfunction and/or symptoms are an increased LV size (LV end-diastolic dimension equal to or greater than 70 mm and an LV end-systolic dimension equal to or greater than 50 MM). Sudden death in asymptomatic patients appears to occur only in those with a massively dilated left ventricle (LV end-diastolic dimension of $\geq$ 80 mm). The 5-year mortality of symptomatic patients with severe AR is about 25 percent, and the 10-year mortality averages 50 percent.

Medical Therapy

All patients with AR need antibiotic prophylaxis to prevent infective endocarditis (Chap. 21). Patients with AR of rheumatic origin require antibody prophylaxis to prevent recurrence of rheumatic carditis. Patients with syphilitic AR should receive a course of antibiotics to treat syphilis. Patients with mild AR need no specific therapy. Patients with moderate AR also usually need no specific therapy but are advised to avoid heavy physical exertion, competitive sports, and isometric exercise. Patients with *severe AR* who are asymptomatic and have normal LV function should be treated with a vasodilator (long-acting nifedipine or calcium channel blocking agent) to decrease the progression of disease unless there is contraindication to its use. Vasodilators are also indicated for short-term therapy in patients awaiting valve replacement to optimize their hemodynamics. Vasodilators are of considerable short-term benefit in patients with functional class III and IV heart failure. All such patients also need digitalis, diuretics, and ACE inhibitors.

Surgical Therapy

Patients with severe chronic AR need valve surgery. The correct timing of surgical therapy is now better defined but not fully clarified. Valve replacement should be performed before irreversible LV dysfunction occurs. Decisions about surgery in AR should be based on the clinical functional class and on the

LV ejection fraction. Patients with chronic, severe AR who are symptomatic need valve replacement. The benefit of valve replacement has been demonstrated even when the LV ejection fraction is reduced to 0.25 or less. Recent data indicate that patients with severe AR, a LV end-diastolic dimension on echocardiography of > 80 mm, and mild to moderate reduction of LV ejection fraction can often benefit from valve replacement. Postoperatively, they are symptomatically improved, the LV ejection fraction increases, and the LV size is reduced; the 5- and 10-year survivals are 87 and 71 percent respectively.

While this is controversial, most authorities recommend valve replacement for patients who are asymptomatic and who have a reduced LV ejection fraction at rest. Surgery is also recommended for asymptomatic patients who have severe obstructive coronary artery disease or need surgery for other valve disease.

Aortic valve replacement with or without associated coronary bypass for obstructive coronary artery disease can be performed at many surgical centers with an operative mortality of 5 percent or less. In patients without associated coronary artery disease or reduced LV systolic function, the operative mortality may be in the range of 1 to 2 percent. If aortic valve replacement is successful and uncomplicated, LV volumes and hypertrophy regress toward or return to normal. Impaired LV systolic pump function improves postoperatively in 50 percent or more of patients. The 5-year survival of patients undergoing aortic valve replacement in severe AR is 85 percent. The 5-year survival of patients with LV ejection fraction is equal to or greater than 0.45 is 87 percent versus 54 percent in patients with an ejection fraction less than 0.45. New techniques of aortic valve repair are being developed and evaluated; the early results are encouraging in selected subgroups. It is possible that, eventually, selected patients may do well with aortic valve repair rather than aortic valve replacement for AR.

SUGGESTED READING

Bonow RO: Asymptomatic aortic regurgitation: indications for operation. *J Card Surg* 1994; 9:170–173.

Bonow RO, Lakatos E, Maron BJ, Epstein SE: Serial long-term assessment of the natural history of asymptomatic patients with chronic aortic regurgitation and normal left ventricular systolic function. *Circulation* 1991; 84:1615–1635.

Bonow RO, Carabello B, de Leon AC Jr, et al: ACC/AHA guidelines for the management of valvular heart disease. *J Am Coll Cardiol* 1998. In press.

Bonow RO, Dodd JT, Maron BJ, et al: Long-term serial changes in left ventricular function and reversal of ventricular dilatation after valve replacement for chronic aortic regurgitation. *Circulation* 1988; 78:1108–1120.

Carabello BA, Usher BW, Hendrix GH, et al: Predictors of outcome for aortic valve replacement in patients with aortic regurgitation and left ventricular dysfunction: a change in the measuring stick. *J Am Coll Cardiol* 1987; 10: 991–997.

Connolly HM, Oh JK, Orszulak TA, et al: Aortic valve replacement for aortic stenosis with severe left ventricular dysfunction: prognostic indicators. *Circulation* 1997; 95: 2395–2400.

Gaasch WH, Sundaram M, Meyer TE: Managing asymptomatic patients with chronic aortic regurgitation. *Chest* 1997; 111:1702–1709.

Hammermeister KE, Sethi GK, Henderson WG, et al: A comparison of outcomes in men 11 years after heart-valve replacement with a mechanical valve or bioprosthesis. Veterans Affairs Cooperative Study on Valvular Heart Disease. *N Engl J Med* 1993; 328:1289–1296.

Horstkotte D, Loogen F: The natural history of aortic valve stenosis. *Eur Heart J* 1988; (9 suppl E): 57–64.

Otto CM, Pearlman AS, Gardner CL: Hemodynamic progression of aortic stenosis in adults assessed by Doppler echocardiography. *J Am Coll Cardiol* 1989; 13:545–550.

Pellikka PA, Nishimura RA, Bailey KR, Tajik AJ: The natural history of adults with asymptomatic, hemodynamically significant aortic stenosis. *J Am Coll Cardiol* 1990; 15: 1012–1017.

Rahimtoola SH: Aortic valve disease. In: Alexander RW, Schlant RC, Fuster V, O'Rourke RA, Roberts R, Sonnenblick EH (eds): *Hurst's the Heart*, 9th ed. New York, McGraw Hill, 1998:1753–1788.

Ross JJ, Braunwald E: Aortic stenosis. *Circulation* 1968; 36(suppl 5):61–67.

Sasayama S. Ross JJ, Franklin D, et al: Adaptations of the left ventricle to chronic pressure overload. *Circ Res* 1976; 38: 172–178.

Scognamiglio R, Rahimtoola SH, Fasoli G, et al: Nifedipine in asymptomatic patients with severe aortic regurgitation and normal left ventricular function. *N Engl J Med* 1994; 331:689–694.

Wood P: Aortic stenosis. *Am J Cardiol* 1958; 1:553–571.

13

MITRAL VALVE DISEASE

Robert A. O'Rourke

MITRAL STENOSIS

Definition, Etiology, and Pathology

Mitral stenosis (MS), an obstruction of blood flow between the left atrium (LA) and the left ventricle (LV), is caused by abnormal mitral valve function. In almost all adult cases, the cause of MS is previous rheumatic carditis. About 60 percent of patients with rheumatic mitral valve disease do not give a history of rheumatic fever or chorea, however, and about 50 percent of patients with acute rheumatic carditis do not eventually have clinical valvular disease. Other causes of MS are uncommon, but obstruction to LV inflow can be congenital or due to active infective endocarditis, neoplasm, massive annular calcification, systemic lupus erythermatosus, carcinoid, methysergide therapy, Hunter-Hurler syndromes, Fabry's disease, Whipple's disease, and rheumatoid arthritis. Other causes include a left atrium myxoma, massive left atrial ball thrombus, and cor atriatum, in which a congenital membrane is present in left atrium.

Rheumatic valvulitis results in scarring and contracture of the leaflets and chordae tendineae. There are often adhesions and fusions of the commissures. This process results in a funnel-shaped structure of the leaflets.

Acute rheumatic carditis is a pancarditis because it involves the pericardium, myocardium, and endocardium. In temperate climates and developed countries, there is usually a long interval (10 to 20 years) between an episode of rheumatic fever and the clinical presentation of symptomatic MS. In tropical and subtropical climates and in less developed countries, the latent period is often shorter and MS may occur during childhood or adolescence.

Pathophysiology

The pathophysiologic features of MS result from obstruction of the flow of blood between the LA and the LV. With reduction in valve area, energy is lost to friction during the transport of blood from the LA to the LV. Accordingly, a pressure gradient is present across the stenotic valve. The increase in LA pressure and volume is reflected back into the pulmonary venous system. Chronic elevation of the LA and pulmonary venous pressures causes hyperplasia and hypertrophy of pulmonary arteries. Pulmonary arterial (PA) hypertension eventually results. About 15 to 30 percent of patients develop a marked disproportionate elevation of pulmonary vascular resistance, much of which rapidly decreases following relief from mitral obstruction. This chronic mitral stenosis imposes a pressure overload in the left atrium and ultimately in the right ventricle. The relationship between valve area, cardiac output, flow period, and average diastolic gradient between the LA and the LV is defined by the formula of Gorlin and Gorlin.

Pulmonary venous hypertension alters lung function in several ways. Distribution of blood flow in the lung is altered, with a relative increase in flow to the upper lungs and therefore in physiologic dead space. Pulmonary compliance generally decreases with increasing pulmonary capillary pressure, thus increasing the work of breathing, particularly during exercise. Chronic changes in the pulmonary capillaries and arteries include fibrosis and thickening.

In some patients with high pulmonary vascular resistance and right ventricular (RV) dysfunction, cardiac output may be low. The body maintains oxygen consumption by extracting more oxygen from the arterial blood, and the mixed venous oxygen content falls. The reduced cardiac output may result in a surprisingly small pressure gradient across the mitral valve despite severe stenosis.

Long-standing MS with severe PA hypertension and RV dysfunction may be accompanied by chronic systemic venous hypertension. Tricuspid regurgitation is frequently present, even in the absence of intrinsic diseases of this valve. Chronic passive congestion in the liver often leads to central lobular necrosis and eventually to cardiac cirrhosis.

Clinical Manifestations

HISTORY

An asymptomatic interval is usually present between the initiating event of acute rheumatic fever and the presentation of symptomatic MS. Initially there is little or no gradient at rest, but with increased cardiac output, LA pressure rises and exertional dyspnea develops. As mitral obstruction increases, dyspnea occurs at lower workloads. The progression of disability is often so protracted that patients may adapt by circumscribing their lifestyles. It is important, therefore, to document what activities a patient can perform without symptoms and at what activity level symptoms begin. As obstruction progresses, patients note orthopnea and paroxysmal nocturnal dyspnea that apparently results from redistribution of blood to the thorax in the supine position. With severe MS and elevated pulmonary vascular resistance, fatigue rather than dyspnea may be the predominant symptom. Dependent edema, nausea, anorexia, and right-upper-quadrant pain reflect systemic venous congestion resulting from elevated systemic venous pressure.

Palpitations are a frequent complaint of patients with MS and may represent frequent premature atrial contractions or paroxysmal atrial fibrillation or flutter. Fifty percent or more of patients with severe symptomatic MS have chronic atrial fibrillation. Symptoms of RV failure (hepatomegaly, edema, and ascites) may predominate in patients with severe pulmonary hypertension. Hemoptysis, hoarseness, and exertional chest pain are dramatic but infrequent manifestations of mitral stenosis.

Systemic embolism, a frequent complication of MS, may result in stroke, occlusion of an extremity's arterial supply, occlusion of the aortic bifurcation, and visceral or myocardial infarction.

PHYSICAL FINDINGS

During the latent, presymptomatic interval, incidental physical findings may be normal or may provide evidence of mild MS. Often, the only characteristic finding noted at rest will be a loud S_1 and a presystolic murmur. A short diastolic decresendo rumble may be heard only with exercise. In patients with symptomatic stenosis, the findings are more obvious and a careful physical examination usually leads to the correct diagnosis.

The jugular venous pressure may be normal or may show evidence of elevated right atrial pressure. A prominent a wave is a result of RV hypertension or associated tricuspid stenosis. A prominent v wave is caused by tricuspid regurgitation. Atrial fibrillation produces an irregular venous pulse with absent a waves. The chest findings may be normal or may reveal signs of pulmonary congestion with rales or pleural fluid. Marked left atrial enlargement may produce egophony at the tip of the left scapula.

The precordium is usually unremarkable on inspection. On palpation, the apical impulse is normal or tapping. An abnormal LV impulse suggests disease other than isolated MS. A diastolic thrill is usually appreciated only when the patient is examined in the left lateral decubitus position. When PA hypertension is present, a sustained RV lift along the left sternal border and the sound of pulmonic valve closure (P_2) may be palpable. On auscultation in the supine position, the only abnormality appreciated may be the accentuated S_1, which is caused by flexible valve leaflets and the wide closing excursion of the valve leaflets. Failure to examine the patient in the left lateral decubitus position accounts for most failures to diagnose symptomatic MS. The diastolic rumble is heard best with the bell of the stethoscope applied at the apical impulse. Nevertheless, the murmur may be localized and the region around the apical impulse also should be auscultated. The open snap (OS) occurs when the movement of the domed mitral valve into the LV is suddenly stopped. It is heard best with the diaphragm of the stethoscope and is often most easily heard midway between the apex and the left sternal border. In this intermediate region, the S_1, P_2, and OS can be identified.

The OS occurs after the LV pressure falls below LA pressure in early diastole. When LA pressure is high, as in severe MS, the snap occurs earlier in diastole. Although the OS is present in most cases of MS, it is absent in patients with stiff, fibrotic, or calcified leaflets. Thus, absence of the OS in severe MS suggest that mitral valve replacement rather than commissurotomy may be necessary.

The low-pitched diastolic rumble follows the OS and is best heard with the bell of the stethoscope. In some patients with low cardiac output or mild MS, brief exercise, such as situps or walking, is adequate to increase flow and accentuate murmur. The murmur is low-pitched, rumbling, and decrescendo. In general, the more severe the MS, the longer the murmur. Presystolic accentuation of the murmur occurs in sinus rhythm and has been reported even in atrial fibrillation.

The two most important auscultatory signs of severe MS are a short A_2–OS interval and a pandiastolic rumble. The diastolic murmur may not be pandiastolic in severe MS if the stroke volume is low and there is no tachycardia.

Systolic murmurs may also be heard in association with the murmur of MS. A blowing, holosystolic murmur at the apex suggests associated mitral regurgitation, whereas a systolic blowing murmur heard best at the lower left sternal border that increases with inspiration usually signifies tricuspid regurgitation. The Graham Steell murmur is a high-pitched diastolic decrescendo murmur of pulmonic regurgitation caused by severe PA hypertension. In most patients with MS, such a murmur usually indicates aortic regurgitation. In general, with the possible exception of concomitant severe aortic regurgitation and/or significant LV systolic dysfunction, a left-sided third heart sound (S_3) is not compatible with severe MS. If an S_3 and a rumble are present, mitral regurgitation (MR) is usually the predominant lesion.

ELECTROCARDIOGRAM

Patients in sinus rhythm may have a widened P wave caused by interatrial conduction delay and/or prolonged LA depolarization. Classically, the P wave is broad and notched in lead II and biphasic, with an accentuated terminal negatively in lead V_1; it measures 0.12 s or more. Atrial fibrillation is common. LV hypertrophy is not present unless there are associated lesions. RV hypertrophy may be present if PA hypertension is marked.

CHEST ROENTGENOGRAM

The posteroanterior and lateral chest films are often so typical that they are diagnostic to clinicians. The thoracic cage is normal. The lung fields show evidence of elevated pulmonary venous pressure. Blood flow is more evenly redistributed to the upper lobes, resulting in apparent prominence of upper lobe vascularity. Increased pulmonary venous pressure results in transudation of fluid into the interstitium. Accumulation of fluid in the interlobular septa produces linear streaks in the bases that extend to the pleura (Kerley B lines). Interstitial fluid may also be seen as perivascular or peribronchial cuffing (Kerley A lines). With transudation of fluid into the alveolar spaces, alveolar pulmonary edema is seen. These changes indicate a long-standing elevated LA pressure. PA hypertension often results in enlargement of the main PA and right and left main pulmonary arteries.

The cardiac silhouette usually does not show generalized cardiomegaly, but the LA is always enlarged. In the posteroanterior film there is a density behind the RA border (double atrial shadow), prominence of the LA appendage on the left heart border between the main PA and LV apex, and elevation of the left main bronchus. The lateral film shows the LA bulging posteriorly. The RV may be enlarged if PA hypertension has been present. The combination of a normal-sized LV, enlarged LA, and pulmonary venous congestion should always suggest MS. Mitral valve calcification is occasionally noted on the plain chest x-ray.

Laboratory Studies

ECHOCARDIOGRAPHY

Echocardiography is the most reliable means of diagnosing mitral stenosis. The characteristic M-mode echocardiographic features are a decreased EF slope of the anterior mitral leaflet, anterior diastolic motion of the posterior leaflet, and thickened dense echoes indicative of calcification. Two-dimensional (2D) echocardiography will demonstrate the valve orifice and allow calculation of the mitral valve area. Doppler echocardiography will provide estimates of the gradient across the valve and of the pulmonary artery pressures.

The maximal gradient across the mitral valve can be calculated from the peak diastolic LV inflow velocity utilizing the Bernoulli equation. In addition, Doppler echocardiography may provide estimates of mitral valve area by calculating the half-time of the mitral velocity pressure. As the severity of mitral stenosis increases, the rate of deceleration decreases, prolonging the pressure half-time. Dividing an empiric constant of 220 by the pressure-time yields an estimate of the mitral valve area that correlates with values obtained during cardiac catheterization. If tricuspid regurgitation can be documented by continuous-waveform Doppler, the RV systolic pressure can be estimated from the following formula:

$$(4 \times \text{peak tricuspid velocity}^2)$$
$$+ \text{ the estimated right atrial pressure}$$

Transesophageal echocardiography is a useful technique for detecting left atrial thrombus, determining the anatomy of the mitral valve and subvalvular apparatus, and assessing the suitability of the patient for catheter balloon mitral commissurotomy or surgical valve repair.

CARDIAC CATHETERIZATION AND ANGIOGRAPHY

In most patients with symptoms from presumed MS, right and left heart catheterization should be performed as part of a preoperative assessment. Simultaneous measurement of cardiac output and the gradient between the LA and LV and the subsequent calculation of the valve area remains the "gold standard" for assessing the severity of MS. LV angiography assesses the competence of the valve, an important determination of operability for mitral commissurotomy. Quantification of LV function provides a useful prognostic indicator of operative late survival and of the expected functional result. Aortic valve function should be evaluated in all patients. Selective supraventricular angiography should be performed in all patients unless there is a contraindication. Tricuspid valve function can be assessed when there is a question of coexisting lesions. In certain circumstances, dynamic exercise in the catheterization laboratory with measurement of mitral valve gradient, cardiac output, and LA and PA pressures can be extremely useful. Selective coronary arteriography establishes site, severity, and extent of coronary artery disease and should be performed in patients with angina, in those with multiple risk factors for coronary artery disease, and in those 35 years of age and older who are being considered for interventional therapy.

OTHER LABORATORY STUDIES

In most clinical situations, other laboratory tests are not needed. Occasionally, a treadmill exercise test to evaluate functional capacity may be very useful clinically, particularly when a patient denies symptoms in spite of severe hemodynamic abnormalities.

Natural History and Prognosis

In the United States, most cases of acute rheumatic fever occur in the early teenage years. Commonly, another 10 years elapse before a murmur is audible. Symptoms will then begin in another 10 years and will progress to the point that relief of mitral valve obstruction is required in the fourth to fifth decade of life. Approximately 50 percent of patients develop symptoms gradually; the rest experience a sudden deterioration in their clinical condition. Atrial fibrillation, fever, emotional stress, or pregnancy can produce this deterioration. Atrial fibrillation develops in about 40 percent of patients with mitral stenosis. An important threat to patients with mitral stenosis who develop atrial fibrillation is systemic embolization.

The 10-year survival of patients with asymptomatic MS is approximately 84 percent, and that of those who are mildly symptomatic is 34 to 42 percent. Patients in the New York Heart Association functional class IV have a very poor survival without treatment. Forty-two percent survive for 1 year and 10 percent or less survive for 5 years. All are dead by 10 years.

Medical Treatment

Prophylaxis against recurrence of rheumatic fever (Table 13-1) should be continued until the age of 35 or longer if the patient is in frequent contact with children in the "strep throat" age group. Endocarditis prophylaxis (Chap. 21) is a lifelong requirement. If atrial fibrillation develops, the preferred drug for rate control is digoxin. Rate control during exercise is frequently not adequate when digitalis is used alone, however, and it is usually necessary to add a rate-slowing calcium antagonist, a beta blocker, or amiodarone. Digoxin and diltiazem or digoxin and low-dose amiodarone are probably the two best combination regimens. Diuretics reduce pulmonary congestion and peripheral edema and allow most patients freedom from severe salt restriction. For the patient with mild symptoms, the maintenance of sinus rhythm is desirable. In patients who need interventional therapy, cardioversion is usually performed at the completion of the procedure. Anticoagulation with warfarin is usually begun about 3 weeks prior to cardioversion and continued for 4 weeks after the procedure. Patients with chronic atrial fibrillation and those with a previous history of embolism should receive anticoagulation with warfarin (INR of 2 to 3) unless there is a specific contraindication. An alternative approach that enables earlier cardioversion is to exclude left atrial thrombi by echocardiography, institute 2 to 3 days of intravenous heparin therapy, cardiovert the patient to sinus rhythm, and continue warfarin therapy for at least 4 weeks. Systemic embolization necessitates permanent anticoagulation. A single systemic embolic episode is *not* an absolute indication for mitral valve surgery; emboli can occur in milder forms of mitral stenosis.

TABLE 13-1

PROPHYLAXIS AGAINST RHEUMATIC FEVER

Penicillin V, 250 mg twice daily PO
If allergic to penicillin, sulfadiazine 1 g daily PO
If allergic to both penicillin and sulfadiazine, erythromycin 250 mg twice daily PO
Benzathine penicillin G, 1.2 million units intramuscularly every 4 weeks (every 3 weeks for high-risk
subjects)

Interventional Therapy

Unless there is a contraindication, surgery or catheter balloon commissurotomy (CBC) should be recommended to an MS patient with functional class III or IV symptoms. For younger patients with pliable, noncalcified valves and without important mitral regurgitation, this means CBC or surgical valve repair. The hemodynamic results of CBC or surgical commissurotomy are excellent. Because of the low mortality and morbidity of mitral commissurotomy/valve repair, CBC or surgery is also offered to patients with functional class II symptoms. The results of successful surgical commissurotomy are excellent; in centers with experienced and skilled staff, surgical mortality is less than 1 percent. Late mortality at 10 years is less than 5 percent, the thromboembolic rate is 2 percent per year or less, and the reoperation rate ranges from 0.5 to 4.5 percent per year.

For the older patient with a stiff or calcified valve or when moderate mitral regurgitation is present, mitral valve replacement is usually performed. Valve replacement carries a higher operative mortality than does commissurotomy (up to 5 percent) as well as the morbidity associated with valve prostheses. Hemodynamic results of mitral valve replacement are often not ideal. Nevertheless, survival at 10 years after mitral valve replacement for functional class III and IV patients is still better than 60 percent.

Use of the double-balloon technique or the Inoue balloon with CBC produces immediate and 3-month hemodynamic and clinical results comparable to those obtained by surgical commissurotomy. The mitral valve area increases from a mean of 1.0 to 2.0 cm. There are reductions of LA and PA pressures at rest and exercise and an increase of exercise capacity. The immediate results of CBC are greatly influenced by the characteristics of the valve and its supporting apparatus; these are best determined by 2D echocardiography (transthoracic and/or transesophageal). Repeat CBC or mitral valve replacement is needed in 20 percent of patients within 5 to 7 years. Late survival is poorer in patients in whom functional class IV, higher echocardiographic score, higher left ventricular end-diastolic pressure, or higher PA systolic pressure is present prior to CBC.

Specifically, mitral CBC is recommended for *symptomatic* patients with moderate or severe MS (mitral valve area equal to or less than 1.5 cm^2) and a valve morphology favorable for percutaneous balloon valvulotomy in the absence of left atrial thrombus or moderate to severe MR. It is also frequently recommended for *asymptomatic* patients with moderate or severe MS and a valve morphology favorable for percutaneous balloon valvulotomy when there is objective evidence of hemodynamic compromise (PA systolic pressure greater than 50 mmHg at rest or 60 mmHg with exercise in the absence of left atrial thrombus or moderate to severe mitral regurgitation). CBC has also been recommended by some for *asymptomatic* patients with moderate to severe MS, favorable valve morphology, and the new onset of atrial fibrillation. In the symptomatic patient with a valve morphology favorable for repair who has a left atrial thrombus present despite anticoagulation, surgical mitral valve repair is usually recommended instead. In patients with moderate to severe MS who have class III to IV symptoms and are not considered candidates for CBC or mitral valve repair, valve replacement is recommended.

MITRAL REGURGITATION

Definition, Etiology, and Pathology

Mitral regurgitation (MR) is characterized by an abnormal reversal of blood flow from the LV to the LA due to abnormalities in the mitral apparatus. While rheumatic valvulitis remains an important cause of mitral regurgitation, it is no longer the most common etiology. Mitral valve prolapse is probably now the most common cause. Among patients requiring mitral valve replacement for isolated mitral regurgitation, myxomatous transformation of the leaflets is the most common finding. Many of these patients have associated rupture of the chordae tendineae.

Other common causes of MR are papillary muscle ischemia or infarction secondary to coronary artery disease, dilatation of the left ventricle, and infective endocarditis. Additional causes of MR include mitral annular calcification, hypertrophic cardiomyopathy, trauma, inherent connective tissue disorders, and congenital deformities. The competence of the mitral valve depends on the normal structure and function of every part of the mitral apparatus including the leaflets, chordae tendineae, annulus, left atrium, papillary muscles, and left ventricular myocardium surrounding the papillary muscles. All cases of MR are associated with an abnormality in at least one of these components. Recently, there has been a reported association of MR as well as aortic regurgitation and tricuspid regurgitation with the use of several anorectic drugs, including fenfluramine, dexfenfluramine, and the combination of either drug with phenteramine. The frequency of this association, its mechanism, and its natural history are undergoing further study.

Pathophysiology

In *chronic* MR, the portion of the LV stroke volume ejected into the LA determines the degree of LA and LV enlargement. The LV responds to the chronic volume overload by dilatation. The dilatation is accompanied by increased wall thickness and hypertrophy, which help maintain mechanical function. The diastolic volume of the LV in MR includes the systolic output of the right ventricle and the volume regurgitated into the LA during the previous systole. The elastic recoil of the atrium and the increased compliance of the LV often allow rapid diastolic filling of the LV without a significant increase in the LV, LA, and pulmonary capillary diastolic pressures.

In *acute* MR the hemodynamic burden is different. The sudden regurgitation imposed on normal chambers does not allow compensatory dilatation of the left atrium and the left ventricle. Consequently, marked elevations of the LA and pulmonary capillary pressures are produced. The pressure overload on the pulmonary vascular circulation produces acute pulmonary edema.

LV dysfunction is a frequent and dismal complication of MR. Although interstitial fibrosis is present in advanced LV failure, the exact mechanism of LV dysfunction remains obscure. The changes in myofiber contractility parallel the changes in global LV function and are associated with reduced myofiber content. During diastole, LV relaxation is frequently abnormal, but chamber stiffness is usually reduced. Age and decreased systolic function are associated with increased chamber stiffness.

Clinical Manifestations

HISTORY

Patients with MR usually have no symptoms. Severe MR may be associated with no or minimal symptoms for years. Fatigue due to low cardiac output and mild dyspnea on exertion are the most usual symptoms and are rapidly improved by rest. Symptoms of fatigue and mild dyspnea progress to orthopnea, paroxysmal nocturnal dyspnea, and peripheral edema. Often the history provides clues to the cause of the MR. A history of myocardial infarction or angina suggests papillary muscle involvement. Sudden appearance or sudden worsening of symptoms in a middle-aged male suggests chordal rupture. A patient with an enlarged heart and chronic symptoms of heart failure prior to the detection of mitral regurgitation probably has dilated cardiomyopathy. A remote history of rheumatic fever suggests rheumatic heart disease; a recent history of fever raises the possibility of infective endocarditis. Severe dyspnea on exertion, paroxysmal nocturnal dyspnea, frank pulmonary edema, or even hemoptysis is observed later in the course of the disease. Such symptoms may be triggered by the new onset of atrial fibrillation, an increase in the extent of regurgitation, endocarditis or ruptured chordae, or change in LV compliance or function.

In patients with severe regurgitation of *acute onset*, symptoms are usually more dramatic; they include pulmonary edema or congestive heart failure but may progressively subside with administration of diuretic and vasodilators and increased compliance of the left atrium.

PHYSICAL FINDINGS

The blood pressure is usually normal. The arterial pulse upstroke is brisk, particularly when there is a markedly reduced ejection time in patients with MR. The LV cardiac apex impulse is often laterally displaced, diffuse, and brief in patients with an enlarged LV. An apical systolic thrill is characteristic of severe regurgitation. A left sternal border lift is observed in patients with RV dilatation in early systole and may be difficult to distinguish from an LA lift in later systole due to the marked filling and expansion of the left atrium that is associated with severe MR. The first heart sound is often obscured by the murmur and is usually normal in intensity. The second heart sound is usually normal, but it may be paradoxically (P_2, A_2) split with expiration if the LV ejection time is markedly shortened. The presence of a third heart sound (S_3) is directly related to the volume of the regurgitation in patients with organic MR. MR often is associated with an early diastolic rumble due to the increased mitral flow in diastole; at times the rumble is prolonged even in the absence of MS. The S_3 sound and diastolic rumble are low-pitched and may be difficult to detect without careful auscultation performed with the patient in the left lateral decubitus position. The S_3 may be increased with expiration. In patients with ischemic and/or functional MR, the S_3 corresponds more often to restrictive LV filling. A fourth heart sound (S_4) is heard mainly in MR of recent onset in patients in sinus rhythm. Midsystolic clicks are markers of mitral valve prolapse (discussed later) and are due to the sudden tension of the chordae. The hallmark of MR is the systolic murmur. In regurgitation of at least moderate degree, the murmur is holosystolic, including the first and second heart sounds. If an OS or S_3 is mistakenly interpreted as being the second heart sound, the murmur may appear to be midsystolic. The murmur is usually high-pitched and blowing but may be harsh, especially in mitral valve prolapse. The maximum loudness of the murmur is usually at the apex; it may radiate to the axilla in patients where the anterior leaflet results in greater regurgitation and to the left sternal border when the posterior leaflet is primarily affected. With prolapse or ruptured chordae of the anterior leaflet, the murmur can only be heard in the back, in the neck, and sometimes on the skull. In cases where the murmur radiates to the base, it may be difficult to distinguish from the murmurs of aortic stenosis or hypertrophic cardiomyopathy. With increases in LV afterload such as handgrip exercise, the murmur is usually louder. By contrast, with reductions of afterload or LV size, the murmur becomes softer (e.g., vasodilators). The murmur does not increase with postextrasystolic beats, probably because the regurgitant flow is unchanged due to the combination of a decreased regurgitant orifice and increased regurgitant gradient. Murmurs of shorter duration usually correspond to mild regurgitation; these may be mid- or late systolic in patients with mitral valve prolapse or early systolic in patients with functional MR.

ELECTROCARDIOGRAM

Chronic mitral regurgitation produces LA or LV enlargement typically manifest by increased amplitude of the P waves and QRS complex. If atrial fibrillation is present, the left atrial enlargement is usually associated with a coarse fibrillatory pattern. When papillary muscle ischemia or infarction is the cause of the regurgitation, evidence of previous inferior or posterior myocardial infarction is often present. Right ventricular hypertrophy is uncommon.

CHEST ROENTGENOGRAM

In chronic severe MR, the chest roentgenogram reveals left atrial and left ventricular enlargement. With rheumatic disease, calcification of the leaflets may be visible; with degenerative disease, a calcified mitral annulus often is evident. Acute, severe MR is usually manifest by a normal cardiac size and pulmonary edema.

Laboratory Studies

ECHOCARDIOGRAPHY/DOPPLER

The echocardiogram is usually helpful in defining the etiology of the MR (e.g., flail leaflets, severe prolapse, mitral annulus calcification, systolic anterior motion of anterior leaflet, and endocarditis vegetation) and determining its consequences. The echo Doppler technique provides an estimate of the severity of the regurgitation by assessing the velocity, width, and length of the regurgitant jet. Color-flow imaging demonstrates the origin and direction of the jet. Accordingly, the jet length, the ratio of the jet area to the left atrial area, or more simply the jet area have been suggested as good indices of the severity of MR. Small jets such as those seen in normal subjects consistently correspond to mild regurgitations. Color-flow imaging for defining regurgitant lesions has significant limitations that are intrinsically related to the nature of regurgitant jets. The extent of a jet is determined by its momentum, and thus as much by regurgitant velocity as by regurgitant flow. Also, jets are constrained by the LA and expand more in large atria. The eccentric jets of valvular prolapse depend on the left atrial wall and tend to underestimate regurgitation. In contrast, the central jets of ischemic and functional MR expand markedly in a large atrium and tend to overestimate regurgitation. Transesophageal echocardiography usually shows larger jets but does not eliminate these limitations of color-flow imaging. The pulmonary venous velocity profile is useful to assess the degree of regurgitation. Systolic reversal of flow in the pulmonary veins is a strong argument for severe MR. Several quantitative methods have been used to measure parameters that reflect the degree of MR; however, the reliability of these techniques for the quantitative assessment of MR remains to be demonstrated.

CARDIAC CATHETERIZATION AND ANGIOGRAPHY

Cardiac catheterization is utilized to assess the hemodynamic status, severity of MR, left ventricular function, and coronary artery anatomy. It confirms the diagnosis of mitral regurgitation as well. A large v wave in the pulmonary capillary wedge pressure tracings suggest MR, but its absence does not exclude MR. A balloon flotation catheter, inserted at the bedside to determine oxygen saturation in the right heart chambers and the presence or absence of v wave in the pulmonary capillary pressures, is helpful in establishing the cause of a new systolic murmur that develops in a patient after acute myocardial infarction. The assessment of the degree of regurgitation can be obtained by LV contrast angiography and can be qualitatively graded in three or four grades on the basis of the degree and persistence of opacification of the left atrium. The assessment of LV function can be performed using quantitative angiography. LV volumes are determined by the regurgitant volume, duration of regurgitation, etiology of regurgitation, and LV function. The most frequently utilized indices of LV function are the end-systolic volume and the LV ventricular ejection fraction. Both have been shown to be useful prognostically. The hemodynamic response to exercise (e.g., cardiac output, pulmonary artery pressure) often indicate the need for valve replacement.

Regional wall motion abnormalities have been observed in patients with MR even in the absence of coronary lesions. Selected coronary angiography is at present the only technique for defining the coronary artery anatomy. It is usually performed in patients above 35 years of age or in those with angina or multiple risk factors of coronary artery disease.

OTHER LABORATORY STUDIES

Radionuclide angiography can be used to estimate the left ventricular end-diastolic and end-systolic volume as well as the right and left ventricular ejection fractions. The detection of exercise-induced LV dysfunction is frequent; however, the significance of such measurements on the long-term prognosis has not been analyzed in large series of patients. The comparison of the counts measured over the RV and LV allows the calculation of the mitral valve regurgitant fraction. Exercise testing is often useful for determining the patient's exercise capacity, particularly in patients who appear relatively asymptomatic despite severe MR.

Natural History

Because of the qualitative and imprecise assessment of the degree of regurgitation, the natural history of MR is poorly defined. Patients with mild rheumatic MR appear to have a good prognosis. The prognosis of patients with mitral valve prolapse and no or mild regurgitation is usually excellent. Some deaths may occur in patients with murmurs of MR and more often when LV function is markedly decreased.

The predictors of poor outcome in patients with MR who are treated medically include severe symptoms (classes III to IV) even if the symptoms are transient, pulmonary hypertension, markedly increased LV end-diastolic volume, decreased cardiac output, and reduced LV ejection fraction. A comparison of the outcome of medically and surgically treated patients shows a trend in favor of the surgical treatment, especially early surgery, with a definite improvement of outcome with surgery in patients who have decreased systolic LV function.

Medical Treatment

Prevention of infective endocarditis is necessary in patients with MR. Young patients with rheumatic MR should receive rheumatic fever prophylaxis. In patients with atrial fibrillation, rate control is achieved using digoxin and/or beta blockers, diltiazem, and amiodarone. Long-term maintenance of sinus rhythm after cardioversion in patients with severe MR or enlarged LA is usually not possible in those who are treated medically. Oral anticoagulation should be used in patients with atrial fibrillation.

Afterload reduction decreases the amount of regurgitation not only by reducing the LV systolic pressure but also by decreasing the effective regurgitant orifice area. The acute utilization of sodium nitroprusside in unstable patients with severe MR, especially in the context of myocardial infarction, may be lifesaving in patients being prepared for mitral valve surgery. Chronic afterload reduction is more controversial. Diuretic treatment is extremely useful for the control of heart failure and for the chronic control of symptoms, especially dyspnea.

Surgical Treatment

Mitral valve reconstruction for MR is often possible. The frequency with which valve repair can be used in patients with MR varies with the experience of the operating team and the spectrum of underlying valve disease; repair is more often feasible in patients with degenerative valve disease than in those with regurgitation caused by rheumatic valvulitis or endocarditis. LV systolic function and late survival in general are better with mitral valve repair than with mitral valve replacement because of the lesser decline or maintenance of normal LV function when the chordae are preserved at the time of surgery.

In patients whose mitral valves cannot be repaired, mitral valve replacement with chordal preservation is less likely to depress LV function than mitral valve replacement without preservation of the chordae tendineae. Patients with severe symptoms due to MR should be treated surgically even if symptoms are markedly improved by medical treatment. Patients who are functional class I or II but with signs of overt LV dysfunction (LV ejection fraction <60 percent, end-systolic diameter >45 mm) should be treated surgically particularly if they are candidates for valve repair or valve replacement with chordal preservation. In patients with *severe* mitral regurgitation who have no or minimal symptoms and no signs of LV dysfunction, surgery is a reasonable option when it is likely that the mitral valve can be repaired with chordal preservation. This pertains to patients with a low operative risk of 1 to 2 percent and valvular lesions that can be repaired as indicated by echocardiography. Intraoperative transesophageal echo should be performed by experienced physicians to monitor the repair procedure and help with decisions warranted by an imperfect result.

Patients who have no symptoms due to severe MR and normal LV systolic function who are candidates for mitral valve repair and have severe pulmonary hypertension at rest or with exercise, atrial fibrillation, or recurrent thromboemboli despite anticoagulation therapy are commonly recommended for early surgery if mitral valve repair with preservation of the chordae is the likely procedure.

MITRAL VALVE PROLAPSE

Definition, Etiology, and Pathology

Mitral valve prolapse (MVP) refers to the systolic billowing of one or both mitral leaflets into the left atrium with or without mitral regurgitation. MVP often occurs as a clinical entity with no or only mild MR; it is frequently associated with unique clinical characteristics when compared with the other causes of MR.

In primary MVP there is interchordal hooding due to leaflet redundancy involving both the rough and clear zones of the involved leaflets. The basic microscopic features of primary MVP is a marked proliferation of the spongiosa, the delicate myxomatous connective tissue between the atrials (a thick layer of collagen and elastic tissue forming the atrial support of the leaflets) and the fibrosa or ventricularis, which is composed of dense layers of collagen and forms the basic support of the leaflet. In primary MVP, myxomatous proliferation of the mucopolysaccharide-containing spongiosa tissue causes local interruption of the fibrosa. Secondary effects include fibrosis of the surface of the mitral valve leaflets and/or elongation of the chordae tendineae as well as ventricular friction lesions. Fibrin deposits often form at the mitral valve–left atrial angle. The primary form of MVP may occur in families where it appears to be inherited as an autosomal dominant trait with varying penetrance. Primary MVP has been found with increased frequency in patients with Marfan's syndrome and in other inheritable connective tissue diseases.

A second form of MVP occurs in which myxomatous proliferation of the spongiosa portion of the mitral valve is absent. Echocardiographic evidence of MVP, often with mitral regurgitation, can be produced in closed-chest dogs undergoing transient coronary artery occlusion. Occasionally, unequivocal MVP has been detected de novo in patients after acute coronary syndromes; however, in most patients with coronary artery disease and MVP, the two entities are coincident but unrelated.

Several recent studies indicate that valvular regurgitation caused by MVP may result from postinflammatory changes, including those following rheumatic fever. Mitral valve prolapse has also been observed in patients with hypertrophic cardiomyopathy and posterior MVP may result from a disproportionally small left ventricular cavity, altered papillary muscle alignment, or a combination of factors. Patients with primary and secondary MVP must be distinguished from those with normal variations on cardiac auscultation or echocardiography that are misinterpreted to represent the MVP syndrome.

Pathophysiology

In patients with MVP, there is frequently LA and LV enlargement, depending upon the presence and severity of MR. The supporting apparatus is often involved. In a patient with connective tissue diseases, the mitral annulus is usually dilated, sometimes calcified, and does not decrease the circumference by the usual 30 percent during LV systole. The effects of mild to moderate MR or cardiac function are similar to those from other causes of mitral regurgitation.

Clinical Manifestations

HISTORY

The diagnosis of MVP is most commonly made by cardiac auscultation in asymptomatic patients or by echocardiography being performed for some other purpose. A patient may be elevated because of a family history of cardiac disease or occasionally by being referred because of an abnormal resting electrocardiogram. The most common presenting complaint is palpitations, which are usually due to ventricular premature beats. Supraventricular arrhythmias are also frequent, and the most common sustained tachycardia is paroxysmal reentry supraventricular tachycardia.

Chest pain is a frequent complaint of patients with MVP. It is atypical in most patients without coexistent ischemic heart disease and rarely resembles classic angina pectoris. Dyspnea and fatigue are frequent symptoms in patients with MVP, including many without severe MR. Objective exercise testing often fails to show an impairment in exercise tolerance, and some patients exhibit distinct episodes of hyperventilation. Neuropsychiatric complaints are not uncommon in patients with MVP. Some patients have panic attacks and others frank manic-depressive syndromes. Transcerebral ischemic episodes occur with increased incidence in patients with MVP, and some patients develop stroke syndromes.

PHYSICAL FINDINGS

The presence of thoracic skeletal abnormalities—the most common being scoliosis, pectus excavatum, straightened thoracic spine, and narrowed anteroposterior diameter of the chest—may suggest the diagnosis of MVP.

The principal cardiac auscultatory feature of this syndrome is the midsystolic click, a high-pitched sound of short duration. The click may vary considerably in intensity and location in systole according to LV loading conditions and contractility. It results from the sudden tension of the mitral valve apparatus as the leaflets prolapse into the LA during systole. Multiple systolic clicks may be generated by different portions of the mitral leaflets prolapsing at various times during systole. The major differentiating feature of the midsystolic click of MVP from that due to other causes is that its timing during systole may be altered by maneuvers that change hemodynamic conditions.

Dynamic auscultation is often useful for establishing the clinical diagnosis of the MVP syndrome. Changes in the LV end-diastolic volume lead to changes in the timing of the midsystolic click and murmur. When end-diastolic volume is decreased, the critical volume is achieved earlier in systole and the click-murmur complex occurs shortly after the first heart sound. In general, any maneuver that decreases the end-diastolic LV volume, increases the rate of ventricular contraction, or decreases the resistance to LV ejection of blood causes the MVP to occur early in systole and the systolic click and murmur to move toward the first heart sound. This occurs when the patient suddenly stands, performs submaximal handgrip exercise, or performs the Valsalva maneuver.

ELECTROCARDIOGRAM

The electrocardiogram (ECG) is usually normal in patients with MVP. The most common abnormality in the MVP syndrome is the presence of ST-T-wave depression or T-wave inversion in the inferior leads. MVP is associated with an increased incidence of false-positive exercise ECG results in patients with normal coronary arteries, especially females.

Although arrhythmias may be observed on a resting ECG or during treadmill or bicycle exercise, they are detected more frequently by continuous ambulatory ECG recordings. Most of the arrhythmias detected, however, are not life-threatening and often do not correlate with the patient's symptoms.

CHEST ROENTGENOGRAM

Posteroanterior and lateral chest films usually show normal cardiopulmonary findings. The skeletal abnormalities described above can be seen. When severe MR is present, both LA and LV enlargement often result. Various degrees of pulmonary venous congestion are evident when LV failure results. Acute chordal rupture with a sudden increase in the amount of mitral regurgitation may present as pulmonary edema without obvious LV or LA dilation. Calcification of the mitral annulus may be seen, particularly in adults with Marfan's syndrome.

Laboratory Studies

ELECTROCARDIOGRAPHY

2D Doppler echocardiography is the most useful noninvasive test for defining MVP. The M-mode echocardiographic definition of MVP includes equal to or greater than 2 mm of posterior displacement of one or both leaflets or holosystolic posterior "hamocking" equal to or greater than 3 mm. On 2D echocardiography, systolic displacement of one or both mitral leaflets in the parasternal long-axis view, particularly when they coapt on the left atrial side of the annular plane, indicates a high likelihood of MVP. There is disagreement concerning the reliability of echocardiographic diagnosis of MVP when observed only in the apical four-chamber view (see below). The diagnosis of MVP is even more certain when the leaflet thickness is >5 mm. Leaflet redundancy is often associated with an enlarged mitral annulus and elongated chordae tendineae. On Doppler velocity recordings, the presence or absence of MR is an important consideration, and MVP is more likely when the MR is detected as a high-velocity eccentric jet in late systole, midway or more posterior in the LA.

At present, there is no consensus on the 2D echocardiographic criteria for MVP. Since echocardiography is a tomographic cross-sectional technique, no single view should be considered diagnostic. The parasternal long-axis view permits visualization of the medial aspect of the anterior mitral leaflet and middle scallop of the posterior leaflet. If the findings of prolapse are localized to the lateral scallop in the posterior leaflet, they would be best visualized by the apical four-chamber view. All available echocardiographic views should be utilized, with the provision that billowing of the anterior leaflet alone in the four-chamber view is not evidence of prolapse; however, a displacement of the posterior leaflet or the coaptation point in any view including the apical views suggests the diagnosis of prolapse. The echocardiographic criteria for MVP should include structural changes such as leaflet thickening, redundancy, annular dilatation, and chordal elongation.

Patients with echocardiographic evidence for MVP but without evidence of thickened/redundant leaflets or definite MR are more difficult to classify. If such patients have auscultatory findings of MVP, then the echocardiogram usually confirms the diagnosis.

CARDIAC CATHETERIZATION AND ANGIOGRAPHY

Cardiac catheterization is rarely used as a diagnostic technique for MVP. Also, contrast ventriculography is unnecessary for determining LV function, since it can usually be determined by 2D echocardiography or radionuclide ventriculography. While contrast cine ventriculography is often useful for assessing the severity of MR, cardiac catheterization and angiography are most commonly used in patients with MVP to exclude the possibility of coronary artery disease.

Intracardiac pressures and cardiac output are usually normal in uncomplicated MVP; however, these measurements become progressively more abnormal as MR becomes more severe. LV angiography usually confirms the presence of prolapse of the mitral valve. LV wall motion is usually normal in patients with primary MVP, but some patients show abnormal contraction patterns in the absence of coronary artery disease.

OTHER LABORATORY STUDIES

Exercise myocardial perfusion imaging with thallium or technetium sestamibi has been recommended as an adjunct to exercise ECG for determining the presence or absence of coexisting mitral ischemia in patients with MVP. Most MVP patients with clinical evidence of coronary artery disease have an abnormal exercise scintigram. Blood pool gated radionuclide studies have been used for the initial and serial assessment of LV function, but this is usually determined at the time of echocardiography.

The indications for electrophysiologic testing in a patient with MVP are similar to those in general practice. The upright tilt test with monitoring of blood pressure and rhythm may be valuable in patients with light-headedness or syncope and in diagnosing autonomic dysfunction.

Natural History and Prognosis

In most patient studies, the MVP syndrome is associated with a benign prognosis. The age-adjusted survival rate for both males and females with MVP is similar to that in patients without this common clinical entity. The gradual progression of MR in patients with mitral prolapse, however, may result in progressive dilatation of the LA and LV. LA dilatation often results in atrial fibrillation, and moderate to severe MR eventually results in LV dysfunction and the development of congestive heart failure in certain patients. Several long-term prognostic studies suggest that complications of MVP occur most commonly in patients with a mitral systolic murmur, thickened redundant mitral valve leaflets, or increased LV or LA size. Sudden death is an uncommon but obviously the most serious complication of mitral valve prolapse. While sudden death is infrequent, its highest incidence has been reported in the familial form of MVP. Infective endocarditis is a serious complication of MVP, and MVP is the leading predisposing cardiovascular lesion in most series of patients reported with endocarditis. Progressive MR occurs frequently in patients with long-standing MVP. Fibrin emboli are responsible in some patients for visual problems consistent with involvement of the ophthalmic or posterior cerebral circulation.

Medical Therapy

There is general agreement that patients with the characteristic systolic click-murmur complex undergoing procedures associated with bacteremia should be treated with antibiotic endocarditis prophylaxis. Such prophylaxis is also usually recommended for patients who have an isolated systolic click with echocardiographic evidence of MVP and significant MR. Some recommend endocarditis antibiotic prophylaxis for patients with isolated systolic click who have echo evidence of high-risk MVP even in the absence of MR. In general, antibiotic endocarditis prophylaxis is not recommended routinely for patients with an isolated systolic click who have no or equivocal evidence of MVP by echocardiography; however, this remains controversial.

Daily aspirin therapy is recommended for MVP patients with documented focal or neurologic events who are in sinus rhythm with no atrial thrombi. Such patients should also avoid cigarettes and oral contraceptives. Warfarin therapy is usually recommended for patients with mitral valve prolapse who have atrial fibrillation and by some for patients who have transient ischemic attacks despite aspirin therapy or who have paroxysmal atrial fibrillation. Aspirin therapy is indicated in poststroke MVP patients with contraindications to anticoagulants and is recommended by some for echocardiographic high-risk patients in sinus rhythm.

Surgical Therapy

Management of patients with MVP may require valve surgery, particularly in those who develop a flail mitral leaflet due to ruptured chordae tendineae or their marked elongation. Most such valves can be repaired successfully by surgeons experienced in mitral valve repair, especially when the posterior leaflet of the mitral valve is predominantly affected. Surgery is less frequently recommended for patients with severe MR and mild symptoms who are likely to require valve replacement. When feasible, valve repair or replacement with preservation of the chordae is often recommended for asymptomatic patients with severe MR and recurrent or chronic atrial fibrillation. There are less data to support surgery for *asymptomatic* patients with severe mitral regurgitation and a resting pulmonary artery pressure above 50 mmHg or an exercise pulmonary artery pressure above 60 mmHg. However, symptomatic class III to IV patients with these same findings are usually sent to surgery. Patients with severe mitral regurgitation and recurrent thromboemboli despite anticoagulant therapy are often recommended for mitral valve surgery when mitral repair is tenable. Surgery is not recommended for patients with severe mitral regurgitation and recurrent ventricular arrhythmias despite medical therapy.

SUGGESTED READING

Bonow RO, Carabello B, De Leon AC Jr, et al: ACC/AHA guidelines for valvular heart disease. *J Am Coll Cardiol* 1998. In press.

Committee on Rheumatic Fever, Endocarditis, and Kawasaki Disease of the Council on Cardiovascular Disease in the Young of the American Heart Association: Treatment of streptococcal pharyngitis and prevention of rheumatic fever: a statement for health professionals. *Pediatrics* 1995; 96:758–764.

Crawford MH, Souchek J, Oprian CA, et al: Determinants of survival and left ventricular performance after mitral valve replacement. Department of Veterans Affairs Cooperative Study on Valvular Heart Disease. *Circulation* 1990; 81:1173–1181.

Currie PJ, Seward JB, Chan KL, et al: Continuous wave Doppler determination of right ventricular pressure: a simultaneous Doppler-catheterization study in 127 patients. *J Am Coll Cardiol* 1985; 6:750–756.

Devereux RB, Frary CJ, Framer-Fox R, et al: Cost-effectiveness of infective endocarditis prophylaxis for mitral valve prolapse with or without a mitral regurgitant murmur. *Am J Cardiol* 1994; 74:1024–1029.

Hatle L, Angelsen B, Tromsdal A: Noninvasive assessment of atrioventricular pressure half-time by Doppler ultrasound. *Circulation* 1979; 60:1096–1104.

Lembo NJ, Dell'Italia LJ, Crawford MH, et al: Mitral valve prolapse in patients with prior rheumatic fever. *Circulation* 1988; 77:830–836.

Manning WJ, Silverman DI, Keighley CS, et al: Transesophageal echocardiographically facilitated early cardioversion from atrial fibrillation using short-term anticoagulation: final results of a prospective 4.5 year study. *J Am Coll Cardiol* 1995; 256:1354–1361.

Nishimura RA, Rihal CS, Tajik AJ, Holmes DR Jr: Accurate measurement of the transmitral gradient in patients with mitral stenosis: a simultaneous catheterization and Doppler echocardiographic study. *J Am Coll Cardiol* 1994; 24:152–158.

O'Rourke RA: Mitral valve prolapse syndrome. In: Alexander RW, Schlant RC, Fuster V, O'Rourke RA, Roberts R, Sonnenblick EH (eds): *Hurst's The Heart*, 9th ed. New York, McGraw-Hill, 1998:1821–1831.

O'Rourke RA: The mitral valve prolapse syndrome. In: Chizner MA (ed): *Classic Teaching in Clinical Cardiology*. Cedar Grove, NJ, Laennec, 1996:1049–1070.

Palacios IF, Tuzcu ME, Weyman AE: Clinical follow-up of patients undergoing percutaneous mitral balloon valvotomy. *Circulation* 1995; 91:671–676.

Patel JJ, Shama D, Mitha AS, et al: Balloon valvuloplasty versus closed commissurotomy for pliable mitral stenosis: a prospective hemodynamic study. *J Am Coll Cardiol* 1991; 18:1318–1322.

Rahimtoola SH, Enriquez-Sarano M, Schaff HV, Frye E: Mitral valve disease. In: Alexander RW, Schlant RC, Fuster V, O'Rourke RA, Roberts R, Sonnenblick EH (eds): *Hurst's The Heart*, 9th ed. New York, McGraw-Hill, 1998:1789–1819.

Reid CL, McKay CR, Chandraratna PA, et al: Mechanisms of increase in mitral valve area and influence of anatomic features in double-balloon, catheter balloon valvuloplasty in adults with rheumatic mitral stenosis: a Doppler and two-dimensional echocradiographic study. *Circulation* 1987; 76:628–636.

Shah PM: Echocardiographic diagnosis of mitral valve prolapse. *J Am Soc Echocardiogr* 1994; 7:286–293.

Tei C, Sakamaki T, Shah PM, et al: Mitral valve prolapse in short-term experimental coronary occlusion: a possible mechanism of ischemic mitral regurgitation. *Circulation* 1983; 68:183–189.

Turi ZG, Reyes VP, Raju BS, et al: Percutaneous balloon versus surgical closed commissurotomy for mitral stenosis: a prospective, randomized trial. *Circulation* 1991; 83:1179–1185.

Tuzcu EM, Block PC, Griffin BP, et al: Immediate and long-term outcome of percutaneous mitral valvotomy in patients 65 years and older. *Circulation* 1992; 85:963–971.

Wisenbaugh T, Skudicky D, Sareli P: Prediction of outcome after valve replacement for rheumatic mitral regurgitation in the era of chordal preservation. *Circulation* 1994; 89:191–197.

Wood P: An appreciation of mitral stenosis: part II. *Br Med J* 1954; I:1113–1124.

14

TRICUSPID AND PULMONIC VALVE DISEASE

Robert A. O'Rourke

TRICUSPID VALVE DISEASE

Definition, Etiology, and Pathology

Tricuspid stenosis results from obstruction to diastolic flow across the valve during filling of the right ventricle. *Tricuspid regurgitation* occurs when the tricuspid valve allows blood to enter the right atrium during a right ventricular (RV) contraction.

The most common cause of *tricuspid stenosis* (TS) is rheumatic fever. This is usually associated with concomitant mitral stenosis. Isolated stenosis of the tricuspid valve can be caused by the carcinoid syndrome, infective endocarditis, endocardial fibroelastosis, endomyocardial fibrosis, and systemic lupus erythematosus. TS has also been reported to occur in patients with Fabry's disease, Whipple's disease, and in those receiving methysergide therapy. Mechanical obstruction of the valve can occur with right atrium myxoma, tumor metastases, and thrombi in the right atrium, each resulting in the hemodynamic abnormalities of TS. Additionally, RV inflow tract obstruction can be due to thrombosis, endocarditis, degeneration, or calcification affecting a prosthetic tricuspid valve.

In rheumatic tricuspid valve disease, alterations in the valve are characterized by fibrosis with contracture of the leaflets and commissural fusion. The latter leads to TS, which is often minor and would go unrecognized clinically if it were not for the high flow across the valve caused by the coexistent regurgitation. Whenever the tricuspid valve is affected by rheumatic disease, there is also involvement of left-sided valves.

Diseases causing *tricuspid regurgitation* (TR) are more numerous than those causing TS. Importantly, the *normal* tricuspid valve commonly does not completely coapt in systole, as is shown by the frequent occurrence of TR jets on Doppler ultrasound. Usually the volume of regurgitant flow is so small that the TR is silent; this finding of trivial TR occurs in 24 to 96 percent of normal individuals and must be considered a variant of normal by Doppler ultrasound.

Pathologic TR is most commonly due to the diseases that cause RV dilatation and failure; LV failure or pulmonary hypertension can result in TR. Primary diseases of the tricuspid valve apparatus that include the tricuspid annulus, leaflets, chordae, papillary muscle, and RV wall also cause TR. The most common etiology of isolated TR is infective endocarditis in drug addicts. Other causes include RV myocardial infarction, trauma, carcinoid, leaflet prolapse, and congenital abnormalities such as atrial septal defect and Ebstein's anomaly. TR has been reported to occur in patients with rheumatoid arthritis and Marfan's syndrome and in those who have undergone radiation therapy. Primary involvement of the tricuspid valve due to rheumatic fever results in TS usually accompanied by TR. Recently, TR has been reported as one of the valvular abnormalities found in patients taking the anorectic drugs fenfluramine and dexfenfluramine. The cause-and-effect relationship remains to be defined.

Carcinoid heart disease is seen in up to 53 percent of patients with malignant carcinoid tumor (usually originating in the ileum) with extensive metastases. Carcinoid usually causes TR and TS and less often pulmonic stenosis and regurgitation. In infective endocarditis, the TR results from improper coaptation of the leaflets because of interposed vegetations.

The most common variety of TR is the secondary type that results from the enlargement of the orifice and annulus resulting from congestive heart failure with RV dilatation due to LV disease. TR may diminish when the heart failure is treated successfully but can be permanent with long-standing RV dilatation.

Various degrees of prolapse of the tricuspid valve are commonly present in the general population and may occur in 3 to 54 percent of patients with mitral valve prolapse. Reported cases of severe TR from prolapse have been relatively uncommon.

Traumatic tricuspid regurgitation usually results from rupture of one or more of the components of the tensor apparatus, with disruption of the papillary muscle occurring more often than rupture of the chordae.

Pathophysiology

TS decreases diastolic flow across the valve, elevates right atrial pressure, and reduces cardiac output. There is stiffening of the valve by fibrosis or disease and commisural fusion, both of which narrow the effective valvular orifice. Flow from systemic veins or the right atrium into the RV is obstructed and a pressure gradient develops in diastole between the right atrium and the right ventricle. The normal area of the tricuspid valve is 7 cm^2; impairment of RV filling occurs when the valve area is reduced to less than 1.5 cm^2. Elevation of the mean right atrial pressure above 10 mmHg usually results in peripheral edema. Development of atrial fibrillation produces a higher right atrial pressure in TS than when sinus rhythm and normal right atrial contraction are present.

In TR, the systolic blood flow into the right atrium (RA) elevates the mean right atrial pressure. Regurgitant flow produces a prominent CV wave reflected to the venous system. Diastolic volume overload of the RV causes further dilatation of the RV and movement of the interventricular septum toward the LV during diastole. RV failure further raises the mean RA and vena caval pressures and result in systemic venous congestion and signs of RV failure.

Clinical Manifestations

HISTORY

The most frequent symptoms in TS are dyspnea and fatigue. When mitral stenosis (MS) coexists, the development of significant TS can diminish the paroxysmal symptoms of dyspnea, pulmonary congestion, and pulmonary hypertension. Occasionally, patients with TS complain of prominent pulsations in the neck veins, which can precede the development of peripheral edema.

Since TR usually accompanies LV failure or MS, the presenting symptoms include dyspnea, orthopnea, and peripheral edema. Even though LV failure is usually present, paroxysmal nocturnal dyspnea is often absent. TR under these conditions may occasionally improve pulmonary symptoms and provide a physiologic basis for the alleviation of left-sided heart failure by the development of right-sided heart failure. Some patients also have less pulmonary edema due to the development of pulmonary arteriolar disease. If the TR is produced by infective endocarditis, symptoms of a febrile illness may be accompanied by fatigue and peripheral edema.

PHYSICAL FINDINGS

TS is frequently associated with lesions of the mitral valve. When sinus rhythm is present, the jugular veins will display the prominent A wave indicative of impaired RV diastolic filling with atrial systole. The A wave in the neck may be of moderate height and sometimes reaches the mandible. Simultaneous auscultation of the heart is often required to confirm that the rise in the venous A wave is simultaneous with the first heart sound. The CV wave is small and the y descent is slow and insignificant. A murmur due to TS, which is a low-pitched, middiastolic and/or presystolic rumble, may be heard at the left sternal border and increases with inspiration.

In patients with primary TR not due to pulmonary hypertension, there are large v waves in the jugular venous pulse. There is a dilated RV with a left parasternal precordial lift and right-sided third or fourth heart sounds. There is commonly a long systolic murmur in the third or fourth intercostal space at the left sternal border that increases with inspiration. This murmur is often confined to early and midsystole or may not be heard at all when there is a small gradient between the right ventricle and the right atrium during systole and a large regurgitant orifice. A short systolic murmur is frequently present in patients with tricuspid regurgiation in the absence of pulmonary hypertension; such a murmur is common in patients with infective endocarditis affecting the tricuspid valve, especially drug addicts. When a large amount of blood returns to the RV in diastole, a short diastolic rumble along the left sternal border may be heard and usually is louder during inspiration. When TR is due to pulmonary hypertension, there is an accentuated pulmonic component (P_2) of the second heart sound; a high-pitched decrescendo diastolic murmur of pulmonic regurgitation (PR) is often heard in the second and third left interspaces. In patients with TR and atrial fibrillation, there is a prominent CV wave in the jugular veins produced by the regurgitation flow into the RV. The characteristic physical findings of TR due to pulmonary hypertension is a holosystolic murmur at the left sternal border that increases during inspiration; there is a right ventricular–right atrial pressure gradient throughout systole.

ELECTROCARDIOGRAM

The characteristic electrocardiographic (ECG) finding in tricuspid stenosis is a large P wave of right atrial enlargement in the absence of RV hypertrophy that is present in patients with sinus rhythm. Atrial fibrillation is frequent in patients with TR. When TR results from myocardial infarction, acute or chronic ECG changes will be seen in the inferior ECG leads, and ST-segment elevation, indicating RV infarction may be present in right-sided precordial ECG leads.

CHEST ROENTGENOGRAM

In TS, the most characteristic radiographic finding is prominence of the right atrium without significant pulmonary arterial enlargement or changes due to the pulmonary hypertension. TR may produce some degree of RA enlargement, but there will usually be accompanying RV enlargement. When multiple fluffy infiltrates are seen on chest x-ray involving several portions of the lung, a diagnosis of tricuspid valve endocarditis is highly likely.

Laboratory Studies

ECHOCARDIOGRAPHY/DOPPLER

The characteristic pattern of TS can often be recorded with the echocardiogram. Fibrosis and calcification of the valve can be identified. Obstructive lesions such as myxoma, thrombus, or other tumors can be recognized echocardiographically. The echo Doppler technique can be used to estimate the diastolic gradient across the valve, but there are not many studies confirming its accuracy.

In TR, there may be echocardiographic evidence of systolic prolapse, rupture of the chordae or papillary muscle, or vegetative lesions on the valve. Increased RV dimensions indicate impaired RV function and the likelihood of secondary TR. Contrast echocardiography with peripheral venous injection can identify the back-and-forth flow across the valve. The echo Doppler technique can estimate the severity of the regurgitation and the systolic pressure in the right ventricle. Color-flow Doppler imaging can delineate the patterns and sites of regurgitation across the valve apparatus.

CARDIAC CATHETERIZATION AND ANGIOGRAPHY

If TS is clinically suspected, simultaneous pressures should be recorded in the RA and RV in order to measure the gradient across the valve accurately. Since the normal gradient across the tricuspid valve is less than 1 mmHg, small gradients may be missed if pullback pressure is recorded from the right ventricle to the right atrium. The area of the tricuspid valve in significant stenosis is usually less than 1.5 cm^2; in severe TS it is less than 1 cm^2.

Accurate angiographic documentation of TR is difficult to obtain because the catheter overrides the tricuspid valve, and ventricular irritability during a RV contrast injection can induce TR. A prominent *CV* wave in the right atrium suggests TR; an intracardiac phonocardiogram may record a regurgitant murmur.

OTHER LABORATORY STUDIES

A radionuclide ventriculogram can delineate the dimensions of the RA and RV and measure the RV ejection fraction. The determination of right heart volumes may help differentiate between stenosis and regurgitation of the tricuspid valve.

Natural History and Prognosis

With TS, the symptoms are usually those of mitral stenosis in the absence of pulmonary congestion. The presence of peripheral edema should raise the possibility of underlying TS. Significant TS may slow the development of characteristic symptoms of mitral stenosis and result in underestimation of the severity of mitral valve obstruction.

With TR due to RV hypertension, the symptoms and clinical course are primarily related to the left-sided heart conditions that produce a pressure-volume overload on the RV. TR always develops eventually with severe RV failure. In infective endocarditis of the tricuspid valve, the type of organism may significantly influence the course and the response to antibiotics.

Medical Management

In TS, the usually precautionary measures of antibiotic coverage and prevention of endocarditis apply. Peripheral edema may not respond well to the administration of digitalis, diuretics, and vasodilator therapy, thus emphasizing the clinical importance of detecting underlying TS. Tricuspid catheter balloon valvuloplasty has been used successfully in patients with predominant TS.

With TR, treatment of RV failure requires digitalis and diuretics; vasodilating agents are also required for the management of LV failure. If failure of the right heart is caused by MS, early intervention to enlarge or replace the mitral valve is appropriate.

Antibiotic prophylaxis against endocarditis is appropriate for patients with either TS or TR. If rheumatic fever is the likely etiology of the tricuspid valve disease, rheumatic fever prophylaxis is indicated (Table 13-1).

If atrial fibrillation develops, chronic anticoagulation with low-dose warfarin (INR of 2.0 to 3.0) is warranted, since the accompanying incidence of systemic and cerebral emboli is estimated at 10 to 20 percent.

Surgical Management

The decision to proceed with valvular heart surgery is usually based on the severity of the aortic and mitral valve disease rather than on the severity of the disease of the tricuspid valve. If there are signs of TS and if TS is demonstrated by cardic catheterization and two-dimensional (2D) echocardiography, the tricuspid valve is directly visualized at operation with the anticipation of performing commissurotomy or valve replacement. When there are signs of severe TR secondary to MS, it is important to document the duration of the regurgitation and the severity and duration of pulmonary hypertension. If the TR is severe and long-standing and if there is chronic pulmonary artery hypertension, it is unlikely that the TR will resolve in the early postoperative period after mitral valve surgery. If the TR and pulmonary hypertension are of short duration, mitral valve replacement or repair will usually reduce pulmonary artery pressure in the early postoperative period, with a resulting decrease in the tricuspid valve regurgitation. In this circumstance, the surgeon usually decides after mitral valve surgery whether or not a procedure to reduce TR is indicated.

TS may be treated successfully by commissurotomy, which is usually performed under direct vision. The procedure may be combined with annuloplasty to correct valve regurgitation. Valve replacement is occasionally necessary if the changes in the leaflets and subvalvular structures are advanced or if severe regurgitation cannot be relieved by annuloplasty.

In general, the early and late results of tricuspid annuloplasty have been superior to those of valve replacement and valve replacement should be avoided when possible.

PULMONIC VALVE DISEASE

Definition, Etiology, and Pathology

Pulmonic stenosis (PS) is caused by obstruction to the systolic flow across the valve and is most commonly congenital (see Chap. 22). Sarcomas and myxomas can sometimes extend to the pulmonic valve, causing PS. Previous cardiac surgery on a congenital pulmonic valve lesion can result in pulmonic regurgitation (PR). The carcinoid syndrome with cardiac involvement can create mild PS and associated PR. Compression of the pulmonary artery can stimulate valvular stenosis and is occasionally produced by tumor, aneurysm, or even constrictive pericarditis.

Acquired lesions of the pulmonic valve generally lead to PR. On rare occasions, an inflammatory process can create stenosis and regurgitation of the valve. Pulmonary hypertension from any cause can produce PR. Inflammatory diseases—such as endocarditis, rheumatic fever, and on rare occasions tuberculosis—can result in PR.

Pathophysiology

PR is the most frequently acquired lesion of the pulmonic valve. It may be secondary to pulmonary hypertension or may be caused by primary abnormalities in the leaflets. PR imposes a volume overload on the RV; if pulmonary hypertension preexists, the overload is superimposed on hypertrophied myocardium. Volume overload of the RV may cause an increase in diastolic volume of the chamber, an increase in RV stroke volume, and subsequent RV failure, resulting in TR. Fortunately, isolated PR can usually be tolerated for a long time without cardiac decompensation.

Clinical Manifestations

HISTORY
Clinical manifestations of acquired pulmonic valvular lesions depend on the severity of the hemodynamic impairment as well as the extent of the underlying disease. Isolated PR can be tolerated without symptoms. Severe pulmonary hypertension may cause syncope in addition to shortness of breath and fatigue. With inflammatory lesions of the pulmonic valve, febrile

manifestations and pulmonary infection may be present. The carcinoid syndrome is characterized by episodes of facial flushing, increased intestinal activity, diarrhea, and bronchospasm. Tumors involving the pulmonic valve may exert pressure from expansion and metastases that affect the heart and lungs.

PHYSICAL FINDINGS

If RV failure and tricuspid regurgitation have developed as a result of PR, a prominent *CV* wave will be present in the jugular venous pulse. Increased right ventricular activity may be visible with a systolic lift palpable along the left sternal border. If pulmonary hypertension is present, the pulmonic second heart sound (P_2) will be accentuated over the left upper sternal border. The murmur of acquired PR is a high-pitched diastolic blowing murmur along the left sternal border. This may be difficult to differentiate from the murmur of the aortic regurgitation, but the absence of peripheral findings of aortic regurgitation is useful in identifying regurgitation of the pulmonic valve as the source of the abnormal diastolic flow. Congenital PR characteristically is associated with a low-pitched, decrescendo murmur along the left sternal border; the peak of the murmur occurring shortly after the pulmonic component of the second heart sound.

ELECTROCARDIOGRAPHY

Although there are no characteristic changes with pulmonic valvular lesions, preexisting pulmonary hypertension will produce RV hypertrophy, QRS, right axis deviation, and changes in the P wave suggesting RA enlargement. If pulmonary hypertension is secondary to mitral stenosis, P mitrale with characteristic notches will be present in lead II.

CHEST ROENTGENOGRAM

Patients with pulmonic valve regurgitation have pulmonary prominence along with an increase in the RV dimension. If stenosis of the pulmonary valve is acquired, there may be poststenotic dilatation or prominence of the main pulmonary artery.

Laboratory Studies

ECHOCARDIOGRAPHY/DOPPLER

Echocardiography can delineate the anatomy of the pulmonic valve as well as intrinsic or extrinsic lesions impinging on the valve apparatus. Sometimes the vegetative lesion or tumor can be detected in the pulmonary valve area. The echo Doppler technique can estimate both the severity of the regurgitation and the stenosis of the pulmonic valve. An analysis of the echo Doppler recordings can provide estimates of the pulmonary artery pressure. Color-flow imaging can further confirm the patterns of regurgitation in the RV outflow tract.

CARDIAC CATHETERIZATION AND ANGIOGRAPHY
PR is not readily demonstrated angiographically, but a right-sided injection can outline the pulmonary valve as well as poststenotic dilatation. A normal aortic root contrast injection helps eliminate AR as the etiology of a diastolic murmur along the left sternal border. Nevertheless, the distinction between PR and AR is usually best made by echo Doppler studies. Intracardiac phonocardiography has been employed to detect the diastolic murmur of PR in the RV outflow tract.

OTHER LABORATORY STUDIES
A radionuclide ventriculogram can be used to estimate RV size and function in patients with stenotic and regurgitant lesions of the pulmonic valve. Since combined lesions of the aortic and mitral valves often create pulmonary hypertension and RV dysfunction, radionuclide ventriculography is useful in estimating the RV ejection fraction.

Natural History and Prognosis

The clinical history is imporant in delineating causes of left-sided heart failure that can lead to pulmonary hypertension and regurgitation of the pulmonic valve. Mild to moderate pulmonary regurgitation can be relatively well tolerated, and the natural history in patients with pulmonic regurgitation really depends upon the severity of left heart failure or of the other syndromes involving the pulmonic valve. The natural history and prognosis of congenital PS are discussed in Chap. 22.

Medical Treatment

Patients with congenital pulmonic valve stenosis are usually best treated by catheter balloon valvulotomy. Antibiotic prophylaxis against endocarditis is appropriate for patients with pulmonic valve regurgitation. If pulmonary emboli contribute to the pulmonary hypertension, anticoagulation is indicated. Further treatment of pulmonary hypertension may require management of failure of the left side of the heart, correction of mitral stenosis or the use of vasodilating agents that can lower pulmonary artery pressure. Vasodilating agents are often ineffective in treating primary pulmonary hypertension.

Surgical Management

Pulmonic valve surgery for acquired disease is performed infrequently. PS on an acquired basis is rare. Although there are a variety of causes of PR, this hemodynamic condition is relatively well tolerated if pulmonary vascular resistance is normal. Pulmonic valve replacement may be performed for acquired conditions such as the carcinoid heart syndrome and effective endocarditis, but it is usually limited to cases where RV dysfunction has become severe after surgery for congenital heart disease.

Although PR is well tolerated for several years after correction of malformations such as tetralogy of Fallot, it may become hemodynamically significant especially if pulmonary artery hypertension is present or develops. In this situation, the placement of a pulmonic valve prosthesis may significantly improve the patient's functional status. In general, bioprosthetic valves have been preferred because of the tendency for mechanical valve thrombosis to develop in this position. Pulmonic valvulectomy in combination with antibiotic therapy is often the most effective treatment for patients with isolated pulmonic valve endocarditis.

SUGGESTED READINGS

Ansari A: Isolated pulmonary valvular regurgitation: current perspectives. *Prog Cardiovac Dis* 1991; 33:329–334.

Bonow RO, Carabello B, De Leon AC Jr, et al: ACC/AHA guidelines for valvular heart disease. *J Am Coll Cardiol* 1998. In press.

Cheitlin MD, Alpert JS, Armstrong WF, et al: ACC/AHA guidelines for the clinical application of echocardiography. *Circulation* 1997; 95:1686–1744.

DeSimone R, Lange R, Tanzeem A, et al: Adjustable tricuspid valve annuloplasty assisted by intraoperative transesophageal color Doppler echocardiography. *Am J Cardiol* 1993; 71:926–931.

Duran CMG: Tricuspid valve surgery revisited. *J Card Surg* 1994; 9:242–247.

Herrmann HC, Hill JA, Krol J, et al: Effectiveness of percutaneous balloon valvuloplasty in adults with pulmonic valve stenosis. *Am J Cardiol* 1991; 68:1111–1113.

Holmes JC, Fowler ND, Kaplan S: Pulmonary valvular insufficiency. *Am J Med* 1968; 44:851–862.

Orbe LC, Sobrino N, Arcas R, et al: Initial outcome of percutaneous balloon valvuloplasty in rheumatic tricuspid valve stenosis. *Am J Cardiol* 1993; 71:353–354.

O'Rourke RA, Rackley CE, Edwards JE, et al: Tricuspid valve, pulmonic valve, and multivalvular disease. In: Alexander RW, Schlant RC, Fuster V, O'Rourke RA, Roberts R, Sonnenblick EH (eds): *Hurst's The Heart*, 9th ed. New York, McGraw Hill; 1998:1833–1850.

Pellikka PA. Tajik AJ, Khandheria BK, et al: Carcinoid heart disease: clinical and echocardiographic spectrum in 74 patients. *Circulation* 1993; 87:1188–1196.

Van Son JA, Danielson GK, Schaff HV, Miller FA Jr: Traumatic tricuspid valve insufficiency: experience in thirteen patients. *J Thorac Cardiovasc Surg* 1994; 108:893–898.

Waller BF, Howard J, Fess S: Pathology of tricuspid valve stenosis and pure tricuspid regurgitation—part I. *Clin Cardiol* 1995; 18:97–102.

Waller BF, Howard J, Fess S: Pathology of tricuspid valve stenosis and pure tricuspid regurgitation—part II. *Clin Cardiol* 1995; 18:167–174.

Waller BF, Howard J, Fess S: Pathology of tricuspid valve stenosis and pure tricuspid regurgitation—part III. *Clin Cardiol* 1995; 18:225–230.

Waller BF, Moriarty AT, Eble JN, et al: Etiology of pure tricuspid regurgitation based on annular circumference and leaflet area: analysis of 45 necropsy patients with clinical and morphologic evidence of pure tricuspid regurgitation. *J Am Card Coll* 1986; 7:1063–1074.

Wooley CF, Fontana ME, Kilman JW, Ryan JM: Tricuspid stenosis. Atrial systolic murmur, tricuspid opening snap, and right atrial pressure pulse. *Am J Med* 1985; 78: 375–384.

15

PROSTHETIC HEART VALVES

Robert A. O'Rourke

HEART VALVE PROSTHESES

A heart valve prosthesis consists of an orifice through which blood flows and an occluding mechanism that closes and opens the orifice. There are two types of heart valves: *mechanical prostheses*, with rigid, manufactured occluders, and *biological* or tissue valves, with flexible leaflet occluders of animal or human types. Among the mechanical valves there are three basic types depending on whether the occluding mechanism is a reciprocating ball, a tilting disk, or two semicircular hinged leaflets. The biological valves include those whose origin is from the patient, another human, or another species.

Mechanical Valves

Ball valves appeared in the early 1960s, disk valves in the early 1970s, and bileaflet valves predominantly during the 1980s.

BALL VALVES
The first successful valve replacement devices, which led to long-term survivors and a design that has endured until today, used a ball-in-cage design. Several modifications of this design have been used, but only the Starr-Edwards valve has endured; it has been used about 200,000 times.

DISK VALVES
The first successful low-profile design was the Björk-Shiley tilting-disk valve, which was introduced in 1969. It evolved through several design refinements and about 360,000 valves were implanted. These refinements also introduced a structural failure caused by strut fracture in the Convexo-Concave model, and it is no longer used.

Tilting valves employ a circulating disk as an occluder. It is retained by wirelike arms or close loops that project into the orifice. The Medtronic Hall valve has a titanium housing and a carbon-coated disk with a unique central hole.

BILEAFLET VALVES

Current development in mechanical valves is based on the bileaflet design, introduced by St. Jude Medical in 1977. Unlike the free-floating occluders in ball and disk valves, the two semicircular leaflets of the bileaflet valve are connected to the orifice housing by a hinge mechanism. During opening, there are three flow areas: one central and two peripheral. The St. Jude bileaflet valve has leaflets that open to an angle of 85 degrees from the plane of the orifice and travel from 55 to 60 degrees to the fully closed position, depending on valve size.

Biological Valves

Biological valves include as wide a variety of models as do the mechanical valves.

AUTOGRAFT

The pulmonary autograft procedure consists of an autotransplant of the patient's own pulmonary valve to the aortic position; the pulmonary valve is then replaced by an aortic or pulmonary homograft. This procedure, first described in 1967, is called the *Ross procedure*.

AUTOLOGOUS PERICARDIAL VALVE

This is a new category of valve developed in an attempt to combine the reproducibility and ease of insertion of a commercial, stented heterograft valve with the benefits of autologous tissue. The valve is fashioned from the patient's own nonvalvular tissue. It is a frame-mounted autologous pericardial valve, which is assembled from a kit in the operating room. From an immunologic standpoint, autologous tissue might be the best material when a tissue transfer is required.

HOMOGRAFT

The homograft or allograft valve is considered a preferred substitute for aortic valve replacement, especially in younger patients. In this situation there is a transplantation from a donor of the same species, such as a donor's aortic or pulmonic valve, into a recipient's aortic or pulmonic position. Several methods of procurement, sterilization, and preservation have been used.

PORCINE HETEROGRAFT

This utilizes transplantation of a valve from another species, either as an intact valve, such as the porcine aortic valve, or a valve fashioned from heterogenous tissue, such as the bovine pericardial valve. In the porcine heterograft (or xenograft), the valve tissue is sterilized by glutaraldehyde, which renders it bioacceptable due to destruction of antigenicity and stabilizes the collagen cross-links for durability. There are two major porcine heterograft valves: the Hancock and the Carpentier-Edwards porcine valves. Most porcine valves are mounted on rigid or flexible stents, to which the leaflets in the sewing ring are attached.

BOVINE PERICARDIAL VALVE

Pericardial valves that are tailored and sewn into a valvular configuration using bovine pericardium as a fabric result in a valve that opens more completely than a porcine valve and has better hemodynamics. Unfortunately, the Ionescu-Shiley, the first commercially available pericardial valve, had a higher failure rate than porcine valves.

The Carpentier-Edwards pericardial bioprosthesis has a method of construction that does not require stitches to pass through the leaflets, as did the Ionescu-Shiley pericardial valve, and during its use since 1982 the results have been excellent.

American Association for Thoracic Surgery/Society of Thoracic Surgeons Guidelines for Clinical Reporting

The complications that were determined to be of critical importance by these guidelines established in 1988 and revised in 1966 include structural valvular deterioration, nonstructural dysfunction, valve thrombosis, embolism, bleeding events, and operative valvular endocarditis.

The consequences of the above morbid events include re-operation, valve-related mortality, sudden unexpected un-explained death, cardiac death, total deaths, and permanent valve-related impairment.

Valve Selection Criteria

Because of the wide variation and results among and between various valve models, it is important to rank valves within valve types on the basis of complication rates. Some general recommendations, however, can be made with regard to valve selection.

A biological valve should be used when the patient cannot or will not take anticoagulants, desires pregnancy, or has a short life expectancy. A mechanical valve should be used if the patient needs anticoagulant therapy (e.g., for atrial fibrillation), has a mechanical valve in another position, previously had a stroke, requires double valve replacement, or has a long-life expectancy. Mechanical valves should be considered for double valve replacement because the risk of structural deterioration for two porcine valves is additive, whereas the thromboembolic risk of two mechanical valves is not.

MANAGEMENT OF PATIENTS WITH PROSTHETIC VALVES

All patients with prosthetic valves need appropriate antibiotic prophylaxis against infective endocarditis (Chap. 21). Patients with rheumatic heart disease continue to need antibiotics as prophylaxis against recurrence of rheumatic carditis (Table 13-1). Adequate antithrombotic therapy is needed for appropriate patients (Chap. 46).

During the first 4 to 6 weeks after surgery, the physician and surgeon jointly manage the patient, directing their attention to relieving postoperative discomfort, readjusting cardiac medications, and instituting anticoagulation if not contraindicated. A graduated plan of activity is started and in most cases enables the patient to return to full activity in 4 to 6 weeks.

Multiple noninvasive tests have emerged for assessing valvular and ventricular function. Fluoroscopy can reveal abnormal rocking of the dehiscing prosthesis or limitation of the occluder if the latter is opaque as well as strut fracture of a Björk-Shiley valve.

Echocardiography/Doppler ultrasound is the most useful noninvasive test. It provides information about prosthesis stenosis/regurgitation, valve area, assessment of other valve disease(s), pulmonary hypertension, atrial size, left ventricular hypertrophy, left ventricular size and function, and pericardial effusion/thickening. An echo/Doppler study is essential at the first postoperative visit because it provides an assessment of the effects and results of surgery and serves as a baseline for comparison should complications and/or deterioration occur later. Subsequently, it is performed as needed in symptomatic patients and in asymptomatic patients at 2- to 5-year intervals.

In patients with a bioprosthesis in the mitral position, echo/ Doppler ultrasound should be performed yearly after 5 years and in the aortic position annually after 8 years because of the increasing incidence of bioprosthetic structural valve deterioration.

Antithrombotic Therapy

All patients with mechanical valves require warfarin therapy. Even with the use of warfarin, the risk of thromboemboli in these patients 1 to 2 percent per year and the risk is considerably higher without treatment with warfarin. Almost all studies have shown that the risk of embolism is greater with a valve in the mitral position as compared with a valve in the aortic position.

MECHANICAL VALVES

All patients with mechanical valves require warfarin, and the INR should be maintained between 2.0 and 3.5. The addition of low-dose aspirin (50 to 100 mg/day) to warfarin therapy may further decrease the risk of thromboembolism. Many authors recommend the addition of aspirin to warfarin unless there is a contraindication to its use. The combination is particularly appropriate in patients who have had an embolus while on warfarin therapy and/or who are known to be particularly hypercoagulable. Importantly, the thromboembolic risk increases early after the insertion of the prosthetic valve; this is a reason to initiate heparin therapy within the first 24 to 48 h of surgery, with maintenance of the aPPT between 60 and 80 s until warfarin therapy has achieved an INR of 2.0 to 3.5.

BIOLOGICAL (TISSUE) VALVES

Since there is an increased risk of thromboemboli during the first 3 months after implantation of a biological prosthetic valve, initial anticoagulation with warfarin is indicated. The risk is particularly high in the first few days after surgery, and heparin therapy should be started within 24 to 48 h, with maintenance of the aPPT between 60 and 80 s until an International Normalized Ratio (INR) of 2.0 to 3.0 is achieved with warfarin. After 3 months, the tissue valve can be treated like native valve disease, and warfarin can be discontinued in approximately two-thirds of patients with biological valves. Associated atrial fibrillation or a left ventricular ejection fraction of less than 0.30 are reasons for lifelong warfarin therapy.

SUGGESTED READING

Barratt-Boyes B: Homograft aortic valve replacement in aortic incompetence and stenosis. *Thorax* 1964; 19:131–135.

Bjork VO: A new tilting disc valve prosthesis. *Scand J Thorac Cardiovasc Surg* 1969; 3:1–10.

Cannegieter SC, Rosendaal FR, Briet E: Thromboembolic and bleeding complications in patients with mechanical heart valve prostheses. *Circulation* 1994; 89:635–641.

Carpentier A, Dubost C: From xenograft to bioprosthesis. In: Ionescu MI, Ross DN, Wooler GH (eds): *Biological Tissue in Heart Valve Replacement*. London, Butterworth, 1971:515–541.

Edmunds LH, Clark RE, Cohn LH, et al: Guidelines for reporting morbidity and mortality after cardiac valvular operations: Ad Hoc Liaison Committee for Standardizing Definitions of Prosthetic Heart Valve Morbidity of the American Association for Thoracic Surgery and the Society of Thoracic Surgeons. *J Thorac Cardiovasc Surg* 1996; 112:708–711.

Grunkemeier GL, Starr A, Rahimtoola SH: Clinical performance of prosthetic heart valves. In: Alexander RW, Schlant RC, Fuster V, O'Rourke RA, Roberts R, Sonnenblick EH (eds): *Hurst's The Heart*, 9th ed. New York, McGraw-Hill, 1998:1833–1850.

Grunkemeier G, Starr A, Rahimtoola SH: Replacement of heart valves. In: O'Rourke RA, (ed): *The Heart: Update I*. New York, McGraw-Hill, 1996:98–123.

Grunkemeier GL, Starr A, Rahimtoola SH: Prosthetic heart valve performance: long-term follow-up. *Curr Probl Cardiol* 1992; 26:355–406.

Guidelines for reporting morbidity and mortality after cardiac valvular operations. *Ann Thorac Surg* 1988; 46:257–259.

Kearon C, Hirsh J: Management of anticoagulation before and after elective surgery. *N Engl J Med* 1997; 336:1506–1511.

McAnulty JH, Rahimtoola SH: Antithrombotic therapy and valvular heart disease. In: Alexander RW, Schlant RC, Fuster V, O'Rourke RA, Roberts R, Sonnenblick EH (eds): *Hurst's The Heart*, 9th ed. New York, McGraw-Hill, 1998:1833–1850.

Ross DN: Homograft replacement of the aortic valve. *Lancet* 1962; 2:487.

Starr A, Edwards M: Mitral replacement: clinical experience with a ball valve prosthesis. *Ann Surg* 1961; 154:726–740.

Stein PD, Alpert JS, Copeland J, et al: Antithrombotic therapy in patients with biological and prosthetic heart values. *Chest* 1995; 108(suppl):371S–379S.

Stinson EB, Griepp RB, Oyer PE, Shumway NE: Long-term experience with porcine aortic valve xenografts. *J Thorac Cardiovasc Surg* 1977; 73:54–63.

Turpie AG, Gent M, Laupacis A, et al: A comparison of aspirin with placebo in patients treated with warfarin after heart-valve replacement. *N Engl J Med* 1993; 329:524–529.

16

MYOCARDITIS AND OTHER SPECIFIC HEART MUSCLE DISEASES

Marlo F. Leonen
John B. O'Connell

The specific heart muscle diseases are secondary cardiomyopathies that are either attributable to an infective, metabolic, heredofamilial, allergic or toxic etiology or occur as a cardiac manifestation of abnormalities of other systems. This chapter deals with myocarditis and a few of the specific heart muscle diseases that are not otherwise discussed elsewhere (Table 16-1).

MYOCARDITIS

Etiology

Although pathologically defined as any inflammation or degeneration of the heart, myocarditis is believed to be caused by a variety of infectious, toxic, and allergic causes. The most common infectious agent associated with myocarditis is the Coxsackie B virus. The true prevalence of the disease is most likely underestimated, since serologic studies of infection with one of the cardiotropic viruses and autopsy data demonstrate histological evidence in patients who do not manifest symptoms (Table 16-2).

Pathogenesis

Increasing evidence shows that the disease process following myocardial viral infection is not caused by a direct effect but more likely results from the host inflammatory response to the causative infectious agent. It has been virtually impossible to culture the virus from myocardial tissue, although molecular identification of enteroviral RNA has recently been demon-

TABLE 16-1

WHO CLASSIFICATION OF CARDIOMYOPATHY

Cardiomyopathy of unknown cause
    Dilated cardiomyopathy
    Hypertrophic cardiomyopathy
    Restrictive cardiomyopathy
    Unclassified cardiomyopathy
Specific heart muscle disease
    Infective
    Metabolic
    General systems disease
    Heredofamilial
    Sensitivity and toxic reactions

strated. The specific immune responses that lead to the myocardial injury are incompletely defined. In humans, macrophages and T lymphocytes are likely involved. The myocardial injury may either be irreversible through perforin-mediated cytotoxicity or reversible through a mechanism involving cytokines. Autoantibodies to membrane, contractile, and energy transport proteins of the myocardium have also been postulated to play a role in the pathogenesis. Because most infections are subclinical and progressive cardiac disease is unusual, host immunoregulatory mechanisms likely mediate the disease process.

Clinical Manifestations

In the minority of cases that myocarditis becomes clinically apparent, an antecedent viral syndrome is common. A significant proportion of patients with active myocarditis present with left ventricular dysfunction. Fulminant heart failure may ensue, although varying degrees of systolic or diastolic dysfunction of varying chronicity may occur. Chest pain is present in approximately one-third of patients, and an occasional patient may present with a syndrome identical to an acute myocardial infarction with angiographically normal coronaries. Coronary arteritis and vasospasm have been reported. Syncope or palpitations may be secondary to atrioventricular block or ventricular arrhythmia. In some unfortunate patients, this may lead to sudden death.

Laboratory Diagnosis

The current "gold standard" for clinical diagnosis is endomyo-cardial biopsy, although application of this procedure has been tempered by the limitation of specific treatment of the disease. The usual technique involves the percutaneous intravascular sampling of the right ventricular septum under fluoroscopic or echocardiographic guidance. A minimum of four to six samples

TABLE 16-2

IMPORTANT CAUSES OF MYOCARDITIS

Infection
  Viral
    Coxsackie (A,B)
    Influenza (A,B)
    ECHO
    Cytomegalovirus
    Epstein-Barr virus
    Human immunodeficiency virus
    Hepatitis (B,C)
  Bacterial, rickettsial, spirochetal
    Corynebacterium diphtheriae
    Beta-hemolytic streptococci
    Mycoplasma pneumoniae
    Coxiella burnetti (Q fever)
    Rickettsia rickettsii (Rocky Mountain spotted fever)
    Borrelia burgdorferi (Lyme disease)
  Protozoal
    Trypanosoma cruzi (Chagas' disease)
    Toxoplasma gondii
  Metazoal
    Trichinosis
    Echinococcosis
  Fungal
    Candidiasis
    Cryptococcosis
Toxic
  Anthracyclines
  Catecholamines
Hypersensitivity
  Antibiotics (penicillins, sulfonamides)
  Methyldopa

are obtained for histologic analysis. A working standard, the Dallas criteria, defines active myocarditis as "an inflammatory infiltrate of the myocardium with necrosis and/or degeneration of adjacent myocytes not typical of the ischemic damage associated with coronary artery disease."

Other blood and cardiac studies are less specific. An elevated erythrocyte sedimentation rate (ESR) occurs in 60 percent, while leukocytosis occurs in 2 percent. Serologic evidence of acute infection with one of the cardiotropic viruses may be demonstrated by either an elevated IgM antibody titer or a fourfold rise in IgG titer over a 4- to 6-week period. Heart-specific antibodies may occur in peripheral blood, although these have also been shown in dilated cardiomyopathes in the absence of myocarditis. Creatine kinase MB, along with other cardiac-specific enzymes, may be elevated, but this unfortunately lacks specificity. The electrocardiogram shows sinus tachycardia in the majority of cases. Conduction defects are common, and so are nonspecific ST- and T-wave abnormalities, QTc prolongation, and low voltage. A myocardial infarction pattern has been reported in a few patients with active myocarditis. Echocardiography and cardiac catheterization may reveal a pattern of dilated, restrictive, or hypertrophic cardiomyopathy that may persist following resolution of inflammation. Regional wall motion abnormalities are surprisingly common.

Treatment

Restriction of physical activity in patients with active myocardial inflammation is prudent. The pharmacologic treatment of patients who present with left ventricular dysfunction is similar to the standard therapy for congestive heart failure, with the use of digoxin, diuretics, and angiotensin-converting enzyme (ACE) inhibitors. Because of the reported higher risk of mural thrombosis with subsequent emboli in those with active myocarditis, oral anticoagulation has been suggested. The use of antiarrhythmic therapy may be indicated, although the insertion of permanent pacemakers or implantable cardioverter/defibrillator (ICD) devices must be tempered against the usual resolution of arrhythmias and heart block following histologic healing of the inflammatory process. An aggressive approach to the management of cardiogenic shock in fulminant cases must be employed, using intravenous inotropic and vasodilator agents, left ventricular mechanical assistance, and balloon counterpulsation as necessary. Cardiac transplantation should

TABLE 16-3
TREATMENT GUIDELINES FOR MYOCARDITIS

Optimize conventional therapy for heart failure (salt restriction, diuretics, digitalis, and angiotensin-converting enzyme inhibitors)

Restrict physical activity

Consider anticoagulation

Treat life-threatening arrhythmias

Immunosupression may be considered if the condition deteriorates despite conservative treatment

Consider transplantation only after all other therapeutic options have failed

be considered only when optimal medical therapy has been administered and adequate time for recovery has been allowed. High early postoperative mortality due to rejection has been seen in transplant patients with active immune-mediated myocardial injury. The routine administration of immunosuppression is not supported by the Myocarditis Treatment Trial, which revealed a neutral effect on mortality and left ventricular function in patients receiving a 24-week regimen of prednisone and either azathioprine or cyclosporine (Table 16-3).

SPECIFIC CAUSES OF MYOCARDITIS

Chagas' Disease

Endemic to rural South and Central America, Chagas' disease is secondary to the protozoan *Trypanosoma cruzi*. Cardiac manifestations are believed to be at least in part due to cellular and humoral host responses, leading to myocarditis that typically presents with heart failure, heart block, or arrhythmia. Nifurtimox affords specific treatment. In some cases, permanent pacemaker treatment and/or chronic antiarrhythmic therapy with amiodarone may be indicated.

Toxoplasmosis

Acute infection by *Toxoplasma gondii* may result in active myocarditis, particularly following cardiac transplantation, when the donor heart may transmit the infection. Diagnosis rests on finding toxoplasma cysts within myocytes by endomyocardial biopsy. Specific treatment consists of administering pyrimethamine and sulfadiazine.

Cytomegalovirus

Cytomegalovirus myocarditis is usually self-limited and asymptomatic except in the cardiac transplant recipient, where infection is associated with either fulminant heart failure or allograft coronary artery disease. Intravenous ganciclovir effectively eradicates the virus.

Lyme Carditis

Lyme disease, caused by the spirochete *Borrelia burgdorferi,* may progress in its chronic phase to Lyme carditis, which usually presents as heart block. Left ventricular dysfunction is rare. Tetracycline is useful in the early phase of Lyme disease and glucocorticoids have been reported to produce cardiac improvement in the later stages.

Eosinophilic Myocarditis

Eosinophilic infiltration of the heart muscle may lead to acute necrotizing myocarditis, which may occur in the absence of peripheral eosinophilia. Prompt improvement of cardiac function has been reported following administration of glucocorticoids in uncontrolled trials.

Giant-Cell Myocarditis

Demonstration of giant cells by endomyocardial biopsy identifies an aggressive form of myocarditis that is characterized by progressive disease unabated by medical therapy. This form of myocarditis may be associated with autoimmune diseases such as myasthenia gravis, autoimmune hemolytic anemia, or polymyositis. Recurrences have been reported following cardiac transplantation.

PHEOCHROMOCYTOMA AND THE HEART

Cardiac manifestations in pheochromocytoma are a result of elevated levels of circulating catecholamines and general sympathetic stimulation. Clinically, presentation may be systolic or diastolic heart failure, cardiac arrhythmias, ischemic heart disease in the absence of epicardial coronary disease, and complications of hypertension. Contraction-band necrosis has been reported in histologic sections.

CARCINOID HEART DISEASE

Cardiac involvement occurs in about half of patients with carcinoid syndrome. The characteristic lesion is myofibromatous involvement of the right (tricuspid and pulmonary) heart valves, although in some patients involvement of the mural endocardium may present with restrictive heart disease. Five-year survival rates of 70 to 80 percent have been achieved using interferon and somatostatin analogs. Surgical management to correct stenotic and regurgitant right heart valves is encouraging.

INFLAMMATORY CARDIAC DISEASE

Hyperoxaluria

In both primary and secondary oxalosis, oxalate crystals may deposit in the conducting system of the heart, causing heart block, and occasionally in the myocardium and the coronary arteries. Congestive heart failure and cardiac arrhythmias may occur. Interestingly, the cardiac dysfunction seen in primary oxalosis has been reported to reverse after combined kidney/liver transplantation.

Gout

Uric acid crystal deposition may be found in the walls of blood vessels, myocardial interstitium, pericardium, conducting system, mitral annulus, and heart valves, eliciting a foreign-body type of granulomatous response. Nevertheless, heart muscle disease associated with hyperuricemia is rare.

HEMOCHROMATOSIS

Iron deposition in the heart as a result of a genetic abnormality or an acquired defect in iron handing may result in a mixed dilated/restrictive cardiomyopathy with both systolic and diastolic dysfunction. Iron deposits are more prominent in the ventricles and are more epicardial than endocardial in location. Conduction system involvement is uncommon and the coronary vasculature appears to be uninvolved. Iron removal by either phlebotomy or chelation therapy with deferoxamine has been associated with improvement in both resting and exercise left ventricular dysfunction.

AMYLOID HEART DISEASE

Amyloid deposition in the interstitium of the myocardium results in increased wall thickness, early diastolic dysfunction, and late development of systolic heart failure. Amyloid heart disease is categorized according to the specific protein deposited. Amyloid AL, myeloma-associated or primary systemic amyloidosis, consists of immunoglobulin light chains. Amyloid AA, or secondary amyloidosis, results from the deposition of amyloid A protein and is seen in such chronic inflammatory disorders as rheumatoid arthritis, Crohn's disease, or familial Mediterranean fever. Familial amyloidosis, inherited as an autosomal dominant disorder, is an expanding group characterized by protein amyloid AF, a variant form of prealbumin (or transthyretin) with a single amino acid transmutation specific for the disease. Amyloid SSA, known as senile cardiac amyloid, is also associated with an abnormal transthyretin protein deposition and may only affect the atria, only the aorta, or multiple organ systems in senile systemic amyloidosis.

The diagnosis of amyloid heart disease rests on a compatible clinical history and a characteristic echocardiographic picture. Cardiac ultrasound may reveal a striking granular, or speckled appearance of a thickened myocardium, along with biatrial enlargement, thickened atrial septa and valves, and, in some cases, a small pericardial effusion. Therapy is limited to symptomatic treatment of heart failure. Patients are unusually sensitive to digitalis glycosides. Once failure symptoms are recognized, the prognosis is grim, with a reported survival of less than 5 years.

SARCOIDOSIS

The characteristic lesion is the sarcoid granuloma, which preferentially involves the cephalad portion of the ventricular septum or the left ventricular papillary muscles. Although cardiac involvement is found in about one-third of autopsy cases of patients with sarcoidosis, clinical cardiac manifestations occur in less than 10 percent. Presenting symptoms vary from different degrees of heart block, arrhythmias, and heart failure. Nuclear scintigraphy with the use of thallium or sestamibi has been reported to be helpful, although the diagnosis is most commonly inferred from the cardiac symptoms in association with evidence of sarcoidosis elsewhere. Administration of glucocorticoids seems to improve cardiac symtoms, reverse electrocardiographic abnormalities, and normalize thallium scan defects in about half of the patients treated.

SUGGESTED READINGS

Aretz HT, Billingham ME, Edwards WD, et al: Myocarditis. *Am J Cardiovasc Pathol* 1986; 1:3–14.

Dabestani A, Child JS, Perloff JK, et al: Cardiac abnormalities in primary hemochromatosis. *Ann NY Acad Sci* 1988; 526:234–243.

Higuchi MDL, DeMorais CF, Barreto ACP, et al: The role of active myocarditis in the development of heart failure in chronic Chagas' disease: a study based on endomyocardial biopsies. *Clin Cardiol* 1987; 10:665–670.

Lundin L, Norheim I, Landelius J, et al: Carcinoid heart disease: relationship of circulating vasoactive substances to ultrasound detectable cardiac abnormalities. *Circulation* 1988; 77:264–269.

Mason JW, Billingham ME, Ricci DR: Treatment of acute inflammatory myocarditis assisted by endomyocardial biopsy. *Am J Cardiol* 1980; 45:1037–1044.

Mason JW, O'Connell JB: Clinical merit of endomyocardial biopsy. *Circulation* 1989; 79:971–979.

Mason JW, O'Connell JB, Herskowitz A, et al: A clinical trial of immunosuppressive therapy for myocarditis. *N. Engl J Med* 1995; 333:269–275.

O'Connell JB, Mason JW: Diagnosing and treating active myocarditis. *West J Med* 1989; 150:431–435.

O'Connell JB, Mason JW: The applicability of results of streamlined trials to clinical practice: The Myocarditis Treatment Trial. *Stat Med* 1990; 9:193–197.

O'Connell JB, Renland DG: Myocarditis and specific cardiomyopathies. In Alexander RW, Schlant RC, Fuster V, O'Rourke RA, Roberts R, Sonnenblick EH (eds): *Hurst's*

The Heart, 9th ed. McGraw-Hill, New York, 1998:2089–2107.

Olson LJ, Okafor EC, Clements IP: Cardiac abnormalities of Lyme disease: manifestations and management. *Mayo Clin Proc* 1986; 61:745–749.

Report of the 1995 World Health Organization/International Society and Federation of Cardiology Task Force on the definition and classification of cardiomyopathies. *Circulation* 1996; 93:841–842.

Satoh M, Tamura G, Segawa I, et al: Expression of cytokine genes and presence of enteroviral genomic RNA in endomyocardial biopsy tissues of myocarditis and dilated cardiomyopathy. *Virchows Arch* 1996; 427:503–509.

Spry, CJF, Tai PC: The eosinophil in myocardial disease. *Eur Heart J* 1987; 8(suppl J):81–84.

Tracy S, Wiegand V, McManus B, et al: Molecular approaches to enteroviral diagnosis in idiopathic cardiomyopathy and myocarditis. *J Am Coll Cardiol* 1990; 15:1688–1694.

Valentine H, McKenna WJ, Nihoyannopoulos P, et al: Sarcoidosis, a pattern of clinical and morphological presentation. *Br Heart J* 1987; 57:256–263.

Young LHY, Joag SV, Zheng L-M, et al: Perforin-mediated myocardial damage in acute myocarditis. *Lancet* 1990; 336:1019–1021.

17

IDIOPATHIC DILATED CARDIOMYOPATHY

Edward M. Gilbert
Michael R. Bristow

DEFINITION

Idiopathic dilated cardiomyopathy (IDC) is a disease of unknown etiology that principally affects the myocardium. The diagnosis of IDC is established by the presence of left ventricular dilation and systolic dysfunction in the absence of congenital, coronary, valvular, or pericardial heart disease. In some patients the development of IDC is associated with clinical factors such as alcoholism, pregnancy, or a family history of cardiomyopathy. However, IDC is distinct from "secondary myocardial diseases," which occur with a specific systemic disorder that may be metabolic, collagen-vascular, infiltrative, neuromuscular, inflammatory, or neoplastic in origin. There are currently no specific gross anatomic, histologic, or ultrastructural morphologic features that can unequivocally differentiate IDC from other causes of heart failure.

PATHOLOGY

The major morphologic feature of IDC on postmortem examination is dilatation of the cardiac chambers. There is an increase in total weight, muscle mass, and myocyte cell volume in IDC, but left ventricular wall thickness is usually not increased because of the marked dilatation of the ventricular cavities. Grossly visible scars may be present in either ventricle. Scarring occurs in the absence of significant narrowing of the epicardial coronary arteries. Intracardiac thrombi and mural endocardial plaques (from the organization of thrombi) are present at necropsy in more than 50 percent of patients with IDC.

301

The characteristic findings of IDC on microscopy are marked myocyte hypertrophy and very large, bizarre-shaped nuclei. Myocyte atrophy and myofilament loss is also seen. These morphometric changes do not correlate with the severity of illness. Ultrastructural abnormalities—such as mitochondrial changes, T-tubular dilatation, and intracellular lipid droplets—may be observed in IDC but can also be observed in other forms of heart disease. There may be interstitial parenchymal and perivascular focal infiltrates of small lymphocytes, but adjacent myocyte damage is not observed. Fibrosis is nearly always present in IDC; its pattern is quite variable, from a fine perimyocytic distribution to coarse scars indistinguishable from those present in chronic ischemia. However, small intramural arteries and capillaries are normal in IDC.

ETIOLOGY

The etiology of IDC has not been established. Since IDC is defined clinically, it is likely that several specific etiologies are responsible for this condition. Apparently "idiopathic" dilated cardiomyopathy may be familial and presumably genetic in etiology. Immune mechanisms may play an etiologic role in other cases. A number of immune regulatory abnormalities have been identified in IDC, including humoral and cellular autoimmune reactivity against myocytes, decreased natural-killer cell activity, and abnormal suppressor cell activity. These findings are not universally present in patients with IDC, however, and some are also present in other heart diseases. HLA associations have also been identified in IDC; the frequency of HLA-B27, -A2, -DR4, and -DQ4 are increased compared with controls and the frequency of HLA-DRw6 is decreased compared with controls. The association in IDC with specific HLA antigens suggests a possible immunologic etiology for this disease. These specific HLA antigens are present in less than 50 percent of patients with IDC, however, and the heterogeneity of these antigens does not point to a unique site for a putative disease-associated gene. Further investigation will be necessary to elucidate the significance of these findings.

Most cases of IDC are believed to be sporadic. A recent study of the first-degree relatives of a sequential series of patients found that 12 of 59 index patients had familial disease. This relatively high percentage of familial dilated cardiomyopathy suggests a genetic basis for at least some of the cases.

Dystrophin gene abnormalities have been described in patients with either X-linked or sporadic cardiomyopathy. More recently, missense mutations in the cardiac actin gene have been identified in two unrelated families with hereditary idiopathic dilated cardiomyopathy. Thus, it appears that genetic abnormalities of contractile proteins may result in heart failure. In addition, alterations in the relative expression of normal genes has been observed within the myocardium of subjects with heart failure (see "Pathophysiology," below).

A clinical and pathologic syndrome that is similar to IDC may develop after resolution of viral myocarditis in animal models and biopsy-proven myocarditis in human subjects. This has led to speculation that IDC may develop in some individuals as a result of subclinical viral myocarditis. Theoretically, an episode of myocarditis could initiate a variety of autoimmune reactions that injure the myocardium and ultimately result in the development of IDC. The abnormalities in immune regulation and the variety of antimyocardial antibodies present in IDC are consistent with this hypothesis; however, it is generally not possible to isolate infectious virus or to demonstrate the presence of viral antigens in the myocardium of patients with IDC. Enteroviral RNA sequences are found in hearty biopsy samples in IDC, but in only approximately one-third of patients. Furthermore, active myocardial inflammation is usually not detected in IDC. Finally, corticosteroid therapy in patients with IDC does not result in significant clinical improvements. Thus, while the viral infection–autoimmune hypothesis is an attractive candidate for the etiology of some cases of IDC, it remains unproven.

CLINICAL MANIFESTATIONS

History

Patients of all ages may develop IDC, with the highest incidence found in middle age. Symptoms often develop gradually, and some patients may be asymptomatic for months or years. Other patients present with a sudden onset of symptoms, frequently after some clinical event that has increased cardiac demands, such as surgery or systemic infection. In young adults, this initial acute presentation may be incorrectly diagnosed as "pneumonia," since heart disease is usually not expected in this age group. While it is attractive to attribute the rapid onset of heart failure to an acute process such as myocarditis, it is uncommon to find specific myocardial diseases on endomyocardial biopsy.

Most symptoms of IDC are caused by heart failure. These symptoms include pulmonary congestion (e.g., dyspnea on exertion, orthopnea, paroxysmal nocturnal dyspnea, and dyspnea at rest), systemic congestion (e.g., peripheral edema, nausea and abdominal pain from hepatic congestion, and nocturia), and low cardiac output (e.g., fatigue and weakness). Angina pectoris may be present even though the coronary arteries are normal. Symptoms of light-headedness, dizziness, or syncope may occur from arrhythmias or drug-related orthostatic hypotension.

Physical Examination

The findings on physical examination are nonspecific and related to the hemodynamic abnormalities of heart failure. Blood pressure may be normal when heart failure is well compensated; but with progression of left ventricular dysfunction, the pulse pressure narrows, pulsus alternans appears, and hypotension develops. Resting tachycardia is common. Peripheral edema develops in the presence of salt and water retention. Cool, pale, or cyanotic extremities with delayed capillary refill and constricted peripheral veins suggest poor peripheral perfusion. Although less impressive to the novice observer than edema, physical signs of poor peripheral perfusion are of greater clinical importance, since such findings indicate that cardiac output is severely reduced. When pulmonary congestion is present, patients may develop tachypnea, breathlessness, pulmonary crackles, wheezing, and physical signs of pleural effusion. Jugular venous distention and hepatomegaly are found when right heart pressures are elevated. The left ventricular impulse may be laterally displaced and a right ventricular impulse may be present on palpation. Gallop sounds and the murmurs of mitral and tricuspid regurgitation are frequently audible.

Electrocardiogram

There are no "typical" electrocardiographic findings in IDC. Sinus tachycardia is common, and atrial and ventricular arrhythmias or atrioventricular conduction abnormalities may be observed. Intraventricular conduction abnormalities (including bundle branch blocks), as well as ST-segment and T-wave changes are common. Some patients will present with poor precordial R-wave progression and even with Q waves that mimic old myocardial infarction. Arrhythmias are frequently recorded on the ambulatory electrocardiogram.

Chest X-ray

Cardiomegaly is the typical feature of IDC on the chest roentgenogram. With the development of pulmonary venous hypertension, there may be pulmonary venous redistribution, interstitial edema, and alveolar edema. Dilation of the azygos vein and superior vena cava may be observed with the development of systemic venous hypertension. Pleural effusions may also be present.

SPECIAL LABORATORY STUDIES

Echocardiography

The echocardiogram is an excellent tool for the evaluation of IDC. Two-dimensional study may show left-sided or four-chamber dilatation. Mitral and tricuspid regurgitation are frequently seen with Doppler evaluation. Left ventricular and septal wall thickness is usually normal. A pericardial effusion may be present. In most cases of IDC, there is global left ventricular hypokinesis, but some individuals will have segmental wall motion abnormalities.

Radionuclide Imaging Techniques

Radionuclide ventriculography can also be used to demonstrate left ventricular dilatation and hypokinesis. This method is particularly useful in the serial assessment of left ventricular function.

Cardiac Catheterization

Left ventricular dilatation and hypokinesis are present on left ventricular angiography. Hemodynamic findings are typical for left ventricular or biventricular failure. Coronary arteriography generally shows normal coronary arteries, but older patients may have "incidental" coronary artery lesions (stenosis < 50% of luminal diameter). The diagnosis of IDC, however, cannot be made in the presence of significant coronary artery disease.

Endomyocardial Biopsy

Endomyocardial biopsy is frequently performed in patients with unexplained heart failure, since this procedure is the only accurate technique for the diagnosis of myocarditis and infiltrative cardiomyopathies. This interest in diagnosing myocarditis is based on anecdotal reports that patients with biopsy-proven myocarditis may respond to immunosuppressive therapy. But until clinical trials have established that immunosuppression is effective for the treatment of myocarditis, the role of endomyocardial biopsy in the evaluation of IDC remains undefined. Although infiltrative cardiomyopathies such as sarcoidosis or hemosiderosis may respond to specific treatment, the results from endomyocardial biopsy will alter therapy in less than 10 percent of unexplained dilated cardiomyopathy.

PATHOPHYSIOLOGY

The primary pathophysiologic abnormality in IDC is a decrease in myocardial contractility. This results in a reduction in left ventricular ejection fraction and an increase in left ventricular end-diastolic volume as well as the typical pathophysiologic changes that occur with heart failure of any etiology. Because of the nonlinear nature of the structure-function relationship, some patients with severely reduced left ventricular ejection fraction may have very few symptoms.

Significant alterations occur in IDC in the myocardial adrenergic receptor–G protein–adenylate cyclase complex. In severe heart failure, there is a 60 to 70 percent reduction in $beta_1$-adrenergic receptor density and $beta_1$-receptor mRNA expression. Myocardial $beta_2$-receptor density and mRNA expression are not changed in IDC, but $beta_2$-agonist responsiveness is mildly reduced (approximately 30 percent) due to $beta_2$-receptor uncoupling. A 30 to 40 percent increase in the activity of the inhibitory Gi protein is also present in heart failure. Preliminary observations suggest that $beta_2$-receptor uncoupling may be related to the increase in Gi activity. The VIP receptor is decreased by approximately 60 percent, but it exhibits increased affinity for VIP, which tends to counteract the effects of VIP-receptor downregulation. There is a small increase in $alpha_1$-receptor concentration. Gs protein function, the H_2-histamine receptor, M_2-muscarinic receptor, and adenylate cyclase are not significantly altered in IDC. Several

mechanisms related to contractility are normal in end-stage
IDC, including the density of calcium-antagonist binding sites,
maximal tension response to calcium in vitro or in vivo, SR
calcium ATPase gene expression, and creatine kinase activity.
We recently found that myocardial gene expression for the
alpha-myosin heavy chain is downregulated and beta-myosin
heavy chain expression is upregulated in subjects with heart
failure from IDC. The relative expression of the genes for the
alpha- and beta-myosin heavy chain isoforms returns toward
normal in subjects who experience an improvement in left ventricular function.

NATURAL HISTORY AND PROGNOSIS

The incidence of IDC is estimated at 3 to 10 cases/100,000,
and approximately 20,000 new cases are diagnosed annually in
the United States. Patients with IDC usually develop progressive deterioration of ventricular function, resulting in death
from either pump failure or arrhythmia. Currently, the 5-year
mortality is estimated at 40 to 80 percent. In the majority of
reports, 1-year mortality is about 25 percent and 2-year mortality about 35 to 40 percent. However, some authors have
reported that patients who survive more than 2 or 3 years have
a relatively good long-term prognosis, and survival curves of
such long-term survivors may even parallel survival curves of
the general population. Stabilization and even spontaneous
improvement have been observed in 20 to 50 percent of
patients, but complete recovery of cardiac function is rare.

What accounts for the large variability in patient survival in
IDC reported by these different authors? Several potential factors can be suggested, including differences in underlying
pathogenic mechanism for IDC, differences in duration or
severity of disease, as well as differences in medical therapy.
Currently, patients with IDC are being identified earlier in the
course of their disease because of technologic advances in noninvasive cardiac imaging and a greater awareness of the
disease among clinicians. Recent advances in medical therapy
may also produce a favorable effect on long-term outcome.
Thus, the patient with IDC in the 1990s may have a better
prognosis than patients reported in the literature. Several
authors have analyzed the impact of several clinical, hemodynamic, laboratory, and histologic parameters on survival in
IDC. The prognostic significance of these parameters is summarized in Table 17-1.

TABLE 17-1

PREDICTING PROGNOSIS IN IDIOPATHIC DILATED CARDIOMYOPATHY

	Predictive	Possibly Predictive	Not Predictive
Clinical factors	Symptoms	Alcoholism Peripartum presentation Family history	Age Duration of illness Preceding viral illness
Hemodynamic factors	LV* ejection fraction Cardiac index	LV chamber size Atrial pressures	
Dysrhythmia	LV conduction delay Complex ventricular ectopy	Atrioventricular block Atrial fibrillation	Simple ventricular ectopy
Histologic morphology	Atrial natriuretic factor† Hyponatremia†	Myofibril volume	Other histologic findings
Neuroendocrine factors		Plasma norepinephrine†	

*Left ventricular.

†Identified as prognostic factors for all causes of heart failure.

308

DIFFERENTIAL DIAGNOSIS

IDC must be distinguished from other causes of heart failure. All patients should be evaluated with a complete history and physical examination, chest x-ray, electrocardiogram, and echocardiogram with Doppler. These studies will identify most known causes of heart failure. Patients older than 40 years of age or who have a history suggestive of ischemia or a high coronary risk profile should also undergo coronary angiography to exclude coronary disease.

TREATMENT

Medical Therapy

Medical therapy of IDC is directed at controlling the heart failure state. This is accomplished by controlling salt and water retention, reducing the workload of the heart, and improving its pumping performance (see also Chap. 1). Patients should be counseled to limit their physical activities based on their functional capacity and to reduce their dietary intake of sodium.

Diuretic therapy must be individualized to account for variations in intravascular volume and renal function. Inadequate diuretic therapy will result in symptomatic pulmonary and systemic venous congestion. Overdiuresis will result in azotemia and electrolyte disorders and will ultimately decrease cardiac output. Patients should be instructed to follow their weights daily so that diuretic doses can be changed in response to weight changes, which usually occur prior to the development of frank symptoms. Serum electrolytes, including magnesium, should be carefully monitored with diuretic therapy. Loop diuretics such as furosemide, torsemide, or bumetanide are the agents of choice when dietary measures are not adequate for controlling congestive failure. Thiazide diuretics are frequently ineffective at the low glomerular filtration rates that occur with heart failure. Most patients receiving diuretics will also require potassium supplementation and/or the addition of a potassium-sparing diuretic to maintain serum potassium > 4.5 meq/dL. When patients are refractory to loop diuretic therapy alone, the addition of metolazone can result in marked diuresis. Metolazone is a potent long-acting diuretic whose site of action is primarily the cortical diluting segment and secondarily the proximal tubule. Metolazone is unique among the thiazide class of diuretics because it is the only one that is effective when the glomerular filtration rate is less than 25 mL/min. Combined therapy with furosemide and metolazone should be initiated with caution because of the potential for excessive diuresis and electrolyte loss (see also Chaps. 1 and 42).

Vasodilator therapy reduces the workload of the heart and improves pumping performance. The angiotensin converting enzyme (ACE) inhibitors and the combination of hydralazine and isosorbide dinitrate have been established to be effective vasodilator regimens. The ACE inhibitors also have important neuroendocrine effects in addition to their vasodilator activity. Patients treated with ACE inhibitors have improved survival as compared with those on placebo or the combination of hydralazine and isosorbide dinitrate. Vasodilator therapy should be initiated at low dose and titrated upward, with careful observation for orthostatic hypotension. Hypotension is most frequently observed in volume-depleted patients. Therefore it is necessary to carefully assess intravascular volume status and, when necessary, to correct relative hypovolemia prior to initiation of vasodilator therapy.

Digitalis is a complex drug and observed beneficial clinical responses may be related to a variety of properties, not just its positive inotropic effects. Digitalis is the undisputed drug of choice for the chronic control of ventricular rate in atrial fibrillation. Several recent randomized drug trials have shown that digitalis also improves ejection fraction, exercise tolerance, and clinical symptoms compared to placebo, even in sinus rhythm. In a recently completed muticentered trial (DIG trial), there was no difference in survival between heart failure patients treated with digoxin or placebo. Because digoxin improves symptoms but does not improve survival, it is our practice to use digoxin in sinus rhythm only when cardiomegaly and left ventricular systolic dysfunction are present and heart failure symptoms have not been readily controlled with diuretics and ACE inhibitors. It is important to follow digoxin and serum electrolyte concentrations to avoid lethal arrhythmias of digitalis toxicity.

Because of the risk for pulmonary and systemic emboli, it is common practice to initiate warfarin anticoagulation in patients with left ventricular mural thrombi, atrial fibrillation, or a history of systemic or pulmonary embolic disease. Some clinicians advocate the use of anticoagulation in patients with severe heart failure symptoms [New York Heart Association (NYHA) class IV] or severely reduced left ventricular ejection fraction (less than 20 percent). It should be recognized, however, that the efficacy of anticoagulation in IDC has not been established by prospective, controlled clinical trials. Anticoagulant therapy must be carefully monitored, with warfarin dose adjusted to maintain an International Normalized Ratio (INR) of 2 to 3.

Treatment of arrhythmias in IDC is a difficult problem, and the roles of antiarrhythmic drugs, the use of electrophysiologic testing to guide therapy, and use of the automatic implantable defibrillator still need to be defined. The most important goals in the management of arrhythmia are to correct hypoxia, as well as electrolyte and acid-base abnormalities, and to treat heart failure. Antiarrhythmic medications have not been shown to improve survival in IDC. All antiarrhythmic drugs have at least some negative inotropic effects. In addition, many antiarrhythmic drugs are also potentially proarrhythmic, particularly in patients with severe heart failure. Because of these problems, it is our practice to reserve antiarrhythmic therapy to patients with sustained ventricular tachycardia, syncope, or sudden death. These patients should be referred to a practitioner with expertise in the management of arrhythmias. Studies of amiodorone in this setting have shown encouraging results. An automatic implantable defibrillator should also be considered in such patients. The relative merits of these two therapeutic choices are currently under prospective investigation.

Beta-Adrenergic Blocking Agents

Several studies have demonstrated that long-term beta blockade improves hemodynamic function and clinical symptoms in patients with IDC. Therapy with beta-adrenergic blockers can be successfully initiated in most patients with IDC and heart failure. Clinical experience suggests that beta$_1$-selective antagonists (such as metoprolol or bisoprolol) or nonselective beta-antagonists with vasodilator properties (such as bucindolol or carvedilol) are better tolerated than nonselective beta-antagonists such as propranolol. In our experience, significant improvements in left ventricular ejection fraction and clinical symptoms occur in greater than 50 percent of patients treated with beta-blocking agents. Beta-blocker therapy must be initiated at very low doses, however, and these must be tritrated upward very slowly. In two prospective clinical trials, carvedilol reduced disease progression in heart failure patients when it was added to therapy with diuretics, digitalis, and ACE inhibitors. A stratified analysis of the U.S. multicenter carvedilol trials program demonstrated a 65 percent reduction in mortality in patients with heart failure. Survival was the primary endpoint for the CIBIS II trial. In this trial, bisoprolol reduced mortality by 32 percent as compared with placebo. These results suggest that beta-blocker therapy should be considered for all patients with idiopathic dilated cardiomyopathy without specific contraindications. At present, carvedilol is the only beta-blocker approved for the treatment of heart failure by the U.S. Food and Drug Administration.

Experimental Therapies

Several other classes of drugs are currently being studied for the treatment of heart failure. In the PRAISE trial, the calcium-channel blocker amlodipine reduced mortality in subjects with a nonischemic cardiomyopathy and NYHA class III or IV heart failure when added to standard triple therapy. This mortality benefit was not observed in patients with ischemic cardiomyopathy. Mortality was not reduced in trials of other calcium-channel blockers such as felodipine and mebefridil. Angiotensin receptor blockers are also under investigation. These agents have short-term hemodynamic effects that are similar to those of ACE inhibitors. In one small study of elderly patients with heart failure (ELITE), mortality was lower in the group receiving the angiotensin receptor blocker losartan than in the group receiving the ACE inhibitor captopril. Larger survival trials are currently under way to test amlodipine and losartan. Pending their results, these agents should not be considered substitutes for ACE inhibitors and beta-blockers. Angiotensin receptor blockers may be considered for patients who are intolerant of ACE inhibitors because of cough. This use of these agents for the treatment of heart failure has not been approved by the U.S. Food and Drug Administration (see also Chap. 1).

Cardiac Transplantation

With improved understanding of donor and recipient selection, the establishment of an aggressive approach to the treatment of infections, the development of endomyocardial biopsy to accurately diagnose acute cardiac allograft rejection, and improved immunosuppression, the 1-year survival of cardiac transplant recipients now exceeds 85 percent. The survival of carefully selected transplant recipients far exceeds the survival of similar patients who are not transplanted. Transplantation is an important option in the management of IDC, since so many patients are young or middle-aged adults and patients with advanced NYHA class III or IV symptoms refractory to medical management (including therapy with beta-blockers) should be considered for transplantation.

SUGGESTED READING

Baandrup U, Olsen EG: Critical analysis of endomyocardial biopsies from patients suspected of having cardiomyopathy. *Br Heart J* 1981; 45:475–486.

Bristow MR, Bohlmeyer TJ, Gilbert EM: Dilated cardiomyopathy. In: Alexander RW, Schlant RC, Fuster V, O'Rourke RA, Roberts R, Sonnenblick EH (eds): *Hurst's The Heart*, 9th ed. New York, McGraw-Hill, 1998:2039–2055.

Bristow MR, Ginsburg R, Fowler M, et al: β_1- and β_2-adrenergic receptor subpopulations in normal and failing human ventricular myocardium. *Circ Res* 1986; 59:297–309.

Costanzo-Nordine MR, O'Connell JB, Engelmier RS, et al: Dilated cardiomyopathy: functional status, hemodynamics, arrhythmias, and prognosis. *Cathet Cardiovasc Diagn* 1985; 11:445–453.

Fuster V, Gersh BJ, Giuliani ER, et al: The natural history of idiopathic dilated cardiomyopathy. *Am J Cardiol* 1981; 47:525–531.

Gilbert EM, Abraham WT, Olsen S, et al: Comparative hemodynamic LV functional, and antiadrenergic effects of chronic treatment with metoprolol vs. carvedilol in the failing heart. *Circulation* 1996; 94:2817–2825.

Gilbert EM, Anderson JL, Deitchman D, et al: Long-term β-blocker vasodilator therapy improves cardiac function in idiopathic dilated cardiomyopathy: a double-blind, randomized study of bucindolol versus placebo. *Am J Med* 1990; 88:223–229.

Lowes BD, Minobe W, Abraham WT, et al: Changes in gene expression in the intact human heart: down-regulation of α-myosin heavy chain in hypertrophied, failing human ventricular myocardium. *J Clin Invest* 1997; 100:2315–2324.

Olson TM, Michels VV, Thibodeau SN, et al: Actin mutations in dilated cardiomyopathy, a heritable form of heart failure. *Science* 1998; 280:750–752.

Packer M, Bristow MR, Cohn JN, et al: Effect of carvedilol on morbidity and mortality in chronic heart failure. *N Engl J Med* 1996; 334:1349–1355.

18

HYPERTROPHIC CARDIOMYOPATHY

Barry J. Maron

DEFINITION

The cardinal feature of *hypertrophic cardiomyopathy* (HCM) is an unexplained increase in left ventricular wall thickness without ventricular dilatation. Left ventricular systolic function is usually normal but often hypercontractile.

ETIOLOGY

HCM is usually familial and genetically transmitted as an autosomal dominant trait. A myriad of mutations in five genes encoding proteins of the cardiac sarcomere have been identified as disease-causing.

PATHOLOGY

Characteristic abnormalities at necropsy include (1) an asymmetric pattern of left ventricular hypertrophy in which the ventricular septum is thicker than portions of the left ventricular free wall; (2) a small or normal-sized left ventricular cavity; (3) a fibrous mural endocardial plaque on the ventricular septum in the outflow tract in apposition to the anterior mitral leaflet; (4) mitral valve enlargement and elongation with or without secondary thickening; (5) atrial dilatation; (6) abnormal intramural coronary arteries with thickened walls and narrowed lumina; (7) interstitial and replacement fibrosis; and

(8) disorganized left ventricular architecture with myocardial cell disarray.

The abnormal intramural coronary arteries are regarded as a form of "small vessel disease" and are probably responsible for myocardial ischemia and replacement fibrosis. Areas of cellular disorganization, often diffusely distributed throughout the myocardium, are thought to potentially represent the arrhythmogenic substrate in this disease, possibly serving as a nidus for ventricular arrhythmias.

CLINICAL MANIFESTATIONS

History

Exertional dyspnea and chest pain are the most common symptoms. Patients also frequently experience impaired consciousness (syncope, near-syncope, or dizziness), palpitations, or orthopnea or paroxysmal nocturnal dyspnea in more advanced stages of heart failure. Dyspnea probably results from abnormal ventricular compliance and increased pulmonary venous pressure and is accentuated by tachycardia. Chest pain may be characteristic of angina pectoris (in the absence of coronary artery disease) but is usually atypical in character.

Physical Examination

In the presence of a left ventricular outflow tract pressure gradient, the characteristic physical findings include (1) systolic ejection murmur along the lower left sternal border and at the apex, decreasing with squatting and handgrip but intensified with standing, the Valsalva maneuver, and following a premature ventricular beat; (2) abrupt upstroke to the arterial pulse, at times demonstrating a bisferiens (bifid) configuration with initial brisk percussion wave and a subsequent tidal wave; (3) a bifid apical systolic impulse, often preceded by a prominent presystolic impulse giving a triple-beat character; and (4) an *a* wave in the jugular venous pulse associated with a prominent S_4 at the cardiac apex. These findings are absent in patients with the much more common nonobstructive form of the disease; the latter patients have no or only a soft systolic ejection murmur.

DIAGNOSTIC TESTS

Electrocardiogram

The electrocardiogram (ECG) is abnormal in 90 to 95 percent of patients with HCM and a wide variety of patterns occur; however, none of these abnormalities is either diagnostic or specific. Patterns of left ventricular hypertrophy with increased QRS voltage and ST-T wave abnormalities are common. Also frequent are left anterior hemiblock, deep inferior Q waves (which may simulate healed myocardial infarction), prominent T-wave inversion, left atrial enlargement, and "pseudoinfarction" patterns with dimished R-wave voltages in the right precordial leads.

Chest X-ray

The cardiac silhouette is often enlarged, and left atrial enlargement and interstitial pulmonary edema may be present. The assessment of cardiac size on x-ray usually adds little to the clinical evaluation, since cardiac dimensions are best assessed with echocardiography.

SPECIAL DIAGNOSTIC EVALUATION

Echocardiography

The two-dimensional echocardiogram is the most useful noninvasive diagnostic test for hypertrophic cardiomyopathy. The anatomic and echocardiographic diagnosis of HCM is based on the identification of a hypertrophied, nondilated left ventricle in the absence of another cardiac or systemic disease capable of producing the degree of wall thickening evident in that patient. A wide variety of patterns of left ventricular wall thickening may be present in patients with HCM—i.e., commonly involving diffuse hypertrophy of both septum and free wall, although not infrequently localized thickening of relatively small portions of the left ventricular wall. The mitral valve (which is often elongated) is displaced anteriorly toward the septum within the small left ventricular outflow tract. Systolic anterior motion (SAM) of the mitral valve with ventricular septal contact is responsible for dynamic obstruction to left ventricular outflow and is also associated with premature partial closure of the aortic valve. Doppler echocardiography may assess the diastolic filling and relaxation abnormalities commonly present in this disease, dynamic characteristics of the outflow pressure gradient, and also the severity of mitral regurgitation (with color-flow imaging).

Ambulatory Electrocardiography

The Holter ECG may be of value in identifying short bursts of nonsustained ventricular tachycardia, markers for the risk of sudden cardiac death when multiple and repetitive.

Radionuclide Imaging Technique

Technetium-99m labeling of the blood pool and gated blood pool scanning can define cavity size and offer quantitation of the left ventricular ejection fraction and filling abnormalities; myocardial perfusion imaging with thallium-201 may demonstrate reversible or fixed perfusion defects suggestive of areas of myocardial ischemia or scarring.

Hemodynamics

Cardiac catheterization and angiography are usually performed only when surgery is contemplated in order to measure (1) intracardiac pressures and the presence of resting or provocable obstruction (with Valsalva maneuver, isoproterenol infusion, or amyl nitrite inhalation); (2) the presence or absence of coronary artery disease; and (3) left ventricular size and contractility. There is no generally accepted clinical indication for endomyocardial biopsy.

Left Ventricular Function

The predominant functional abnormality in most patients with HCM is impaired ventricular compliance with abnormal diastolic filling and relaxation, presumably due to the distorted myocardial architecture and resultant asynchronous myocardial contraction and relaxation. Of note, the heart failure that patients with HCM manifest characteristically occurs in the presence of a nondilated chamber and normal (or even enhanced) left ventricular systolic function. Therefore, the congestive symptoms so common in HCM are usually a consequence of impaired filling rather than pump failure. Only in the "end-stage" phase of the disease, associated with cavity enlargement and wall thinning, is left ventricular systolic dysfunction evident.

NATURAL HISTORY

HCM should be regarded as a highly complex disease process capable of having important clinical consequences and causing premature death in some patients; however, it should also be recognized that other patients experience little or no disability and even achieve normal life expectancy, often without the aid of major therapeutic interventions. The natural history of HCM is highly variable and unpredictable; sudden cardiac death may occur in young people (usually 12 to 35 years of age), and HCM is the most common cause of sudden death in young competitive athletes. However, sudden catastrophes may also occur in midlife and beyond. Progressive symptoms of heart failure may dominate after age 35 to 40, although other patients may live their entire lives without incurring important functional limitation. Elderly patients with HCM may also suddenly experience severe heart failure after being free of symptoms for many decades. Both patients with and patients without outflow obstruction may die suddenly.

The onset of atrial fibrillation, with loss of the atrial contribution to ventricular filling, may result in severe clinical deterioration, heart failure, and peripheral embolization. Infective endocarditis may occur, the anterior mitral leaflet being the most common site. Pregnancy with vaginal delivery is usually well tolerated.

DIFFERENTIAL DIAGNOSIS

HCM has been regarded clinically as a "great masquerader." The murmur of obstructive HCM can mimic that of aortic valvular stenosis, ventricular septal defect, or mitral regurgitation. Some children with HCM have marked right ventricular outflow obstruction with a murmur similar to that of pulmonary valve stenosis. Chest pain in HCM may be typical of coronary artery disease (angina pectoris), and some ECG patterns suggest a prior healed myocardial infarction.

TREATMENT

Treatment for HCM (Figure 18-1) is focused initially on drug therapy to reduce cardiac symptoms and improve functional limitation. Surgery (ventricular septal myotomy-myectomy;

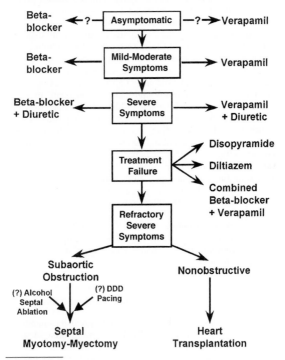

Management of HCM

FIGURE 18-1

Therapeutic strategies for congestive symptoms in patients with HCM. Question marks indicate treatment recommendations that are largely unresolved. (Adapted with permission from Maron BJ, Epstein SE: Heart disease: update. In: Braunwald E (ed): *Textbook of Cardiovascular Medicine*, 3d ed. London, Saunders, 1990.)

Morrow procedure) has been reserved for those patients who fail to benefit from pharmacologic treatment.

Asymptomatic patients may not require specific drug therapy, although prophylactic treatment with beta blockers or verapamil has been utilized selectively by some investigators in an effort to retard disease progression.

Exertional dyspnea (and chest pain) often responds to a variety of beta-blocking agents. The calcium channel blocker verapamil has been shown to reduce symptoms and improve exercise capacity in a large proportion of patients who do not benefit from beta blockers, probably by improving ventricular filling. Some centers have utilized disopyramide (instead of verapamil) to relieve limiting symptoms.

In those patients judged to be at high risk for sudden death (particularly patients with prior cardiac arrest or family history of multiple premature HCM-related deaths), amiodarone and particularly the implantable cardioverter-defibrillator have been employed.

Anticoagulation is indicated in patients with atrial fibrillation and should be continued indefinitely once this arrhythmia has been documented. Vasodilators should be avoided and diuretics used judiciously in treating patients with outflow obstruction. Digitalis may increase the outflow gradient and therefore is contraindicated in patients with obstructive HCM. In those patients with progressive congestive heart failure secondary to impaired systolic function (i.e., "end-stage" HCM), however, the therapeutic strategy is similar to that utilized for pump failure in other diseases, ultimately including the possibility of cardiac transplantation. Antibiotic prophylaxis is indicated to protect against infective endocarditis in patients with outflow obstruction (see Chap. 21)

Surgical treatment with muscular resection from the basal ventricular septum (myotomy-myectomy; Morrow procedure) is presently the "gold standard" for severely symptomatic patients with a peak systolic left ventricular outflow tract gradient of at least 50 mmHg who become refractory to drug treatment. As a result of this operation, the outflow gradient and mitral regurgitation are generally abolished or greatly reduced, and symptoms are relieved in the vast majority. While operation can improve the quality of life, it does not eliminate subsequent risk for disease progression or sudden cardiac death. In major centers the operative risk is now acceptably low (equal to or less than 1 to 2 percent).

In place of myotomy-myectomy, mitral valve replacement has been performed when severe mitral regurgitation due to intrinsic mitral valve disease (such as mitral valve prolapse) is present or in those selected patients for whom the risk of performing myotomy-myectomy is judged to be relatively high due to the particular distribution of ventricular septal hypertrophy. Pacemaker implantation is utilized for associated bradycardia or high-grade atrioventricular block and recently has been proposed as an alternative to surgery to reduce outflow gradient and symptoms. Partial septal ablation by injection of alcohol into the first major septal perforator coronary artery is a new experimental procedure designed to reduce the outflow gradient.

SUGGESTED READING

Klues HG, Schiffers A, Maron BJ: Phenotypic spectrum and patterns of left ventricular hypertrophy in hypertrophic cardiomyopathy: morphologic observations and significance as assessed by two-dimensional echocardiography in 600 patients. *J Am Coll Cardiol* 1995; 26:1699–1708.

Maron BJ: Hypertrophic cardiomyopathy. *Lancet* 1997; 350: 127–133.

Maron BJ: Hypertrophic cardiomyopathy. In: Alexander WR, Schlant RC, Fuster V, O'Rourke RA, Roberts R, Sonnenblick EH (eds): *Hurst's The Heart*, 9th ed. New York, McGraw-Hill, 1997:2057–2074.

Maron BJ, Bonow RO, Cannon RO III, et al: Hypertrophic cardiomyopathy: interrelations of clinical manifestations, pathophysiology, and therapy. *N Engl J Med* 1987; 316: 780–789, 844–852.

McKenna WJ, Deanfield JE, Faroqui A, et al: Prognosis in hypertrophic cardiomyopathy: role of age and clinical, electrocardiographic and hemodynamic features. *Am J Cardiol* 1981; 47:532–538.

Morrow AG, Reitz BA, Epstein SE, et al: Operative treatment in hypertrophic subaortic stenosis: techniques and the results of pre- and postoperative assessments in 83 patients. *Circulation* 1975; 52:88–102.

Spirito P, Seidman CE, McKenna WJ, Maron BJ: The management of hypertrophic cardiomyopathy. *N Engl J Med* 1997; 336:775–785.

Wigle ED, Sasson Z, Henderson MA, et al: Hypertrophic cardiomyopathy: the importance of the site and extent of hypertrophy—a review. *Progr Cardiovasc Dis* 1985; 28: 1–83.

19

RESTRICTIVE CARDIOMYOPATHY

Ralph Shabetai

Restrictive cardiomyopathy characterizes a small but important subset of patients with diastolic heart failure. It is defined as a primary or secondary disease of the myocardium that creates diastolic dysfunction clinically indistinguishable from constrictive pericarditis. Therefore, it is characterized by absence of ventricular dilatation or hypertrophy, raised systemic and pulmonary venous pressures, and ventricular diastolic pressure identical to that of constrictive pericarditis. The venous pressure pulse is characterized by prominent x and y descents, and the heart is usually small. However, the atria are enlarged—sometimes massively, in which case the chest radiograph shows cardiomegaly. Ventricular systolic function is normal.

ETIOLOGY

Some cases are of unknown etiology (idiopathic), others are due to infiltrative myocardial disease such as amyloidosis. Hemochromatosis usually causes dilated cardiomyopathy but less commonly causes restrictive cardiomyopathy. Less common causes include myocarditis, cardiac transplantation, sarcoidosis, pseudoxanthoma elasticum, Löffler's eosinophilic endomyocardial fibrosis, and, in children, Gaucher's disease.

HISTORY

The disease is often far advanced when the patients first present. The manifestations are of left- and right-sided heart failure.

DIAGNOSIS

Clinical Examination

Prominent features are the increased jugular venous pressure with prominent x and y descents, evidence of pulmonary congestion, edema, and often ascites.

Electrocardiogram (ECG)

This is almost always abnormal. Left bundle branch block and left atrial enlargement are common. Amyloidosis is characterized by low voltage in the ECG in contrast to increased wall thickness of the left ventricle by echocardiography. Cardiac arrhythmias, especially atrial fibrillation, are common, especially in amyloidosis.

Chest Radiograph

Characteristically, the heart is of normal size except when there is massive biatrial enlargement. Pulmonary congestion is severe.

Echo Doppler Cardiography

The ventricles are of normal size and have normal systolic function. Pericardial effusion is sometimes present. Major structural disease of the heart valves is absent in most cases. Infiltrative myocardial disease may be mistaken for ventricular hypertrophy. The left ventricular diameter increases in early diastole but does not increase further during mid- or late diastole.

Left ventricular inflow velocity is characterized by a very prominent E wave that has a severely reduced deceleration time.

Other Imaging Modalities

Computed tomography (CT) and magnetic resonance imaging (MRI) indicate absence of abnormal thickness of the pericardium. MRI may also disclose the abnormal myocardial structure.

Cardiac Catheterization

This procedure is frequently done to differentiate restrictive cardiomyopathy from constrictive pericarditis and also to obtain an endomyocardial biopsy to establish the cause of restrictive cardiomyopathy. The right atrial pressure tracing is identical to that observed in constrictive pericarditis. Likewise, a typical dip-and-plateau configuration of ventricular diastolic pressure is present. Pulmonary hypertension is usually modest, with a peak systolic pressure in the right ventricle around 40 mmHg. The plateaus of right and left ventricular diastolic pressure may be equal in amplitude, just as they are in constrictive pericarditis. On the other hand, when the plateau of pressure is significantly higher in the left as compared with the right ventricle, the patient is much more likely to have restrictive cardiomyopathy than constrictive pericarditis.

Left ventriculography confirms normal systolic function and the absence of significant regional wall motion abnormalities. Endomyocardial biopsy is an integral part of the hemodynamic study of restrictive cardiomyopathy.

Distinguishing Restrictive Cardiomyopathy from Constrictive Pericarditis

A history of tuberculosis, trauma, prior pericarditis, or systemic disease that may involve the pericardium favors constrictive pericarditis. A history of prior radiation suggests constrictive pericarditis or combined constrictive pericarditis and restrictive cardiomyopathy. Prior cardiac surgery suggests constrictive pericarditis. Cardiac transplantation or a history or evidence of amyloidosis, hemochromatosis, or other infiltrative disease suggests restrictive cardiomyopathy.

In constrictive pericarditis, the velocity of mitral inflow declines greatly with inspiration. Respiratory variation in mitral inflow velocity in restrictive cardiomyopathy is normal—that is, virtually absent. If, after clinical, imaging, and invasive hemodynamic evaluation, the differential diagnosis remains in doubt, endomyocardial biopsy is almost always abnormal in restrictive cardiomyopathy and normal in constrictive pericarditis.

Clinical Examination

By definition, the clinical examination is not helpful in distinguishing between constrictive pericarditis and restrictive cardiomyopathy.

Electrocardiography

Depolarization abnormalities such as left bundle branch block favor restrictive cardiomyopathy. Atrioventricular conduction disturbance may also be found in restrictive cardiomyopathy. Isolated repolarization abnormalities are equally consistent with constrictive pericarditis and restrictive cardiomyopathy. Atrial fibrillation may occur in both conditions.

Chest Radiograph

Calcification of the pericardium strongly supports constrictive pericarditis, but its absence is equally compatible with constrictive pericarditis or restrictive cardiomyopathy. A CT or MRI scan showing abnormal thickness of the pericardium favors constrictive pericarditis. Absence of this finding is strongly suggestive of restrictive cardiomyopathy but is not diagnostic.

TREATMENT

Treatment in Patients with Greatly Elevated Ventricular Filling Pressures

Diuretics may be useful. Excessive use of these agents can be detrimental. Drugs with positive inotropic action are not helpful, and vasodilation should be applied very cautiously lest it further decrease ventricular filling. Calcium channel blocking agents may possibly increase ventricular diastolic compliance, but the evidence supporting this is not robust.

Infiltrative Cardiomyopathies

AMYLOID HEART DISEASE

This is the most commonly encountered infiltrative cardiomyopathy in adults who reside in nontropical zones. Cardiac amyloidosis may be isolated or may be part of a more generalized amyloidosis. Cardiac involvement may present as typical restrictive cardiomyopathy or as a combination of systolic and diastolic heart failure without the dip-and-plateau configuration of ventricular diastolic pressure. Cardiac arrhythmias are common. The prognosis is poor and the condition may recur following cardiac transplantation (see also Chap. 16).

CARCINOID HEART DISEASE

When this condition creates subacute tricuspid regurgitation, the picture of restrictive cardiomyopathy may resemble that seen in some cases of infarction of the right ventricle.

Hemochromatosis

Dilated heart failure is much more common than restrictive cardiomyopathy, but the latter has been reported and may precede systolic heart failure. Excellent results can be expected from repeated venesection to lower the iron load. The patients usually manifest hepatic dysfunction, diabetes, and brown pigmentation (see also Chap. 16).

Pseudoxanthoma Elasticum

Endocardial fibroelastosis, sometimes with calcification, is one of the many lesions that may be found in this rare, but striking, inherited disorder of elastic tissue metabolism. Surgical resection of calcified bands of endomyocardium combined with mitral valve replacement has been successful in the treatment of some cases (see also Chap. 26).

Sarcoidosis

Sarcoidosis, when it affects the heart, causes arrhythmia and atrioventricular conduction disturbances. Restrictive cardiomyopathy is much less common, but does occur. In addition to the granulomata, fibrosis may be present. The condition should not be confused with giant cell myocarditis, in which granulomata are absent. The prognosis is poor, mainly because of conduction disturbance and arrhythmia. Patients with biopsy-proven disease are treated with prednisone. The value of this treatment is unknown at present. Cardiac transplantation has been successful in a number of reported cases (Chap. 16).

Glycogen Storage Diseases of the Heart

These diseases mainly affect infants and children. The best known is Pompe's disease (type II glycogenesis), an autosomal recessive disorder caused by acid maltase deficiency. Glycogen is deposited in massive amounts in the cardiac and skeletal musculature. Manifestations appear in the first few months of life. The myocardium becomes so infiltrated that the disease process may resemble hypertrophic cardiomyopathy. More commonly, the heart is massively dilated and the picture of restrictive cardiomyopathy does not appear. The diagnosis can be made from skeletal muscle biopsy, which confirms the absence of alpha-glucosidase.

A number of other rare abnormalities of glycogen metabolism also may affect the heart. Nodular glycogenic infiltration occurs in the pediatric age group and is associated with tuberous sclerosis in half the cases.

Endomyocardial Fibrosis: Hypereosinophilic Syndrome

Two forms of the eosinophilic syndrome are recognized: tropical endomyocardial fibrosis and nontropical eosinophilic endomyocardial disease (Löffler's). The major characteristics are eosinophilia with eosinophilic infiltration of the endomyocardium. Fibrosis is extensive and eventually virtually occludes the ventricular cavity.

The eosinophils are highly toxic, causing necrosis of adjacent myocardial cells. There is not a clear distinction between Löffler's syndrome and endomyocardial involvement in eosinophilic leukemia.

Hemodynamically, the findings are typical of restrictive cardiomyopathy. End-diastolic ventricular volume is greatly reduced; therefore, sinus tachycardia is common. Atrial fibrillation also is common and is associated with left atrial dilation.

Nontropical eosinophilic endomyocardial fibrosis is uncommon. The prognosis is poor, perhaps justifying resection of the endocardium, which has been performed in a few cases with some degree of success.

SUGGESTED READINGS

Benotti JR, Grossman W, Cohn PF: Clinical profile of restrictive cardiomyopathy. *Circulation* 1980; 61:1206–1212.

Grossman W: Diastolic dysfunction in congestive heart failure. *N Engl J Med* 1991; 325:1557–1564.

Hwang R, Menh CC, Lin CY: Clinical analysis of five infants with glycogen storage disease of the heart—Pompe's disease. *Jpn Heart J* 1986; 27:25–34.

Kern MJ, Lorell BH, Grossman W: Cardiac amyloidosis masquerading as constrictive pericarditis. *Cathet Cardiovasc Diagn* 1982; 8:629–635.

Kushwaha SS, Fallon JT, Fuster V: Restrictive cardiomyopathy. *N Engl J Med* 1997; 336:267–276.

Meaney E, Shabetai R, Ghargava V, et al: Cardiac amyloidosis, constrictive pericarditis and restrictive cardiomyopathy. *Am J Cardiol* 1976; 38:547–556.

Shabetai R: Restrictive, obliterative, and infiltrative cardiomyopathies. In: Alexander RC, Schlant RC, Fuster V, O'Rourke RA, Roberts R, Sonnenblick EH (eds): *Hurst's The Heart*, 9th ed. New York, McGraw-Hill, 1988:2075–2088.

20

DISEASES OF THE PERICARDIUM

Ralph Shabetai

ACUTE VIRAL AND IDIOPATHIC PERICARDITIS

The inflammation is fibrinous, and pericardial effusion is sometimes present. The illness presents with malaise, fever, and chest pain, which may be either pleuritic or retrosternal and crushing. The pain sometimes is relieved by sitting up. Trapezoid ridge radiation is characteristic. The leukocyte count and erythrocyte sedimentation rate are elevated.

Pericardial friction rub is the diagnostic sign. It is superficial, scratchy or creaky, and best appreciated using firm pressure with the stethoscope.

Diffuse ST-segment elevation is the classic electrocardiographic (ECG) finding, but the ST is depressed in a V_R and V_1. The T wave remains upright. PR depression is specific but insensitive.

Treatment

The illness is usually brief and responsive to nonsteroidal anti-inflammatory agents, including colchicine. The patient must be observed for increasing pericardial effusion (echocardiogram) and tamponade.

RECURRENT PERICARDITIS

In some patients an immunologic abnormality develops that causes violent attacks of pericarditis, with or without effusion, recurring over many months or years.

Treatment

Attempt to treat recurrences with nonsteroidal anti-inflammatory agents. When that is not possible, prednisone should be given, usually starting at 60 mg/day, supplemented with an anti-inflammatory agent such as ibuprofen. Azathioprine also can be used as an adjunct. Colchicine may be effective. Relapses may recur for many years. Try to prevent steroid dependence in such patients. Pericardiotomy should be considered only when the patient is dependent on steroids and has serious side effects.

PERICARDIAL EFFUSION

Etiology

The common causes are acute viral or idiopathic pericarditis and neoplasm—most commonly bronchogenic, mammary, or lymphomatous. Drugs such as procainamide can also cause pericardial effusion. AIDS is now an important and quite frequent cause (see Chap. 29).

Diagnosis

There are no specific symptoms; physical examination is helpful only when there is tamponade. Whenever pericardial effusion is a reasonable possibility, it is essential to obtain an echocardiogram. It is important to assess cardiac function as well to rule out underlying cardiac dysfunction.

Nature of the Fluid

Pericardial fluid is quite often bloody, almost regardless of the cause. Pericardiocentesis for diagnosis has a remarkably low yield and should be avoided. Pericardiocentesis indicated for treatment, paradoxically, has a significant diagnostic yield. Thus, in patients who undergo the procedure for cardiac tamponade or where there is a strong suspicion of purulent pericarditis, diagnosis of the underlying cause is often possible. Pericardioscopy has been used in a few centers to aid in the selection of biopsy sites from the pericardium and epicardium. Malignant pericardial effusion is increasingly managed by balloon pericardiotomy.

CHRONIC LARGE IDIOPATHIC EFFUSION

Most will not recur after tap, but those that do can be tapped again. Further recurrence is rare and usually indicates pericardiectomy.

THE PERICARDIAL COMPRESSIVE SYNDROMES

Cardiac tamponade and *constrictive pericarditis* share important features but also have a number of important differentiating characteristics. The pathophysiology in both is impaired diastolic filling. In both, systemic and pulmonary venous pressures are elevated, cardiac output may be low, and there is equalization of left and right ventricular diastolic pressure.

In *tamponade*, the raised venous pressure is characterized by respiratory variation and a dominant x descent, coincident with the carotid pulse. In *constrictive pericarditis,* the characteristic rapid y descent of venous pressure immediately follows the carotid pulse.

Pulsus paradoxus occurs in moderate and severe cardiac tamponade, but not in mild tamponade. Pulsus paradoxus is less common in constrictive pericarditis except in elastic constrictive pericarditis, in which the pericardium can stretch a little.

Cardiac Tamponade

Common causes include trauma (which may be iatrogenic), neoplasm, and viral or idiopathic pericarditis.

In the acute form, often due to trauma, the heart is small, the venous pressure is extremely high, and hypotension is prominent. The effusion may be small because the pericardium cannot be stretched fast enough to accommodate a large effusion. Pulsus paradoxus is often prominent.

In medical cases, effusion usually develops more slowly and the effusions are usually larger. In moderately severe cases, cardiac output is reduced about 25 percent but arterial blood pressure is maintained near normal. Venous pressure is elevated.

As cardiac tamponade worsens, hypotension occurs, cardiac output is considerably lower, and pulsus paradoxus usually is evident. Venous pressure is high and the x descent prominent. A pericardial friction rub may or may not be present, and the ECG may not show signs of acute pericarditis but may show pulsus alternans.

ECHOCARDIOGRAPHY

This is a singularly helpful examination. The diagnosis cannot be entertained if the echocardiogram does not confirm pericardial effusion. Right atrial compression and right ventricular diastolic collapse are highly sensitive and specific signs of moderate tamponade. These echocardiographic signs appear before hypotension or pulsus paradoxus and when the reduction of cardiac output remains modest. False-positive and false-negative results occur but are uncommon and usually explicable. If untreated, tamponade progresses to the decompensated state marked by hypotension, low cardiac output, oliguria, and altered consciousness.

PREEXISTING HEART DISEASE

Preexisting heart disease may alter the manifestations of cardiac tamponade. Particularly important is elevation of left ventricular end-diastolic pressure, as may occur in patients receiving hemodialysis. In these cases, pulsus paradoxus characteristically is absent. Pulsus paradoxus is also absent in aortic regurgitation and atrial septal defect.

TREATMENT

In mild cases, especially idiopathic or viral, it may be unnecessary to remove pericardial fluid. Such patients should be observed frequently and monitored by serial echocardiograms while undergoing anti-inflammatory treatment. If the pericardial effusion becomes smaller and signs of cardiac tamponade diminish, the patient usually will not require pericardiocentesis. Failure of such resolution is an indication for removal of pericardial fluid.

Others, including all with significantly compromised hemodynamics, must undergo pericardiocentesis, open drainage, or balloon pericardiotomy. Except in drastic emergencies, pericardiocentesis should always be preceded by echocardiography.

The administration of isoproterenol or a vasodilator and volume infusions are improvising measures that precede pericardiocentesis and are not substitutes for removal of fluid.

Constrictive Pericarditis

ETIOLOGY

The common causes are neoplasm, radiation, and infection. Tuberculosis is less common in recent series. Many cases are idiopathic and some posttraumatic. A small number are complications of cardiac surgery.

SYMPTOMS
The symptoms are indistinguishable from those of severe right-sided heart failure, but there may be a history of pericarditis. Dyspnea, fatigue, weight gain, edema, and ascites are the principal features.

PHYSICAL EXAMINATION
Signs suggesting congestive heart failure without apparent etiology should raise the possibility of constrictive pericarditis. The venous pressure is elevated and dominated by a prominent y descent. The apex beat may be impalpable or systolic retraction may be less evident. A third heart sound, the pericardial knock, is more common than it used to be when constrictive pericarditis was a more chronic process. The ECG usually shows nonspecific ST- and T-wave changes.

HEMODYNAMICS
Cardiac output may be low. The atrial pressure shows a deep y descent corresponding with the early diastolic dip of ventricular pressure. The left and right ventricular diastolic pressures show a dip-and-plateau (square root) configuration, and the two plateaus are equal in amplitude.

Effusive Constrictive Pericarditis

This syndrome occurs when there is not just pericardial effusion, often causing cardiac tamponade, but also pericardial *pathology*. When pericardial fluid is removed, the classic clinical and hemodynamic findings of cardiac tamponade give way to those of constrictive pericarditis. The syndrome should be suspected in cases of tamponade when right atrial pressure does not return to normal after removal of pericardial fluid.

DIFFERENTIAL DIAGNOSIS
Constrictive pericarditis is sometimes mistaken for *cirrhosis of the liver* because of edema, ascites, and hepatic dysfunction. In cirrhosis, however, venous pressure is normal or only slightly elevated.

Tricuspid valve disease may simulate right-sided heart failure or constrictive pericarditis. Patients should be auscultated for the systolic murmur of tricuspid regurgitation or the diastolic rumble of tricuspid stenosis, both of which become louder during inspiration. Doppler echo of tricuspid blood flow velocity in tricuspid regurgitation yields spectral and color evidence of tricuspid regurgitation with absent respiratory variation of inflow when tricuspid regurgitation is severe. Tricuspid stenosis is shown by increased velocity of inflow.

Right ventricular infarction may also simulate right-sided heart failure or constrictive pericarditis. In right ventricular infarction, however, there are usually other features of ischemic heart disease, commonly inferior infarction.

Restrictive cardiomyopathy simulates constrictive pericarditis. The differential diagnosis can be straightforward or difficult. Prior pericarditis or calcification of the pericardium favors constrictive pericarditis. ST-T wave changes favor cardiomyopathy. The finding of equal diastolic pressures in the ventricles is consistent with either diagnosis, but left ventricular diastolic pressure considerably above the right favors cardiomyopathy. A thickened pericardium by computed tomography (CT) or magnetic resonance imaging (MRI) is helpful, but a normal-appearing pericardium does not rule out constrictive pericarditis. In constrictive pericarditis but not in restrictive cardiomyopathy, mitral inflow velocity falls during inspiration while tricuspid velocity increases. If doubt still remains, endomyocardial biopsy should be performed. In the very small minority of cases that remain undiagnosed, exploratory thoractomy is justified (see Chap. 19).

Other Syndromes of Constrictive Pericarditis

Occult constrictive pericarditis may be disclosed by infusion of a large volume of fluid, which causes the previously normal right atrial and pulmonary wedge pressures to become equal and develop the waveform of constrictive pericarditis.

Localized constrictive pericarditis can occur, especially in postoperative cases.

Treatment

The usual treatment is *pericardiectomy*. Venous pressure below 7 or 8 mmHg can be followed. With the availability of cardiopulmonary bypass, surgical mortality has been greatly reduced. Heavily calcific constrictive pericarditis still poses surgical risk, in the range of 5 to 10 percent. When constriction is by the visceral pericardium, it is difficult to separate the planes; the operation is risky, and it is seldom possible to perform a complete pericardiectomy.

SOME SPECIFIC PERICARDIAL DISORDERS

Dialysis-Related and Uremic Pericardial Disease

Pericarditis is a late finding in *untreated chronic renal disease* and often predicts short survival.

Patients undergoing *hemodialysis* are subject to pericarditis at lower levels of nitrogen retention. This may take the form of pericardial effusion, or a friction rub. Effusions frequently lead to cardiac tamponade.

Diagnosis is difficult; the venous pressure is often high, because these patients retain fluid. Therefore, the combination of a pericardial effusion by echocardiogram and raised venous pressure may or may not indicate cardiac tamponade. These patients do not have pulsus paradoxus. Right atrial pressure may be lower than pulmonary wedge pressure. Thus, cardiac tamponade develops when pericardial pressure becomes equal to right atrial pressure but is not as high as pulmonary wedge pressure. This difference explains the absence of pulsus paradoxus. The echocardiogram should be scanned for right atrial compression or right ventricular collapse.

In some hospitals, a large persistent pericardial effusion, and certainly cardiac tamponade, is an indication for subxyphoid drainage of the pericardial fluid. In others the policy is less aggressive. The patient with effusion but without tamponade may respond to intensification of dialysis. Pericardiocentesis is done in some dialysis units for early cardiac tamponade but not in others, where surgical drainage is preferred. Mortality from cardiac tamponade in dialysis patients is considerable, but this complication of dialysis has become much less common in recent years (see also Chap. 30).

Pericardium and Myocardial Infarction

Pericardial friction rub is common in the first day or two of acute myocardial infarction. It is not an indication to stop anticoagulants or thrombolytic treatment.

Asymptomatic pericardial effusion is common in the course of acute myocardial infarction.

Dressler's syndrome probably is an autoimmune response to myocardial injury. There may be an infective component, as suggested by the epidemiology, which shows swings from frequent occurrence to long periods when it is not encountered.

POSTPERICARDIOTOMY SYNDROME

This is a syndrome of fever, leukocytosis, chest pain, and, frequently, pericardial effusion in patients who have undergone operation involving the opening of the pericardium. The syndrome may appear a week or two after operation or may be delayed. Sometimes it exhibits features of recurrence. Treatment should be with nonsteroidal anti-inflammatory drugs, but if the syndrome cannot be suppressed, prednisone is needed.

Neoplastic Pericardial Disease

The most common neoplasms are secondaries from the lung or from breast lymphoma, and leukemia. Mesothelioma is the most common primary tumor (see also Chap. 32).

Neoplasm is an important cause of pericardial effusion. Cardiac tamponade and constrictive pericarditis are fairly frequent. When a long remission or cure has not been secured, treatment should be directed toward the patient's comfort. Balloon pericardiotomy is often a good option. Patients in whom the outlook for the neoplasm is favorable can be treated more radically.

Radiation-Induced Pericardial Disease

Radiation of the mediastinum can produce both constrictive pericarditis and restrictive cardiomyopathy. The onset may be early but can be delayed several years. Endomyocardial biopsy is useful in deciding whether or not the patient would improve with pericardiectomy.

Hypersensitivity and Collagen Vascular Pericardial Disorders

Rheumatoid arthritis can cause subacute constrictive pericarditis and be associated with other abnormalities such as heart block and valve disease. Pericardiectomy is often required.

Pericardial disease develops in nearly all patients with *lupus erythematosus* when life is prolonged by steroid treatment. The usual lesion is fibrinous pericarditis, but a large pericardial effusion may develop. Late complications include cardiac tamponade and constrictive pericarditis (see also Chap. 26).

Drug-Induced Pericardial Disease

Pericardial abnormalities may develop in response to hydralazine, procainamide, and isoniazid and less commonly to other drugs.

Infectious (Nonviral) Pericarditis

TUBERCULOSIS PERICARDITIS
This disease is now quite uncommon, though it is encountered in patients with impaired immunologic status—e.g., in AIDS patients and in the hemodialysis population. Tuberculosis should be suspected if no other etiology is readily apparent, if pericardial disease does not subside spontaneously or respond readily to anti-inflammatory treatment, or if pericardial disease occurs in a patient who has had immunosuppression.

Tuberculous pericarditis usually is treated with a triple drug regimen such as rifampin 600 mg, isoniazid 300 mg daily, and ethambutol 50 mg/kg body weight daily.

BACTERIAL PERICARDITIS
Infectious pericarditis has become less common. However, it is a life-threatening disease with a high mortality. *Straphylococcus* can affect the pericardium, especially in the elderly and in the immunocompromised. Most frequently, treatment is by open surgical drainage.

SUGGESTED READINGS

Callahan JA, Seward JB, Nishimura RA, et al: Two dimensional echocardiographically guided pericardiocentesis: experience in 117 consecutive patients. *Am J Cardiol* 1985; 55:476–479.

Hurrell DG, Nishimura RA, Higano ST, et al: Value of dynamic respiratory changes in left and right ventricular pressures for the diagnosis of constrictive pericarditis. *Circulation* 1996; 93:2007–2013.

Leimgruber PP, Klopfenstein HS, Wann LS, et al: The hemodynamic derangement associated with right ventricular diastolic collapse in cardiac tamponade. *Circulation* 1983; 68:612–620.

Oh JK, Tajik AJ, Appleton CP, et al: Preload reduction to unmask the characteristic Doppler features of constrictive pericarditis: a new observation. *Circulation* 1997; 95: 796–799.

Palacios IF, Tuzcu EM, Ziskind AA, et al: Percutaneous balloon pericardial window for patients with malignant pericardial effusion and tamponade. *Cathet Cardiovasc Diagn* 1991; 22:244–249.

Schoenfeldt MH, Supple EW, Dec GW, et al: Restrictive cardiomyopathy versus constrictive pericarditis. *Circulation* 1987; 75:1012–1017.

Shabetai R: Diseases of the pericardium. In: Alexander RA, Schlant RC, Fuster V, O'Rourke RA, Roberts R, Sonnenblick EH (eds): *Hurst's The Heart*, 9th ed. New York, McGraw-Hill, 1998:2169–2203.

Shabetai R, Fowler NO, Guntheroth WG: The hemodynamics of cardiac tamponade and constrictive pericarditis. *Am J Cardiol* 1970; 26:480–489.

Spodick DH: Acoustic phenomena in pericardial disease. *Am Heart J* 1971; 81:114–124.

Spodick DH: Diagnostic electrocardiographic sequences in acute pericarditis. *Circulation* 1973; 48:575–580.

21

INFECTIVE ENDOCARDITIS

David T. Durack

DEFINITIONS

Infective endocarditis (IE) is caused by microbial infection of the endothelial lining of the heart. The characteristic lesion is a vegetation, which usually develops on a heart valve. *Acute bacterial endocarditis* (ABE) is caused by virulent organisms and runs its course over days to weeks. *Subacute bacterial endocarditis* (SBE) is caused by organisms of low virulence and runs its course over weeks to months. Sterile vegetations are called *nonbacterial thrombotic endocarditis* (NBTE).

THE SUSCEPTIBLE HOST

IE can occur at any age, but today is most common in older adults. The median age of onset is about 50 years. The male/female ratio is approximately 2:1 overall, but is higher in elderly patients. Most patients who develop IE have a preexisting cardiac condition that affects the valves. The lesions pose different degrees of risk for development of IE (Table 21-1). Intravenous drug abusers are at high risk for IE even if their heart valves are normal; they have a high frequency of tricuspid valve infection caused by *Staphylococcus aureus*.

PATHOGENESIS AND PATHOLOGY

Figure 21-1 shows the pathogenesis of IE. NBTE may develop on heart valves in a wide variety of clinical conditions. Small aggregates of platelets have been found occasionally on normal valves, but they occur frequently on the surfaces of valves

TABLE 21-1

ESTIMATES OF THE RELATIVE RISK FOR INFECTIVE ENDOCARDITIS POSED BY VARIOUS CARDIAC LESIONS

Relatively high risk	Intermediate risk	Very low or negligible risk
Prosthetic heart valves	Mitral valve prolapse with regurgitation	Mitral valve prolapse without regurgitation
Previous infective endocarditis	Pure mitral stenosis	Trivial valvular regurgitation by echocardiography without structural abnormality
Cyanotic congenital heart disease	Tricuspid valve disease	
Patent ductus arteriosus	Pulmonary stenosis	
Aortic regurgitation and/or stenosis*	Asymmetric septal hypertrophy	Isolated atrial septal defect (secundum)
Mitral regurgitation	Bicuspid aortic valve or calcific aortic sclerosis with minimal hemodynamic abnormality	Arteriosclerotic plaques
Mitral stenosis with regurgitation		Coronary artery disease

342

Ventricular septal defect Coarctation of the aorta	Degenerative valvular disease in elderly patients	Cardiac pacemaker
Surgically repaired intracardiac lesions with residual hemodynamic abnormality	Surgically repaired intracardiac lesions with minimal or no hemodynamic abnormality, less than 6 months after operation	Surgically repaired intracardiac lesions with minimal or no hemodynamic abnormality, more than 6 months after operation

*Includes tricuspid, bicuspid, and unicuspid valves.

SOURCE: Adapted from Durack DT: Infective endocarditis. In Alexander RW, Schlant RC, Fuster V, O'Rourke RA, Roberts R, Sonnenblick EH (eds): *Hurst's The Heart*, 9th ed. New York, McGraw-Hill, 1998:2205–2239, with permission from the publisher.

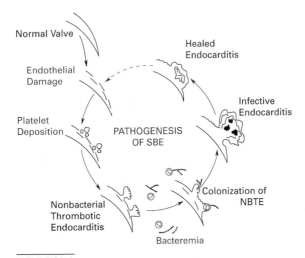

FIGURE 21-1

The main events in pathogenesis of nonbacterial thrombotic endocarditis (NBTE) and subacute bacterial endocarditis (SBE).

damaged by congenital or rheumatic disease or by previous
infective endocarditis. The common factor leading to platelet
deposition is endothelial damage. This exposes subendothelial
connective tissue containing collagen fibers, which in turn
causes platelets to aggregate at the site. These microscopic
platelet thrombi may harmlessly embolize away, or they may
be stabilized by fibrin to form NBTE. The vegetations of
NBTE are friable masses, usually situated along the lines of
valve closure. They vary greatly in size, sometimes being
rather large and causing infarctions when they embolize. There
is little inflammatory reaction at the site of attachment. Histologically, the vegetations of NBTE consist of degenerating
masses of platelets interwoven with strands of fibrin.

The essential event leading to infective endocarditis is
attachment of microorganisms circulating in the bloodstream
onto an endocardial surface, especially NBTE. Once lodged on
the endocardium, bacteria multiply rapidly. The vegetation
provides an ideal environment for the growth of microbial
colonies. The presence of bacteria stimulates further thrombosis; layers of fibrin are deposited around growing bacteria,

causing the vegetations to enlarge. Histologically, colonies of microorganisms are found embedded in a fibrin-platelet matrix. Although the inflammatory reaction at the site of attachment may be extensive, even progressing to form a frank abscess, the vegetations themselves characteristically contain relatively few leukocytes. Even these few are prevented from reaching bacteria by layers of fibrin, which form protective barriers around colonies.

Formation of an abscess is one of the most important complications of valvular infection. Abscesses often develop by direct extension of valvular infection into the fibrous cardiac skeleton supporting the valves. From here, abscesses can extend farther into the adjacent myocardium. Abscesses are uncommon in SBE unless a valvular prosthesis is present. They occur more often in ABE, and are found in the majority of patients who die with active prosthetic valve infection (PVE). Because valve-ring abscesses are located close to the conduction system, they can cause conduction disturbances and arrhythmias.

ETIOLOGIC ORGANISMS

Many different species of microbes can cause infective endocarditis, but gram-positive cocci predominate. Streptococci or staphylococci cause more than 80 percent of IE on native valves. Among the streptococci, alpha-hemolytic (viridans) streptococci from the mouth cause most cases of SBE. The approximate frequencies of the main organisms causing IE are shown in Table 21-2.

CLINICAL MANIFESTATIONS

The clinical and laboratory manifestations of infective endocarditis can be grouped under three headings (see Table 21-3):

- Manifestations of a systemic infection.
- Manifestations of an intravascular lesion.
- Manifestations of an immunologic reaction to infection.

The symptoms of SBE can develop insidiously and with great variability. Fevers, chills, rigors, night sweats, general malaise, anorexia, fatigue, and weakness are typical. The patient often loses weight. Headaches and musculoskeletal complaints, including myalgias, arthralgias, and back pain, are

TABLE 21-2

FREQUENCY OF VARIOUS ORGANISMS CAUSING INFECTIVE ENDOCARDITIS*

Organism	NVE, %	IV Drug Abusers, %	Early PVE, %	Late PVE, %
Streptococci	60	15–25	5	35
Viridans, alpha-hemolytic	35	5	<5	25
Strep. bovis (group D)	10	<5	<5	<5
Enterococcus faecalis (group D)	10	10	<5	<5
Other streptococci	<5	<5	<5	<5
Staphylococci	25	50	50	30
Coagulase-positive	23	50	20	10
Coagulase-negative	<5	<5	30	20
Gram-negative aerobic bacilli	<5	5	20	10

Fungi	<5	5	5
Miscellaneous bacteria	<5	5	5
Diphtheroids, propionibacteria	<1	<5	<5
Other anaerobes	<1	<1	<1
Rickettsia	<1	<1	<1
Chlamydia	<1	<1	<1
Polymicrobial infection	<1	5	<5
Culture-negative endocarditis	5–15	<5	<5

*These are representative figures collated from the literature; wide local variations in frequency are to be expected. NVE = native valve endocarditis; PVE = prosthetic valve endocarditis.

SOURCE: Adapted from Durack DT: Infective endocarditis. In: Alexander RW, Schlant RC, Fuster V, O'Rourke RA, Roberts R, Sonnenblick EH (eds): *Hurst's The Heart*, 9th ed. New York, McGraw-Hill, 1998:2205–2239, with permission from the publisher.

TABLE 21-3

SUMMARY OF THE MAJOR CLINICAL MANIFESTATIONS OF INFECTIVE ENDOCARDITIS

Manifestation	History	Examination	Investigations
Systemic infection	Fever, chills, rigors, sweats, malaise, weakness, lethargy, delirium, headache, anorexia, weight loss, backache, arthralgia, myalgia Portal of entry: Oropharynx, skin Urinary tract Drug abuse Nosocomial bacteremia	Fever Pallor Weight loss Asthenia Splenomegaly	Anemia Leukocytosis (variable) Raised ESR Blood cultures positive Abnormal CSF
Intravascular lesion	Dyspnea, chest pain, focal weakness, stroke, abdominal	Murmurs Signs of cardiac failure	Hematuria Chest x-ray

	pain, cold and painful extremities	Petechiae—skin, eye, mucosae Roth spots, Osler's nodes Janeway lesions Splinter hemorrhages Stroke Mycotic aneurysm Ischemia or infarction of viscera or extremities	Echocardiography Arteriography Liver-spleen scan Lung scan, brian scan, CT scan Histology, culture of emboli
Immunologic reactions	Arthralgia, myalgia, tenosynovitis	Arthritis Signs of uremia Vascular phenomena Finger clubbing	Proteinuria, hematuria, casts, uremia, acidosis Polyclonal increases in gamma globulins Rheumatoid factor, decreased complement, and immune complexes in serum Antistaphylococcal teichoic acid antibodies

SOURCE: Adapted from Alexander RW, Schlant RC, Fuster V, O'Rourke RA, Roberts R, Sonnenblick EH (eds): *Hurst's The Heart*, 9th ed. New York, McGraw-Hill, 1998:2205–2239, with permission of the publisher.

349

common. This symptom complex is often described by the patient or physician as a "flu-like illness." Symptoms usually persist over several weeks, worsening intermittently, before the diagnosis is made. In ABE, the symptoms are both accelerated and accentuated in severity. Patients experience hectic fevers, rigors, and prostration, usually leading to admission to hospital within a few days.

Evidence of an intravascular lesion is provided by symptoms of left- or right-sided heart failure and by such manifestations of embolization as focal neurologic injury, chest pain, flank pain, left-upper-quadrant pain, hematuria, or ischemia of an extremity. Symptoms of cardiac failure may develop or worsen suddenly in either acute or subacute disease because of mechanical complications such as perforation of a valve leaflet, rupture of chordae tendineae, or development of functional stenosis from obstruction of blood flow by large vegetations. Alternatively, heart failure may develop insidiously, or preexisting chronic heart failure may worsen because of progressive damage to the valves or associated structures. Myocardial infarction due to coronary artery embolism may contribute to heart failure

IE should be considered in any patient who presents with the classic triad of *fever, anemia, murmur*. Important physical signs of IE include new or changed murmurs (especially if they indicate new aortic or mitral regurgitation), heart failure, splenomegaly, signs of embolization, and peripheral signs (see Tables 21-3 and 21-4).

DIAGNOSTIC TESTS

The *chest x-ray* may be normal, or may reveal congestive heart failure or other manifestations of valvular heart disease. Multiple small, patchy opacities can be due to septic pulmonary emboli from tricuspid valve vegetations. *Electrocardiography* may show evidence of hypertrophy, or of myocardial infarction due to emboli. Conduction abnormalities indicate the possibility of a valve-ring abscess near the conduction system. Pericarditis is rare in SBE except in patients with renal failure, but it can occur in patients with ABE, and occasionally as a complication of valve ring abscess.

Anemia and an elevated *erythrocyte sedimentation rate* are usual, but nonspecific findings. The *white blood cell count* may be normal or elevated and is seldom helpful in the diagnosis of IE. *Urinalysis* can reveal microscopic hematuria and proteinuria due to microemboli or gross hematuria due to embolic infarction. Immune-complex glomerulonephritis, a common complication of IE, can produce red cell casts, white cell casts, and protein in the urine.

The antigenic stimulus due to organisms in the vegetation can raise polyclonal antibodies, resulting in positive *rheumatoid factor* in about 30 percent of patients with SBE, circulating immune complexes, and occasional false-positive tests for syphilis.

CT scanning and MRI can reveal cerebritis, embolic infarction, or hemorrhage in the brain and infarcts or abscess formation in the spleen or other sites.

Blood Culture

This is the most important test in the diagnosis of IE. In more than 90 percent of cases, three sets of blood cultures suffice to yield the infecting organisms. Venous blood samples for these should be drawn several hours apart, before antibiotic therapy is started. If the patient has ABE, indicating that antibiotic therapy must be started immediately, the three sets should be drawn with minimal delay.

Echocardiography

Echocardiography is essential in the diagnosis and assessment of IE. It can detect vegetations, valve leaflet rupture, chordal rupture, valve-ring abscess, and myocardial abscess. Two-dimensional echocardiography combined with a Doppler flow study is particularly useful in evaluating endocarditis, especially if a transesophageal sensor is used. Two-dimensional echocardiography is 80 to 95 percent sensitive for larger vegetations but cannot detect very small vegetations; therefore, a negative echocardiogram does not rule out IE. It is somewhat more difficult to assess IE echocardiographically when the tricuspid valve, the pulmonic valve, or prosthetic valves are involved. Serial echocardiograms may be helpful in assessing cardiac function and the need for valve replacement. The disappearance or persistence of vegetations on echocardiograms during treatment are not reliable criteria for the success or failure of antibiotic therapy, but enlargement during therapy can indicate treatment failure.

Cardiac Catheterization

Cardiac catheterization is generally well tolerated and safe during IE. It can define anatomic abnormalities such as valvular lesions, congenital defects, coronary artery disease, asymmet-

TABLE 21-4
CHARACTERISTICS OF SOME PERIPHERAL SIGNS OF INFECTIVE ENDOCARDITIS

	Petechiae	Splinter hemorrhages	Roth spots	Osler's nodes	Janeway lesions	Clubbing
Appearance	Tiny red hemorrhagic spots	"Splinters" under nails; red when fresh, then brown or black	Small bright red patches with white centers	Pea-sized red or purplish nodules	Red macules	Curvature of the nails in two planes, with swelling of the terminal phalanges
Distribution	Anywhere, especially above clavicles, in mouth, and in conjunctivae	Distal third of nails	Retinae	Fingers and toes, occasionally hands and feet	Palms and soles; occasionally on flanks, forearms, ankles, feet, ears	Fingers and/or toes

Incidence	Common, in both SBE and ABE	Common in both SBE and ABE	Infrequent; usually in SBE	Infrequent; usually in SBE	Infrequent; usually in ABE	Rare; in SBE only
Pathology	Increased capillary permeability; microemboli	Blood in avascular squamous epithelium under nail; due to microemboli or increased capillary fragility	Inflammation and hemorrhage	Intracutaneous local vasculitis; bacteria rarely found; occasional abscess formation; probably embolic in origin	Origin uncertain; possibly embolic or allergic in origin	Soft-tissue proliferation, occasionally periosteal new bone formation
Pain	None	None	None	Mild to moderately severe	None	Usually none, sometimes painful
Duration	Days	Weeks	Days	Days	Several hours to day	Weeks to months
Diagnostic significance	Nonspecific; also found in septicemia, after cardiac surgery, and in many other disorders	Nonspecific; found in up to 10% of normal people and up to 40% of patients with mitral stenosis	Strongly suggestive of endocarditis but not diagnostic	Almost pathognomonic for endocarditis	Suggestive of endocarditis	Nonspecific; found in many cardiopulmonary disorders; can be congenital

SOURCE: Adapted from Durack DT: Infective endocarditis. In: Hoeprich PD, Jordan MC, Ronald AR (eds.): *Infectious Diseases*, 5th ed. Philadelphia, Lippincott, 1994:1240–1248, with permission of the author and publisher.

rical septal hypertrophy, coarctation of the aorta, or mycotic aneurysm and provide physiologic measurements that aid management. It is especially useful when surgery is being considered or when antibiotic treatment seems to be failing.

Microbiological Tests

The etiologic organism should be accurately identified. Then, if possible, the minimal inhibitory concentrations (MIC) of appropriate antibiotics for the organism should be determined. The results will guide the choice of an appropriate antibiotic. Measurement of serum inhibitory and bactericidal titers (SIT, SBT) for the organism while the patient is being treated is not essential in routine management of IE.

DIAGNOSTIC CRITERIA

1. Definite infective endocarditis:
   a. Pathologic criteria:
      Microorganisms: demonstrated by culture or histology in a vegetation, or in a vegetation that has embolized, or in an intracardiac abscess, *or*
   b. Clinical criteria (using specific definitions listed below):
      i. Two major criteria *or*
      ii. One major and three minor criteria, *or*
      iii. Five minor criteria
2. Possible infective endocarditis:
   Findings consistent with infective endocarditis that fall short of *definite*, but not *rejected*.
3. Rejected:
   a. Firm alternate diagnosis explaining evidence of infective endocarditis, *or*
   b. Resolution of endocarditis syndrome, with antibiotic therapy for 4 days or less, *or*
   c. No pathologic evidence of IE at surgery or autopsy, after antibiotic therapy for 4 days or less.

Definition of Terms

MAJOR CRITERIA
1. Positive blood culture for infective endocarditis:
   a. Typical microorganism for infective endocarditis from two separate blood cultures:
      i. *Viridans* streptococci (including nutritional variant strains), *Streptococcus bovis*, HACEK group, *or*

      ii. Community-acquired *Staph. aureus* or enterococci, in the absence of a primary focus, *or*

  b. Persistently positive blood culture, defined as recovery of a microorganism consistent with infective endocarditis from:

      i. Blood cultures drawn more than 12 h apart, *or*

      ii. All of three or majority of four or more separate blood cultures, with first and last drawn at least 1 h apart.

2. Evidence of endocardial involvement:

  a. Positive echocardiogram for infective endocarditis:

      i. Oscillating intracardiac mass, on valve or supporting structures, or in the path of regurgitant jets, or on implanted material, in the absence of an alternative anatomic explanation, *or*

      ii. Abscess, *or*

      iii. New partial dehiscence of prosthetic valve, *or*

  b. New valvular regurgitation (increase or change in pre-existing murmur not sufficient)

3. Positive serology for *Bartonella* or *Coxiella* (Q fever)

MINOR CRITERIA

1. Predisposition: predisposing heart condition or intravenous drug use

2. Fever: $\geq 38.0°C$ (100.4°F)

3. Vascular phenomena: major arterial emboli, septic pulmonary infarcts, mycotic aneurysm, intracranial hemorrhage, conjunctival hemorrhages, Janeway lesions

4. Immunologic phenomena: glomerulonephritis, Osler nodes, Roth spots, rheumatoid factor

5. Microbiologic evidence: positive blood culture but not meeting major criterion above (excluding single positive cultures for coagulase negative staphylococci and organisms that do not cause endocarditis) *or* serologic evidence of active infection with organism consistent with infective endocarditis

6. Echocardiogram: consistent with infective endocarditis but not meeting major criterion above

DIFFERENTIAL DIAGNOSIS

Because the clinical manifestations of endocarditis are numerous and often nonspecific, the differential diagnosis of this disease is very wide. ABE shares many clinical features with primary septicemias due to *Staph. aureus*, *Neisseria*, pneumo-cocci, and gram-negative bacilli. Findings in ABE may suggest pneumonia, meningitis, brain abscess, stroke, malaria, acute pericarditis, vasculitis, and disseminated intravascular coagulation. SBE must be considered during the workup of every patient with fever of unknown origin. Its manifestations can mimic those of rheumatic fever, osteomyelitis, tuberculosis, meningitis, intraabdominal infections, salmonellosis, brucellosis, glomerulonephritis, myocardial infarction, stroke, endocardial thrombi, atrial myxoma, connective tissue diseases, vasculitis, occult malignancy (especially lymphomas), chronic cardiac failure, pericarditis, and even psychoneurosis.

NATURAL HISTORY AND PROGNOSIS

IE is almost always fatal if untreated. Optimal treatment regimens can give cure rates of 95 percent or better for streptococcal native-valve endocarditis and for *Staph. aureus* IE in intravenous drug abusers. However, cure rates are much lower when unfavorable prognostic factors are present. Heart failure and central nervous system (CNS) complications are the most important adverse prognostic factors. Others include gram-negative or fungal infection, prosthetic valve infection, development of abscesses in the valve ring or myocardium, old age, renal failure, and culture-negative IE.

Eradication of the etiologic organisms (microbiologic cure) can be achieved in a high proportion of patients with bacterial endocarditis. However, both early and long-term mortality rates remain significant because of preexisting disease and because of damage done before the infection is eradicated. Survival curves after admission with IE show a significant number of late deaths over time, despite microbiologic cure.

RECURRENCE

Recurrent endocarditis is a general term that includes both *relapses* and *reinfections*. Occasional relapses occur even after optimal treatment, so follow-up clinical evaluation, including blood cultures if there is any fever or other sign of relapse, should be meticulously performed. Most relapses occur within a few weeks of the end of treatment. Reinfection is a new episode of endocarditis occurring after cure of a previous episode. Patients remain permanently at risk of infection after cure of infective endocarditis because of residual valve scarring (see Fig. 21-1). Further episodes are fairly common, being recorded in 5 to 30 percent of cases. This wide variation in reported incidence of recurrence is partly due to variable duration of follow-up. Intravenous drug abusers and patients with severe periodontitis are at highest risk for reinfection. Some patients have suffered three or more separate episodes of IE. Patients who have previously had native-valve endocarditis are at higher risk for prosthetic valve infection, for reasons not yet understood.

TREATMENT

Antibiotic therapy should be started immediately after blood cultures have been drawn if the patient has ABE or is deteriorating rapidly. If the patient has SBE and is stable, antibiotics may be withheld for 1 or 2 days while investigations and blood cultures are in progress. Appropriate empirical treatment for SBE is ampicillin 2.0 g IV every 4 h *plus* gentamicin 1.0 mg/kg IV every 8 h. For ABE, *add* nafcillin 1.5 g IV every 4 h *or* vancomycin 1.0 g IV every 12 h.

Standard treatment regimens for the most common organisms causing IE are given in Table 21-5. The choice of treatment regimens for gram-negative ABE, fungal IE, and other uncommon forms of endocarditis requires specialist consultation.

Surgical treatment usually consists of the removal of the infected native valve and insertion of a prosthetic valve. Clear indications should be present before proceeding with valve replacement for IE, because prosthetic valves cause significant long-term morbidity. When surgery is indicated, however, it should not be delayed, because the patient's condition may deteriorate rapidly. Valve replacement can be successful even if the duration of antimicrobial therapy has been too short to kill the organisms. Other procedures include excision of vegetations, valvuloplasty, valve repair, and closure of an abscess cavity. Indications for surgery include moderate or severe heart failure not responding to medical therapy, presence of a valvering abscess, valvular obstruction, unstable prosthesis, repeated major emboli, and failure of antimicrobials to control infection. Surgical intervention may also be indicated for management of major systemic emboli, removal of splenic abscess, or treatment of mycotic aneurysms.

PREVENTION

Antibiotics may prevent IE if given just before dental and other procedures that can cause transient bacteremia. This is appropriate when both the preexisting cardiac lesion and the

TABLE 21-5

TREATMENT REGIMENS FOR INFECTIVE ENDOCARDITIS CAUSED BY GRAM-POSITIVE COCCI

Organism	Regimen	Duration, weeks	Comments
Alphahemolytic (viridans) streptococci; Strep. bovis	1. Penicillin G 4 million units IV every 6 h *plus* gentamicin 1.0 mg/kg every 12 h IV, *or*	2	Standard regimen, for patients less than 65 years old without renal failure eighth-nerve defects, or serious complications
	2. Penicillin G 4 million units IV every 6 h *plus* gentamicin 1.0 mg/kg every 12 h IV (for first 2 weeks only), *or*	4	For patients with complicated disease—e.g., CNS involvement, shock, moderately penicillin-resistant streptococci, failed previous treatment
	3. Penicillin G 4 million units IV every 6 h IV, *or*	4	For patients more than 65 years old, with renal failure on eighth-nerve defects
	4. Ceftriaxone 2g IV or IM once daily, *or*	4	For patients allergic to penicillins
	5. Vancomycin 10 mg/kg IV every 12 h	4	For patients allergic to penicillins and cephalosporins
Enterococcus faecalis and other penicillin-resistant organisms	1. Ampicillin 2 g IV every 4 h *plus* gentamicin 1.0 mg/kg IV every 8 h, *or*	4–6	4 weeks should be adequate for most cases with symptoms present for less than 3 months.

358

		Weeks	
	2. Vancomycin 15 mg/kg IV every 12 h IV *plus* gentamicin 1.0 mg/kg IV every 8 h (not to exceed 80 mg) every 8 h	4–6	For patients allergic to penicillins; 4 weeks should be adequate for most cases. Serum levels should be monitored.
Staph. aureus	1. Nafcillin 1.5 g IV every 4 h, *or*	4–6	Standard regimen
	2. Nafcillin as above, *plus* gentamicin 1.0 mg/kg IV every 8 h for the first 3–5 days only, *or*	4–6	For patients with severe disseminated staphylococcal disease, synergy may be advantageous during early stages of treatment
	3. Cefazolin 2 g IV every 8 h, *or*	4–6	For patients allergic to penicillins
	4. Vancomycin 15 mg/kg IV every 12 h	4–6	For patients allergic to penicillins and cephalosporins; for methicillin-resistant strains
Group A streptococci, *Strep. pneumoniae*	1. Penicillin G 2 million units IV every 6 h, *or*	2–4	
	2. Cefazolin, 1 g IV every 8 h	2–4	
HACEK* group	1. Ampicillin 2 g IV every 4 h *plus* gentamicin 1.0 mg/kg IV every 12 h IV, *or*	4	Gentamicin may be discontinued if organism is fully sensitive to ampicillin.
	2. Ceftriaxone 1–2 g IV or IM once daily	4	For patients allergic to penicillins; suitable for home therapy after stabilization in hospital.

*HACEK: *Haemophilus* spp., *Actinobacillus* spp., *Cardiobacterium* spp., *Eikenella* spp., *Kingella* spp.

SOURCE: Adapted from Alexander RW, Schlant RC, Fuster V, O'Rourke RA, Roberts R, Sonnenblick EH (eds): *Hurst's The Heart,* 9th ed. New York, McGraw-Hill, 1998:2205–2239, with permission of the publisher.

TABLE 21-6

REGIMENS FOR PROPHYLAXIS OF INFECTIVE ENDOCARDITIS

Standard regimen	For dental procedures; oral or upper respiratory tract surgery; minor GI or GU tract procedures	Amoxicillin, 2.0 g orally 1 h before
Special regimens	Oral regimen for penicillin-allergic patients (oral and respiratory tract only)	Clindamycin, 600 mg orally 1 h before
	Parenteral regimen for high-risk patients; also for GI or GU tract procedures	Ampicillin, 2.0 g IM or IV, *plus* gentamicin 1.5 mg/kg IM or IV, 0.5 h before
	Parenteral regimen for penicillin-allergic patients	Vancomycin, 1.0 g IV slowly over 1 h, starting 1 h before; *add* gentamicin; 1.5 mg/kg IM or IV. If GI or GU tract involved
	Cardiac surgery including implantation of prosthetic valves	Cefazolin, 2.0 g IV, at induction of anesthesia, repeated 8 and 16 h later *or* vancomycin 1.0 g IV slowly over 1 h, starting at induction, then 0.5 g IV 8 and 16 h later

SOURCE: Adapted from Dajani AS et al., 1997, and Durack DT, 1995 and 1998 (see Suggested Reading list), with permission of the publishers.

procedure seem to pose significant risk for IE—e.g., tooth extraction in a patient with a bicuspid aortic valve. Suggested regimens are listed in Table 21-6. Antibiotic prophylaxis for IE is unnecessary for many low-risk procedures such as diagnostic cardiac catheterization, pacemaker insertion, intubation, flexible bronchoscopy, uncomplicated gastrointestinal diagnostic procedures, dental fillings above the gum line, and adjustment of orthodontic appliances.

SUGGESTED READING

Dajani AS, Taubert KA, Wilson WR, et al: Prevention of bacterial endocarditis: recommendations by the American Heart Association. *JAMA* 1997; 277:1794–1801.

Durack DT: Prevention of infective endocarditis. *N Engl J Med* 1995; 332:38–44.

Durack DT: Prophylaxis of infective endocarditis. In: Mandell GL, Douglas RG, Jr., Dolin R (eds): *Principles and Practice of Infectious Diseases*. New York, Churchill Livingstone, 1995:793–813.

Durack DT: Infective endocarditis. In: Alexander RW, Schlant RC, Fuster V, O'Rourke RA, Roberts R, Sonnenblick EH (eds): *Hurst's The Heart*, 9th ed. New York, McGraw-Hill, 1998:2205–2239.

Durack DT, Bright DK, Lukes AS, Duke Endocarditis Service: New criteria for diagnosis of infective endocarditis: utilization of specific echocardiographic findings. *Am J Med* 1994; 96:200–209.

Kaye, D (ed): *Infective Endocarditis*. New York, Raven Press, 1992.

Scheld WM, Sande MA: Endocarditis and intravascular infections. In: Mandell GL, Douglas RG Jr, Dolin R (eds): *Principles and Practice of Infectious Diseases*. New York, Churchill Livingstone, 1995:740–783.

Threlkeld MG, Cobbs CG: Infectious disorders of prosthetic valves and intravascular devices. In: Mandell GL, Bennett JE, Dolin R (eds): *Principles and Practice of Infectious Diseases*. New York, Churchill Livingstone, 1995:783–793.

Wilson WR, Karchmer AW, Dajani AS, et al: Antibiotic treatment of adults with infective endocarditis due to viridans streptococci, enterococci, staphylococci and HACEK microorganisms. *JAMA* 1995; 274:1706–1713.

C H A P T E R

22

CONGENITAL HEART DISEASE

William H. Plauth, Jr.
Michael D. Freed

ATRIAL SEPTAL DEFECT

Definition

An *atrial septal defect* is any opening in the interatrial septum other than a patent foramen ovale.

Pathology

Most atrial septal defects are large enough to allow free communication between the atria. They are classified by anatomic location as follows:

- *Ostium secundum defect*, the most common type, in the area of the fossa ovalis. Mitral valve prolapse may be associated.
- *Ostium primum defect*, a defect with a complex known as *common atrioventricular canal defect*, in the area inferior to the fossa ovalis and frequently involving the atrioventricular valves and interventricular septum (see also "Ventricular Septal Defect," below.)
- *Sinus venosus defect*, in the septum superior to the fossa ovalis near the orifice of the superior vena cava and almost invariably associated with partial anomalous drainage of the right pulmonary veins.

All types of atrial septal defect may have dilated right heart chambers and pulmonary arteries. Pulmonary hypertension and pulmonary vascular disease may develop, but usually not before the patient's third decade (see "Eisenmenger's Syndrome," toward the end of this chapter).

Pathophysiology

Usually there is no resistance to blood flow or a pressure difference between the atria even in the face of a large left-to-right shunt. Development of pulmonary hypertension and pulmonary vascular disease will lead to shunt reversal and cyanosis. Mitral regurgitation is common with ostium primum defects.

Clinical Manifestations

HISTORY

Most children are asymptomatic except with primum defects complicated by mitral regurgitation. Mild fatigue and dyspnea often appear in late adolescence and are present in most adults. Congestive heart failure and arrhythmias are common in the fourth decade and beyond. Atrial septal defects are more common in females than in males and are, excluding a bicuspid aortic valve or mitral valve prolapse, the most common form of congenital heart disease among adults.

PHYSICAL EXAMINATION

Many patients are slender. A right ventricular lift to palpation along the lower left sternal border, a widely split S_2 with the split fixed throughout the respiratory cycle, and a midsystolic murmur at the left upper sternal border reflecting increased right ventricular stroke volume are characteristic. There may be a diastolic flow murmur over the lower left sternal border denoting increased flow across the tricuspid valve during ventricular diastole. A holosystolic murmur of mitral regurgitation is common with primum defects. With pulmonary hypertension and pulmonary vascular disease, the flow murmurs disappear, the pulmonary component of S_2 becomes accentuated, and cyanosis develops. At that stage, tricuspid and pulmonary regurgitation may be present.

Usual Diagnostic Tests

CHEST X-RAY

Mild to moderate cardiac enlargement, prominence of the main pulmonary artery and its branch, and increased pulmonary arterial flow are characteristic. Absence of left atrial enlargement helps distinguish uncomplicated atrial septal defects from other left-to-right shunts. Left atrial and left ventricular enlargement may occur with primum defects accompanied by significant mitral regurgitation. With pulmonary vascular disease, the pulmonary flow pattern becomes diminished but the central pulmonary arterial enlargement remains.

ELECTROCARDIOGRAM

The rsR pattern in the anterior precordial leads is characteristic and indicates mild right ventricular conduction delay or mild hypertrophy. Most patients with secundum defects have a right inferior QRS vector in the frontal plane. With primum defects, the QRS vector is usually superior. They may also have left atrial and left ventricular hypertrophy. Atrial fibrillation and flutter are not uncommon among adults.

ECHOCARDIOGRAPHY

Echocardiography demonstrates right atrial and right ventricular enlargement and flat or paradoxical ventricular septal motion. Most secundum and primum defects in children are easily visualized by two-dimensional imaging with color Doppler, whereas sinus venosus defects are somewhat more difficult to see. Associated defects, such as mitral abnormalities with the primum defect, can be identified. Transesophageal study offers excellent images for those in whom the transthoracic approach is inadequate, particularly in adults.

CARDIAC CATHETERIZATION

In young patients with the typical clinical and echocardiographic findings of an uncomplicated secundum defect, cardiac catheterization is not necessary. It is indicated for those with primum defects and in adults in whom pulmonary hypertension or coronary artery disease are more likely.

The atrial septum is usually easily traversed from the femoral approach. There is a significant step-up in oxygen saturation at the right atrial level. Similar pressures are recorded in both atria with a mean gradient between them of less than 3 mmHg. The right ventricular and pulmonary arterial pressures should be assessed and pulmonary vascular resistance calculated. A systolic pressure of up to 20 mmHg across the right ventricular outflow tract can be attributed to increased flow without representing true pulmonary stenosis. The site of pulmonary venous connections and the presence of associated defects can be demonstrated by selective biplane cineangiography.

Natural History and Prognosis

Most children and adolescents are asymptomatic. Because of lack of symptoms and unimpressive findings, it is common for this diagnosis to go unsuspected until early adult years. Thereafter symptoms and complications occur with increasing frequency and severity. The most important problems are pul-

monary hypertension and pulmonary vascular disease, atrial
arrhythmias, and congestive heart failure. Pulmonary vascular
disease may progress rapidly during pregnancy and with the
use of contraceptive drugs. These complications lead to high
morbidity and mortality with surgical correction and following
it. Congestive heart failure, pulmonary embolism, paradoxical
emboli, brain abscess, and infection are the causes of death in
those who do not have surgical correction. Congestive heart
failure is frequently associated with atrial fibrillation. Patients
with primum defects or mitral abnormalities are subject to
infective endocarditis.

Treatment

Surgery is recommended just prior to school age or whenever
the diagnosis is made thereafter. Surgery is recommended if
the pulmonary/systemic blood flow ratio is 1.5:1 or greater
providing there is no serious malfunction of the left side of the
heart causing this. If pulmonary hypertension is present,
surgery is recommended for those with a flow ratio of 1.5:1 or
greater if the systemic arterial saturation is at least 92 percent
and the pulmonary vascular resistance is less than 15 Wood
units (mmHg/L/min) or 1200 dynes $\cdot$ s $\cdot$ cm^{-5}, corrected for
body surface area. With more severe pulmonary vascular disease, surgery offers no benefit.

Surgical closure is accomplished by direct suture repair or
the use of a pericardial or Dacron patch. This choice is determined by the location, size, and shape of the defect. Pulmonary
venous connections should be inspected so that anomalous
veins may be redirected to the left atrium. In patients with primum defects, valvuloplasty of the mitral valve may be
necessary. The risk of surgery in childhood is very low. There
is greater morbidity and less complete resolution of abnormalities in adults. Postoperative follow-up is necessary to
determine adequacy of repair, document reversal, or disappearance of previous abnormalities and detect and manage any
arrhythmias.

VENTRICULAR SEPTAL DEFECT

Definition

A *ventricular septal defect* (VSD) is an opening in the interventricular septum that permits communication between the
two ventricles.

Pathology

There is a wide range of size among VSDs. They are classified by their anatomic location as follows:

- Perimembranous defects, the most common type, in the superior septum immediately below the crista supraventricularis of the right ventricle.
- Conal, supracristal, or subarterial doubly committed defects in the right ventricular outflow tract and adjacent to the aortic valve, which tends to prolapse into these defects, with resulting aortic regurgitation.
- Inlet or atrioventricular canal defects in the posterior septum beneath the septal leaflet of the tricuspid valve.
- Muscular defects commonly with multiple openings, usually in the trabecular and apical portions of the septum.

Cardiac chamber enlargement and hypertrophy depend on shunt volume and the degree of pulmonary hypertension.

Pathophysiology

Functional alterations depend on the defect's size and the status of the pulmonary vascular bed. A small defect (less than $0.5 \text{ cm}^2/\text{m}^2$) offers a large resistance to flow. The left-to-right shunt is small and there is no pulmonary hypertension. There is little additional work for the heart. A defect of moderate size (0.5 to $1.0 \text{ cm}^2/\text{m}^2$) still permits a gradient between left and right ventricular systolic pressures, and the pulmonary pressure is generally less than 80 percent of the systemic systolic pressure. A large left-to-right shunt may be present, however, which imposes a significant volume overload on the left side of the heart. Pulmonary vascular disease can result but is uncommon (see "Eisenmenger's Syndrome," below).

If the defect is large, approximately equal to or greater than the aortic valve orifice (at least $1.0 \text{ cm}^2/\text{m}^2$), it offers no resistance to blood flow. The systolic pressures in both ventricles are the same. The relative proportion of flow to each circulation is governed by the relative resistances in the pulmonary and systemic vascular beds. These patients are at high risk for the development of pulmonary vascular disease.

At birth, the pulmonary vascular resistance is high and there is little left-to-right shunting across even a large VSD. The resistance decreases over the first few days and weeks of life, with a corresponding increase in the magnitude of shunting. Congestive heart failure then appears with the larger defects.

Clinical Manifestations

HISTORY
Infants with larger defects develop heart failure usually by 3 months of age. Parents note tachypnea, increased work of breathing, fatigue with feedings, poor weight gain, and excessive sweating. Infants and children with small ventricular defects are asymptomatic. Their defects are ordinarily detected early by routine examination because of the loudness of the murmur. Diagnosis is rarely delayed beyond early childhood because of the prominence of the murmur even when the defect is small.

PHYSICAL EXAMINATION
Characteristically there is a loud, harsh holosystolic murmur at the lower left sternal border, which is frequently accompanied by a palpable thrill. In patients with significant left-to-right shunts (pulmonary/systemic flow ratios $\geq 2:1$), there is a mid-diastolic rumble at the apex caused by high flow across the mitral valve. The right and left ventricular impulses are prominent to palpation. The pulmonary component of the second heart sound (P_2) is increased with pulmonary hypertension. There may also be signs of congestive heart failure such as dyspnea, tachypnea, and enlargement of the liver. Pulmonary rales and peripheral edema are uncommon and reflect severe and long-standing heart failure.

With advanced pulmonary vascular disease, these findings regress. The second heart sound is loud and single and murmurs of pulmonary and tricuspid regurgitation may become audible. Systemic arterial desaturation becomes detectable and is progressive.

Signs of decreasing left-to-right shunting may also be obvious if the defect becomes smaller or if infundibular pulmonary stenosis develops. The murmur of aortic regurgitation may also appear in a few children secondary to prolapse of an aortic leaflet into the defect.

Usual Diagnostic Tests

CHEST X-RAY
Heart size—especially the size of the left atrium and left ventricle—and pulmonary vascularity reflect the magnitude of left-to-right shunting. Small defects cause no radiographic abnormalities.

ELECTROCARDIOGRAM

With a small defect, the electrocardiogram may be normal or demonstrate slightly augmented left ventricular voltage. With larger defects, biventricular and left atrial hypertrophy are usual. Right atrial and isolated right ventricular hypertrophy are seen only with severe pulmonary vascular disease or when severe pulmonary stenosis develops.

ECHOCARDIOGRAPHY

Two-dimensional echocardiographic imaging permits visualization of most defects of significant size. With the addition of color flow mapping, even tiny muscular or multiple defects can be identified and localized. A defect's position and diameter as well as chamber size, ventricular function, and the presence and/or absence of pulmonary stenosis can be determined. The RV systolic pressure can be estimated accurately by Doppler measurement of the systolic pressure gradient across the defect itself or, if tricuspid regurgitation exists, the systolic gradient across the tricuspid valve.

CARDIAC CATHETERIZATION

There is a step-up in oxygen saturation at the right ventricular level. The pulmonary arterial pressures can be measured and the pulmonary vascular resistance calculated. Selective biplane cineangiograms in the left ventricle will outline the defect.

Natural History and Prognosis

Small VSDs impose only the risk of infective endocarditis. The majority of these become smaller or close spontaneously during childhood. Although diminution in size is also a characteristic of many large defects, the appearance of congestive heart failure, severe pulmonary stenosis (a clinical picture indistinguishable from tetralogy of Fallot), or the threat of pulmonary vascular disease almost invariably requires surgical intervention during infancy or early childhood. Once there are irreversible pulmonary vascular changes, the outlook is bleak, although survival beyond the third decade is common. A few children will develop progressive aortic regurgitation.

TREATMENT

It is important to identify those patients with moderate or large defects as early as possible, since they are at risk for developing the complications described above. Congestive heart failure is treated with digoxin and diuretics as required. Cardiac catheterization is recommended for all infants with heart failure, failure to thrive, or evidence of pulmonary hypertension.

Surgery is recommended in infancy if heart failure is difficult to manage medically or if the pulmonary arterial systolic pressure is greater than one-half the systemic systolic pressure beyond 6 months of age. Surgery is also recommended if there is any degree of pulmonary hypertension by age 2 years. In the remainder of cases, it is advisable to correct the defect before the child enters school if there is still a large shunt (pulmonary/systemic flow ratio greater than 1.8:1) or cardiac enlargement persists. If pulmonary vascular disease is already present at the time the diagnosis is made, surgery is still recommended if the pulmonary vascular resistance is less than 11 Wood units (880 dynes $\cdot$ s $\cdot$ cm^{-5}) or if the pulmonary/systemic vascular resistance ratio is less than 0.7:1 provided the pulmonary/systemic flow ratio is at least 1.5:1. Beyond this level, the risk of surgery more than offsets any possible benefit. The same criteria apply to adults.

Primary patch closure of the defect is the procedure of choice, even in young infants. Perimembranous and inlet defects should be repaired through the right atrium and tricuspid valve if possible. Supracristal defects usually require right ventriculotomy, whereas multiple apical defects may require both approaches and even a posterior left ventriculotomy. Avoidance of injury to the AV conduction system requires careful attention when the perimembranous or inlet types are being repaired.

PATENT DUCTUS ARTERIOSUS

Definition

Patent ductus arteriosus is persistent patency of the fetal vessel that normally connects the aorta and pulmonary artery.

Pathology

The ductus arteriosus connects the origin of the left pulmonary artery to the aortic arch just beyond the left subclavian artery. Normally it closes and becomes a ligament in the first few weeks of life.

Pathophysiology

The diameter and length of the ductus determine the severity of the hemodynamic abnormalities, which are similar to those described above for ventricular septal defects. The flow, however, occurs in diastole as well as in systole.

Clinical Manifestations

HISTORY
Patent ductus arteriosus is more common in premature infants and in those exposed to rubella during the first trimester of pregnancy of a nonimmune mother; it is also more common in situations associated with hypoxia. Symptoms are related to significant left-to-right shunting with poor growth and/or overt heart failure. Only rarely is diagnosis delayed beyond childhood.

PHYSICAL EXAMINATION
Peripheral pulses are bounding, especially with a large shunt. The typical finding is a continuous or machinery murmur best heard at the left upper sternal border and below the left clavicle. The murmur peaks at S_2. The length of the diastolic portion of the murmur may be abbreviated in newborns and older patients with pulmonary hypertension. A diastolic rumble at the apex is present with larger shunts. The pulmonary closure sound is increased when pulmonary hypertension is present. Findings of pulmonary vascular disease may predominate in the older child or adult, in which case systolic arterial desaturation or frank cyanosis may be found in the left arm and hand and the lower segment of the body.

Usual Diagnostic Tests

CHEST X-RAY
Findings are similar to those described for ventricular septal defects except that the ascending aorta may be enlarged.

ELECTROCARDIOGRAM
This is the same as for ventricular septal defect.

ECHOCARDIOGRAPHY
Larger defects can frequently be imaged. Color flow Doppler echocardiography is extremely valuable for detecting the shunt.

CARDIAC CATHETERIZATION

Catheterization is not necessary in children with a typical picture of uncomplicated patent ductus and no pulmonary hypertension, but it is usually recommended in adults. At catheterization there is a step-up in oxygen saturation at the pulmonary arterial level. The catheter usually crosses the ductus easily to enter the descending thoracic aorta from the pulmonary artery. Pulmonary arterial pressures and resistance should be ascertained. Aortography will outline the ductus. In specialized laboratories, the ductus may be closed by catheter introduction of an occluding device or by coil embolization.

Natural History and Prognosis

Infective endarteritis can occur regardless of the size of the ductus. The risk increases with length of survival. Heart failure, pulmonary hypertension, and pulmonary vascular disease can occur in those with significant shunts. Infants with heart failure can die suddenly. Irreversible pulmonary vascular disease is rarely seen in the first year or two of life. Calcification of the ductal wall is common in adults.

Treatment

Antibiotic coverage at times of risk to prevent infective endarteritis is appropriate in all patients. Indomethacin is often successful in causing closing for the ductus in the premature infant. Diuretics and volume control are helpful in the treatment of heart failure, but patients unresponsive to medical treatment should have the ductus closed surgically without delay. Closure is recommended for all with continued patency beyond the first 6 months of life. This is usually done electively at 1 to 2 years of age in the asymptomatic patient. In skillful hands, transcatheter coil occlusion is an alternative for the ductus of small diameter. Surgical closure by division or ligation remains the procedure of choice in most instances. In adults with calcification, placement of a Dacron patch over the aortic orifice of the ductus may be advisable. Care is necessary to avoid injury to the adjacent left recurrent laryngeal nerve.

COARCTATION OF THE AORTA

Definition

Coarctation of the aorta is a discrete narrowing of the distal segment of the aortic arch due to a deformity of a medial layer of the aortic wall.

Pathology

An infolding of the aortic wall in its anterior, superior, and posterior aspects creates a narrowed and eccentric lumen either opposite the ductus or just proximal to it. Prominent collateral vessels form in the older child and adult to carry blood to the distal aortic segment. A bicuspid aortic valve is present in approximately one-half of patients with coarctation.

Pathophysiology

Arterial pressures above the coarctation are elevated. There is a significant systolic and mean pressure gradient between the upper and lower extremities. Exercise increases these pressure gradients.

Clinical Manifestations

HISTORY

In infants, the symptoms are those of heart failure. There is often difficulty with feeding and poor weight gain. Older children are usually asymptomatic, but they may have mild fatigue or claudication in their legs when running. Rarely, diagnosis is delayed until adolescence or early childhood, when hypertension is discovered or complications occur (see "Natural History," below).

PHYSICAL EXAMINATION

Signs of heart failure are common in infants. If a murmur is present, it is usually midsystolic and nondescript unless best heard over the interscapular region. Prominent upper segment pulses and a left ventricular lift are present. Pulses in the lower extremities are much weaker (or absent) and delayed in timing. Blood pressure measurements in both arms and a leg will confirm the diagnosis. The frequency and degree of hypertension in the upper segment increase with age beyond infancy. Findings are particularly prominent in adults. An early systolic click at the apex suggests an associated bicuspid aortic valve.

Usual Diagnostic Tests

CHEST X-RAY
Generalized cardiomegaly and pulmonary venous congestion are common in infants. Many children have a normal heart size. Left atrial and left ventricular enlargement are common in the adult. A "figure 3" configuration of the left aortic margin at the coarctation may be seen in overpenetrated films, and an "E sign" or "reverse 3" seen with a barium swallow. Rib notching due to the presence of collateral vessels is evident in older children and adults.

ELECTROCARDIOGRAM
Infants have right or biventricular hypertrophy. T-wave inversion in the left precordial leads is common. Children may have normal electrocardiograms, but left ventricular hypertrophy becomes progressively more common with age.

ECHOCARDIOGRAPHY
The coarctation can often be imaged from the suprasternal approach, with the Doppler flow pattern and gradient estimates offering diagnostic confirmation. The precordial and sub-xiphoid views are of great value to detect the presence and severity of associated lesions, which are common in the infant.

CARDIAC CATHETERIZATION
Cardiac catheterization may not be indicated if other imaging techniques are diagnostic and there are no associated defects. If the cardiac index is normal, there will be left atrial and left ventricular hypertension and significant systolic and mean pressure differences across the coarctation. Associated defects should be sought in symptomatic infants. Aortography will demonstrate the site and length of the coarctation.

Natural History and Prognosis

Congestive heart failure is the most common problem associated with coarctation of the aorta in infancy. Of all individuals born with coarctation, approximately half will present within the first month or two of life with this complication. Persistent hypertension leads to complications in adults with uncorrected coarctation in the second and third decades. Premature death in the form of dissecting aortic aneurysm, cerebral arterial rupture, heart failure, or myocardial infarction are most common in those who are not operated on or whose surgery is delayed into the third decade. There is a significant risk of endocarditis or endarteritis both before and after operation.

Recoarctation is common in patients who have undergone surgery in infancy. Residual or recurrent hypertension without recoarctation is most common in those who are operated on after 6 years of age.

Treatment

Vigorous treatment of heart failure is indicated. This includes temporary palliation with prostaglandin E_1 in the newborn to maintain ductal patency. Surgery is recommended in all infants with significant complicating defects and in those who do not respond dramatically to decongestive measures. Elective surgery between 1 and 4 years of age is indicated in the remainder. Older children and adults should undergo surgical correction without delay.

Repair by subclavian flap angioplasty is generally recommended in infancy. In older patients, resection and direct end-to-end anastomosis is most commonly employed. Occasionally in children and commonly in adults, bypass of coarctation using a tubular vascular prosthesis may be required because of the length of the obstruction or changes in the arterial wall.

Balloon angioplasty is an alternative to surgery and has been demonstrated to be particularly beneficial in patients with recurrent coarctation.

VALVAR AORTIC STENOSIS (INCLUDING BICUSPID AORTIC VALVE)

Definition

In *valvar aortic stenosis* there is subtotal obstruction to the left ventricular outflow due to a congenital aortic *valvar* deformity.

Pathology

Most commonly, the aortic valve is bicuspid. The degree of commissural fusion varies from case to case. Occasionally the valve is unicuspid and severely obstructive. Uncommonly the valve is tricuspid, again with variable commissural fusion. The valve leaflets are thickened and may be severely dysplastic. Calcification occurs in adults. Left ventricular hypertrophy is present in patients with significant obstruction.

Pathophysiology

The hemodynamics of congenital valvar aortic stenosis are similar to those of the acquired form (see Chap. 12, "Aortic Valve Disease"). A bicuspid aortic valve may be unobstructive during early years but tends to become thickened and calcified in middle to late adult years.

Clinical Manifestations

HISTORY
Most infants and children are asymptomatic. Symptoms of heart failure occur in infants with critical stenosis or in older adults, especially when this condition is complicated by atherosclerotic cardiovascular disease. Syncope, angina, and sudden death can also occur with severe disease at any age. Patients usually come to medical attention because of the murmur.

PHYSICAL EXAMINATION
Most patients have a systolic thrill at the right upper sternal border and over the carotid arteries. Arterial pulse amplitude may be diminished in severe disease. Paradoxical splitting of S_2 occurs almost exclusively in adults. A systolic click at the apex is characteristic and differentiates valvar stenosis from other forms of left ventricular outflow obstruction. A fourth heart sound in older children and adults signifies severe obstruction. There is a harsh crescendo-decrescendo murmur, which is loudest at the right upper sternal border. Louder and longer systolic murmurs generally indicate more important obstruction except in the presence of heart failure, in which, with decreased cardiac output, the intensity of the murmur may be markedly diminished. Some children and adults also develop the murmur of aortic regurgitation.

Usual Diagnostic Tests

CHEST X-RAY
Significant cardiac enlargement with or without pulmonary edema indicates severe disease. Poststenotic dilatation of the

ascending aorta is characteristic. Calcification of the valve may be seen in adults.

ELECTROCARDIOGRAM

Left ventricular hypertrophy seldom helps in judging severity. Diminished anterior forces, a deep SV_1 ($\geq$ 30 mm), or absence of the Q wave in lead V_6 suggest severe stenosis. The appearance of ST-segment depression in the left precordial leads with exercise is a reliable method for identifying young patients with at least a moderate gradient.

ECHOCARDIOGRAPHY

Echocardiographic imaging permits recognition of abnormal aortic valve motion and identification of the number of cusps as well as differentiation from other types of aortic stenosis. Left ventricular function can be assessed. Doppler interrogation can accurately predict the systolic and mean gradients as well. Transesophageal study is of particular value in adults when transthoracic imaging is less than ideal.

CARDIAC CATHETERIZATION

Symptomatic patients and those with echocardiographic and Doppler evidence of severe obstruction do not require catheterization. Those with clinical evidence of only moderate obstruction should have this confirmed by cardiac catheterization to determine the proper method of management. In the absence of congestive heart failure (i.e., in the presence of a normal cardiac output), the peak left ventricular–aortic systolic pressure gradient or calculated valve area is used to assess the severity. Moderate obstruction is commonly defined as a gradient between 50 and 75 mmHg or a valve area between 0.5 and 0.8 cm²/m² (see also Chap. 12).

Natural History and Prognosis

Infants with severe obstruction develop heart failure and require emergency surgical or balloon valvuloplastic intervention. Mortality is significant with or without intervention and is often related to endocardial fibroelastosis, papillary muscle necrosis, or a small left ventricle. Most infants and children with milder aortic stenosis will show a gradual progression in severity during childhood and adolescence. This is complicated by calcification and associated atherosclerosis in the adult years. Stenosis tends to recur even after successful palliation. Sudden death can occur in those with at least a moderate gradient. Infective endocarditis poses a serious threat due to systemic embolization and the development of severe aortic regurgitation.

Treatment

Medical treatment of the symptomatic patient, whether infant or adult, should be instituted, but relief of the obstruction should also be obtained without delay. Surgery or balloon valvuloplasty is recommended for any patient with a gradient of 75 mmHg or more or a valve area of less than $0.5 \ cm^2/m^2$. Intervention is also recommended for those with symptoms or ischemic electrocardiographic findings but with a lower gradient of 40 to 75 mmHg. Exercise restrictions are recommended for all patients except for those with mild disease. Careful evaluation and follow-up of all patients with aortic stenosis is mandatory.

Surgical relief of valve aortic stenosis is accomplished by a carefully placed incision in the fused but well supported true commissure. A conservative approach is essential to prevent an intolerable degree of aortic regurgitation. Restenosis in those patients who have undergone valvotomy in infancy or childhood or calcification in adults with significant stenosis eventually makes unavoidable aortic valve replacement.

VALVAR PULMONARY STENOSIS

Definition

In *valvar pulmonary stenosis* there is subtotal obstruction of right ventricular outflow due to a congenital deformity of the pulmonary valve.

Pathology

The most common type is the dome-shaped stenosis of commissural fusion with the valvar tissue appearing as a cone- or dome-shaped structure perforated at its distal end. There is poststenotic dilation of the pulmonary trunk and the ventricular septum is intact. Less commonly, there is severe dysplasia of the valve leaflets, usually with a degree of annular narrowing, with three markedly thickened leaflets of limited mobility. The degree of right ventricular hypertrophy is a function of the severity of the obstruction. Muscular infundibular stenosis may occur secondarily.

Pathophysiology

There is a systolic pressure difference between the right ventricle and the pulmonary artery. If there is right ventricular failure, this pressure difference may be low because of the low flow. In infants, severe obstruction often causes cyanosis from left-to-right shunting via the patent foramen ovale.

Clinical Manifestations

HISTORY

Most infants and children are asymptomatic, though easy fatigue with exercise may be present in some. Exercise intolerance is common in adults with important obstruction. Only infants and adults with severe pulmonary stenosis tend to develop right ventricular failure.

PHYSICAL EXAMINATION

Physical findings are similar in all age groups. A prominent *a* wave may be seen on examination of the jugular venous pulse in older children with at least moderate obstruction. A systolic thrill at the upper left sternal border and in the suprasternal notch is characteristic. The right ventricular impulse becomes increasingly forceful with more severe obstruction. An early systolic ejection click accentuated with expiration is audible at the left upper sternal border and is characteristic of commissural fusion but not of valvar dysplasia. As severity increases, the P_2 becomes progressively softer and more delayed until it finally becomes inaudible. There is a harsh systolic crescendo-decrescendo murmur best heard at the left upper sternal border. With mild or moderate stenosis, the murmur peaks in midsystole and ends at or before the aortic component of S_2. In severe stenosis, the murmur peaks late in systole and extends beyond the aortic closure sound. Right-sided heart failure with hepatomegaly and tricuspid regurgitation occurs in infants with very critical obstruction and in adults with long-standing stenosis. If there is an atrial communication, left-to-right shunting is the rule unless obstruction is severe, in which case the shunt will be right-to-left with resulting cyanosis.

Usual Diagnostic Tests

CHEST X-RAY

Poststenotic dilatation of the main and proximal left pulmonary arteries is characteristic of commissural fusion but not valvar

dysplasia. Significant cardiac enlargement signals critical
obstruction; pulmonary flow may be diminished in the patient
with a right-to-left shunt at atrial level.

ELECTROCARDIOGRAM

Right ventricular forces in the anterior precordial leads correlate reasonably well with severity of stenosis. Right axis
deviation and right atrial hypertrophy occur with significant
obstruction.

ECHOCARDIOGRAPHY

Abnormal pulmonary valve motion and thickness may be seen,
and Doppler evaluation provides an excellent measure of
severity.

CARDIAC CATHETERIZATION

There is elevated right ventricular systolic pressure and a distinct systolic pressure difference across the valve. If right
ventricular output is normal, the gradient defines the severity.
An atrial shunt can be detected. Right ventricular biplane
cineangiography demonstrates the valvar anatomy as well as
any associated infundibular hypertrophy.

Natural History and Prognosis

In most patients with mild to moderate pulmonary stenosis, the
clinical course is favorable without a significant tendency for
the gradient to increase with time. A few patients, particularly
infants, may show significant progression in early childhood.
Growth or the development of infundibular stenosis may be a
factor. The prognosis for those with unrelieved critical obstruction in infancy or severe obstruction of many years duration is
poor. Right ventricular dysfunction ensues. Heart failure and
arrhythmias can lead to premature death in adults. Infective
endocarditis is a risk.

Treatment

Heart failure, cyanosis, or a right ventricular pressure at or
above systemic levels is an indication for prompt intervention.
In asymptomatic patients, elective intervention is usually recommended when the right ventricular systolic pressure is near
70 mmHg or the gradient is near 50 mmHg. Balloon valvuloplasty is the initial treatment of choice of most patients; in

skilled hands it is very successful. Results with dysplastic leaflets are much less impressive. Prophylaxis against infective endocarditis is recommended indefinitely.

Surgery is recommended for those who have had an unsatisfactory result with balloon valvuloplasty, those with severe valvar dysplasia or annular hypoplasia, or those in whom significant infundibular stenosis does not regress after successful valvuloplasty. Fused commissures are excised, very dysplastic leaflets are removed, subvalvar muscle bundles are excised, and patch augmentation of a small annulus can be provided.

TETRALOGY OF FALLOT

Definition

Tetralogy of Fallot is characterized by a large ventricular septal defect, overriding aorta, pulmonary stenosis, and right ventricular hypertrophy.

Pathology

The aorta is enlarged; it straddles the ventricular septal defect and arises to a varying degree from both ventricles. Fibrous continuity of the aortic and mitral valves is maintained. The infundibulum is the dominant site of obstruction to pulmonary flow, although the pulmonary valve is often malformed and the pulmonary arteries are often small. A right aortic arch is common.

Pathophysiology

Because the ventricular septal defect is large, the systolic pressure is equal in both ventricles and the aorta. The degree of right ventricular obstruction determines the direction and volume of shunting. In most cases this is greater than systemic resistance, causing a right-to-left shunt and cyanosis.

Clinical Manifestations

HISTORY

Most cases are recognized in the first few months of life because of cyanosis. In some patients, the pulmonary stenosis slowly becomes more severe and may not cause cyanosis until later in childhood or, rarely, even in adulthood. Dyspnea with exertion is common. Attacks of suddenly increasing cyanosis and hyperpnea, called *hypoxic spells*, occur in infants and young children. Squatting with exercise is almost pathognomonic of tetralogy.

PHYSICAL EXAMINATION
The degree of cyanosis varies. Clubbing is seen in those with cyanosis of more than 3 months duration. Signs of congestive heart failure occur only in adults unless there is severe superimposed illness during childhood. A right ventricular lift is present. A crescendo-decrescendo murmur is audible at the left midsternal border. It usually is harsh but it may be quite soft, or it may become inaudible with very severe stenosis or hypoxic spells. The S_2 is single.

Usual Diagnostic Test

CHEST X-RAY
Right ventricular enlargement, a concave pulmonary arterial segment, and diminished pulmonary flow are characteristic. The *coeur en sabot* (boot-shaped contour of the heart) is classic during childhood. A right aortic arch may be seen.

ELECTROCARDIOGRAM
The mean QRS axis in the frontal plane is right and inferior with right ventricular hypertrophy.

ECHOCARDIOGRAPHY
All the structural components of tetralogy of Fallot can be demonstrated, including the level(s) of right ventricular outflow tract obstruction. Color-flow Doppler will confirm the direction of shunting across the ventricular septal defect.

OTHER LABORATORY STUDIES
The *hemoglobin* and *hematocrit* should be evaluated frequently so that the development of relative anemia or severe polycythemia may be avoided. Because bleeding may be associated with polycythemia, *platelet and coagulation studies* are advisable if surgery is planned. Elevated uric acid levels and gout can occur if cyanosis is severe and of long standing.

CARDIAC CATHETERIZATION

The diagnosis is confirmed by establishment of the presence of a nonrestrictive ventricular septal defect with right-to-left shunting and by delineation of the severity and location of the right ventricular outflow tract obstruction. Congenital coronary arterial anomalies can occur that will influence the surgical approach.

Natural History and Prognosis

The severity of right ventricular outflow obstruction is progressive. Untreated, the prognosis is poor because of complications from hypoxemia and polycythemia, such as stroke and brain abscess. Sudden death can occur; survival beyond age 30 years is unusual. Infective endocarditis is common.

Treatment

Hypoxic spells are treated initially with the knee-chest position, oxygen, and morphine. If the spell is severe, a beta-adrenergic blocking drug, volume supplementation, and sodium bicarbonate may be required. In newborns, administration of prostaglandin E_1 may improve cyanosis temporarily by opening the ductus.

Complications such as endocarditis, dehydration, and anemia should be prevented or appropriately treated. Prompt surgical intervention is indicated at any age for patients with hypoxic spells, progressive symptoms, or with polycythemia with the hematocrit approaching 65 percent. For the remainder of patients, elective surgical correction should not be delayed beyond early childhood. Diagnosis after this age warrants prompt surgical intervention.

Infants may be palliated with chronic administration of beta blockers or systemic-to-pulmonary arterial shunt procedures. The latter are preferable, especially when the pulmonary arteries are very small or there are other anatomic abnormalities, such as a major coronary artery crossing the right ventricular outflow tract. Early correction is preferable in most other infants, since it can now be done with an acceptably low mortality if the anatomy is favorable.

Surgical correction requires patch closure of the ventricular septal defect and resection of obstructive infundibular muscle.

Pulmonary valvotomy and patch augmentation of the in-fundibulum, valvar annulus, and central pulmonary arteries are also frequently required.

EBSTEIN'S ANOMALY OF THE TRICUSPID VALVE

Definition and Pathology

In *Ebstein's anomaly*, varying portions of the posterior and septal leaflets of the tricuspid valve are displaced downward and attached to the ventricular wall below the annulus. The proximal part of the right ventricle is thin-walled and continuous with the right atrium. The papillary muscles and chordae are usually malformed, with multiple direct attachments of valvar tissue to the mural endocardium. An interatrial communication—a stretched patent foramen ovale—is present in most cases.

Pathophysiology

There is obstruction to right ventricular filling because of the decreased size of the functional right ventricle. At least mild tricuspid regurgitation is the rule. A right-to-left shunt occurs through the stretched foramen ovale. The severity of the hemodynamic abnormalities varies widely.

Clinical Manifestations

HISTORY
Approximately half of the reported cases present with cyanosis and right-sided heart failure in early infancy. The remainder present at all ages with symptoms of dyspnea on exertion due to hypoxemia and palpitations due to arrhythmia being the most common. Presentation as an adult is not rare, as symptoms and physical findings can appear deceptively benign.

PHYSICAL EXAMINATION
The newborn may have severe cyanosis initially due to high pulmonary vascular resistance. Cyanosis and clubbing in older patients tends to be mild. Only the rare patient with an intact atrial septum is not cyanotic. The liver may be enlarged and the jugular venous pulse elevated. The precordium is quiet, even in

the presence of striking cardiomegaly. The holosystolic murmur of tricuspid regurgitation is commonly heard and may be accompanied by a "scratchy" diastolic murmur of tricuspid stenosis. The S_1 is split and loud. The S_2 is widely and persistently split. S_3 and S_4 sounds are audible in older patients.

Usual Diagnostic Tests

CHEST X-RAY
Heart size varies; ordinarily it is very large owing to a very dilated right atrium. Pulmonary blood flow is diminished in proportion to the degree of cyanosis.

ELECTROCARDIOGRAM
Giant, spiked P waves, a prolonged PR interval, and right ventricular conduction delay or complete right bundle branch block are common. In approximately 10 percent of cases, the pattern of preexcitation, or Wolff-Parkinson-White syndrome, is present.

ECHOCARDIOGRAPHY
The exact anatomy of the valve, its function, and its attachments as well as associated defects are clearly demonstrable.

CARDIAC CATHETERIZATION
In most cases two-dimensional echocardiography with color Doppler evaluation is sufficient; catheterization is now less commonly performed than previously. Caution should be exercised, as the risk of rhythm disturbance is greater than usual. Right atrial pressures are elevated. The characteristic right ventricular pressure recording is not obtained until the catheter reaches the apex or outflow tract. An electrode catheter can demonstrate the area of atrial pressures, but with a ventricular electrogram the pulmonary artery may be difficult to enter. Biplane cineangiography will demonstrate the anomaly and associated defects.

Natural History and Prognosis

The natural history varies greatly with the severity of the anomaly. Infants and those with associated defects are at highest risk of early death; survival to old age has been reported with mild disease. Symptoms tend to progress in most patients. Severe symptoms, cardiomegaly, and cyanosis correlate with mortality. Death is usually the result of heart failure, complications of cyanosis, arrhythmia, or low cardiac output.

Treatment

Medical management includes treatment of heart failure, arrhythmias, and complications of cyanosis as well as prevention of endocarditis.

Surgery is generally recommended only for those with severe disease. Reconstruction or replacement of the tricuspid valve is necessary with closure of the atrial communication. In a small number of selected patients, the bidirectional Glenn or the modified Fontan operation has been used successfully. Interruption of accessory conduction pathways can also be accomplished if needed.

EISENMENGER'S SYNDROME

Pulmonary arterial hypertension is usually the result of transmission of systemic arterial pressure to the pulmonary artery via a larger communication. Less commonly, it is due to obstruction in the left side of the heart. *Pulmonary vascular obstructive disease* is the process of structural changes in the smaller muscular arteries and arterioles of the lungs that gradually diminishes their ability to transport blood. Patients with defects that cause both increased pulmonary blood flow and pressure are most likely to develop this feared complication.

Eisenmenger's syndrome consists of a large communication between the left and right sides of the heart, pulmonary arterial pressures at systemic level, and pulmonary vascular obstructive disease that reverses the original left-to-right shunt and causes cyanosis.

Clinical Manifestations

HISTORY

The age at which cyanosis and its symptoms appear varies greatly; those defects associated with both hypoxia and congestive failure cause earlier presentations. Exertional fatigue and dyspnea are common. Syncope from inadequate cardiac output or arterial oxygen desaturation, arrhythmias, hemoptysis, and chest pain occur when the disease progresses to a very severe level.

PHYSICAL EXAMINATION

A right ventricular lift, loud pulmonary closure sound, and cyanosis with clubbing are typical findings. The murmurs of pulmonary and tricuspid regurgitations may be audible.

Usual Diagnostic Tests

CHEST X-RAY

The right ventricle is hypertrophied, but the overall heart size may not be large. The central pulmonary arteries are enlarged, with rapid tapering or "pruning" to small vessels distally.

ELECTROCARDIOGRAM

Right axis deviation and right atrial and right ventricular hypertrophy are present.

ECHOCARDIOGRAPHY

The underlying cardiac defect can be outlined along with supporting evidence of the presence of pulmonary artery hypertension.

CARDIAC CATHETERIZATION

The presence of pulmonary hypertension and elevated vascular pulmonary resistance is established and the underlying defects are outlined. Measurements should be repeated after inhalation of 100% oxygen and again after pulmonary arterial administration of tolazoline or the inhalation of nitrous oxide to seek any reversible component of the disease.

Natural History and Prognosis

The disease is a progressive one with a poor prognosis of survival beyond the young- to midadult years. Sudden death, heart failure, brain abscess, pulmonary hemorrhage, and stroke all contribute to this. Pregnancy is associated with a very high mortality and fetal wastage.

Treatment

Medical therapy is usually not effective in combating the complications of severe pulmonary vascular disease, but it should be tried. Pregnancy and the use of oral contraceptives should be strenuously avoided. Prevention of infective endocarditis is important.

Corrective surgery is contraindicated when the pulmonary vascular resistance is 11 Wood units (880 dynes $\cdot$ s $\cdot$ cm^{-5}) or more or the pulmonary/systemic resistance ratio is 0.7:1 or greater with a pulmonary/systemic flow ratio less than 1.5:1. Lung transplantation with repair of the cardiac defect or combined heart and lung transplantation are the only surgical means of treatment for these patients.

LONG-TERM POSTOPERATIVE FOLLOW-UP

With the dramatic advances that have occurred in the surgical treatment for congenital heart defects, more patients are reaching adulthood.

It must be recognized that there are residua, sequelae, and complications resulting from this type of surgery. A residual part of a defect may deliberately have been avoided at operation. Sequelae are unavoidable consequences; complications are unexpected but related events that occur after surgery. Of surgery for all congenital heart defects, only surgical correction of patent ductus arteriosus is likely to pose no long-term problems.

Patients frequently have residual murmurs. Careful evaluation of the hemodynamic abnormalities that these represent is very important. This will help recognize and anticipate potential complications. Doppler and two-dimensional echocardiography are particularly helpful and noninvasive. Significant residual shunts or defects should have been repaired in childhood. Continued cyanosis usually signals complex uncorrected defects or pulmonary vascular disease.

Sick sinus syndrome with bradytachyarrhythmias can occur after repair of an atrial septal defect. The most serious late complications after repair of ventricular septal defect or tetralogy of Fallot are development of complete heart block and serious ventricular arrhythmias. Dysrhythmias are particularly important late problems, especially among patients who have had complex operations. Serious ventricular dysfunction also may occur late after surgery for complex defects. Ambulatory electrocardiographic monitoring and stress exercise testing are very helpful in guiding management. Valvar aortic stenosis tends to recur and continues to progress over the years despite successful surgery or valvuloplasty in childhood. Progressive aortic regurgitation may also appear, with or without surgery. Pulmonary regurgitation following relief of RV outflow tract obstruction, once thought benign, can produce progressive right ventricular dilatation, dysfunction, arrhythmias, and tricuspid regurgitation with right atrial dilatation.

As a rule, the risk of infective endocarditis is not diminished after surgery with the exception of those patients who have undergone patent ductus ligation or repair of VSD or of secundum ASD in whom there is no residual shunt or murmur. Patients in whom it has been necessary to place an artificial valve are at increased risk.

In general, all patients who have had surgery for congenital heart defects with the exception of a patent ductus should be followed longitudinally throughout life for the development of problems.

SUGGESTED READING

Deanfield JE, Gersh BJ, Warnes CA, Mair DD: Congenital heart disease in adults. In: Alexander RW, Schlant RC, Fuster V, O'Rourke RA, Roberts R, Sonnenblick EH, (eds): *Hurst's The Heart*, 9th ed. New York, McGraw-Hill, 1998:1995–2027.

Emmanouilides GC, Riemenschneider TA, Allen HD, Gutgesell HP (eds): *Moss and Adams Heart Disease in Infants, Children, and Adolescents*. 5th ed. Baltimore, Williams & Wilkins, 1995.

Freed MD, Plauth WH Jr: The pathology, pathrophysiology, recognition, and treatment of congenital heart disease. In: Alexander RW, Schlant RC, Fuster V, O'Rourke RA, Roberts R, Sonnenblick EH (eds): *Hurst's The Heart*, 9th ed. New York, McGraw-Hill, 1998:1925–1993.

Fyler DC (ed): *Nadas' Pediatric Cardiology*. Philadelphia, Hanley & Belfus, 1992.

Garson A Jr, Bricker JT, Fisher DJ, Neish SR (eds): *The Science and Practice of Pediatric Cardiology*, 2nd ed. Baltimore, Williams & Wilkins, 1998.

Perloff JK: *The Clinical Recognition of Congenital Heart Disease*, 4th ed. Philadelphia, Saunders, 1994.

Perloff JK, Child JS (eds): *Congenital Heart Disease in Adults*, 2nd ed. Philadelphia, Saunders, 1998.

Roberts WC (ed): *Adult Congenital Heart Disease*. Philadelphia, Davis, 1987.

Snider AR, Serwer GA, Ritter SB: *Echocardiography in Pediatric Heart Disease*, 2nd ed. St. Louis, Mosby, 1997.

23

PULMONARY EMBOLISM

Joseph S. Alpert
James E. Dalen

Each year an estimated 50,000 to 100,000 deaths occur in the United States due primarily to pulmonary embolism; in another 100,000 patients who die with other major diseases, pulmonary embolism is a significant contributory cause. Most of these deaths occur in patients who are not treated because the diagnosis is not established. The mortality of untreated pulmonary embolism is 20 to 30 percent. If the diagnosis is established and appropriate treatment instituted, mortality is less than 10 percent.

In the overwhelming majority of cases, pulmonary embolism originates as deep venous thrombosis (DVT) in the proximal deep venous system in the lower legs. Some of the factors that predispose to DVT are shown in Table 23-1. The optimal strategy for preventing fatal pulmonary embolism is to recognize patients who are at increased risk of DVT and to institute appropriate prophylactic treatment. In order to reduce the incidence of pulmonary embolism, DVT must be prevented or, at the very least, recognized and treated.

DEEP VENOUS THROMBOSIS

Pathophysiology and Natural History

Three factors underlie the development of venous thrombosis: stasis, injury to the intima of the vein, and a hypercoagulable state. All three factors need not be present simultaneously for DVT to occur. It appears that stasis and intimal injury are the most important.

Venous thrombosis seems to originate as a platelet nidus in the vicinity of the venous valves in the lower extremities. Platelet aggregation and activation initiate the clotting cascade,

TABLE 23-1

RISK FACTORS FOR DEEP VENOUS THROMBOSIS

Venous stasis, vascular injury, or secondary
hypercoagulable states

Surgery (especially involving lower extremities or
  pelvis)
Trauma (especially fractured hip in the elderly and
  acute injury of the head and spinal cord)
Bed rest, immobility, stroke, or paralysis
Congestive heart failure, low-cardiac-output states
Malignancy
Pregnancy
Oral contraceptive agents, estrogen therapy
Obesity
Varicose veins
Inflammatory bowel disease
Advanced age
Prior thromboembolism
Nephrotic syndrome
Sepsis

Primary hypercoagulable state

Antithrombin III deficiency
Protein C deficiency
Protein S deficiency
Disorders of plasminogen and plasminogen activation
Paroxysmal nocturnal hemoglobinuria
Myeloproliferative disorders
Polycythemia vera
Heparin-induced thrombocytopenia
Lupus anticoagulant/anticardiolipin antibodies

with subsequent formation of a red fibrin thrombus. The
thrombus gradually grows as further platelets, fibrin, and red
cells are incorporated. The intrinsic fibrinolytic system is also
activated, attacking the fibrin framework of the thrombus. In
some instances, the intrinsic fibrinolytic system is successful in
dissolving the thrombus; in other situations, pieces of the
thrombus break off and embolize to the pulmonary circulation.
Residual venous thrombus becomes organized and incorporated into the venous wall. It has long been recognized that

patients with clinically evident malignancy are often at increased risk for venous thromboembolism.

More than 90 percent of pulmonary emboli originate in the deep veins of the lower extremities. Occasionally, DVT in pelvic veins or thrombi in the right heart give rise to pulmonary emboli. Thrombi localized to the deep venous system of the calf are of limited risk with respect to their tendency to embolize. Once such thrombi propagate into the popliteal vein and/or the veins of the thigh, however, the risk for pulmonary embolism increases markedly. Thus, symptomatic pulmonary embolism usually stems from recurrent episodes of embolization that arise from thrombi in the proximal veins of the lower extremities.

Prevention

There are multiple techniques to prevent DVT, as shown in Table 23-2. In medical patients with a factor for increased risk or in surgical patients over 40 years of age who are undergoing major operations without contraindications, low-dose heparin, 5000 units subcutaneously every 8 to 12 h, is very effective in preventing DVT. In patients at higher risk, e.g., in those undergoing urologic procedures, low-dose subcutaneous heparin does not provide adequate protection. In these circumstances adjusted-dose heparin [activated partial thromboplastin time (APTT) $1\frac{1}{2}$ to 2 times control 6 h after injection], low-dose warfarin [International Normalized Ratio (INR) 2.0 to 3.0], or intravenous dextran provides more appropriate prophylaxis. In patients at increased risk of bleeding, such as those undergoing neurosurgical procedures, intermittent pneumatic compression (IPC) boots have been shown to be effective. In patients undergoing elective surgery, these prophylactic therapies should begin prior to the induction of anesthesia because DVT often begins while patients are anesthetized in the operating room. Aspirin and sulfinpyrazone have not been shown effective in preventing DVT. Hirudin and especially low-molecular-weight heparin appear very promising.

Diagnosis

If the prophylactic therapies listed above are instituted, the risks of DVT and pulmonary embolism, its potentially lethal complication, are greatly diminished. If prophylactic treatment is not used or if it fails, the clinician must recognize DVT as soon as possible in order to institute treatment to prevent pul-

TABLE 23-2

PROPHYLAXIS AGAINST DEEP VENOUS THROMBOSIS PULMONARY EMBOLISM FOR SPECIFIC PATIENT GROUPS

Patient Group	Prophylaxis
Medical or surgical patients under 40 years of age with no clinical risk factors	Early ambulation
Medical patients with one or more risk factors (Table 23-1) or surgical patients over 40 years of age undergoing major operations but with no additional risk factors	GCS; LDH every 8–12 h, fixed dose LMWH, or IPC
Surgical patients over 40 years of age undergoing major operations and with additional risk factors	GCS; adjusted-dose subcutaneous unfractionated heparin or fixed-dose LMWH (IPC is an alternative in patients prone to hematomas or infection)
Very high risk general surgery patients with multiple risk factors	GCS; IPC and adjusted-dose subcutaneous unfractionated heparin or fixed-dose LMWH; in selected patients, perioperative warfarin (INR 2.0–3.0)
Total hip replacement	GCS; adjusted doses of warfarin (INR 2.0–3.0) or unfractionated heparin (APTT 1.5–2.5 times control 6 h after injection); when available, LMWH (without laboratory control)

Hip fractures	GCS; warfarin (INR 2.0–3.0) or LMWH
Knee surgery, neurosurgery	GCS; IPC; LMWH
Acute spinal cord injury with paralysis	GCS; LDH; LMWH; IPC
Multiple trauma	GCS; IPC; warfarin (INR 2.0–3.0); LMWH
Myocardial infarction	GCS; LDH (IVH if anterior infarct or increased risk factors); IPC if heparin is contraindicated
Ischemic stroke with lower extremity paralysis	GCS; LDH (alternative: LMWH, IPC, warfarin)
Long-term indwelling central vein catheter	GCS; warfarin, 1 mg/day
Hip or knee surgery in high-risk patients with history of serious, previous pulmonary embolism	GCS; warfarin; consider prophylactic inferior vena cava filter

Key: GCS, graded compression stockings; LDH, low-dose subcutaneous heparin; LMWH, low-molecular-weight heparin; IPC, intermittent pneumatic compression; INR, International Normalized Ratio; APTT, activated partial thromboplastin time; IVH, intravenous heparin.

Source: Modified from Dalen and Hirsh, 1995, with permission.

monary embolism. The classic signs of DVT—unilateral leg swelling and tenderness—occur only in the minority of cases; most episodes of DVT are clinically silent. Venography or noninvasive tests of venous thrombosis should usually be performed when DVT is suspected.

The most sensitive and specific test for the recognition of DVT is venography; it is invasive and uncomfortable.

Impedance plethysmography (IPG) is extremely useful in evaluating patients with suspected DVT; it can be performed in 15 to 20 min. The correlation between unilaterally positive IPG and venography exceeds 90 percent. A bilaterally normal IPG nearly excludes proximal (above the knee) DVT. The test is less sensitive to DVT confined to the distal leg (below the knee). If the IPG is positive in both legs, venography should be performed (if available), because DVT may or may not be present.

Scanning the legs after the injection of fibrinogen labeled with iodine 125 (^{125}I) is a very sensitive test for the diagnosis of DVT. The disadvantage of this test is that the ^{125}I must be injected prior to the development of DVT.

The Doppler technique is another accurate, noninvasive test used for the diagnosis of DVT; its accuracy is dependent on the expertise of the operator: the sensitivity and specificity of this technique in detecting DVT can vary in different medical centers.

Treatment

If proximal DVT is detected, treatment includes intravenous heparin for 7 to 10 days, followed by oral warfarin therapy begun on day 2 to 3. Heparin is given as an intravenous bolus of 5000 units, followed by an infusion of 1000 units per hour; the dosage is adjusted to produce an APTT between 1.5 and 2.5 times control. Warfarin anticoagulation is started simultaneously with the initiation of heparin. Without treatment, pulmonary embolism occurs in 50 percent of patients with proximal DVT. The probability of clinically significant pulmonary embolism in patients with DVT limited to the distal lower extremity is less than 10 percent; in this circumstance in an asymptomatic patient, many physicians would not institute heparin treatment. If DVT limited to the distal lower extremity is not treated with heparin, however, repeat IPG or Doppler study is indicated to make certain that there has not been extension to the proximal venous circulation.

Some advocate thrombolytic treatment with urokinase, streptokinase, or tissue plasminogen activator (t-Pa) in patients with symptomatic proximal DVT in order to prevent the postphlebitic syndrome. The incidence of the postphlebitic syndrome in patients treated with heparin is uncertain, however; at present, it is controversial whether or not fibrinolytic therapy prevents the postphlebitic syndrome in patients with DVT.

PULMONARY EMBOLISM

Pathophysiology and Natural History

The major hemodynamic consequences of pulmonary embolism result from reduction in the cross-sectional area of the pulmonary vascular bed, with associated increases in pulmonary vascular resistance and right ventricular afterload. Because the reserve capacity of the pulmonary vascular bed is substantial, a single large embolic thrombus (or repeated smaller thrombi) must impact the pulmonary vascular bed before pulmonary vascular resistance and right ventricular afterload increase. In general, more than 50 percent of the pulmonary vascular tree must be occluded before pulmonary arterial pressure increases significantly. Patients with advanced cardiopulmonary disease, e.g., marked left ventricular failure, occasionally develop marked hemodynamic abnormalities following modest episodes of pulmonary embolism.

Increased right ventricular afterload leads to right ventricular dilatation and eventually failure, with decreased cardiac output and systemic arterial hypotension.

Arterial hypoxemia is quite common in patients with acute pulmonary embolism. Ventilation/perfusion mismatches and reduced cardiac output contribute to shunting of deoxygenated blood across the pulmonary vascular bed.

Hyperventilation usually produces arterial hypocarbia and respiratory alkalosis. An occasional patient with massive embolism may develop hypercarbia despite marked hyperventilation.

Pulmonary infarction or hemorrhage results when distal emboli totally occlude a small pulmonary artery. Infarction following more proximal pulmonary embolism is often prevented, since the pulmonary parenchyma has four sources of oxygen: the airways, the pulmonary arteries, bronchial arterial collateral circulation, and back-diffusion from the pulmonary veins. Some or all of these compensatory mechanisms may be compromised in patients with heart failure or intrinsic lung disease. Consequently, pulmonary infarction is more common under these circumstances.

Patients who survive an initial episode of pulmonary embolism usually do well if the diagnosis is made and appropriate therapy is instituted. The intrinsic fibrinolytic system of the pulmonary vascular bed initiates dissolution of thrombi. With heparin therapy alone, 36 percent of perfusion lung scan defects resolve within 5 days. At 2 weeks, 52 percent of defects have resolved; at 3 weeks, 73 percent; and at 1 year, 76 percent. Arterial hypoxemia and chest roentgenographic abnormalities improve as the thromboemboli resolve. Persistent pulmonary hypertension and chronic cor pulmonale from unresolved embolism are unusual.

Diagnosis of Acute Pulmonary Embolism

One reason that the diagnosis of acute pulmonary embolism is frequently missed is that it may present as one of three different clinical syndromes: pulmonary infarction (or hemorrhage), acute cor pulmonale, or acute "unexplained" dyspnea.

PULMONARY INFARCTION OR HEMORRHAGE

This is the most common presentation of pulmonary embolism. More than 50 percent of all patients in whom pulmonary embolism is diagnosed have signs or symptoms of pulmonary infarction or hemorrhage. The classic symptom is the abrupt onset of pleuritic chest pain, with or without dyspnea. Hemoptysis occurs in a minority of patients.

Pulmonary hemorrhage or infarction usually causes a pulmonary infiltrate on the chest radiograph. Other roentgen abnormalities include an elevated diaphragm due to splinting of respiration and a small pleural effusion that is usually unilateral and may or may not be bloody.

On physical examination, tachypnea with a respiratory rate greater than 20 per minute is nearly always present. Signs of right ventricular failure are absent. Examination of the lungs usually reveals rales, wheezes, or evidence of a pleural effusion. A pleural friction rub may be present. Evidence of deep venous thrombosis on physical examination is present in a minority of cases.

The most useful routine laboratory tests in patients with suspected pulmonary infarction or hemorrhage include the chest x-ray, while blood cell count (WBC) and differential, Gram's stain of sputum if available, and arterial blood gas analysis. The WBC, differential, and sputum examination help to diagnose bacterial bronchitis or pneumonia. Analysis of arterial blood gases in patients with pulmonary infarction or hemorrhage will nearly always demonstrate hypocapnia and respiratory alkalosis secondary to tachypnea.

The most useful screening test for pulmonary embolism is the ventilation/perfusion ($\dot{V}/\dot{Q}$) lung scan. The most specific finding for acute pulmonary embolism is the presence of large segmental perfusion defects that ventilate normally. Spiral CT is another noninvasive and accurate modality for the identification of pulmonary embolism. It visualizes pulmonary emboli only in the first four to five branching subdivisions of the pulmonary arterial tree. Selective pulmonary angiography may be indicated for a definitive diagnosis.

ACUTE COR PULMONALE

Acute cor pulmonale occurs when pulmonary embolism obstructs more than 60 to 75 percent of the pulmonary circulation.

In addition to marked, acute dyspnea, patients with acute cor pulmonale may present with signs of decreased cardiac output: hypotension, syncope, or cardiac arrest. On physical examination, tachypnea, tachycardia, and possibly hypotension are found. Signs of acute right ventricular failure—distended neck veins, right-sided S_3 gallop, and parasternal heave—are usually present. The lungs may be clear, but signs of DVT may be noted.

The most useful diagnostic tests in patients with suspected acute cor pulmonale are the electrocardiogram (ECG), measurement of central venous pressure, and arterial blood gas analysis. The ECG in patients with acute cor pulmonale often will show a new $S_1Q_3T_3$ pattern, incomplete right bundle branch block, and/or signs of right ventricular ischemia. The ECG helps to exclude acute myocardial infarction. Measurement of central venous pressure may be critical in diagnosing acute cor pulmonale in patients who present with systemic hypotension or cardiovascular collapse. If systemic arterial hypotension is due to acute cor pulmonale, right atrial and central venous pressure will be elevated. Arterial blood gas analysis in patients with acute cor pulmonale usually but not always demonstrates hypoxemia as well and hypocapnia. The chest radiograph is not helpful in the diagnosis of acute cor pulmonale and often is unremarkable.

Once the diagnosis of acute cor pulmonale is suspected, it should be confirmed by $\dot{V}/\dot{Q}$ lung scar, spiral CT, or pulmonary angiography. In unstable patients in whom surgical or thrombolytic therapy may be indicated, it usually is wise to proceed promptly to pulmonary angiography for definitive diagnosis.

ACUTE UNEXPLAINED DYSPNEA

The diagnosis of pulmonary embolism is most difficult in a patient with a submassive lesion who does not sustain pulmonary infarction. The ECG remains normal, and there are no physical signs of right ventricular failure. If pulmonary infarction or hemorrhage does not occur, pleuritic pain is absent and there are no specific abnormalities on chest radiograph. In this circumstance, the primary symptom of acute pulmonary embolism is the sudden onset of dyspnea. The only abnormalities on physical examination are tachypnea, possibly tachycardia, and anxiety. The lungs are clear; there are no signs of acute right ventricular failure. Signs of DVT may be present or absent. Thus, many of the commonly obtained clinical variables associated with pulmonary embolism are not abnormal.

The principal differential diagnoses in patients with acute dyspnea due to pulmonary embolism are left ventricular failure, pneumonia, and the hyperventilation syndrome. Left ventricular failure and pneumonia usually can be excluded by further history, physical examination, and chest radiographs. Patients presenting with dyspnea due to acute pulmonary embolism usually but not always have significant hypoxemia. Measurement of arterial blood gases while the patient breathes room air may allow one to distinguish between potentially lethal pulmonary embolism and benign hyperventilation.

As with other syndromes of acute pulmonary embolism, IPG or venous Doppler examination is very helpful, and the diagnosis should be confirmed by $\dot{V}/\dot{Q}$ lung scan, spiral CT, or pulmonary angiography.

Treatment of Acute Pulmonary Embolism

The therapeutic modalities available for patients with acute pulmonary embolism can be divided into two groups: prophylactic and definitive therapy. *Prophylactic therapy* is based on

the concept that the body's intrinsic fibrinolytic system will
dissolve thromboembolic material that finds its way into the
pulmonary vascular bed. Such dissolution usually leads to resolution of the pathophysiological changes associated with
acute pulmonary embolism over a period of 7 to 14 days. Thus,
prophylactic therapy aims at preventing further embolic episodes. Examples of prophylactic therapy include anticoagulation with heparin, low-molecular-weight heparin, or warfarin
and inferior vena cava interruption (Table 23-3). Prophylactic

TABLE 23-3

**HEPARIN REGIMENS FOR THE TREATMENT OF DEEP
VENOUS THROMBOSIS AND PULMONARY EMBOLISM**

Diagnosis	Heparin dosage
Deep venous thrombosis without pulmonary embolism or with minor pulmonary embolism	5000-U IV loading dose followed by 1000–1500 U IV per hour; check activated partial thromboplastin time (APTT) 4–6 h after initiating infusion and adjust heparin dose to prolong APTT to 1.5–2.5 times control or low-molecular-weight heparin 1 mg/kg bid subcutaneously. No APTT determinations required. Warfarin is started on day 2 to 3. Heparin is discontinued on days 7 to 10.
Major pulmonary embolism with or without right ventricular failure and hypotension	10,000-U IV loading dose followed by 1000–2000 U IV per hour; check APTT 4 h after initiating infusion and adjust heparin dose to pro-long APTT to 1.5–2.5 times control (employ smaller loading and infusion dosage for smaller individuals or patients with hepatic and/or renal insufficiency). Warfarin is started on day 2 to 3 (if patient is stable) and heparin is discontinued on days 7 to 10.

therapy with anticoagulants is initiated as soon as the clinician has a high index of suspicion for an increased risk of DVT or pulmonary embolism.

Definitive therapy focuses on thromboemboli that have already arrived in the pulmonary vascular bed. Definitive therapy attempts to remove or dissolve such emboli in order to effect a more rapid resolution of the pathophysiologic sequelae of pulmonary embolism. Examples of definitive treatment for pulmonary embolism include thrombolytic agents and pulmonary embolectomy.

The incidence of bleeding during heparin therapy is determined not by the dose of heparin but by defects in the walls of blood vessels. Therefore, contraindications to anticoagulants include conditions that predispose to bleeding—e.g., active peptic ulcer disease, esophageal varices, hemorrhagic diatheses, severe liver or kidney disease, severe hypertension, intracranial disease, and recent surgery on brain, spinal cord, joints, or genitourinary tract.

Heparin is continued for 7 to 10 days and should overlap oral warfarin therapy until a prothrombin time of INR = 2.0 to 3.0 (approximately 1.3 to 1.5 times control, using rabbit brain thromboplastin) is achieved. Warfarin is usually started at a dose of 5 mg/day.

Thereafter, warfarin dosage is adjusted according to the results of the prothrombin time. After the prothrombin time has been in the therapeutic range for 1 to 2 days, heparin is discontinued and fine regulation of the prothrombin time is achieved by altering the dosage of warfarin. Warfarin should be continued for as long as the patient has an underlying predisposition to thromboembolism, e.g., bed rest. In the patient with a fracture, this period of time should be for 2 months after the cast or traction is removed and the individual is ambulatory. If predisposition to thromboembolism is transitory, patients who have suffered an episode of acute pulmonary embolism should receive oral anticoagulation for 3 to 6 months. In the individual with permanent predisposition to thromboembolism, anticoagulation should be lifelong.

An alternative to daily oral warfarin is injection of subcutaneous heparin every 8 to 12 h for a minimum of 12 weeks. Subcutaneous heparin can be either fixed dose (i.e., 5000 units every 8 to 12 h) or adjusted dose to main the predose APTT at $1\frac{1}{2}$ times the control value. Low-molecular-weight heparin can be substituted for unfractionated heparin. The dose is 1 mg/kg bid and APTT testing is not required.

Venous interruption is also a form of prophylactic therapy in that it is performed to prevent additional venous thromboemboli from reaching the pulmonary vascular bed. Interruption of the vena cava is highly effective in preventing further episodes of thromboembolism. Interruption of the inferior vena cava is indicated (1) when embolism occurs in patients receiving appropriate anticoagulant therapy, (2) when anticoagulants are contraindicated, (3) when diseases predisposing to venous thrombosis and pulmonary embolism are prominent and persistent, (4) when septic embolism occurs, (5) when paradoxical embolism occurs, and (6) in some patients with massive embolism in whom a further episode of embolism would be fatal. At present, the most common form of venous interruption is not surgery but rather the insertion of a filter or umbrella into the inferior vena cava.

As noted earlier, most patients with pulmonary embolism have an excellent prognosis if the diagnosis is suspected and confirmed and treatment is initiated. Individuals with massive embolism, right ventricular failure, and hypotension form a subgroup with a high in-hospital mortality (32 percent). One would expect these patients to benefit from a direct, definitive attack on the pulmonary vascular thromboemboli, and, indeed, this small minority of patients appears to benefit from thrombolytic dissolution of thromboemboli or even embolectomy. Pulmonary embolectomy is associated with high surgical mortality (approximately 30 to 50 percent) because the operation is usually performed on patients who are in profound shock.

The primary indications for embolectomy are the presence of right ventricular failure and systemic arterial hypotension requiring vasopressors in a patient with bilateral central pulmonary emboli documented by pulmonary angiography or spiral CT. In this setting, embolectomy can be lifesaving. Contraindications to pulmonary embolectomy include recurrent pulmonary embolism without angiographic evidence of occluded central pulmonary arteries, pulmonary arterial systolic pressure in excess of 70 mmHg, severe underlying heart disease complicated by heart failure, and marked pulmonary insufficiency secondary to severe pulmonary disease, i.e., chronic obstructive lung disease. An alternative to operative pulmonary embolectomy, which requires a thoracotomy, is percutaneous pulmonary embolectomy employing a special catheter.

The search for an agent to dissolve thromboemboli in human beings has been long and complex. The agents studied most extensively are streptokinase, urokinase, and t-PA. Thrombolytic therapy is potentially most useful in patients with documented massive pulmonary embolism and hemodynamic abnormalities, such as right ventricular failure and/or hypotension. A multicenter randomized trial has documented improved right ventricular function and possibly a reduced number of adverse clinical events in patients with acute pulmonary embolism treated with t-PA as compared with intravenous heparin.

The dosage of urokinase for patients with acute pulmonary embolism is an initial intravenous infusion of 4400 IU/kg body weight dissolved in 15 mL of sterile water given over 10 min. Maintenance therapy is then initiated: 4400 IU/kg per hour for a total of 12 h. Streptokinase is given as an initial loading dose of 250,000 IU dissolved in normal saline solution or 5% dextrose in water and administered over 30 min. Maintenance therapy consists of 100,000 IU/h for 24 h.

In selected patients, thrombolytic therapy with t-PA or reteplase can rapidly lyse pulmonary embolism; these agents act more rapidly and may cause less bleeding than urokinase. The usual dosage of t-PA is 100 mg given intravenously over 2 h. It has also been administered at a rate of 0.6 mg/kg intravenously over 2 min.

At the present time, thrombolytic therapy is generally reserved for patients with hemodynamic embarrassment, right ventricular dysfunction, anatomically large pulmonary embolism, or extensive deep venous thrombosis.

FAT EMBOLISM

Fat emboli to the lungs produce a fat embolism syndrome (FES): dyspnea (respiratory insufficiency), confusion (neurologic dysfunction), and petechiae.

The entrance of free globules of fat into systemic veins most often occurs after fractures of long bones, especially fractures of the tibia and femur in automobile accidents. Fat also may enter the circulation following direct injury to subcutaneous fat tissue by contusion, concussion, burns, childbirth, poisoning, the use of a pump oxygenator, or high-altitude flights. Patients can develop the acute respiratory distress syndrome (ARDS), with extensive intrapulmonary hemorrhage and damage to pulmonary vascular endothelium and parenchyma. Clinically, it is frequently difficult to distinguish FES from ARDS.

Patients often have a lucid interval of 6 h to several days (typically, 24 to 40 h) following trauma before the first symptoms or signs of fat embolism are recognized. Most features of the syndrome result from fat emboli either to the lungs or to the brain. Cardiorespiratory manifestations of pulmonary fat emboli include tachypnea, dyspnea, sinus tachycardia, hypoxemia, and pyrexia to 39.4°C (103°F). Individuals with severe respiratory distress may become cyanotic. Patients with FES often have copious bronchial secretions, which may be hemorrhagic. The cerebral symptoms, which may occur simultaneously with or after the pulmonary symptoms, include headache, increasing irritability, disturbances of consciousness, disorientation, delirium, confusion, restlessness, convulsions, apathy, stupor, and coma. Focal cerebral syndromes also may occur. As noted above, patients with fat embolism may present with or may develop the full clinical pictures of ARDS. Oliguria and even anuria may develop.

Signs of systemic fat embolism include petechiae, especially on the anterior chest, axillary folds, neck, fundi, and conjunctivae. Rarely, fat emboli are seen in the retinal vessels. In some patients, the petechiae, whether spontaneous or induced, may be related to the associated thrombocytopenia, although small fat emboli may be found in biopsies of cutaneous capillaries adjacent to the petechiae. The prothrombin time and partial thromboplastin time may be increased, and plasma fibrinogen may be reduced. The serum calcium level often is decreased, presumably because of the interaction between increased serum fatty acids and calcium; serum lipase and tributyrinase concentrations usually are elevated. A frozen selection of clotted blood examined for fat may be of some diagnostic value early in the course of the FES, particularly in patients with an arterial P_{O_2} less than 60 mmHg. The finding of fat droplets in sputum or urine is also suggestive but not diagnostic of fat embolism. Arterial hypoxemia is one of the earliest and most important laboratory findings. The chest roentgenogram usually shows extensive fluffy infiltrates; occasionally, only hazy, diffuse, fine stippling is seen throughout both lungs. The chest radiograph also may be compatible with pulmonary edema.

There is no specific therapy for fat embolism; the most important principle is maintenance of pulmonary oxygenation and function. It is important to correct the arterial hypoxemia that is usually present and occasionally quite marked. Supplemental inspiratory oxygen usually is required. Occasionally, assisted ventilation, with or without positive end-expiratory pressure (PEEP), is needed.

It has been suggested that massive doses of glucocorticoids may decrease alveolar damage, although adequate clinical trials are lacking. Doses usually employed are hydrocortisone, 1 to 2 g/day, or methylprednisolone, 13 mg/kg per day for 3 to 5 days. Low-dose heparin was formerly recommended in order to decrease platelet adhesiveness. The stimulatory effect of heparin upon lipase activity in the lung is theoretically detrimental, however, since it might increase the amounts of toxic fatty acids in the lungs. Other former therapies no longer used include low-molecular-weight dextran, intravenous ethyl alcohol, hypothermia, and various detergents.

SUGGESTED READING

Alpert JS, Smith R, Carlson J, et al: Mortality in patients treated for pulmonary embolism. *JAMA* 1976; 236: 1477–1480.

Alpert JS, Dalen JE: Pulmonary embolism. In: Alexander RW, Schlant RC, Fuster V, O'Rourke RA, Roberts R, Sonnenblick EH, (eds): *Hurst's The Heart*, 9th ed. New York, McGraw-Hill, 1998:1719–1738.

Becker DM, Philbrick JT, Bachhuber TL, et al: D-Dimer testing and acute venous thromboembolism: a shortcut to accurate diagnosis? *Arch Intern Med* 1996; 156:939–946.

Dalen JE, Alpert JS: Natural history of pulmonary embolism. *Prog Cardiovasc Dis* 1975; 17:259–269.

Dalen JE, Banas J Jr, Brooks H, et al: Resolution rate of acute pulmonary embolism in man. *N Engl J Med* 1969; 280: 1194–1197.

Dalen JE, Hirsh J (eds): Fourth ACCP Consensus Conference on Anti-thrombotic Therapy. *Chest* 1995; 108(suppl): 225S–522S.

Dalen JE, Alpert JS, Hirsh J: Thrombolytic therapy for pulmonary embolism: Is it effective? Is it safe? When is it indicated? *Arch Intern Med* 1997; 157:2550–2556.

Goldhaber SZ: Thrombolysis for pulmonary embolism. *Prog Cardiovasc Dis* 1991; 34:113–134.

Goldhaber SZ, Haire WD, Feldstein ML, et al: Alteplase versus heparin in acute pulmonary embolism: randomized trial assessing right-ventricular function and pulmonary perfusion. *Lancet* 1993; 341:507–511.

Goldhaber SZ, Morpurgo M: Diagnosis, treatment, and prevention of pulmonary embolism: Report of the WHO/

International Society and Federation of Cardiology Task Force. *JAMA* 1991; 268:1727–1733.

Huisman MV, Bulla HR, TenCate JW, et al: Serial impedance plethysmography for suspected deep venous thrombosis in outpatients. *N Engl J Med* 1986; 314:823–828.

Hull RC, Raskob GE, Pineo GF, et al: A comparison of subcutaneous low-molecular-weight heparin with warfarin sodium for prophylaxis against deep vein thrombosis after hip or knee implantation. *N Engl J Med* 1993; 329: 1370–1376.

Kelley MA, Carson JL, Palevsky HI, et al: Diagnosing pulmonary embolism: new facts and strategies. *Ann Intern Med* 1991; 114:300–306.

Leizorovicz A, Simmoneau G, Decousus H, et al: Comparison of efficacy and safety of low-molecular-weight heparins and unfractionated heparin in initial treatment of deep venous thrombosis: a meta-analysis. *Br Med J* 1994; 309:299–304.

McIntyre KM, Sasahara AA: Hemodynamic and ventricular responses to pulmonary embolism. *Prog Cardiovasc Dis* 1974; 17:175–190.

Meyer G, Tamisier D, Sors H, et al: Pulmonary embolectomy: a 20-year experience at one center. *Ann Thorac Surg* 1991; S1:232–236.

Parakos JA, Adelstein SJ, Smith RE, et al: Late prognosis of acute pulmonary embolism. *N Engl J Med* 1973; 289: 55–58.

PIOPED Investigators: Value of the ventilation/perfusion scan in acute pulmonary embolism: results of the prospective investigation of pulmonary embolism diagnosis (PIOPED). *JAMA* 1990; 263:2753–2796.

Salzman EW: Low-molecular-weight heparin and other new antithrombotic drugs. *N Engl J Med* 1992; 326:1017–1019.

Stein PD, Hull RD, Pineo G: Strategy that includes serial non-invasive leg tests for diagnosis of thromboembolic disease in patients with suspected acute pulmonary embolus based on data from PIOPED. *Arch Intern Med* 1995; 155: 2101–2104.

Urokinase Pulmonary Embolism Study Group: The urokinase pulmonary embolism trial. *Circulation* 1973; 47 (suppl2): 1–108.

24

PULMONARY HYPERTENSION— COR PULMONALE

John H. Newman
James E. Loyd

DEFINITION

Cor pulmonale is a term used to describe right heart failure resulting from pulmonary hypertension that occurs as a consequence of lung disease. Normal mean pulmonary arterial (PA) pressure at rest is 14 ± 3 mmHg. Pulmonary hypertension exists when the mean pulmonary arterial pressure exceeds 20 mmHg at rest or 30 mmHg during moderate exercise. Pulmonary hypertension (PHT) is silent clinically until it becomes severe enough to impair right ventricular (RV) function. RV function is usually impaired when mean PA pressure rises *acutely* to levels greater than 40 mmHg. The RV adapts to *chronic* pulmonary hypertension and can sustain function at mean PA pressures higher than 60 mmHg (80 to 90 systolic). Ultimately, the RV may fail, leading to dyspnea from low cardiac output and to systemic venous engorgement and edema from RV failure.

ETIOLOGIES AND PATHOGENESIS

The pathophysiology of PHT results from one or a combination of three processes: (1) back pressure in the pulmonary vessels, due to mitral valve or left ventricular dysfunction; (2) vasoconstriction from hypoxia and/or hypercarbia; (3) occlusion or obliteration of vessels by emboli, lung parenchymal destruction, or by arterial lesions, as in primary pulmonary hypertension. The following is a list of the most common important causes of pulmonary hypertension.

Acute Pulmonary Hypertension

- Acute pulmonary thromboembolism
- Acute left ventricular (LV) failure
- Acute rupture of the ventricular septum
- Mitral valve/papillary muscle rupture
- Adult respiratory distress syndrome
- High-altitude pulmonary edema
- Fat embolism; amniotic fluid embolism; tumor embolism

Note: Tension pneumothorax and cardiac tamponade can simulate acute cor pulmonale clinically.

Chronic Pulmonary Hypertension

- Chronic left heart failure
- Mitral stenosis; septal defects—atrial or ventricular
- Chronic thrombotic PHT
- Chronic bronchitis and emphysema
- Primary PHT
- Collagen vascular diseases and pulmonary angiitis
- Hypoventilation syndromes: sleep apnea, obesity, hypothyroidism, neuromuscular disease, kyphoscoliosis
- Diffuse interstitial lung diseases (rare)
- High-altitude dwelling ($>10,000$ ft)

Note: The clinically subtle diseases in this group are thrombotic and primary PHT and the hypoventilation syndromes.

CLINICAL DETECTION

Pulmonary hypertension is usually detected because a physician has thought of it. Most of the symptoms attributable to PHT are nonspecific. The following are the most common symptoms in primary PHT: (1) dyspnea on mild exertion; (2) loss of energy; (3) syncope; (4) chest pain (perhaps RV angina); (5) cough; (6) palpitations; (7) cyanosis and hemoptysis. Other symptoms are those that relate to underlying disease, such as cough and sputum production in chronic obstructive pulmonary disease (COPD), or daytime somnolence and obstructive snoring in sleep apnea. Often, PHT is detected by identifying suggestive signs on a chest radiograph or electrocardiogram (ECG). Occasionally a good observer will deduce the presence of PHT by such clues as neck vein distension or a fixed split S_2 with a loud pulmonic component in a patient with few signs but troublesome symptoms.

Physical Signs

Physical signs are highly variable. This is a list of common findings in PHT. No patient has all of them. Some patients have none of them.

- Blood pressure: Normal, or mild hypotension when cardiac output is reduced. Narrow pulse pressure. Severe systemic hypotension when cardiac output (CO) is grossly impaired. Sleep apnea may present with systemic hypertension.
- Pulse: Normal, or tachycardia if CO is reduced. Pulse >100 beats per minute indicates severe disease.
- Cardiac: RV heave; RV S_3; S_2 loud and widely split; $P_2 > A_2$. Occasionally tricuspid regurgitation (TR), murmur, rarely a pulmonary regurgitation (PR) murmur. The most common finding is a loud, widely split, or fixed split S_2.
- Veins: Jugular vein distension only when there is overt right ventricular (RV) failure or severe TR, indicating elevated right atrial (RA) pressure.
- Others: Liver congestion, peripheral edema, acral cyanosis, tachypnea, abdominal tenderness due to congestion, ascites, hepatojugular reflux.

Diagnostic Tests

CHEST RADIOGRAPH

Figure 24-1 depicts the classic features found in severe PHT, including Eisenmenger's syndrome due to atrial septal defect (ASD), some cases of thrombotic PHT, and PPH. The enlarged pulmonary arteries may be mistaken for hilar adenopathy. The large main pulmonary artery (PA) can be mistaken for the aortic knob, which is smaller. The peripheral lung markings are normal or decreased. Right descending pulmonary artery diameter greater than 16 to 18 mm on a standard roentgenogram suggests severe PHT. On the lateral chest radiograph, the RV fills the space between the sternum and anterior heart border, and the left descending PA is enlarged. Other causes of PHT may have their own characteristic radiographic features—e.g., interstitial edema and Kerley B lines in pulmonary venoocclusive disease, a double density effect created by a large left atrium behind the cardiac silhouette in mitral stenosis, or hyperinflation and bullae in emphysema.

ELECTROCARDIOGRAM

Figure 24-2 depicts the classic changes of severe PHT. The mean QRS axis is rightward at 120 degrees. The tall, peaked P

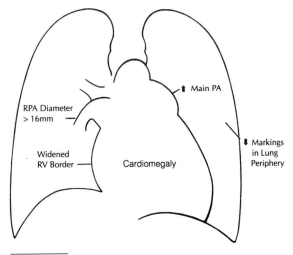

FIGURE 24-1
Radiographic features of pulmonary hypertension.

waves indicate right atrial enlargement. The R>S in V_1–V_3, deep S wave in V_6, and associated ST–T changes indicate RV hypertrophy. Other findings occasionally seen in PHT are S_1, Q_3, T_3, pattern or right bundle branch block (RBBB). Frequently, a significant PHT may exist with few or no changes found on the ECG. Thus, the presence of changes is useful information, but the absence of changes does not exclude disease.

ECHOCARDIOGRAM

Two-dimensional (2D) echocardiography with Doppler estimate of pulmonary artery pressure is the most sensitive noninvasive test to detect significant PHT. Estimates of PA pressure are reasonably accurate when mean PA pressure exceeds 30 mmHg. The technique depends on the presence of a regurgitant tricuspid jet for accuracy. When PHT is suspected, a saline "bubble study" should be added to the request for echocardiography because occasionally it will reveal right-to-left shunt of an atrial septal defect or patent foramen ovale. The echocardiogram also can give information about RV

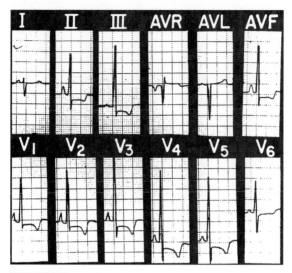

FIGURE 24-2
Classic ECG of pulmonary hypertension.

dimensions and contractility and ventricular septal size and motion.

OTHER TESTS
Other tests depend on the differential diagnosis. Pulmonary embolism should be suspected in all patients with PHT and is best screened for by a ventilation-perfusion lung scan. Doppler leg venograms and pulmonary arteriograms are also helpful in many patients. Hypoventilation syndromes require polysomnography. Right heart and often left heart catheterization is necessary to determine the degree of hypertension and whether or not left ventricular or valvular dysfunction is the cause of PHT. LV diastolic dysfunction may cause significant PHT, even with a normal ejection fraction. Blood oxygen saturation studies should be done to seek intracardiac shunts. Arterial blood gases (ABGs), pulmonary function tests (PFTs) with DLCO and serologic tests are frequently needed to establish the cause of disease. In selected cases, computed tomographic

(CT) scanning may be useful to detect abnormalities such as fibrosing mediastinitis or obstruction of a main pulmonary artery by tumor.

Diagnostic Approach and Management (Selected Diseases)

PRIMARY PULMONARY HYPERTENSION (PPH)

The diagnosis is made by excluding other causes of PHT and requires a normal or low-probability lung scan to rule out thromboembolism, formal right heart catheterization, and serial O_2 saturation measured in RA, RV, and PA with wedge pressure and cardiac output to rule out shunt, valvular lesions, and LV dysfunction. ABGs and PFTs are routinely indicated. Serologic tests, including hepatitis B profile and HIV antibodies, are needed because PPH occurs as a complication of liver cirrhosis or AIDS. Serum antinuclear antibody (ANA) and antiphospholipid antibody should be measured. Patients should receive a vasodilator trial (i.e., prostacyclin, prostaglandin E_1, adenosine) at cardiac catheterization unless pulmonary venoocclusive disease is suspected (in which case catastrophic pulmonary edema may occur). Vasodilation with calcium-channel blockers is highly effective in about 30 percent of cases. A response is usually defined as a 30 percent reduction in pulmonary vascular resistance (PVR) or a 20 percent reduction in PA pressure. Prostacyclin (epoprostenol) given intravenously through a central catheter is now FDA-approved and is the most efficacious therapy for NYH class III/IV patients refractory to other vasodilators. Unless contraindicated, warfarin should be considered for prophylactic anticoagulation. Single-lung transplantation is an option in end-stage disease, with 70 percent 1-year and 60 percent 3-year survival reported.

CHRONIC THROMBOTIC PULMONARY HYPERTENSION

This diagnosis is suggested by a lung scan positive for multiple segmental or lobar defects. Leg vein Doppler studies should be done. Carefully performed selective pulmonary arteriography is both safe and essential to confirm the diagnosis. A vena cava filter is used in most patients proved to have this disease. Patients should receive heparin followed by warfarin (INR 2 to 3) (see Chap. 46). When this diagnosis is considered, serum protein S, C, antithrombin III, and antiphospholipid antibodies should be measured before institution of anticoagulant therapy. The roles of factor V Leiden abnormality and hyperhomocysteinemia are not established in this illness. Many patients with chronic thrombotic PHT develop New York Heart Association functional class IV heart failure. Open-heart surgery with pulmonary thromboendarterectomy is successful in restoring pulmonary perfusion and alleviating symptoms in most patients, and patients should be given anticoagulation drugs for life. Diuretics, digoxin, oxygen, and other supportive therapies are given as needed.

ACUTE PULMONARY THROMBOEMBOLISM

The diagnosis almost always requires a ventilation-perfusion lung scan. A high clinical suspicion for pulmonary embolism coupled with a high-probability scan is > 90 percent accurate for pulmonary embolism. Unfortunately, the majority of patients with acute pulmonary embolism have indeterminate scans, so that pulmonary arteriography is frequently necessary to establish or exclude the diagnosis. Sixteen percent of patients with low-probability lung scans have pulmonary emboli. A positive leg vein study assists in raising the diagnostic certainty. Therapy consists of heparin by bolus plus continuous infusion to raise the PTT > 1.5 times control, followed by warfarin. Recently, low molecular weight heparin has been shown to be efficacious in the therapy of thromboembolism. The role of lytic therapy is currently unclear. Some clinicians use lytic therapy when hemodynamic instability (tachycardia, hypotension) is present; others insert a vena cava filter to impede further emboli and then use heparin. The issue is unsettled. Warfarin should be started the second day of heparinization in a stable daily dose; most clinicians give anticoagulants for only 3 to 6 months (INR 2 to 3) unless there is a persistent condition predisposing to venous stasis and repeat embolism (see also Chaps. 23 and 46).

SUGGESTED READING

Asherson RA, Khamashta MA, Ordi-Ros J, et al: The "primary" antiphospholipid syndrome: major clinical and serological features. *Medicine (Baltimore)* 1989; 68: 366–371.

Becker DM, Philbrick JT, Selby JB: Inferior vena cava filters: indications, safety, effectiveness. *Arch Intern Med* 1992; 152:1985–1994.

Chang CH: The normal roentgenographic measurement of the right descending pulmonary artery in 1,085 cases. *Am J Roentgenol* 1962; 87:929–935.

Coplan N, Shinony R, Ioachim H: Primary pulmonary hypertension associated with human immunodeficiency viral infection. *Am J Med* 1990; 89:96–99.

D'Alonzo GE, Barst RJ, Ayres SM, et al: Survival in patients with primary pulmonary hypertension. *Ann Intern Med* 1991; 115:343–349.

DeCousus H, Leizorovic A, Parent F, et al: A clinical trial of vena cava filters in the prevention of pulmonary embolism in patients with proximal deep-vein thrombosis. *N Engl J Med* 1998; 338:409–415.

Fishman AP: Pulmonary hypertension. In: Alexander RW, Schlant RC, Fuster V, O'Rourke RA, Roberts R, Sonnenblick EH (eds): *Hurst's The Heart*, 9th ed. New York, McGraw-Hill, 1998:1699–1717.

Goldhaber SZ, Meyerovitz MF, Markis JE, et al: Thrombolytic therapy of acute pulmonary embolism: current status and future potential. *J Am Coll Cardiol* 1987; 10:96B–104B.

Grover RF: Chronic hypoxic pulmonary hypertension. In: Fishman AP (ed): *The Pulmonary Circulation: Normal and Abnormal*. Philadelphia, University of Pennsylvania Press, 1990:283–301.

McGinn S, White PD: Acute cor pulmonale resulting from pulmonary embolism, its clinical recognition. *JAMA* 1935; 104:1473–1480.

Moser KM: Venous thromboembolism. *Am Rev Respir Dis* 1990; 141:235–249.

Moser KM, Auger WR, Fedulo PF: Chronic major-vessel thromboembolic pulmonary hypertension. *Circulation* 1990; 81:1735.

Nachman RJ, Silverstein R: Hypercoagulable states. *Ann Intern Med* 1993; 119:819–827.

Newman JH, Ross JC: Chronic cor pulmonale. In: Alexander RW, Schlant RC, Fuster V, O'Rourke RA, Roberts R, Sonnenblick EH (eds): *Hurst's The Heart*, 9th ed. New York, McGraw-Hill, 1998:1739–1749.

Nocturnal Oxygen Therapy Trial Group: Continuous or nocturnal oxygen therapy in hypoxemic chronic obstructive lung disease: a clinical trial. *Ann Intern Med* 1980; 93:391–398.

Palevsky HI, Fishman AP: The management of primary pulmonary hypertension. *JAMA* 1991; 265:1014–1020.

Palevsky HI, Schloo BL, Pietra GG, et al: Primary pulmonary hypertension: vascular structure, morphometry, and responsiveness to vasodilator agents. *Circulation* 1989; 80: 1207–1221.

Perez D, Kramer N: Pulmonary hypertension in systemic lupus erythematosus: report of four cases and review of the literature. *Semin Arthritis Rheum* 1981; 11:177–181.

Pietra GG, Edwards WD, Kay JM, et al: Histopathology of primary pulmonary hypertension: a qualitative and quantitative study of pulmonary blood vessels from 58 persons in The National Heart, Lung and Blood Institute, Primary Pulmonary Hypertension Registry. *Circulation* 1989; 1198–1206.

PIOPED Investigators: Value of the ventilation/perfusion scan in acute pulmonary embolism: results of the prospective investigation of pulmonary embolism diagnosis (PIOPED). *JAMA* 1990; 263:2753–2796.

Rich S, Kaufmann E, Levy PS: This effect of high doses of calcium-channel blockers on survival in primary pulmonary hypertension. *N Engl J Med* 1992; 327:76–81.

Richards DW: The right heart and the lung with some observations on teleology: The J. Burns Amberson Lecture. *Am Rev Respir Dis* 1966; 94:691–702.

Schiller N: Pulmonary artery pressure estimation by Doppler and two-dimensional echocardiography. *Cardiol Clin* 1990; 8:277–287.

Weitzenblum E, Apprill M, Krieger J, et al: Sleep disordered breathing and pulmonary hypertension. *Eur Respir J* 1990; 11(suppl):523S–526S.

Wiedemann H, Matthay R: Cor pulmonale in chronic obstructive pulmonary disease circulatory pathophysiology and management. *Clin Chest Med* 1990; 11:523–545.

THE HEART AND ENDOCRINE DISEASES

Joel Zonszein
Edmund H. Sonnenblick

DIABETES MELLITUS

With both type 1 diabetes (previously known as insulin-dependent diabetes), and type 2 diabetes (previously known as non-insulin-dependent diabetes), the incidence and severity of coronary artery disease (CAD) are increased. Following acute myocardial infarction (MI), short- and long-term mortality is greater, along with an increased development of congestive heart failure (CHF). Hyperglycemia is responsible for the development and progression of microvasculopathy and neuropathy and is a risk factor for coronary events and mortality. In addition to hyperglycemia, the etiology of macrovascular disease is related to several other risk factors (Table 25-1). The high prevalence of CAD found in the "metabolic syndrome X" (abdominal obesity, hypertension, dyslipidemia, glucose intolerance, and insulin resistance) suggests that hyperinsulinemia combined with other risk factors, rather than the severity or duration of hyperglycemia, is responsible for the accelerated rate of arteriosclerosis.

Diabetic Cardiomyopathy

Diabetes mellitus is associated with accelerated obstructive large vessel disease, leading to MI as well as a more diffuse myocyte loss and resulting in dilated cardiomyopathy. With concomitant hypertension, the latter process is more pronounced. Thus, diabetic cardiomyopathy may reflect the end result of large-vessel obstructive atherosclerosis as well as small-vessel vasospastic disease leading to focal areas of tissue loss. Early in this process, diastolic dysfunction may be promi-

TABLE 25-1

CARDIOVASCULAR RISK FACTORS AND MARKERS

Genetic
Adverse environment factors
⇑ Age
Metabolic syndrome "X"
   Hyperinsulinemia
   Impaired glucose tolerance (IGT), resistance to
      insulin-stimulated glucose uptake
   Dyslipidemia
      ⇑ Triglycerides
      ⇓ HDL cholesterol
      ⇑ LDL cholesterol (dense subparticles)
      ⇑ Lipoprotein "little a" [Lp(a)]
      ⇑ Apolipoprotein B
   Central obesity
   Hypertension
Severity of hyperglycemia
Hypercoagulability state
   ⇑ Platelet aggregability
   ⇑ von Willebrand factor procoagulant activity
   ⇑ Fibrinogen
   ⇑ Increased plasminogen activator inhibitor type 1
      (PAI-1) activity
⇑ Leptin
⇑ Tumor necrosis factor (TNF)
Endothelial dysfunction
Hyperhomocysteinemia
Sex steroids (⇑ androgens, ⇓ estrogens,
   ⇓ dehydroepiandosterone (DHEA)
Others
   ⇓ Sex hormone binding protein (SHBP
   ⇓ Insulin-like growth factor binding protein 1
      (IGFP-1)
   ⇑ Uric acid
   Albuminuria
   Cardiac syndrome "X"
   Medications

nent, with elevated ventricular pressures and a normal ejection fraction. Later, systolic dysfunction occurs, with progressive ventricular dilatation and reduced ejection fraction. Diabetic cardiomyopathy is often accompanied by autonomic neuropathy, which contributes to the high morbidity and mortality. Painless myocardial damage in diabetes is also common, due to decreased sensitivity to ischemic pain.

Evolving autonomic neuropathy alters the normal phasic changes of heart rate, termed *heart rate variability* (HRV), characterized by a decrease in the high-frequency rate and an increase in the very low frequency rate. Practical bedside tests for the diagnosis of autonomic cardioneuropathy include decreased beat-to-beat heart rate variability determined during slow respiratory rates; abnormal heart rate response to standing (with an R-R interval ratio equal to or less than 1.00); an abnormal heart rate response (R-R interval ratio equal to or less than 1.10) to a Valsalva maneuver; orthostatic hypotension after standing (with systolic blood pressure falling ≥30 mmHg after 1 min of standing); and an increased resting heart rate (over 100 beats per minute), determined after the patient is resting in a supine position for at least 15 min.

Diet and exercise, while important, fail to normalize hyperglycemia in the majority of patients with type 2 diabetes; medications are thus needed. Treatment of type 2 diabetes has shifted from just treating hyperglycemia to improving insulin resistance, along with early aggressive treatment of other factors (i.e., hypertension, dyslipidemia, smoking cessation, etc.). Even mild elevations of blood sugar should not be tolerated (Table 25-2), since hyperglycemia begets hyperglycemia (known as "glucose toxicity") and aggravates chronic diabetic complications. Newer antidiabetic agents permit more aggressive glycemic control. While chronic administration of high insulin doses in the obese patient with poor metabolic control is not recommended, insulin therapy is the best modality during acute metabolic decompensation and in hospitalized patients.

Management of Myopcardial Infarction

During acute MI, oral antidiabetic agents are often ineffective and insulin is required. While higher insulin doses are needed, insulin is often withheld or given inappropriately at lower doses as "sliding-scale coverage." Tight metabolic control using intensive insulin therapy for short-term therapy has been

TABLE 25-2

INDICES OF GLYCEMIC CONTROL

Biochemical Index	Nondiabetic	Goal	Action Suggested
Preprandial glucose (mg/dL)	<115	80–120	<80 or >140
Bedtime glucose (mg/dL)	<120	100–140	<100 or >160

Note: These values are for nonpregnant individuals. Action suggested depends on the individual patient. Hemoglobin A_{1C} (HbA_{1C}) is referenced to a nondiabetic range of 4 to 6 percent.

SOURCE: From the pharmacological treatment of hyperglycemia in NIDDM. *Diabetes Care* 1995; 18:1510–1518, with permission.

shown to improve long-term survival, particularly in patients with mild disease.

Acutely, timely introduction of thrombolytic therapy to restore flow, along with beta blockers to reduce heart rate and thus limit oxygen needs, may limit the extent of ischemia; such therapy is particularly beneficial in the diabetic population. Angiotensin-converting enzyme (ACE) inhibitors, or angiotensin II receptor blockers, which are beneficial in protecting the diabetic kidney, may improve left ventricular function and reduce the rate of reinfraction. Aspirin is recommended for primary and secondary prevention; however, the correct dose needs to be determined, since the diabetic population has a state of hypercoagulability. Magnesium deficiency is common in diabetics, particularly in those with poor metabolic control and/or on diuretic therapy, and magnesium replacement may be beneficial (see "Magnesium Deficiency," below, for therapeutic recommendations).

Postinfarction, beta-adrenergic blockers, ACE inhibitors, and aspirin have been shown to decrease mortality, particularly in the diabetic population. Vigorous treatment of hypertension and dyslipidemia also reduces recurrent morbid events. Coronary artery bypass graft (CABG) surgery is associated with a high morbidity rate (infections, renal failure, CHF, cardiogenic shock, arrhythmias, and myocardial rupture) and greater mortality in patients with diabetes. On the other hand, after percutaneous transluminal coronary angioplasty (PTCA), restenosis is also increased, and PTCA, as compared with CABG, is associated with an even higher morbidity and mortality rate (BARI study).

HYPERTHYROIDISM

Cardiovascular Manifestations

Hyperthyroidism produces decreased systemic vascular resistance and increased stroke volume, characterized by sinus tachycardia, systolic hypertension, and low diastolic blood pressure. Its most common cardiovascular complications are tachyarrhythmias, thromboembolic accidents, and heart failure. Atrial fibrillation (AF) occurs in 15 percent of thyrotoxic patients. Of those with newly diagnosed AF, 15 percent are hyperthyroid, and about 15 percent of these develop thromboembolism. Cardiovascular abnormalities are often the only manifestations in the elderly ("apathetic hyperthyroidism"). Individuals over the age of 60 with subclinical hyperthyroidism [low serum thyroid-stimulating hormone (TSH) concentration] have a threefold higher risk of developing AF. Congestive heart failure is also more common in the elderly and in those with concomitant intrinsic heart disease.

Diagnosis

The diagnosis is suspected clinically and confirmed by elevated serum levels of T_4 and T_3 and suppressed TSH. A single TSH determination is not reliable in patients who are acutely ill, who are hospitalized, or who have hypothalamic-pituitary disorders.

Treatment

Beta-adrenergic blockers ar very useful in reducing cardiovascular manifestations, while other modalities are used to render the patient euthyroid. Anticoagulation is recommended for AF, especially if it lasts more than 3 days, in the elderly, and in those with CHF, a dilated left atrium, and/or valvulopathy. Anticoagulation is also necessary in patients with spontaneous or induced cardioversion and should be continued for at least 4 weeks after conversion to sinus rhythm and reestablishment of the euthyroid state. Higher-than-normal doses of digitalis and/or beta blockers may be needed in hyperthyroidism due to the increased rate of disposal.

HYPOTHYROIDISM

Cardiovascular Manifestations

Hypothyroidism is characterized by decreased total body oxygen consumption with sinus bradycardia, reduced cardiac output, and elevated total peripheral vascular resistance as well as increased arterial pressure. Uncomplicated hypothyroidism rarely causes overt CHF. A low-voltage electrocardiogram (ECG) and/or enlarged cardiac silhouette are common. Echocardiography is valuable in detecting pericardial effusion, which is common and sometimes massive in hypothyroid states but rarely causes tamponade. The pericardial effusion resolves slowly with thyroid replacement.

Diagnosis

Typical clinical manifestations suggest the diagnosis, which may be overlooked due to their insidious presentation. The diagnosis is confirmed by a low T_4 coupled with an elevated plasma TSH. Routine testing for hypothyroidism has a low yield but is recommended in the elderly. Seriously ill patients may have low total T_4 levels but normal free T_4 and minimally abnormal TSH; hence they are euthyroid ("euthyroid sick syndrome"), and thyroid hormone replacement is not necessary. With recovery from critical illness, T_4 and TSH levels normalize; therefore repeat hormone determinations are recommended. The finding of a normal TSH level is often sufficient to exclude the diagnosis of hypothyroidism.

Treatment

Since T_3 is primarily derived from the extrathyroidal production of T_4, L-thyroxine is the therapeutic preparation of choice. The full daily replacement dose of L-thyroxine is 1.5 to 2.0 μg/kg/day, with elderly patients needing less. During treatment, TSH determinations alone provide a satisfactory and cost-effective assessment of dose adequacy. Thyroid replacement in the elderly and in those with underlying ischemic cardiac disease requires considerable prudence. Vigorous replacement of thyroid hormone may precipitate CHF with underlying valvular disease or cardiomyopathy and may cause MI in patients with CAD. Therefore, low doses of replacement thyroid hormone are recommended (12.5 to 25 μg daily), with very gradual increments. The presence of CAD can be masked by hypothyroidism. Elevated levels of creatine kinase (CK) in serum are commonly from skeletal muscle; this should be differentiated from CK of cardiac origin.

When hypothyroidism is very severe (myxedema coma), hyponatremia, hypoglycemia, cardiorespiratory compromise, and mental changes may be present. While the choice of thyroid hormone, dosage, and route of administration are controversial, an immediate intravenous bolus dose of levothyroxine designed to raise the serum T_4 level to 77 to 90 nmol/L (6 to 7 μg/dL) is recommended. Usually about 500 μg is needed, followed by daily administration of 75 to 100 μg until the patient's vital signs become stable and gastrointestinal function returns to normal, permitting oral administration. Glucocorticoids should be administered until coexistent adrenal insufficiency can be ruled out. Concomitant nitrates, calcium channel blockers, beta-adrenergic blockers, and digitalis may be helpful and should be used with prudence. In patients with unstable coronary disease, bypass surgery or angioplasty is preferable before thyroid replacement therapy is pursued.

AMIODARONE-INDUCED THYROID DYSFUNCTION

Amiodarone can induce hypothyroidism in 20 percent and hyperthyroidism in approximately 10 percent of patients. Amiodarone has a long half-life (25 to 100 days) and is an important source of exogenous iodine (200 mg per tablet). Amiodarone-induced hypothyroidism occurs more often in patients with underlying thyroid disease, many with positive antithyroid antibodies. TSH measurements may be misleading, and identification of high serum levels of reverse T_3 concentration may be helpful with normal or low serum levels, indicating hypothyroidism. Amiodarone withdrawal results in spontaneous remission of thyroid dysfunction in approximately half of these patients.

Hyperthyroidism may be related to iodine overload, particularly in those with preexisting thyroid abnormalities. Amiodarone and desethylamiodarone, its major metabolite, can also exert direct cytotoxic effects, causing a thyroiditis-like syndrome with follicular injury resulting in increased serum interleukin 6 (IL-6) levels. Treatment with thionamides (propylthiouracil and methimazole) can lower thyroid hormone concentrations; in some, the addition of potassium perchlorate (a drug that competitively inhibits iodine uptake) and/or glucocorticoids may also be necessary. When all medical trails fail and amiodorone treatment needs to be continued, a near total thyroidectomy may be necessary.

CALCIUM AND MAGNESIUM IN CARDIAC DISORDERS

Calcium

Hypercalcemia resulting from hyperparathyroidism produces a shortened QT interval as well as increased sensitivity to digitalis and its toxicity. In hypocalcemia, a prolonged QT interval is the typical electrocardiographic finding.

Magnesium Deficiency

Magnesium (Mg) deficiency is commonly unrecognized, undiagnosed, and undertreated. The myocardial irritability related to Mg losses can be reversed by Mg replacement. Hypomagnesemia may induce ventricular arrhythmias, particualrly in the digitalized patient, because of a synergistic adverse effect on the Na^+-K^+-ATPase pump. Mg deficiency is particularly common in hospitalized patients, particularly in those who were poorly nourished or receiving medications such as diuretics and aminoglycoside antibiotics, which may cause further Mg losses.

The diagnosis of Mg deficiency is difficult due to the minimal and nonspecific clinical manifestations and lack of reliable laboratory tests. Tetany is rare, but less specific signs such as tremor, muscle twitching, bizarre movments, focal seizures or generalized convulsions, delirium, or coma may occur. Mg deficiency should be suspected when other electrolyte abnormalities coexist and when electrocardiographic changes such as prolongation of the Q-T and P-R intervals, widening of the QRS complex, ST-segment depression, and low T waves are found. The serum Mg concentration is unreliable, as it does not reflect intracellular Mg. A more effective clinical probe is the "magnesium loading test," which is both therapeutic and diagnostic. It consists of parenteral administration of $MgSO_4$ and the assessment of urinary Mg retention. Individuals with normal Mg balance should eliminate at least 75 percent. Parenteral administration is favored, particularly in medical emergencies. The recommended dose is 2 g of $MgSO_4$ (16.3 meq) given intravenously over 30 min, followed by a constant Mg infusion rate of 1 meq/kg/day for the first 2 to 3 days and 0.5 meq/kg/day for 2 to 3 more days. This approach is recommended where there is a high index of suspicion, particularly in patients with ischemic heart disease or arrhythmias. Mg administration is contraindicated in anuric individuals and in those with significant renal impairment.

ACROMEGALY

Advanced acromegaly, characterized by cosmetic disfigurement and incapacitating arthropathy, is accompanied by a shortened life expectancy resulting from respiratory, cardiovascular, and malignant complications. Hypertension is three times more common, and the extent of cardiomegaly is disproportional to the degree of hypertension. Left ventricular hypertrophy is often associated with CHF. The clinical manifestations and findings of elevated plasma growth hormone (hGH) levels with failure to suppress after a glucose load is diagnostic. Alternatively, high plasma levels of insulin-like growth factor I (IGF-I) are also diagnostic.

Acromegalic tumors are generally removed surgically, but the cure rate remains low. When a surgical approach is not possible or unsuccessful, radiation therapy and/or medical treatment with somatostatin analogs (octreotide acetate) is necessary, since persistent elevation of hGH levels is associated with an increased mortality.

ADRENAL INSUFFICIENCY

Primary destructive adrenal disease (Addison's disease) manifests itself by decreased production of both aldosterone and cortisol. Secondary adrenal disease (pituitary disease) is often manifest as panhypopituitarism. Addison's disease is characterized by small cardiac size, low cardiac output, and hypotension. Since there is deficiency of both cortisol and aldosterone, hyperkalemia and metabolic acidosis may also be present. The most common form of isolated adrenal insufficiency is aldosterone deficiency (hyporeninemic hypoaldosteronism, or type IV renal tubular acidosis). Vascular collapse can take place during stress or acute illnesses. A rapid screening test with synthetic adrenocorticotropic hormone (ACTH) stimulation is valuable, and the diagnosis can be established by determination of cortisol, ACTH, renin, and aldosterone levels.

Following diagnostic tests, the urgent administration of glucocorticoids is crucial and helpful diagnostically, as reflected by a rapid recovery. Glucocorticoid replacement potentiates the vasoconstrictor response to catecholamines and results in a rapid reversal of hypotension.

CUSHING'S SYNDROME

Cushing's syndrome results from elevated plasma levels of cortisol and is most commonly caused by therapy with glucocorticoids for nonendocrine disorders. The most frequent cause of endogenous hypercortisolism is Cushing's disease, caused by a pituitary tumor and bilateral diffused adrenal hyperplasia. Clinical features include central obesity, hypertension, proximal muscle weakness, atrophy of the skin, wide striae, easy bruising, and osteoporosis. Hypokalemic alkalosis favors ectopic ACTH-induced hypercortisolism, which is associated with impaired glucose tolerance or diabetes, hyperinsulinemia, hypertension, obesity and dyslipidemia. Thus, there is a greater risk of accelerated atherosclerosis, which when untreated, can be associated with ischemia-related cardiovascular events with a high mortality rate.

When Cushing's syndrome is suspected, an overnight dexamethasone suppression screening test can be obtained by giving 1 mg of dexamethasone orally at midnight; normal individuals will have a plasma cortisol levels of less than 5 μg/dL the following morning. In those failing to suppress, a more extensive evaluation is needed to confirm the diagnosis and determine the cause. This is done by dynamic studies of the hypothalamic-pituitary-adrenal axis. Localization of the source of hypercortisolism is established by contrast imaging studies. The source of hypercortisolism needs to be identified in order to address its treatment (by surgery, radiation, medical, therapy, or a combination). The major cardiac manifestations reflect systemic arterial hypertension and hypokalemia and need to be treated accordingly. CHF and cerebrovascular accidents are common causes of death in untreated patients.

SEX STEROIDS AND THE HEART

Sex steroids affect the vascular system, primarily through changes in lipid metabolism, but since estrogen receptors are present in the vasculature, a direct effect is likely. Estrogen decreases low-density lipoprotein (LDL) cholesterol and raises high-density lipoprotein (HDL) cholesterol. In contrast, exogenous androgens, especially those that cannot be metabolized to estrogen (i.e., nonaromatizable androgens), cause unfavorable lipid effects.

Androgens

The indication for androgen therapy is hypogonadism in men. Androgen abuse is common among athletes and body builders, since supraphysiologic doses of testosterone combined with exercise training cause a favorable anabolic effect with increased fat-free mass and muscle size. Similar anabolic effects are found when androgen is given to aging men, where a physiologic decline in testosterone occurs. The extended use of androgens in the elderly as well as in other non-androgen-deficient populations is not advisable due to the potential side effects.

Oral Contraceptives

Complications associated with the use of oral contraceptives—such as venous thromboembolism, myocardial infarction, ischemic and hemorrhagic stroke, as well as worsening of hypertension and diabetes mellitus—have been significantly lessened by decreasing the dosage of estrogen. Most preparations now contain 35 μg of estrogen or less. The risk of adverse cardiovascular events in premenopausal women results not from atherogenesis but from thrombogenesis caused by dyslipidemia, changes in procoagulants, and platelet aggregation. The use of oral contraceptives in smokers is associated with an increased risk of MI, especially when combined with other risk factors such as hypertension, diabetes mellitus, and hyperlipidemia. Since smokers under the age of 35 who use oral contraceptives have morbidity and mortality rates similar to those of women who do not use contraceptives, they may continue to use oral contraceptives only if they stop smoking. Those between the age of 35 and 45 should use oral contraceptives only if they do not smoke and do not have other underlying risk factors for vascular disease.

Estrogen Replacement Therapy

The benefits of estrogen in postmenopausal women include relief of menopausal symptoms, reduction of bone loss, and a 50 percent reduction in the risk of heart disease. Mortality can be reduced by 20 percent by lowering the rate of MI among elderly women with estrogen replacement. Progestins were thought to have unfavorable cardiovascular effects; however, medroxyprogesterone acetate, the most commonly used progestin in the United States, has no adverse effects; when used in combination with estrogen, it has been found to be associated with an even better profile than estrogen alone. Women and their physicians must balance the benefit of hormone replacement therapy against the possible side effects, including a somewhat higher incidence of breast and uterine carcinoma. While the combination of estrogen plus progestin can decrease endometrial carcinoma, it does not affect the risk of breast cancer. While we await further information from prospective clinical trials, such as the Women's Health Initiative Study of the National Institutes of Health, it is recommended that all postmenopausal women be considered candidates for hormone replacement therapy and be educated about its risk and benefits, following the guidelines given by the American College of Physicians.

SUGGESTED READING

Golzari H, Cebul RD, Bahler RC: Atrial fibrillation: restoration and maintenance of sinus rhythm and indications for anticoagulation therapy. *Ann Intern Med* 1996; 125: 311–323.

Harris MI, Cowie CC, Reiber G, et al: *Diabetes in America*, 2d ed. NIH Publication No. 95-1468. Washington DC; U.S. Government Printing Office, 1995.

Hulley S, Grady D, Bush T, et al: Randomized trial of estrogen plus progestin for secondary prevention of coronary heart disease in postmenopausal women. *JAMA* 1998; 280: 605–613.

ISIS-4: a randomized factorial trial assessing early oral captopril, oral mononitrate, and intravenous magnesium sulphate in 58,050 patients with suspected acute myocardial infarction. ISIS-4 (Fourth International Study of Infarct Survival) Collaborative Group. *Lancet* 1995; 345: 669–685.

Jordan RM: Myxedema coma. *Med Clin North Am* 1995; 79: 185–194.

Ladenson PW: Recognition and management of cardiovascular disease related to thyroid dysfunction. *Am J Med* 1990; 88:638–641.

McLean RM: Magnesium and its therapeutic uses: a review. *Am J Med* 1994; 96:63–76.

Melmed S, Ho K, Klibanski A, et al: Clinical review 75: recent advances in pathogenesis, diagnosis, and management of acromegaly. *J Clin Endocrinol Metab* 1995; 80:3395–3402.

Reaven GM: Syndrome X. *Clin Diabetes* 1994; 12:32–36.

Sawin CT, Geller A, Wolf PA, et al: Low serum thyrotropin concentrations as a risk factor for atrial fibrillation in older persons. *N Engl J Med* 1994; 331:1249–1252.

The Bypass Angioplasty Revascularization Investigation (BARI) Investigators: Comparison of coronary bypass surgery with angioplasty in patients with multivessel disease. *N Engl J Med* 1996; 335:217–225.

The Writing Group for the PEPI trial: Effects of estrogen or
    estrogen/progestin regimens on heart disease risk factors
    in postmenopausal women: The Postmenopausal Estro-
    gen/Progestin Interventions (PEPI) Trial. *JAMA* 1995;
    273:199–208.
World Health Organization Collaborative Study of Cardio-
    vascular Disease and Steroid Hormone Contraception:
    Venous thromboembolic disease and combined oral con-
    traceptives: results of international multicentre case-
    control study. *Lancet* 1995; 246:157–182.
Zonszein J, Sonnenblick EH: The heart and endocrine disease.
    In: Alexander RW, Schlant RC, Fuster V, O'Rourke RA,
    Roberts R, Sonnenblick EH (eds): *Hurst's The Heart,* 9th
    ed. New York, McGraw-Hill, 1998:2117–2142.

CHAPTER

26

THE CONNECTIVE TISSUE DISEASES

Robert C. Schlant
William C. Roberts

The connective tissue, or collagen vascular, diseases are systemic disorders of unknown origin characterized by inflammatory lesions of many organs, including the joints, muscles, blood vessels, pleurae, and pericardium. The prognosis is often related to involvement of the kidney, brain, or heart.

The connective tissue diseases can be divided into heritable disorders and nonheritable disorders. The heritable disorders of connective tissue that are associated with cardiovascular defects include Marfan's syndrome (MS), the Ehlers-Danlos syndrome (EDS), pseudoxanthoma elasticum (PXE), osteogenesis imperfecta (OI), annuloaortic ectasia, and familial aneurysms. Nonheritable disorders of connective tissue that are associated with important cardiovascular involvement include systemic lupus erythematosus (SLE), polyarteritis nodosa (PN), rheumatoid arthritis (RA), ankylosing spondylitis, systemic sclerosis (SS), polymyositis/dermatomyositis, giant-cell arteritis, the Churg-Strauss syndrome, the antiphospholipid syndrome, and perhaps syphilis.

HERITABLE CONNECTIVE TISSUE DISORDERS

Marfan's Syndrome

MS is inherited as an autosomal dominant with high penetrance; however, it occurs without a family history in about 25 to 30 percent of patients. It is associated with a wide variety of mutations in the fibrillin-1 gene on chromosome 15.

Characteristically, the ocular, skeletal, and cardiovascular systems are involved. The four major manifestations of MS are a positive family history, ectopia lentis, aortic root dilatation or dissection, and dural ectasia. Many of the minor manifestations occur frequently in the general population, including mitral valve prolapse; early myopia; scoliosis; joint hypermobility; anterior chest deformity, especially asymmetric pectus excavatum or carinatum; long, thin extremities with arachnodactyly; tall stature with increased lower body height; high, narrowly arched palate; fusiform ascending aortic aneurysm ("annuloaortic ectasia"); aortic dissection; mitral regurgitation; mitral annular calcification; papillary muscle dysfunction; spontaneous pneumothorax; cutaneous striae; and inguinal hernia.

The diagnosis requires at least one major manifestation with involvement of the skeleton and at least two other systems if there is no unequivocally affected first-degree relative. If such a relative is documented, there should be involvement of at least two systems. Primary mitral valve prolapse (Chap. 13) is not a forme fruste of MS, although the two conditions may be part of a phenotypic continuum.

Patients with suspected MS should have a careful history and physical examination, including measurements of height, arm span, and floor-to-pelvis distance. A slit-lamp ophthalmic examination, chest film, electrocardiogram, and transthoracic echocardiogram should be obtained. If suggestive widening of the ascending aorta occurs, repeat transthoracic or transesophageal echocardiography should be performed.

Patients with MS should avoid isometric, abrupt, or strenuous exertion; competitive or contact sports; scuba diving; or risk of trauma. Beta-adrenergic blockade, which may retard the rate of dilatation of the aortic root, should be prescribed for patients with MS. In asymptomatic patients with aortic dilatation, surgical repair is generally recommended when the diameter reaches 55 to 60 mm.

Women with MS who become pregnant should be counseled regarding the approximately 50 percent chance that the fetus may have the condition. Such a woman's risk of sudden death is particularly increased if her aortic root diameter exceeds 40 mm or she has moderate or severe aortic regurgitation. Pregnant women with MS should receive a beta-adrenergic blocker and have serial transthoracic echocardiography every 6 to 10 weeks.

Ehlers-Danlos Syndrome

The EDS is a heterogenous group of disorders of connective tissue that are primarily characterized by skin fragility and easy bruising, skin hyperextensibility, multiple ecchymoses, "cigarette paper" scars, and joint hypermobility. A wide variety of cardiovascular abnormalities may be present, including spontaneous rupture of the aorta or large arteries, coronary or intracranial aneurysms, arteriovenous fistulas, mitral and tricuspid valve prolapse, dilation of the aortic root, aortic regurgitation, renal artery aneurysms, systemic arterial hypertension and myocardial infarction.

Pseudoxanthoma Elasticum

PXE is a rare heritable condition that can be transmitted as either an autosomal recessive or an autosomal dominant trait. There are mineral deposits in the elastic fibers, particularly in the skin, Bruch's membrane behind the retina, and blood vessels. The skin lesions, which are described as resembling a "plucked chicken," consist of yellow macules or papules that produce a rough, cobblestone texture.

The skin lesions are typically most prominent in the neck, axillae, antecubital fossae, popliteal spaces, and groins. In the retinae, there may be mottled *peau d'orange* hyperpigmentation and angioid streaks. Retinal hemorrhage and disk drusen may occur.

Calcific deposits may occur in the media of coronary, cerebral, gastrointestinal, renal, and peripheral arteries. Both angina pectoris and myocardial infarction can result. Calcific deposits may occur in the mural endocardium of the ventricles, atria, and atrioventricular valves. Occasionally, mitral stenosis or mitral valve prolapse can be produced as can restrictive cardiomyopathy.

Osteogenesis Imperfecta

This rare heritable disorder of connective tissue is also known as "brittle bone disease" because of the frequent fractures from mild trauma sustained by patients with OI. It is inherited in an autosomal dominant fashion with variable penetrance, but more than 80 mutations have been identified and there is a wide variation in the clinical severity of OI.

Most manifestations of OI are bony, otologic, ocular, dental, and cutaneous. The bony manifestations include short stature, in utero fractures, severe osteoporosis, and severe bone fragility, with bowing of the long bones and repeated fractures. The otologic and ocular manifestations include hearing loss, blue sclerae, and angioid streaks in the retina. Dentinogenesis imperfecta and easy bruising may occur. The cardiovascular manifestations include aortic root dilatation, aortic dissection, aortic regurgitation and mitral regurgitation.

Annuloaortic Ectasia

This condition, which is a pear-shaped enlargement of the sinus and the proximal tubular portions of the ascending aorta, may occur either as part of MS or as a familial condition with no other stigmata of MS.

Familial Aneurysms

A variety of familial aneurysms of cardiovascular structures have been reported, but it is not documented that these are heritable disorders of connective tissue. The structures involved include the aorta, ventricular septum, carotid arteries and intracranial arteries.

NONHERITABLE CONNECTIVE TISSUE DISORDERS

The cardiac manifestations of the six common nonhereditary connective tissue disorders are shown in Table 26-1.

SYSTEMIC LUPUS ERYTHEMATOSUS

Clinical Manifestations

Systemic lupus erythematosus (SLE) occurs most often in the second and third decades of life and affects women more frequently than men. Frequent early signs include fever, arthralgias, arthritis, skin rash, myalgias, and pleuritis.

SLE can produce a *pancarditis* with involvement of the heart valves, endocardium, myocardium, coronary arteries, and pericardium. Pericardial involvement is most frequent; *pericarditis* can be the first manifestation of SLE. Pericardial involvement is usually manifest by effusion or by pericarditis with fibrofibrinous fluid. Rarely, purulent pericarditis, cardiac tamponade, and pericardial constriction may occur. Uremic pericarditis may occur in patients with severe renal failure.

SLE endocarditis and valvulitis is produced by fibrofibrinous sterile vegetations that may occur on any of the heart valves, especially the mitral and aortic. "Atypical verrucous endocarditis" was first described by Libman and Sacks. The lesions are usually clinically silent, but they can produce mur-

TABLE 26-1

PRIMARY CARDIAC MANIFESTATIONS IN NONHEREDITARY CONNECTIVE TISSUE DISEASES

Disease	Pericardium	Myocardium	Endocardium	Coronary Arteries
SLE	++	+	++	+/-
SS	+	++	0	++
PN	+/-	++	0	++
AS	0	+/-	++	0
RA	++	+	+	0
PM/DM	++	++	+/-	+/-

++ = Major site of involvement; + = may be involved; +/- = rarely involved; 0 = not involved.
SLE = systemic lupus erythematosus; SS = systemic sclerosis; PN = polyarteritis nodosa; RA = rheumatoid arthritis; AS = ankylosing spondylitis; PM/DM = polymyositis/dermatomyositis.

SOURCE: Adapted from Schlant RC, Gonzalez EB, Roberts WC: The connective tissue diseases, in Alexander RW, Schlant RC, Fuster V, O'Rourke RA, Roberts R, Sonnenblick EH (eds): Hurst's The Heart, 9th ed. New York, McGraw-Hill, 1998:2271–2294.

murs from either stenosis or regurgitation and, rarely, can produce severe valve dysfunction and require valve replacement. The lesions also predispose to infective endocarditis.

SLE myocarditis is uncommon clinically. A severe form can occur, however, with ventricular arrhythmias, conduction disturbances, and congestive heart failure.

SLE may involve the small intramyocardial arteries and produce fibrinoid necrosis and thromboembolic occlusions. Involvement of the vessels to the SA and AV nodes may produce disturbances of rhythm and conduction, including complete heart block. Systemic hypertension is frequently associated with renal involvement and may result in heart failure. Coronary atherosclerosis may be accelerated by the hypertension, renal failure, and steroid therapy.

Neonatal lupus erythematosus is a rare syndrome that occurs when anti-Ro (SSA) antoantibodies are formed, cross the placenta, and cause a lupus-like syndrome in the newborn, which usually resolves in 3 to 6 months. Complete heart block, usually irreversible, may occur and require a pacemaker. Congenital heart block may be diagnosed by the finding of fetal bradycardia around the 23rd week of gestation.

Special Diagnostic Tests

Serum gamma globulin and cryoproteins frequently are elevated, while serum complement usually is decreased. Lupus erythematosus (LE) cell preparations are characteristic. Antinuclear antibody (ANA), anticytoplasmic, and rheumatoid factors are frequently positive, but are nonspecific. An increased prevalence of cardiac abnormalities has been reported in SLE patients with increased antiphospholipid or anticardiolipin antibody titers.

Natural History and Prognosis

Survival is greater than 80 percent over a 10-year period; most deaths are related to infection, renal failure, cerebritis, or coronary artery disease.

Differential Diagnosis

This includes rheumatic heart disease, infective endocarditis, myocarditis, acute purulent or tuberculous pericarditis, dilated cardiomyopathy, glomerulonephritis, sickle-cell disease, and other connective tissue diseases.

Therapy

Therapy for SLE includes nonsteroidal anti-inflammatory drugs, corticosteroids, and cytotoxic agents such as azthioprine and cyclophosphamide. Systemic hypertension, congestive heart failure, and arrhythmias are managed in the usual ways. Pericardial effusions can produce tamponade and can require pericardiocentesis. SLE valvulitis can require valve replacement.

POLYARTERITIS NODOSA

Polyarteritis nodosa (PN) produces segmental necrotizing arteritis of the medium-sized or small arteries throughout the body, especially in the skin, kidneys, gastrointestinal tract, spleen, lymph nodes, central nervous system, skeletal muscles, and heart.

Diagnosis

The diagnosis is based on the clinical evidence of multisystem involvement and on biopsy evidence of arteritis. The sedimentation rate and serum gamma globulins are frequently elevated, and rheumatoid favor and antinuclear antibodies may be present. The p-antineutrophil cytoplasmic antibody (ANCA) may be present.

Natural History and Prognosis

PN may produce focal myocardial infarction or conduction system abnormalities. Involvement of the renal arteries often leads to renal failure and hypertension, which can produce congestive heart failure. In general, the prognosis in PN is poor.

Differential Diagnosis

This includes giant-cell arteritis, hypersensitivity angiitis, temporal arteritis, Takayasu's arteritis, and arteritis from other connective tissue disorders.

Therapy

Corticosteroids and other anti-inflammatory drugs are used to treat PN. Hypertension, heart failure, and conduction system abnormalities are treated in the usual ways.

RHEUMATOID ARTHRITIS

Clinical Manifestations

Rheumatoid arthritis (RA) is the most common connective tissue disease. It usually affects the joints, with arthralgia and arthritis, fever, anemia, weight loss, subcutaneous nodules, and lymphadenopathy. Pleuritis and necrotizing vasculitis occur less often. Pericarditis may occur in up to 30 percent of patients but usually is overshadowed by other symptoms. Large pericardial effusion can require pericardiocentesis; pericardial constriction can require pericardiectomy. Rheumatoid nodules can involve the myocardium and the heart valves, where they can produce mild-to-severe valvular regurgitation. Very rarely, extensive rheumatoid nodules and myocarditis can produce arrhythmias, conduction disturbances, and congestive heart failure.

Therapy

Arrhythmias, conduction disturbances, heart failure, and acute pericarditis are treated in the usual ways. Therapy for RA usually consists of corticosteroids and nonsteroidal anti-inflammatory agents; penicillamine and gold also are employed.

ANKYLOSING SPONDYLITIS

Clinical Manifestations

Ankylosing spondylitis (AS), or rheumatoid spondylitis, usually affects young men, with inflammation of the spinal joints and, eventually (over 20 or 30 years), immobilization of the spine due to fusion of the costovertebral and sacroiliac joints. Most patients have the HLA-B27 histocompatibility antigen, as do patients with Reiter's syndrome and juvenile arthritis.

The aorta of a patient with ankylosing spondylitis may have an inflammatory sclerosis, just above and below the aortic valve, that produces aortic regurgitation. Extension to the mitral valve can produce mitral regurgitation; extension into the ventricular septum can produce conduction defects. Reiter's syndrome and psoriatic arthritis can each produce a similar inflammatory sclerosis of the aortic root, and aortic regurgitation.

Therapy

Back pain is treated with salicylate or other anti-inflammatory agents such as indomethacin and phenylbutazone. Iritis is treated with corticosteroids. Conduction disturbances and heart failure are treated in the usual ways. Aortic regurgitation can require valve replacement.

SYSTEMIC SCLEROSIS

Clinical Manifestations

Systemic sclerosis (SS) (formerly called *scleroderma*) is associated with fibrous thickening of the skin and fibrous and degenerative changes and diffuse vascular lesions in the fingers, esophagus, small and large bowels, kidneys, heart, and lungs. Systemic hypertension due to renal involvement is common. Most patients have Raynaud's disease of the fingers. A similar phenomenon also occurs in the vasculature of involved internal organs.

The clinical course varies from short and severe to long and benign. A variant of SS is the *CREST syndrome* (*c*alcinosis, *R*aynaud's phenomenon, *e*sophageal abnormality, *s*clerodactyly, *t*elangiectasia), in which symptoms of lung disease and pulmonary hypertension may predominate.

SS may produce patchy myocardial necrosis and fibrosis, possibly due to Raynaud's vasospastic phenomenon of the small coronary vessels. This can result in angina pectoris, myocardial infarction, sudden cardiac death, arrhythmias, and biventricular congestive heart failure. Pericarditis may occur due to SS or, more often, to chronic renal failure. Very rarely, constrictive pericarditis can occur. SS can involve the pulmonary vasculature and produce severe pulmonary vascular disease, with lesions similar to those of primary pulmonary hypertension (see Chap. 24) and with a similar clinical picture, including possible sudden cardiac death.

Therapy

Arrhythmias and congestive heart failure are treated as usual. Calcium-channel blockers, such as nifedipine, may help patients with Raynaud's phenomenon involving the myocardium as well as the fingers or with pulmonary hypertension, but the benefit of such therapy is not proven. Captopril may improve myocardial perfusion. Corticosteroids and D-penicillamine are also of unproven efficacy.

POLYMYOSITIS/DERMATOMYOSITIS

Both polymyositis (PM) and dermatomyositis (DM) are characterized by proximal muscle weakness and periorbital edema. DM also has a characteristic heliotrope rash. Myocarditis and even coronary arteritis can occur. Elevations of serum levels of enzymes reflecting muscle breakdown such as creatine phosphokinase or aldolase are typical. The anti-Jo-1 antibody may be detected. Typical changes may be found in the electromyogram, and biopsy of a proximal muscle such as the deltoid may be confirmatory. Treatment initially is oral prednisone, 40 to 60 mg daily. If this is not successful, treatment may include methotrexate, azathioprine (Imuran), or cyclophosphamide (Cytoxan). Treatment with intravenous immunoglobulin appears promising.

GIANT-CELL (CRANIAL, TEMPORAL, GRANULOMATOUS) ARTERITIS

This systemic inflammatory vasculitis can involve almost any artery or vein in the body but primarily involves extracranial vessels, especially of the external carotid artery. It occurs almost exclusively in patients over 55 years of age. Common symptoms include headache, scalp tenderness, jaw claudication, visual disturbances including blindness and diplopia, weight loss, anemia, and neuromuscular symptoms of polymyalgia rheumatica (PMR). Uncommon presentations include fever of unknown origin, chest pain from aortitis or myocardial infarction, aortic aneurysm, coma, peripheral gangrene, peripheral neuropathies, large vessel involvement with limb claudication, aortic regurgitation, and stroke. Typical findings on physical examination include tenderness of the temporal or occipital arteries, nodulations of the artery, and a tender scalp. The erythrocyte sedimentation rate usually is markedly elevated. The diagnosis is confirmed by a biopsy although this is positive in no more than 60 percent, and a contralateral biopsy may be required. Therapy with oral prednisone, 40 to 60 mg daily, is used to prevent blindness and suppress inflammation with systemic involvement.

CHURG-STRAUSS SYNDROME

The Churg-Strauss syndrome (allergic granulomatosis and angiitis) may be associated with left ventricular dilatation and decreased ejection fraction, as well as mitral regurgitation. Left ventricular systolic function may improve with corticosteroid therapy. Mitral valve replacement may be necessary.

ANTIPHOSPHOLIPID ANTIBODY SYNDROME

The antiphospholipid syndrome (APS) is characterized by the presence of antiphospholipid (aPL) antibody or lupus anticoagulant, usually in high titer, and any or all of the following: recurrent arterial or venous thromboses; recurrent fetal losses; and thrombocytopenia. Spontaneous thromboses occur in small and large arteries in the cerebral and ocular circulations. Livedo reticularis is frequent. SLE is frequently associated. There is an increased incidence of aortic or mitral regurgitation, both in patients with the "primary" APS syndrome and patients with SLE who have aPL antibodies. An increased anticardiolipin antibody is a risk factor for deep venous thrombosis and pulmonary embolus. Treatment has included aspirin, warfarin, heparin, or corticosteroids.

SYPHILIS

Cardiovascular syphilis is not usually considered a connective tissue disorder, although histologically it has many features similar to those of ankylosing spondylitis. Of note, spirochetes have never been identified in the aorta of a patient with cardiovascular syphilis.

Characteristically, syphilis involves the entire tubular portion of the aorta, up to the origin of the innominate artery, which may become diffusely or focally dilated. The aneurysms may press into the sternum or vascular structures. In ankylosing spondylitis, the disease process always involves the basal portions of the aortic valve cusps and always extends into the membranous ventricular septum, the basal portion of the anterior mitral leaflet, or both. In cardiovascular syphilis, the intima of the ascending aorta may have a "tree-bark" appearance of atherosclerotic plaques. Histologically, the aortic lesions of cardiovascular syphilis and ankylosing spondylitis are similar.

SUGGESTED READING

Alexander EL, Firestein GS, Weiss JL: Reversible cold-induced abnormalities in myocardial perfusion and function in systemic sclerosis. *Ann Intern Med* 1986; 105:661–668.

Ansari A, Larson PH, Bates HD: Cardiovascular manifestations of systemic lupus erythematosus. *Prog Cardiovasc Dis* 1985; 27:421–434.

Ellis WW, Baer AN, Robertson RM, et al: Left ventricle dysfunction induced by cold exposure in patients with systemic sclerosis. *Am J Med* 1986; 80:385–392.

Kahan A, Devaux JY, Amor B, et al: Nifedipine and thallium-201 myocardial perfusion in progressive systemic sclerosis. *N Engl J Med* 1986; 314:1397–1402.

Khamashita MA, Cuadrado M, Mujie F, et al: The management of thrombosis in the antiphospholipid-antibody syndrome. *N Engl J Med* 1995; 332:993–997.

Lockshin MD: Antiphospholipid antibody syndrome. *JAMA* 1992; 268:1451–1453.

Pyeritz PE, McKusick VA: The Marfan syndrome: diagnosis and management. *N Engl J Med* 1979; 300:772–777.

Roberts WC, Honig HS: The spectrum of cardiovascular disease in the Marfan syndrome: a clinico-morphologic study of 18 necropsy patients and comparison to 151 previously reported necropsy patients. *Am Heart J* 1982; 104: 115–135.

Royce P, Steinmann B (eds): *Connective Tissue and Its Heritable Disorders: Molecular, Genetic, and Medical Aspects*. New York, Wiley-Liss, 1993.

Schlant RC, Gonzalez EB, Roberts WC: The connective tissue diseases, in Alexander RW, Schlant RC, Fuster V, O'Rourke RA, Roberts R, Sonnenblick EH (eds): *Hurst's The Heart*, 9th ed. New York, McGraw-Hill, 1998:2271–2294.

Wallace DJ, Hahn BH (eds): *Dubois' Lupus Erythematosus*, 4th ed. Philadelphia, Lea & Febiger, 1993.

OBESITY AND HYPERTENSION

An association of obesity with hypertension is clearly indicated by epidemiologic studies, though most obese persons are not hypertensive; even with extreme obesity, one-third are normotensive. A predominantly central or truncal distribution of body fat correlates better with hypertension incidence than does total excess body fat mass. Obesity hypertension differs from essential hypertension in that (1) measurements of blood volume and renin levels do not correlate with the presence of hypertension and (2) systemic vascular resistance is less with higher cardiac output and pulse-wave velocity in obese hypertensives than in lean hypertensive subjects. The clinical triad of obesity, hypertension, and insulin resistance has suggested a hypertensive mechanistic role for the latter, but epidemiologic studies do not demonstrate a consistent association, and chronic hyperinsulinemia is not associated with hypertension. The mechanism of obesity hypertension remains poorly defined. Weight reduction and dietary salt restriction will effect some degree of blood pressure reduction in most cases. In the absence of these first-line measures, angiotensin-converting enzyme (ACE) inhibitors or calcium channel blockers are considered preferable to diuretic or beta-blockade treatment, since the latter agents may have adverse effects on lipids and glucose metabolism in these subjects.

ATHEROSCLEROTIC CORONARY ARTERY DISEASE

Identification of obesity as an independent coronary risk factor—that is, involving pathogenetic factors unique to obesity—

has proved difficult and remains controversial. The presence of obesity, however, with its accompanying atherogenic conditions, has correlated with coronary disease incidence in a number of population studies. Established or potential coronary risk factors accompanying obesity include dyslipidemia, impaired glucose tolerance, insulin resistance, and hypertension. The correlation with coronary disease incidence and the frequency of these accompanying conditions is greater with excess truncal fat distribution than with excess total fat mass. Because of an altered pattern of fatty acid absorption with central obesity, which predisposes to an atherogenic lipid profile, it has been suggested that central obesity may be considered an independent coronary risk factor. Weight reduction in such patients may effect a preferential loss of abdominal visceral fat, with decrements in the measurements of waist-to-hip ratio, blood pressure, plasma glucose and insulin, serum triglycerides, total cholesterol, and low-density lipoprotein (LDL) cholesterol. As yet, however, the potential impact of these salutary changes with weight reduction on the coronary disease process in obese subjects has not been clearly documented or defined.

SLEEP APNEA

More than half of individuals with sleep apnea are obese, and the incidence of sleep apnea is markedly obese subjects approximates 75 percent. Obstructive apnea is the most common type in obese subjects, with transient hypercapneic hypoxemia, pulmonary hypertension, and persistent systemic hypertension. Arrhythmias during apnea in subjects with underlying coronary disease include sinus arrest, atrioventricular (AV) block, ventricular tachycardia, and asystole. Treatment modalities such as tracheostomy, nocturnal positive-pressure breathing, and weight reduction may effect reversal of these arrhythmias as well as decrements in pulmonary and systemic arterial pressure.

CARDIOMYOPATHY

Virtually all morbidly obese people develop left ventricular hypertrophy, in some cases associated with a chronic circulatory congestive state. Clinical manifestations of pulmonary and systemic congestion are associated with increased blood volume and high cardiac output, elevated left ventricular filling pressure, pulmonary hypertension at rest or during exercise, and, in 50 to 60 percent of cases, modest to moderate systemic hypertension. Echocardiographic findings include increased left ventricular wall thickness and left ventricular mass, a modest increase in left ventricular cavity dimension, and enlargement of the left atrium. In some cases left ventricular ejection fraction is depressed, but in others systolic function may be well preserved despite repeated bouts of severe circulatory congestion over a period of many years. Thus, left ventricular systolic function must be assessed by echocardiographic, angiographic, or radionuclide techniques and often cannot be predicted on bedside examination. Left ventricular diastolic dysfunction associated with hypertrophy and diminished chamber compliance is regularly present. In patients with preserved left ventricular systolic function, wall stress—as indicated by the wall thickness–cavity ratio—remains normal, suggesting "adequate" hypertrophy. In those with depressed systolic function, the wall thickness–cavity ratio is reduced, reflecting increased wall stress and "inadequate" hypertrophy. About 5 percent of extremely obese persons develop marked hypoventilation associated with a hypoventilation syndrome of somnolence hypoxemia, cyanosis, respiratory acidosis, and polycythemia. Sleep apnea appears to initiate this syndrome in many but not all of these patients. In this setting, pulmonary vasoconstriction secondary to hypoxemia and acidosis results in an additional component of pulmonary hypertension and a transpulmonary diastolic pressure gradient superimposed upon underlying left ventricular hypertrophy and elevated filling pressure. Biventricular hypertrophy supervenes and occasionally right-sided involvement may predominate.

Frequent electrocardiographic findings in very obese subjects are left axis deviation and low voltage in the standard leads. Despite marked anatomic involvement, electrocardiographic evidences of left ventricular hypertrophy usually are absent. In subjects with hypoventilation, right axis deviation and P pulmonale may be seen. Recurrent bouts of congestion predispose to the development of atrial fibrillation or flutter. Sudden death frequently occurs.

Significant weight loss in markedly obese subjects, usually achieved by gastroplasty or gastrointestinal bypass surgery, results in lowered blood pressure, hemodynamic improvement, reduction in left ventricular mass and chamber size, favorable alteration in left ventricular filling dynamics, and increased ejection fraction if impaired. Substantial improvement in pulmonary function may accompany weight loss in subjects with obesity hypoventilation syndrome.

THERAPEUTIC CONSIDERATIONS

Diuretic therapy is well tolerated in the acute congestive state and, together with dietary sodium restriction, usually is the most effective measure for symptomatic relief (see Chap. 42).

Digitalis and vasodilator therapy have proved useful in some cases involving systolic left ventricular dysfunction. Digitalis dosage should not be based on body weight, as this might lead to toxicity (see Chap. 1).

Antihypertensive therapy is indicated when blood pressure is high. ACE inhibitors, calcium channel blockers, or other agents known to cause regression of myocardial hypertrophy probably are preferable (see Chap. 11).

Low-dose subcutaneous heparin should be considered for patients who present in an acute congestive state as prophylaxis against venous thrombosis and pulmonary embolism (see Chap. 46).

With relief of acute symptoms, medical or surgical measures to achieve long-term weight reduction are indicated. Hemodynamic alterations and left ventricular hypertrophy are reversible with weight reduction.

SUGGESTED READING

Alexander JK: The heart and obesity. In: Alexander RW, Schlant RC, Fuster V, O'Rourke RA, Roberts R, Sonnenblick EH (eds): *Hurst's The Heart*, 9th ed. New York, McGraw-Hill, 1998:2407–2412.

Alexander JK: The cardiomyopathy of obesity. *Prog Cardiovasc Dis* 1985; 26:325–334.

Alpert MA, Lambert CR, Panayiotou H, et al: Relation of duration of morbid obesity to left ventricular mass, systolic function, and diastolic filling and effect of weight loss. *Am J Cardiol* 1995; 76:1194–1197.

Dennis KE, Goldberg AP: Differential effects of body fatness and body fat distribution on risk factors for cardiovascular disease in women: impact of weight loss. *Arterioscler Thromb* 1993; 13:1487–1494.

Hall JE, Summers RL, Brands MW, et al: Resistance to the metabolic action of insulin and its role in hypertension. *Am J Hypertens* 1994; 7:772–778.

Rejala R, Partinen M, Sane T, et al: Obstructive sleep apnea syndrome in morbidly obese patients. *J Intern Med* 1991; 230:125–129.

Reisin E: Weight reduction in the management of hypertension: epidemiologic and mechanistic evidence. *Can J Physiol Pharmacol* 1986; 64:818–824.

28

THE HEART, ALCOHOLISM, AND NUTRITIONAL DISEASE

Charles L. Laham
Timothy J. Regan

ALCOHOLIC CARDIOMYOPATHY

Cardiomyopathy related to long-term alcohol abuse has gained renewed recognition in recent years. Alcoholism has been reported to be the most common identifiable etiologic factor in dilated cardiomyopathy, accounting for up to one-third of cases.

As with other causes of primary myocardial disease, a diffuse abnormality of the myocardium is present, unrelated to coronary atherosclerosis, arterial hypertension, valvular disease, or congenital heart disease. Though symptoms of congestive heart failure may be the most common presentation in primary myocardial disease of multiple etiologies, congestive heart failure may be found on initial examination in fewer than half of patients. In a significant proportion of patients, arrhythmias without congestive heart failure may be the first evident abnormality. Chest pain is not uncommon; occasionally, classic angina pectoris is the only symptom despite normal coronary arteriograms.

Etiology

The evidence that ethanol and/or its metabolites are the etiologic basis for alcoholic cardiomyopathy is circumstantial. The major supportive feature for this idea is the history of ethanol ingestion in intoxicating amounts for many years, frequently marked by periods of spree drinking. The reduced mortality rate in patients who become abstinent further supports the view that ethanol is the major factor in the development of the disease.

In an epidemiologic survey of a working population, the risk of cardiovascular death was increased when chronic alcohol consumption exceeded six drinks a day. As a variable that may contribute to the development of cardiomyopathy, cigarette use is common. In contrast to alcoholic cirrhosis, clinically evident malnutrition is usually not present in the cardiac patient. In females, the disease is rare before menopause.

Diagnosis

Clinical reports of cardiomyopathy have emphasized the problem of denial in attempting to elicit a history of alcoholism. There is male predominance. Suggestive diagnostic aspects include a family history of chronic alcoholism, social disruption, and accident-proneness. The major positive diagnostic feature is the history of ethanol ingestion in intoxicating amounts for many years, frequently marked by periods of spree drinking. Often this information can be obtained only through persistent questioning in multiple visits with the patient or by communication with relatives. Altered blood composition may support the suspicion of addiction. The mean corpuscular red cell volume may be enhanced, serum albumin and transferrin reduced. Platelet adenyl cyclase responses to appropriate stimulation may be reduced; this condition may persist for months to years after the individual becomes abstinent.

Heart Failure

Almost 50 percent of individuals addicted to ethanol show subclinical abnormalities of left ventricular function on noninvasive testing; these abnormalities are commonly diastolic. In those whose cardiac dysfunction progresses to low-cardiac-output heart failure, exertional or nocturnal dyspnea is present. Patients may also complain of weakness and exertional fatigue, presumably because of the reduction in cardiac output. The physical signs of cardiac decomposition seen in these individuals are similar to those observed with other forms of dilated cardiomyopathy—i.e., S_3 and S_4 gallops and cardiomegaly. The murmur of mitral regurgitation is usually well differentiated from that related to rheumatic valvular disease. This murmur is characteristically confined to a portion of systole, is often not holosystolic, and, as a rule, changes as cardiac compensation is restored. A holosystolic murmur due to tricuspid regurgitation may be heard in the third and fourth intercostal spaces along the sternal border. It is also diminished with the amelioration of heart failure.

Since the addicted person often delays seeking medical assistance for weeks to months, evidence of right-sided heart failure is not uncommon, with distended jugular veins; also common are an enlarged, tender liver and edema of the dependent portions of the body.

A common complication of cardiomyopathy is the development of pulmonary or peripheral arterial emboli. Systemic emboli can originate from mural thrombi in the left ventricle and left atrium. Pulmonary emboli are often associated with thrombi in the dilated right heart and with thrombophlebitis in the venous system.

ELECTROCARDIOGRAPHY

The electrocardiogram (ECG) at this time may be relatively normal or may show nonspecific changes. Poor progression of the R wave across the precordium is fairly common, particularly as the disease advances. This is thought to be due to progression of the ventricular disease and conduction delay. Evidence of left ventricular and atrial enlargement is common; left anterior hemiblock occurs in a minority of patients, while left or right bundle branch block appears in approximately 10 percent. A variety of arrhythmias may be present, including atrial or ventricular ectopic beats and atrial fibrillation. These usually are secondary to heart failure.

HEMODYNAMICS

The hemodynamic characteristics of the left ventricle in patients who are compensated from heart failure show a qualitative similarity to those found in patients in the subclinical state. Presentation with noncoronary chest pain rather than heart failure can be associated with a significant increase of end-diastolic pressure but a slight reduction of end-diastolic volume consistent with diminished compliance. Moreover, indices of contractility are frequently diminished in this state. When shortness of breath appears, there is an increase in end-diastolic volume and a further reduction of systolic function as measured by ejection fraction.

PRECIPITATING FACTORS

Usually there are no evident factors other than ethanol that can be readily associated with the episode of heart failure. However, occasional small, isolated outbreaks appear to be associated with the combination of ethanol abuse and the trace metals lead, cobalt, and/or arsenic. Several other factors have to be considered potentially important. Clinically evident malnutrition usually is not present in the cardiac patient, though it is commonly associated with liver disease. In females, the disease is rare before menopause. Systemic hypertension may be present (see below).

Natural History and Prognosis

The course of alcoholic cardiomyopathy is variable, depending largely on the extent of cardiac involvement. The outlook is relatively poor in persons who continue to ingest ethanol in substantial amounts. In one study, a series of 64 alcoholic patients with cardiomyopathy were followed over a 4-year period. One-third remained abstinent, and the mortality rate in this group was 9 percent, though only a minority exhibited clinical improvement paralleling the response in hepatic cirrhosis. Of those who remained actively alcoholic, more than one-half succumbed. Presumably, at certain stages of the disease, the pathogenic mechanisms may continue unabated despite traditional pharmacologic management and abstinence from alcohol.

Management

Pharmacologic therapy for cardiac decompensation depends on the state of cardiac disease when the patient is first encountered. During the first episode of heart failure, if the patient has had symptoms for a relatively short time with only modest cardiomegaly and pulmonary congestion, he or she may be managed initially by diuretics to diminish volume overload. As the disease progresses, there is a role for preload- and afterload-reducing agents. Presumably the latter may provide long-term efficacy, as reported for other causes of heart failure. Digitalis can contribute to the management of congestive failure in the advanced stages of the disease and, assuming the patient's compliance, is most useful in the control of atrial fibrillation or sinus tachycardia.

Arrhythmias

The patient may present with an arrhythmia as the initial manifestation of disease. Supraventricular arrhythmias predominate; atrial fibrillation is the most common. Cardioversion or pharmacologic intervention is frequently required, but sinus rhythm is restored spontaneously in some patients. Plasma electrolytes on admission usually are normal. During a subsequent recurrence, the same arrhythmias may occur as during the original episode.

This entity has been called the *holiday heart syndrome* because of frequent presentation over holidays or weekends. In this syndrome, the acute cardiac rhythm disturbance occurs in association with heavy ethanol consumption in a person who has chronically abused ethanol without other clinical evidence of heart disease. Under unusual circumstances, such as a prolonged period of sleeplessness, the arrhythmia may sometimes be induced acutely without chronic abuse.

Sudden unexpected death in alcoholics has been reported from several countries. This phenomenon has been attributed to cardiac arrest—a conclusion is supported by the observation of a reduced ventricular fibrillation threshold in an animal model of chronic alcoholism. In one series, young and middle-aged persons without coronary disease made up 8 percent of all sudden deaths. A subsequent investigation in the former Soviet Union revealed an incidence of 17 percent in the apparent absence of other toxins. Frequently there was no evidence of cardiac hypertrophy; the liver involvement was often in the form of fatty liver rather than hepatic cirrhosis.

Pathology

In examination of biopsy specimens from living patients and of autopsy tissue preparations, no distinctive features have been revealed in the patients with alcoholic heart disease as compared with those with congestive cardiomyopathy of other causes. Quite early in the prefailure stage, there seems to be dilatation of the sarcoplasmic reticulum and the undifferentiated portion of the intercalated disk, but these changes apparently are obscured at later stages of disease, when considerable myocytolysis may be seen. An increase of fibrous tissue is a usual finding; it may take the form of an increase in the interstitial collagen component or replacement of myocardial fibers. Intramural vessels are usually normal; in the areas of fibrous-tissue accumulation, however, they may show wall thickening.

ALCOHOL AND HYPERTENSION

The diagnosis of alcoholic cardiomyopathy is often obscured when the patient presents with elevated arterial blood pressure.

When seen with other causes of heart failure, systemic hypertension is often considered to be secondary to compensatory peripheral vasoconstriction during cardiac decompensation. Hypertensive episodes may be more frequent in the alcoholic, particularly if measurements are made shortly after a period of ethanol intake. During the late intoxication–early withdrawal period, this response has been reported in up to one-half of noncardiac alcoholics observed in an outpatient setting, usually without the development of classic withdrawal illness. Other evidence of hypertension, such as retinopathy, is usually absent.

Arterial pressure may be moderately elevated for several days, with spontaneous decline to normal thereafter. Substantial elevations may require up to a week for spontaneous normalization. After a short period of abstinence, arterial pressures are normalized in all but 10 percent of noncardiac alcoholics, which represents the incidence of hypertension in a control group. Alcohol is now recognized as a risk factor for the development of systemic arterial hypertension (Chap. 11).

CORONARY ARTERY DISEASE AND MODERATE ALCOHOL INTAKE

Research studies have observed a favorable protective effect in men and women against stroke, coronary artery disease, and sudden cardiac death with consumption of moderate amounts of alcohol. These effects appear equal with intake of wines, beer, or spirits; by extension, given the theoretical additional antioxidant benefits of red wine, they have also been confirmed in patients drinking nonalcoholic grape juice. Unexpected have been the observations of efficacy with alcohol doses of less than one drink per day. An interesting feature in comparison studies involving alcohol intake is that non-drinkers and drinkers are considered to be epidemiologically equivalent in terms of education and social class as related to cardiac risk, an assumption that may not be accurate.

A recently reported prospective study over 9 years involving 490,000 men and women aged 35 to 69 years noted a 30 to 40 percent reduction in cardiovascular death with one drink of alcohol per day. As the level of alcohol intake increased, an exponential increase in accidental injury, cirrhosis, and cancer was noted, potentially negating or surpassing any beneficial mortality effects. Similar studies have concluded, from a risk-benefit perspective, that any positive effect of alcohol is optimized at about one drink daily, particularly in those over 55 to 60 years of age with increased coronary risk factors. Whether or not alcohol use provides an added benefit in patients with cardiac disease on traditional treatment remains an open question. Younger patients with lower cardiac risk are statistically at higher risk from the negative effects of drinking.

ISSUES ON VITAMINS

Thiamine

Beriberi is a high-cardiac-output state associated with arteriolar vasodilatation. Although it has been the classic view that right ventricular failure is dominant when symptoms develop, several studies have documented a significant elevation of left ventricular end-diastolic and pulmonary capillary wedge pressure that is reversible with thiamine therapy.

Folate, Vitamin B_{12}, and Pyridoxine

Emerging data on homocysteine as an independent risk factor for cardiac and peripheral vascular disease have been focused on methods of reducing its blood level as a step in reducing risk. Folate, vitamin B_{12}, and pyridoxine are each cofactors in the metabolism of homocysteine, and decreased levels of these have also been postulated as coronary risk factors. An observational study found decreased levels of folate, vitamin B_{12}, and pyridoxine in 59, 57, and 25 percent of men, respectively, with elevated homocysteine levels (greater than 16 μmol/L). Several randomized studies have shown an average 25 percent reduction in homocysteine levels with folate supplementation, with a mean 7 percent additional benefit with concomitant use of vitamin B_{12} but no independent benefit for pyridoxine despite decreased levels. Suggested doses have been 400 mg/day of folate together with 1 mg of vitamin B_{12}; this appears to lower homocysteine effectively without adverse effects. Recommendations as to the possible use of these compounds as treatments in cardiovascular disease will depend on the results of ongoing placebo-controlled trials.

Vitamin C

Scurvy can be associated with sudden death. Human volunteers on a vitamin C–deficient diet have reported dyspnea and chest pain associated with PR prolongation and ST-segment abnormalities. These symptoms may be reversible with the administration of vitamin C. Preliminary prospective studies of antioxidants have failed to show an independent beneficial cardiovascular effect of vitamin C supplementation, although some studies have suggested a beneficial effect upon endothelial function.

Vitamin E and Selenium

Prospective observational studies in male and female health professionals have suggested an inverse association of coronary artery events and vitamin E intake, particularly in smokers. Of particular importance are the ongoing placebo-controlled trials in subjects with and without cardiovascular disease in whom the dietary intake is supplemented by 400 to 600 IU of the vitamin per day.

Selenium deficiency has been associated with a cardiomyopathy in children. The disease has a regional distribution in agricultural areas where the selenium content of staple grains and soil is reduced. Supplementation of the diet with selenium has been found to be preventative. Although isolated selenium deficiency that produces cardiomyopathy has been described in animal models, combined deprivation of selenium and vitamin E has been shown experimentally to produce patchy myocardial necrosis.

Carotenes

Preliminary interest in beta-carotenes as lipid-soluble antioxidants have not been borne out by several recent placebo-controlled studies. Many failed to show a beneficial effect, and a Finnish study in a population of smokers found a trend toward an increased lung cancer incidence. Thus, current evidence does not support a role for beta-carotenes in cardiovascular disease prevention.

Cachexia

Severe weight loss in individuals who have a relatively normal initial body weight may have important cardiovascular consequence, particularly in infants and children, where primary congestive heart failure resulting from either marasmus or kwashiorkor may occur. In adults, the nutritional deficiency

that characterizes protein-calorie malnutrition is marked by a
significant reduction of heart rate, stroke volume, and cardiac
output associated with a diminished left ventricular end-
diastolic diameter and mass. The ECG is generally unremarkable, but some patients exhibit ST-T wave changes. The
QT interval is usually normal. Patients who recover from
anorexia show improvement in heart rate, posterior wall
dimension, systolic wall stress, and left ventricular internal
dimension. A potential for arrythmias and sudden death has
been postulated.

Hyperalimentation

Therapeutic feeding by the enteric or parenteral route has
assumed increasing importance in the management of a variety
of illnesses. Adults with chronic undernutrition without underlying heart disease may develop heart failure during hyperalimentation; a state resembling congestive heart failure
may develop, characterized by hypermetabolism, ventricular
gallop, augmented cardiac output, and normal ejection fraction. Rapid resolution follows diuretic therapy, slowing of the
rate of hyperalimentation, and a reduction in the daily intake of
sodium.

SUGGESTED READING

Ettinger PO, Wu CF, DeLa Cruz Jr, et al: Arrythmias and the
"holiday heart." *J Am Coll Cardiol* 1983; 1:816–818.
Kupari M, Koskinen P, Soukas A, et al: Left ventricular filling
impairment in asymptomatic chronic alcoholics. *Am J
Cardiol* 1990; 66:1473–1477.
Lonn EM, Yusuf S: Is there a role for antioxidant vitamins in
the prevention of cardiovascular diseases? An update on
epidermiological and clinical trials data. *Can J Cardiol*
1997; 13:957–965.
Regan TJ: The heart, alcoholism, and nutritional disease. In:
Alexander RW, Schlant RC, Fuster V, O'Rourke RA,
Roberts R, Sonnenblick EH (eds): *Hurst's The Heart*, 9th
ed. New York, McGraw-Hill, 1997: 2109–2115.
Robinson K, Kristopher A, Refsum H, et al: Low circulating
folate and vitamin B_6 concentrations: risk factors for
stroke, peripheral vascular disease, and coronary artery
disease. *Circulation* 1998; 97:437–443.

Rosengren A, Wilhelmsen L, Wedel H: Separate and combined effects of smoking and alcohol abuse in middle-aged men. *Acta Med Scand* 1988; 223:111–118.

Thun MJ, Peto R, Lopez AD, et al: Alcohol comsumption and mortality among middle-aged and elderly U.S. adults. *N Engl J Med* 1997; 337:1705–1714.

Vikhert AM, Tsiplenkova VG, Chepachenko NM: Alcoholic cardiomyopathy and sudden cardiac death. *Am Coll Cardiol* 1986; 8:3A–11A.

29

CARDIOVASCULAR INVOLVEMENT IN THE ACQUIRED IMMUNODEFICIENCY SYNDROME (AIDS)

Melvin D. Cheitlin

DEFINITION

Acquired immunodeficiency syndrome (AIDS) is caused by an infection by a virus of the family Retroviridae, first recognized in 1981. The primary virus responsible for AIDS is the HIV-1 organism. The organism, a retrovirus, invades the nucleus of certain cells such as macrophages and immunocytes by means of a specific membrane receptor, and then incorporates its DNA copy into the host genome. After a long incubation period (a mean of 8 to 10 years), the virus expresses itself, killing the host cells and invading other immune cells, usually T-helper lymphocytes. The incubation has been prolonged by the use of antiviral agents such as zidovudine (AZT), dideoxyinosine (DDL), and dedeoxyctidine (DDC) as well as the development of protease inhibitor drugs such as saquinavir, ritonavir, and indinavir. The disease compromises the immune defense mechanisms of the host, and the patient is subject to opportunistic viral, fungal, parasitic, and bacterial infections. These infections mainly involve lung, central nervous system, and gastrointestinal (GI) tract. There is also a predisposition to the development of certain malignancies, especially Kaposi's sarcoma and lymphoma.

Risk factors for the development of AIDS infection are sexual intercourse with a patient with HIV infection, intravenous (IV) drug use, and the intravenous use of blood products.

CARDIOVASCULAR INVOLVEMENT

Cardiovascular involvement was recognized early in the HIV epidemic, but it is not usually an important clinical problem in most patients with AIDS. Cardiovascular involvement is seen in AIDS patients as follows:

1. Pericarditis and pericardial effusion with and without tamponade
2. Pulmonary hypertension leading to right ventricular dilatation and failure
3. Myocardial involvement
   a. Incidental finding of tumor implants
   b. Asymptomatic left ventricular (LV) hypokinesis on echocardiogram
   c. Symptomatic cardiomyopathy
4. Valvular involvement
   a. Mitral valve prolapse
   b. Infective endocarditis
   c. Marantic endocarditis

CLINICAL MANIFESTATIONS

Pericarditis and Pericardial Effusion

In all patients with AIDS, pericarditis and large pericardial effusions are relatively unusual. In patients with HIV disease who have cardiovascular symptoms sufficiently prominent to warrant an echocardiogram, however, pericardial involvement is seen in about one-third. Usually, the presence of pericardial effusion is suspected by finding a large cardiac silhouette on a chest radiograph. Occasionally, patients will present with shortness of breath, tachycardia, an elevated central venous pressure, and even decreased blood pressure—all signs and symptoms of cardiac tamponade. Less frequently, the patients will have chest discomfort, usually with a pleuritic component, and even a pericardial friction rub.

The etiology of the pericarditis is variable *Mycobacterium tuberculosis* as well as fungal and purulent pericarditis can be seen. Also, patients with lymphoma can develop pericardial effusion. In our experience, even with pericardial biopsy, most have no recognizable pathogen and their disease is idiopathic. However, the development of a pericardial effusion is a poor prognostic sign and AIDS patients with an effusion have a shorter life expectancy than those without.

Pulmonary Hypertension

Usually, but not always, pulmonary hypertension is seen in patients who have had multiple *Pneumocystis carinii* pulmonary infections. Frequently, the patient will have a loud pulmonic component of the second heart sound, a right ventricular lift, and even right ventricular hypertrophy (RVH) on electrocardiogram. Occasionally, the patient can present with signs and symptoms of right heart failure, with an increased central venous pressure, peripheral edema, right ventricular S_3 and S_4 gallops, and tricuspid regurgitation.

Myocardial Involvement

At autopsy of AIDS patients without signs or symptoms of cardiovascular involvement during life, 20 to 40 percent of the hearts are found to have focal myocarditis consisting of collections of round cells with and without myocardial necrosis. It is rare to find diffuse myocarditis. Hypokinesis, decreased ejection fraction, and/or dilated left ventricle can be found with echocardiography in 20 percent or more of hospitalized patients with AIDS. Most of these patients have no clinical signs of heart failure. In less than 3 to 5 percent there is a syndrome of left ventricular dilatation and decreased systolic function (cardiomyopathy) with congestive heart failure. Occasional patients with ventricular arrhythmias and even ventricular tachycardia and sudden death have been described.

The etiology of left ventricular dysfunction in AIDS is unclear. The possibilities are:

1. HIV myocarditis.
2. Myocarditis due to opportunistic infections such as toxoplasmosis, coccidiomycosis, cytomegalic viral infection, and coxsackie B viral infection.
3. Autoimmune myocarditis.
4. Drug effects from cardiotoxic drugs causing reversible cardiomyopathy. These include interleukin-2, alpha interferon, and foscarnet. Other known cardiotoxic drugs such as doxorubicin (Adriamycin) are used occasionally. Other possibilities are that zidovudine (Retrovir, AZT) might be responsible for decreased ventricular function in some patients, and certain drug combinations might be the cause of cardiomyopathy. The LV dysfunction might be reversible if the drug is removed.
5. Illicit drugs. The use of such illicit drugs as cocaine and heroin is common in the population at risk for HIV infection. Each of these drugs is capable of causing cardiomyopathy,

although in most patients myocardial involvement is subclinical. Nonillicit drugs such as alcohol, capable of causing cardiomyopathy, also have to be considered.

6. Cytokines, such as tumor necrosis factor (TNF), either paracrine or circulating.

Patients with cardiomyopathy can manifest their disease clinically by the presence of enlarged silhouette on chest radiography, a dilated left ventricle with a displaced apex impulse, S_4 and S_3 gallops, or a murmur of mitral regurgitation due to a dilated left ventricle. Ultimately, congestive heart failure with rales, pulmonary edema, and even right heart failure can be seen. A clinical presentation of congestive heart failure, however, is uncommon in AIDS patients.

Valvular Involvement

Mitral valve prolapse is seen in about 5 percent of the normal population and might, therefore, be expected in a similar percentage of patients with AIDS. The presence of mitral valve prolapse as manifested by a nonejection click or mitral regurgitation murmur, especially a late systolic murmur at the apex, also can be due to cachexia and a decrease in left ventricular end-diastolic volume. This can result in mitral valve prolapse on echocardiography and probably is unrelated to the AIDS infection.

Infective endocarditis of the aortic, mitral, and tricuspid valves is seen, but mostly in patients who are intravenous (IV) drug users. The risk factor is IV drug use, not AIDS per se.

Marantic endocarditis, manifested by sterile fibrinous vegetations on the aortic or mitral valve, can be related to the cachectic state. These vegetations are discovered because of embolization or are seen on echocardiography

CLINICAL WORKUP

Most patients with HIV infection have no clinically important cardiovascular involvement. There is no evidence that finding clinically silent left ventricular dysfunction or small pericardial effusions by echocardiography, or "focal myocarditis" by myocardial biopsy, results in a change in therapy that will alter the course of the disease. Therefore, there is no indication to perform any special tests, such as echocardiography or myocardial biopsy, looking for asymptomatic cardiovascular or myocardial abnormalities.

If the patient has signs and symptoms consistent with cardiovascular disease, however, such as unexplained dyspnea, orthopnea, paroxysmal dyspnea, chest pain, an enlarged cardiac silhouette on chest x-ray, pericardial friction rub, or a heart murmur, appropriate tests to define the nature of the problem may be indicated. A chest x-ray and electrocardiogram (ECG) should be obtained; however, the most useful test is an echocardiogram.

The extent of evaluation and special diagnostic tests or procedures that are appropriate for an individual patient is best determined by the patient's primary physician.

TREATMENT

Pericardial Effusion

If there is a clinical picture of pericarditis with pleuritic chest pain and fever, indomethacin (Indocin) or a nonsteroidal analgesic can be given. If this is unsuccessful in suppressing the symptoms, a short course of steroids can be helpful, although the possibility of disseminating other infective organisms should be recognized.

If a pericardial effusion is detected and the effusion is causing respiratory problems because of lung compression or hemodynamic problems such as an elevated venous pressure or actual tamponade with tachycardia and hypotension, pericardiocentesis usually is indicated. Evacuation of the fluid will eliminate tamponade and decrease the volume of the effusion. Examination of the fluid for cells and stain and culture for organisms can identify treatable etiologies. A catheter can be left in the pericardial cavity to continuous drainage for 1 to 2 days. If the fluid reaccumulates, a repeat pericardiocentesis can be performed, at which time a pericardial window can be formed by means of inflating a balloon catheter as it crosses the pericardium, thereby creating a hole in the pericardium that drains into the left pleural space.

If the effusion is small and the patient is toxic or hemodynamically compromised, open drainage rather than needle pericardiocentesis is the better option. A pericardial biopsy can be obtained for culture and microscopic examination, giving a better chance for identification of a treatable cause, such as lymphoma or infection by a treatable organism.

If the pericardial effusion is not causing toxicity or hemodynamic compromise, and especially if it is small, there is no value in evacuating the fluid, but there is some danger of lacerating the myocardium.

Pulmonary Hypertension

If the patient has a loud second heart sound, especially if the ECG reveals right ventricular hypertrophy, or if the right ventricle is dilated and the left ventricle normal on echocardiogram, an estimation of the pulmonary artery systolic pressure should be obtained by echo-Doppler. Most such patients have a jet of tricuspid regurgitation which, by the modified Bernoulli equation, can give an accurate estimation of right ventricular and pulmonary artery systolic pressure.

The pressure gradient (mmHg) in systole between the right ventricle and the right atrium (G) is given by the formula: $G = 4V^2$, where V equals the velocity of the jet in meters per second. If the estimated venous pressure is added to this, systolic pressure of the right ventricle and pulmonary artery is obtained.

If pulmonary hypertension is present, the patient should be considered for a right heart catheterization and measurement of the pulmonary artery wedge pressure and calculation of pulmonary vascular resistance. If the pulmonary artery pressure and pulmonary vascular resistance are high, an attempt can be made to find a vasodilating drug that will reduce the pulmonary artery pressure without reducing the systemic arterial pressure unacceptably. Various vasodilators can be tried, starting with 100 percent oxygen, followed by hydralazine (Apresoline), isoproterenol (Isuprel), nifedipine (Procardia, Adalat), and prostacycline. If a drug can be found that decreases the pulmonary vascular resistance and pulmonary artery pressure without decreasing cardiac output and without decreasing systemic arterial pressure, this drug should be given chronically. The chances of finding such a drug are small.

Cardiomyopathy and Congestive Heart Failure

If the patient develops the clinical picture of congestive heart failure, a search for some recognized and treatable cause of congestive heart failure should be instituted—for instance, coronary heart disease, valve disease, systemic arterial hypertension, hypertrophic cardiomyopathy, or congenital heart disease. If there is no identifiable cause, drug toxicity should be suspected and any known myocardial depressant should be stopped. Such drugs as Adriamycin, interleukin-2, alpha interferon, and foscarnet have all been shown to cause cardiomyopathy, which in some instances is reversible when the drug is stopped. Even AZT has been implicated as possibly causing myocardial depression in some patients, and therefore a "drug holiday" should be attempted, if possible. If ventricular function improves within 1 week, the drug should be permanently discontinued. If no change occurs, the drug can be reinstituted if it is still indicated.

The patient should receive the usual treatment for congestive heart failure, that is, digoxin, diuretics, and afterload reduction, preferably with an angiotensin converting enzyme inhibitor if it is tolerated (see Chap. 1).

The question of myocardial biopsy is often raised. Although occasional instances of a treatable myocardial lesion are recognized, they are distinctly rare. Still, some authorities consider it reasonable to do a myocardial biopsy in such patients with congestive heart failure and no identifiable cause, looking for these rare occurrences. Finding nonspecific myocarditis by myocardial biopsy is not helpful in management since there is no evidence that corticosteroids or antimetabolites are of any benefit.

VALVE DISEASE

The presence of valvular regurgitation or stenosis should trigger the appropriate workup, including an echo-Doppler study and blood cultures if the patient is febrile. Blood should be cultured for bacteria, both aerobic and anaerobic, as well as for fungi. If there are valvular vegetations and the blood cultures are all negative, especially if the patient is afebrile, marantic endocarditis is very probable and the patient should be given anticoagulants.

If the patient has endocarditis and is a surgical candidate for valve replacement, the status of the HIV disease and especially the prognosis related to the HIV disease must be considered. If the prognosis is good for a relatively prolonged course, then surgery should be offered to the patient without regard to HIV infection. If, on the other hand, the HIV disease is advanced and the prognosis for survival poor, there is no reason to have the patient undergo valve replacement surgery.

Asymptomatic mitral valve prolapse needs no treatment. If mitral regurgitation is present, antibiotics are indicated to prevent infective endocarditis during conditions likely to be associated with bacteremia (see Chap. 21).

RISK TO HEALTH CARE WORKERS

The fear of contracting HIV infection accidentally while caring for a patient with AIDS is a very real concern. The major way in which AIDS is infective in this context is through inadvertent needle stick or skin penetration by knife. Although possible, the risk of infection by a simple splash of blood on mucous membranes is extremely small. As of 1997, a total of 52 health care workers have been reported who seroconverted after needle-sticks in the workplace. Combining the results of several large prospective studies, the risk of HIV-1 infection after percutaneous exposure to blood from an HIV-infected person is approximately 0.3 percent per exposure (95% confidence interval limits 0.13–0.7%).

The greatest risk of infection of health care workers is to be stuck with a contaminated hollow needle. The most likely time for this to occur is during an emergency situation, where there is confusion and rapid activity. A careless act such as grabbing a 4×4 that is hiding a needle is all that is required. Of course, the personnel most at risk are those involved in such activities—nurses and house staff.

The risk to surgical personnel is greater than that to medical personnel. For this reason, cardiovascular operations on patients with HIV are indicated only if the patient will receive substantial benefit. If the patient is HIV+ but without disease or AIDS, life expectancy is long enough in most instances that operation is indicated for the usual reasons. If the patient has AIDS and life expectancy is short, operation is indicated only if the patient will benefit very significantly in terms of symptom relief that cannot be achieved medically.

SUGGESTED READING

Blanchard DG, Hagenhoff C, Clow LC, McCann HA, Dittrich HC: Reversibility of cardiac abnormalities in human immunodeficiency virus (HIV) in infected individuals: a serial echocardiographic study. *J Am Coll Cardiol* 1991; 17:1270–1276.

Cheitlin MD: AIDS and the cardiovascular system, in Alexander RW, Schlant RC, Fuster V, O'Rourke RA, Roberts R, Sonnenblick EH (eds): *Hurst's The Heart*, 9th ed. New York, McGraw Hill, 1998:2143–2152.

Cheitlin MD: Cardiac involvement in AIDS. In: Cohen PT, Sande MA, Volberding PA (eds): *The AIDS Knowledge*

Base. Boston, Massachusetts Medical Society, 1994; 5: 14-5–14-14.

Corallo S, Mutinelli MR, Moroni M, et al: Echocardiography detects myocardial damage in AIDS: a prospective study in 102 patients. *Am Heart J* 1988; 9:887–892.

De Castro S, D'Amati G, Gallo P, et al: Frequency of development of acute global left ventricular dysfunction in human immunodeficiency virus infection. *J Am Coll Cardiol* 1994; 24:1018–1024.

Francis CK: Cardiac involvement in AIDS. *Curr Probl Cardiol* 1990; 15:571–639.

Heidenreich PA, Eisenberg MJ, Kee L, et al: Pericardial effusion in AIDS: incidence and survival. *Circulation* 1995; 92:3229–3234.

Henderson DK, Fahey BJ, Willy M, et al: Risk for occupational transmission of human immunodeficiency virus type 1 (HIV-1) associated with clinical exposures: a prospective evaluation. *Ann Intern Med* 1990; 113: 740–746.

Herskowitz A, Vlahor D, Willoughby S, et al: Prevalence and incidence of left ventricular dysfunction in patients with human immunodeficiency virus infection. *Am J Cardiol* 1993; 71:955–958.

Kaul S, Fishbein MC, Siegel RJ: Cardiac manifestations of acquired immunodeficiency syndrome: a 1991 update. *Am Heart J* 1991; 122:535–544.

Michaels AD, Lederman RJ, MacGregor JS, Cheitlin MD: Cardiovascular involvement in AIDS. *Curr Prob Cardiol* 1997; 22:109–148.

Reilly JM, Cunnion RE. Anderson DW, O'Leary TJ, et al: Frequency of myocarditis, left ventricular dysfunction and ventricular tachycardia in the acquired immunodeficiency syndrome. *Am J Cardiol* 1988: 62:789–793.

Stephen O. Pastan
William E. Mitch

Cardiovascular disease is a major cause of morbidity and death in end-stage renal disease (ESRD) patients treated by chronic dialysis. The cardiovascular mortality rate in ESRD patients is approximately three times higher than in patients without kidney disease, and 45 percent of deaths of dialysis patients in the United States are caused by cardiovascular disease.

Common risk factors for cardiovascular morbidity in dialysis patients include a high prevalence of hypertension, diabetes mellitus, hyperlipidemia, and abnormalities in calcium and phosphate metabolism, with associated hyperparathyroidism and vascular calcification. Additionally, hypotension during hemodialysis, pericardial disease, infective endocarditis, and fluid and electrolyte disturbances can contribute significantly to cardiac morbidity in ESRD patients.

CARDIOVASCULAR RISK FACTORS AND MORBIDITY IN CHRONICALLY UREMIC PATIENTS

Hypertension

The majority of ESRD patients develop systemic hypertension before beginning dialysis therapy. An expanded extracellular fluid volume (ECV) and vasoconstriction are the most important factors causing hypertension in patients with ESRD. Direct evidence that ECV expansion plays a critical role in the pathogenesis of hypertension is found in studies demonstrating rapid resolution of hypertension in most patients after the ECV is reduced substantially by vigorous dialysis.

Patients whose hypertension is resistant to volume removal with dialysis often have vasoconstriction that may be associated with higher levels of plasma renin activity; in this group, drugs inhibiting the renin-angiotensin axis (beta blockers, angiotension converting enzyme inhibitors, or angiotensin$_2$ receptor antagonists) can control hypertension. In some patients, arterial vasoconstriction may be related to overactivity of the sympathetic nervous system or an increase in intracellular calcium in vascular smooth muscle cells. This increase stimulates vascular contraction. Other suggested mechanisms contributing to hypertension in patients with chronic renal failure include a decrease in the production of vasodilator prostaglandins or an increase in the level of the vasoconstrictor endothelin. Finally, the use of recombinant human erythropoietin to correct the anemia associated with ESRD can be accompanied by an increase in blood viscosity and peripheral vascular resistance, resulting in elevation of the blood pressure in approximately 20 to 30 percent of dialysis patients.

The cornerstone of managing hypertension in dialysis patients is reducing the ECV effectively by removing fluid with ultrafiltration during hemodialysis. The addition of antihypertensive drugs may be required for patients who develop symptomatic hypotension during the hemodialysis session but who remain hypertensive between dialysis treatments (see discussion below).

For reasons discussed above, medications that inhibit the renin-angiotension axis appear to be the antihypertensive medications of choice in ESRD patients. Other antihypertensive medications, however, including calcium-channel blockers and minoxidil, can be effective in dialysis patients. Generally, antihypertensive drugs are withheld prior to hemodialysis to minimize the risk of hypotension from occurring during the dialysis treatment. Bilateral nephrectomy is rarely used to eliminate the renin-angiotensin system and to treat dialysis-resistant hypertension, which is actually aggravated by depletion of the ECV (see also Chap. 11).

Diabetes Mellitus and Hyperlipidemia

About 35 percent of all CRF patients who begin maintenance dialysis therapy are diabetic; these patients have a significantly higher mortality rate due to coronary artery diseases than age-matched, nondiabetic patients. The problem is not limited to diabetic complications because hypertension is so frequent.

Close to 50 percent of patients with ESRD have high serum levels of triglycerides, which appear to be related to impaired degradation of very low density lipoprotein by lipoprotein lipase. In patients treated by peritoneal dialysis, the high concentrations of glucose in the dialysate worsen the tendency toward high serum triglyceride levels because glucose increases the production of triglycerides. Another factor is the resistance to insulin action. Other factors that may contribute to hyperlipidemia in patients with chronic renal failure include the use of diuretics and beta blockers. Beside triglycerides, low levels of high-density-lipoprotein (HDL) cholesterol are often documented in dialysis patients. The heparin used to prevent clotting in the dialyzer will cause a transient fall in serum triglycerides by increasing the activity of lipoprotein lipase and thus will complicate interpretation of the lipid profile. Blood samples for plasma lipids should be obtained after a 12-h fast and before the patient is given heparin for the dialysis session.

Strategies for lowering triglycerides and LDL cholesterol include restricting dietary fat, giving fish oil supplements, increasing exercise, and avoiding alcohol or treatment with beta blockers or diuretics. Clofibrate or gemfibrozil may also be effective, but their use has been associated with an increased incidence of myositis and hepatotoxicity in dialysis patients. HMG-CoA reductase inhibitors do seem to be safe in ESRD patients and, at standard doses, will reduce the levels of triglycerides and LDL cholesterol. Although definitive evidence that treatment with antihyperlipidemic drugs prevents atherosclerosis in dialysis patients is lacking, attempts at treatment seem prudent based on results from patients without kidney disease (see also Chap. 7).

Hemodialysis-Associated Hypotension

Clinically significant hypotension occurs in approximately 25 percent of hemodialysis sessions. A common cause of hypotension during dialysis is excessive ultrafiltration, leading to ECV depletion, decreased venous return, and reduced cardiac output. Rapid lowering of plasma osmolality resulting from the removal of urea and other molecules can cause a shift of ECV to the intracellular compartment, because these molecules do not move out of cells as rapidly as dialysis removes them from blood. This leads to an intravascular volume decrease and hypotension.

An important factor governing changes in plasma osmolality is the concentration of sodium in the dialysate. In the past, dialysate with a low sodium concentration was utilized to control hypertension. Today, dialyzers achieve such high rates of fluid removal via ultrafiltration that a higher concentrations of sodium can be used in the dialysate. This practice has reduced the number of hypotensive episodes.

Other factors can increase the tendency toward hemodialysis hypotension and should always be excluded: (1) cardiac dysfunction from long-standing hypertension, ischemic or valvular heart disease, other causes of cardiomyopathy, or pericardial tamponade; (2) a rapid reduction in serum potassium or calcium, causing decreased contractile force; (3) autonomic neuropathy (particularly in diabetic patients); (4) sepsis; (5) occult hemorrhage (e.g., retroperitoneal hemorrhage after femoral vein catheterization); (6) ingestion of food prior to or during dialysis, resulting in splanchnic vasodilation; and (7) use of antihypertensive medications on the day of dialysis.

To reduce the incidence of hypotension, a bicarbonate-buffered dialysate with a sodium concentration of approximately 140 mEq/L should be used and the rate and extent of ultrafiltration monitored carefully. Antihypertensive medications should be avoided immediately prior to hemodialysis. Management of hypotension includes reducing the ultrafiltration rate, placing the patient in the Trendelenburg position, and administering saline through the arteriovenous access. Oxygen administration may be useful in patients with ischemic heart disease.

Hyperparathyroidism

Secondary hyperparathyroidism is common in dialysis patients. Hyperphosphatemia caused by failure to excrete dietary phosphates plays a major role in the pathogenesis of secondary hyperparathyroidism. Hyperphosphatemia results in a reduction in plasma-ionized calcium because of a direct complexing of calcium with phosphates; the low calcium is a potent stimulus for parathyroid hormone (PTH) secretion. In the presence of significant chronic renal failure, there is also decreased activity of proximal renal tubular 1α-hydroxylase activity, leading to limited production of 2,25-dihydroxyvitamin D_3, the active vitamin D metabolite. This results in decreased calcium absorption from the GI tract; 1,25-dihydroxyvitamin D_3 also directly suppresses PTH secretion. Unfortunately, many dialysis patients have a chronically elevated serum phosporus and an increased serum calcium-phosphate product. Morbidity in patients with a product above 60 mg^2/dL^2 includes vascular and soft tissue calcification. Calcification can occur in coronary and peripheral arteries and in the myocardium, and extensive valvular calcification may result in impaired valve function. Finally, it has been proposed that PTH or excessive intracellular calcium can directly impair myocardial function.

Prevention or treatment of secondary hyperparathyroidism is based on correcting hyperphosphatemia. This can be accomplished only if the patient adheres to a diet containing less than 800 mg phosphorus per day; but even so, patients generally need to take phosphate binders with meals, thus limiting the absorption of dietary phosphate. Calcium carbonate or acetate are the preferred agents because they avoid the risk of aluminum toxicity (including osteomalacia, anemia, and encephalopathy), which can occur with aluminum-based binders. If hypocalcemia persists and if there is evidence of secondary hyperparathyroidism, administration of 1,25-dihydroxyvitamin D_3 can be used therapeutically. Subtotal parathyroidectomy is rarely needed to treat severe hyperparathyroidism.

ISCHEMIC HEART DISEASE

Ischemic heart disease is common in the ESRD population; at autopsy, acute myocardial infarctions have been found in over 25 percent of dialysis patients. There are many risk factors for coronary artery disease in this patient population (see above). In addition, factors such as hypertension, ECV overload, anemia, hypotension and hypoxia during hemodialysis—as well as flow through the arteriovenous fistula—can adversely affect the balance between myocardial oxygen supply and demand. For example, myocardial uptake of thallium-201 during exercise was abnormal in more than 50 percent of dialysis patients, many of whom were asymptomatic. Of note, stress-thallium testing has been found to have a positive predictive value of only approximately 70 percent in ESRD patients; some false-positive thallium tests may relate to a decrease in the lateral-to-septal count density ratio reported in patients with renal failure without coronary artery disease. Myocardial ischemia and angina pectoris occur, however, in some dialysis patients with no evidence of significant narrowing of coronary arteries on angiography.

The pharmacologic management of angina pectoris in uremic patients is similar to that used in patients without kidney disease. Nitrates, beta blockers, and calcium channel blockers are well tolerated except that drug dosages often have to be reduced and antianginal medications may need to be withheld before dialysis to avoid hypotension during dialysis. Exercise tolerance has been shown to improve in hemodialysis patients with coronary artery disease following a correction of the anemia with erythropoietin. It is recommended that the hematocrit be maintained above 30 percent by administering erythropoietin. Attention must also be paid to repleting iron stores adequately; transfusion of packed red blood cells may occasionally be needed.

The decision to perform cardiac angiography should be based upon the same criteria as for patients without kidney disease. In patients who are not treated by dialysis, the dose of contrast dye should be minimized to prevent loss of residual renal function, and fluid intake (including the dose of contrast media) should be minimized to prevent fluid overload. In those treated by dialysis, it is not necessary to dialyze patients immediately after angiography unless there is concern about intravascular volume expansion or heart failure.

Coronary artery bypass surgery (CABG) has been shown to lead to a mortality of approximately 10 percent in dialysis patients as well as an increased perioperative morbidity. Despite these high numbers, an improved quality of life post-CABG has made this a frequent operation in dialysis patients. Interestingly, percutaneous transluminal coronary angioplasty (PTCA) has been found to be associated with an unacceptably high rate of restenosis despite good initial angiographic success; the use of PTCA should therefore be reserved only for dialysis patients who are not candidates for CABG. Dialysis should be performed just before cardiac surgery to optimize ECV status and avoid hyperkalemia during surgery from low blood pressure/perfusion, hemolysis, or the use of potassium during cardioplegia.

CONGESTIVE HEART FAILURE

Congestive heart failure accounts for 20 to 30 percent of the mortality in ESRD patients. Echocardiograms reveal a high prevalence of hypertrophic cardiomyopathy characterized by left ventricular hypertrophy (LVH), asymmetrical septal hypertrophy, and/or impaired contractility (uremic cardiomyopathy) as well as dilated cardiomyopathy. Concentric left ventricular hypertrophy occurs in patients with current or previous systemic hypertension. Risk factors for myocardial dysfunction in dialysis patients include hypertension, persistent ECV expansion, anemia, the arteriovenous fistula, ischemic heart disease, metabolic acidosis, electrolyte disturbances (hyperkalemia, hypocalcemia), hyperparathyroidism, and possibly the uremic state itself. Hemodialysis can improve cardiac function dramatically, presumably by controlling hypertension, correcting volume overload, removing uremic toxins, and normalizing blood pH and electrolyte levels.

The prevention of heart failure in dialysis patients requires strict control of hypertension and the ECV. The influence of the arteriovenous fistula can be tested by occluding it and determining if the heart rate slows (Branham's sign). If this is positive, revision of the fistula may be required to decrease excessive blood flow and the oxygen demand of the heart.

Management of heart failure includes bed rest and oxygen therapy plus removal of excess fluid by ultrafiltration. Other causes of heart failure—such as myocardial infarction, arrhythmias, or infective endocarditis—must be excluded. If digitalis is used, appropriate adjustment of dosage plus frequent monitoring of plasma levels are necessary. Left ventricular hypertrophy has been found to improve after correction of anemia in dialysis patients by recombinant human erythropoietin (EPO). Complete normalization of LVH is, however, uncommon.

PERICARDIAL DISEASE

Pericarditis in dialysis patients may be related to inadequate removal of uremic toxins that are coincident to viral or bacterial infections, tuberculosis, or systemic lupus erythematosus or to drugs such as minoxidil. The clinical incidence of pericarditis has decreased from 50 percent in the predialysis era, when it was regarded as a preterminal event, to 5 to 20 percent today. Pericarditis appears to occur less frequently in peritoneal dialysis patients compared to hemodialysis patients, possibly due to a higher clearance of middle molecules by peritoneal dialysis.

The primary treatment for dialysis-associated pericarditis is intensive dialysis (e.g., daily hemodialysis for 1 to 2 weeks).

Heparin is eliminated to avoid pericardial hemorrhage and tamponade, which may be heralded by severe hypotension during dialysis, especially in the absence of volume depletion. Besides intensive dialysis, oral or intrapericardial administration of corticosteroids and indomethacin have been used, but there is limited evidence for the efficacy of these medications. The predominant effect of indomethacin is to reduce fever.

Treatment of a large pericardial effusion by intensification of dialysis may result in improvement, but if there is no improvement or if hemodynamic compromise occurs, surgical drainage of the pericardial effusion via subxiphoid pericardiotomy and creation of a pericardial window is the preferred procedure. Although pericardiectomy has been regarded as definitive in patients with pericarditis and a clinically significant effusion, this more invasive procedure is usually unnecessary. Constrictive pericarditis is rare in dialysis patients, even in patients with pericarditis. Small pericardial effusions are found in 15 to 20 percent of stable, asymptomatic dialysis patients.

INFECTIVE ENDOCARDITIS

The incidence of endocarditis may be as high as 3 to 5 percent in dialysis patients. Predisposing factors include a uremia-associated immunocompromised state, repeated puncture of the arteriovenous fistula, and infection of the arteriovenous access. In addition, calcific aortic or mitral valvular abnormalities related to hyperparathyroidism are present in approximately 10 percent of patients and can serve as a nidus for infection. Note that bacteremia occurs in approximately 10 to 20 percent of dialysis patients. *Staphylococcus aureus* is the most frequent organism, although *Staphylococcus epidermidis*, *Streptococcus viridans*, enterococci, and gram-negative organisms can also occur. The diagnosis of infective endocarditis may be difficult in uremic patients because of the frequency of bacteremia and because systolic and diastolic murmurs are common. Repeated blood cultures, physical examination, and an echocardiographic assessment are mandatory when infective endocarditis is suspected. Treatment of infective endocarditis consists of 4 to 6 weeks of parenteral antibiotics (see Chap. 21).

CARDIAC ARRHYTHMIAS

Risk factors for cardiac arrhythmias in dialysis patients include ischemic heart disease and calcification of the conduction system (from secondary hyperparathyroidism and/or pericarditis), hemodialysis-associated hypotension, dialysis-induced acid-base and electrolyte disturbances (hyper- and hypokalemia, hyper- and hypocalcemia, hypermagnesemia), and hypoxemia. Dialysis patients receiving digitalis have an excessive risk for atrial and ventricular arrhythmias during dialysis because of rapid shifts of potassium. On the other hand, hyperkalemia is believed to be responsible for a significant fraction of the 10 percent death rate from cardiac arrest in dialysis patients.

RENAL FUNCTION IN HEART FAILURE

In heart failure, enhanced sympathetic activity and activation of the renin-angiotensin-aldosterone axis enhance salt reabsorption, while excess vasopressin augments water retention. These responses cause ECV and plasma volume expansion, leading to increased end-diastolic ventricular volume plus peripheral and pulmonary edema. Circulating atrial natriuretic peptide (ANP) levels are increased in heart failure and may modulate the antinatriuretic effects caused by sympathetic and renin system activation. Excessive vasopressin-induced water reabsorption can cause hyponatremia, which is a poor prognostic indicator.

Renal vasoconstriction in heart failure patients can be sufficiently severe to cause prerenal azotemia, which is characterized by an elevated BUN-to-creatinine ratio of greater than 10 to 1, a low urine sodium concentration or low fractional excretion of sodium, and a urinalysis without cellular elements or casts (the presence of granular or cellular casts indicates structural renal damage). Factors that precipitate or exacerbate renal failure include excessive diuresis and worsening cardiac function. When renal perfusion is reduced (e.g., by heart failure), the glomerular filtration rate (GFR) becomes dependent on angiotensin II–induced efferent glomerular arteriolar constriction. Consequently, ACE inhibitors can markedly decrease the GFR; ACE inhibitors and other antihypertensives (e.g., hydralazine) can also reduce glomerular filtration by causing systemic hypotension and reducing renal perfusion pressure. These effects are often reversible when the doses are reduced or the drugs are discontinued. The management of renal failure in heart failure patients is aimed primarily at improving cardiac function.

RENAL FAILURE FOLLOWING CARDIAC CATHETERIZATION

The risk of renal damage following radiocontrast dye rises with diabetes mellitus, multiple myeloma, preexisting renal failure, volume depletion, and heart failure—also with large amounts of contrast dye. The renal failure after contrast dye is typically brief (approximately 5 to 7 days) unless there is substantial underlying renal damage. In high-risk patients, the least amount of contrast dye should be administered and, whenever possible, noninvasive techniques should be used to assess ventricular function and anatomy. ECV expansion with half normal saline (1 mL/kg/h for 12 h prior to and 12 h following the studies) reduces the incidence of contrast nephropathy.

Atheroembolic nephropathy usually occurs in elderly patients with erosive aortic atherosclerosis; they develop cholesterol emboli to the kidneys during arterial catheterization and their serum creatinine rises sharply; it usually does not return to basal levels. At times, renal failure may worsen slowly, leading to ESRD. Hypertension due to activation of the renin-angiotensin system may be present and the urinalysis typically does not reveal casts. Atheroembolization to other locations such as the eyes (cholesterol plaques seen by fundoscopy), pancreas (pancreatitis), and skin (livedo reticularis or gangrene) may be present and suggest the diagnosis. Occasionally, immunologic activation may occur and be signified by an active urinary sediment with hematuria and cellular casts, hypocomplementemia, eosinophilia, and a high sedimentation rate. Biopsy of an affected organ (e.g., skin, kidney with atheroemboli) can help establish the diagnosis. There is no specific treatment.

In contrast to atheroembolic renal disease, thromboembolic renal arterial disease (e.g., in patients with atrial fibrillation or after myocardial infarction) often causes renal infarction. Such patients may present with flank pain, proteinuria, and hematuria; the serum lactate dehydrogenase level is increased and renal failure leads to an increased serum creatinine, particularly if both kidneys are affected. A radioisotope scan or renal arteriography will confirm the diagnosis. Therapy includes anticoagulation, thrombolysis, and—rarely—surgical removal of the thrombus.

CARDIAC DRUGS IN RENAL FAILURE

Many drugs used in the treatment of cardiovascular diseases undergo significant clearance by the kidneys. In order to avoid toxic side effects, the doses of these drugs must be modified depending on the level of the patient's GFR (the best estimate of kidney function). The volume of distribution of digoxin is reduced 30 to 50 percent in ESRD patients, so the loading dose of digoxin should be reduced and the maintenance digoxin dosage should also be decreased, because its primary route of elimination is by glomerular filtration of unmetabolized digoxin. Because of the changes in pharmacokinetics, only general guidelines for maintenance dosages are available: 0.0625 to 0.125 mg digoxin every other day usually results in a therapeutic plasma level, but this should not be assumed, and regular monitoring of the plasma digoxin level is required.

As renal function decreases, so does the ability to eliminate unmetabolized drugs such as procainamide; the half-life increases from approximately 3.5 to 16 h in renal failure. In addition, the half-life of N-acetylprocainamide (NAPA), an active metabolite of procainamide primarily excreted by the kidneys, is markedly prolonged in renal failure. Consequently, maintenance procainamide doses should be reduced or the intervals between dosages prolonged; close monitoring of plasma levels of both procainamide and NAPA is necessary. Since both compounds are removed by dialysis, a dose of procainamide should be administered after hemodialysis.

The beta blockers atenolol and nadolol are eliminated primarily by the kidneys, so a dose reduction of 50 to 75 percent is necessary for CRF patients. Like procainamide, these drugs should be given after dialysis, because a significant fraction is removed by the dialysis procedure. In dialysis or predialysis patients, nitroprusside can result in the accumulation of thiocyanate and neurologic toxicity, manifest as confusion, hyperreflexia, and seizures. Consequently, the dose of nitroprusside should be minimized and the drug given for as short a period as possible.

In dialysis patients, most doses of ACE inhibitors should be reduced by approximately 50 percent, because they and their metabolites are excreted by the kidney. In patients who are not treated by dialysis, these drugs have two major types of toxic effects. First, they can cause hyperkalemia by inhibiting angiotensin-stimulated aldosterone release, resulting in decreased potassium excretion. Second, they can cause rapid loss of renal function in patients with renal artery stenosis or other conditions associated with activation of the renin-angiotensin system, including congestive heart failure. The mechanism for the decrease in GFR is inhibition of angiotensin-induced constriction of the efferent glomerular arteriole. This dilates the arteriole and decreases the hydrostatic pressure across the glomerular capillary wall to reduce filtration.

The use of cyclosporine in heart transplant recipients is often associated with loss of renal function; some patients progress to ESRD. Cyclosporine constricts both afferent and efferent glomerular arterioles, resulting in a reduced GFR. In addition, proximal tubular injury—with vacuolar changes, inclusion bodies, and giant mitochondria—has been noted. Cyclosporine usage has also caused hyperkalemia and renal tubular acidosis; these effects are usually reversible if the dose is reduced or the drug is discontinued.

SUGGESTED READING

Ansari A, Kaupke CJ, Vaziri ND, et al: Cardiac pathology in patients with end-stage renal disease maintained on hemodialysis. *Int J Artif Organs* 1993; 64:560–564.

Batiuk TD, Kurtz SB, Oh JK, et al: Coronary artery bypass operation in dialysis patients. *Mayo Clin Proc* 1991; 66:45–53.

Brown JH, Vites NP, Testa JH, et al: Value of thallium myocardial imaging in the prediction of future cardiovascular events in patients with end-stage renal failure. *Nephrol Dial Transplant* 1993; 8:433–437.

Converse RL, Jacobsen TN, Toto RD, et al: Sympathetic overactivity in patients with chronic renal failure. *N Engl J Med* 1992; 327:1912–1918.

Henrich WL, Hunt JM, Nixon JV: Increased ionized calcium and left ventricular contractility during hemodialysis. *N Engl J Med* 1984; 310:19–23.

Hruska KA, Teitelbaum SL: Renal osteodystrophy. *New Engl J Med* 1995; 33:166–174.

Parfrey PS, Harnett JD, Barre PE: The natural history of myocardial disease in dialysis patients. *J Am Soc Nephrol* 1991; 2:2–12.

Pastan SO, Mitch SE. The kidney in heart disease, in Alexander W, Schlant RC, Fuster V, O'Rourke RA, Roberts R, Sonnenblick EH (eds): *Hurst's The Heart*, 9th ed. New York, McGraw-Hill, 1998:2413–2423.

Radermacher J, Koch KM: Treatment of renal anemia by erythropoietin substitution: the effects on the cardiovascular system. *Clin Nephrol* 1995; 44(suppl 1):S56–S60.

Rostand SG, Brunzell JD, Cannon RO, Victor RG: Cardiovascular complications in renal failure. *J Am Soc Nephrol* 1991; 2:1053–1062.

Solomon R, Werner C, Mann D, et al: Effects of saline, mannitol, and furosemide on acute decreases in renal function induced by radiocontrast agents. *N Engl J Med* 1994; 331:1416–1420.

Thadhani RI, Camargo CA Jr: Atheroembolic renal failure after invasive procedures: natural history based on 52 histologically proven cases. *Medicine* 1995; 74:350–358.

THE HEART AND NONCARDIAC DRUGS, ELECTRICITY, POISONS, AND RADIATION

Andrew L. Smith

This chapter deals with the effects of a variety of noncardiac drugs, electricity, poisons, and radiation on the heart.

NONCARDIAC DRUGS

Chemotherapeutic Agents

Chemotherapeutic agents may result in acute or chronic cardiovascular toxicity. Cardiomyopathy has generally been associated with the anthracyclines (doxorubicin, daunorubicin, epirubicin, idarubicin, mitoxantrone). Cyclophosphamide has been associated with reversible systolic dysfunction and occasionally hemorrhagic myocarditis. Interleukin 2 and interferon alpha may cause hypotension and rarely cardiomyopathy. 5-Flurouracil has been associated with coronary vasospasm. Amsacrine and paclitaxel have been associated with cardiac arrhythmias.

ANTHRACYCLINES

Doxorubicin (Adriamycin) and daunorubicin (Cerubidine) cause dose-related cardiotoxicity possible related to free radical damage. Acute cardiac toxicity may occur after the initial doses and chronic cardiotoxicity within months of therapy; late cardiac toxicity, with delayed development of systolic dysfunction, is becoming increasingly recognized. Diffuse left ventricular dysfunction (cardiomyopathy) occurs in about 7 percent of patients receiving 550 mg/m^2 and in 18 percent after 700 mg/m^2 of doxorubicin. Once cardiomyopathy develops, treatment strategies are similar to those with other forms of systolic dysfunction (Chap. 1). Further cardiac toxins are to be avoided. The clinical course varies from fulminant heart failure to gradually progressive deterioration. Some patients have reversible systolic dysfunction. Serial echocardiography is only of moderate sensitivity and specificity in detecting cardiotoxicity during therapy. Endomyocardial biopsy provides more definitive detection but is not widely used clinically. Dexrazoxane, an iron-chelating agent, is approved as a preventive strategy in women with breast cancer after cumulative doses of doxorubicin 300 mg/m^2.

Psychotropic Agents

TRICYCLIC ANTIDEPRESSANTS

Tricyclic antidepressants have potentially serious cardiovascular effects, including tachycardia, orthostatic hypotension, electrocardiographic (ECG) changes, and depression of ventricular function. These drugs have electrophysiologic properties similar to those of the type IA antiarrhythmics. These drugs are contraindicated in the recovery phase following myocardial infarction. While they may be indicated in the treatment of severely depressed patients, the threshold for their use should rise as the severity of heart disease increases or when there is QT prolongation.

Tricyclic antidepressant overdose is lethal in approximately 2 percent of patients and is generally related to cardiac complications. Initial clinical status and serum drug levels are not predictive of prognosis. QRS prolongation is a sign of toxicity but is an insensitive finding. Gastric lavage, repeat dosing of activated charcoal, and sodium bicarbonate therapy are appropriate treatment strategies. Type I antiarrhythmics should not be used for cardiac rhythm disturbances. Sodium bicarbonate is the initial therapy for ventricular dysrhythmias. Hypotension, which may be refractory to volume loading and bicarbonate therapy, should be treated with vasopressors such as norepinephrine, phenylephrine, or vasopressor doses of dopamine.

Other Antidepressants

The selective serotonin reuptake inhibitors (SSRIs) have rarely been associated with orthostatic hypotension and bradycardia. These drugs may affect the cytochrome P450 system and interfere with other cardiovascular drugs. Case reports of cardiac toxicity are rare.

The monoamine oxidase (MAO) inhibitors have little effect on cardiac conduction or myocardial contractility. Orthostatic hypotension is common. The major concern with these agents is interaction with tyramine-containing substances, resulting in hypertensive crisis.

Lithium may suppress automaticity, particularly of the sinus node. Electrocardiographic changes may simulate hypokalemia, including T-wave inversion, prominent U waves, and QT prolongation. Overdose with lithium may result in sever bradycardia. A low anion gap suggests the presence of lithium toxicity.

Phenothiazine antipsychotic agents including chlorpromazine (Thorazine) and thioridazine (Mellaril) can produce tachycardia, postural hypotension, T-wave changes, QT prolongation, and bundle branch block. Sudden death, presumed to be due to cardiac arrhythmias, has occurred.

Noncardiac Drugs Causing Torsades De Pointes

Torsades de pointes has been associated with tricyclic antidepressants, tetracyclic antidepressants, phenothiazines, haloperidol, chloral hydrate, and probucol. Other implicated drugs include erythromycin, trimethoprime-sulfamethoxazole, terfenadine, astemiozole, cisapride, and pentamidine. Arsenic, organophosphates, and liquid protein diets have also been associated.

Methylxanthines and Beta-Adrenergic Agonists

Caffeine, theophylline, and terbutaline may cause sinus tachycardia and may exacerbate the tendency for cardiac arrhythmias. Controversy exists over the safety of long-term use of these agents in patients with cardiovascular disease.

Ergotamine and Methysergide

Ergotamine and methysergide have similar chemical structures. Valvular heart disease has been reported with both agents. Pericardial, pleural, or retroperitoneal fibrosis may occur. Methysergide is more strongly associated with regurgitant valvular lesions. Therapy should be discontinued if a new murmur is detected. Regression of valvular lesions may occur, although valve replacement is occasionally required. Ergotamine is a vasoconstrictor and may induce coronary artery spasm.

Chloroquine

Chloroquine and hydroxychloroquine can cause skeletal and, rarely, heart muscle disease. When cardiac involvement occurs, features of restrictive cardiomyopathy are most common. Acute chloroquine poisoning results in hypotension, tachycardia, and prolongation of the QRS and is often fatal.

Oral Contraceptive Agents

The cardiovascular risk profile of newer-generation oral contraceptive agents is favorable. Thrombosis resulting in myocardial infarction or deep venous thromboembolism may occur; the risk is significantly increased in smokers. Hypertension is rare. The risk of stroke in otherwise healthy women is only minimally increased. Women who use oral contraceptives should be advised to avoid smoking.

Anabolic Steroids

Illicit use of androgens is a problem in competitive athletes and body builders. Data on human toxicity are limited. Stanozolol and nandrolone have been associated with marked lipid abnormalities and an increase in coronary atherosclerosis. These agents may also cause left ventricular hypertrophy and hypertension.

Cocaine

The cardiovascular complications of cocaine include sudden death, acute myocardial infarction, accelerated atherosclerosis, dilated cardiomyopathy, acute reversible myocarditis, acute severe hypertension, acute aortic dissection, and stroke. Endocarditis may result form intravenous drug use. Electrocardiographic abnormalities include sinus tachycardia, PVCs, ventricular tachycardia, torsades de pointes, prolongation of the QT interval, diffuse ST-segment changes, and ventricular fibrillation.

Chest pain is the most common reason for cocaine users to seek medical attention. The evaluation of cocaine-related chest pain is difficult. Approximately 6 percent of patients presenting to emergency departments with cocaine-related chest pain have myocardial infarction. These patients are often young men without other risk factors for coronary artery disease except for tobacco smoking. The quality or duration of the chest discomfort is often not predictive of infarction. Because young patients often have early repolarization patterns on ECG, ST elevation in leads V_1 to V_3 may be confused with acute infarction. Patients often require monitoring for 12 h until enzymes have excluded infarction.

Treatment strategies for cocaine-induced myocardial ischemia have been developed. Patients presenting with anxiety, tachycardia, or hypertension may respond well to benzodiazepines. Nitroglycerin may reverse coronary vasoconstriction induced by cocaine. Aspirin may prevent thrombus formation. Patients not responding to these measures may benefit from phentolamine or from calcium-channel-blocker therapy with verapamil. Beta-adrenergic antagonists should generally be avoided due to unopposed alpha-mediated vasoconstriction. Combined alpha and beta blockade with labetalol has been used to treat tachyarrhythmias but is not an accepted therapy for myocardial ischemia. In documented cocaine-related myocardial infarction, thrombolytic therapy is highly effective. However, emergent coronary angiography may be necessary to distinguish the patient with acute infarction from one with ST elevation due to early repolarization. Cocaine-related cardiac rhythm disturbances are best managed initially with benzodiazepines. Intravenous sodium bicarbonate and magnesium may be beneficial. Lidocaine should be used cautiously because of concerns regarding a lower seizure threshold. Beta blockade with propanolol or esmolol should be avoided.

Appetite Suppressants

Flenfluramine and dexflenfluramine were recently withdrawn from the market by the FDA because of evidence linking these drugs to heart-valve regurgitation. The incidence of overt valvular disease is low, with less than one case per 1000 patient-years. Risk appears to correlate with duration and possibly with dose of agents used. These drugs have also been associated with primary pulmonary hypertension.

ELECTRICITY

Environmental Accidents

The immediate cardiac effect of lightning or electrical equipment injury may be asystole or ventricular fibrillation. Cardiac arrest may also result form apnea and hypoxia. Atrial and ventricular arrhythmias, conduction abnormalities, and left ventricular dysfunction may occur. Cardiac abnormalities occur from direct myocardial injury or central nervous system injury with intense catecholamine release. Hypertension and tachycardia may be managed with beta-blocking agents.

Cardiopulmonary resuscitation should be continued for a prolonged period after apparent death from lightning, since late recovery may occur. In lightning strikes involving multiple victims, attention should first be directed to those who are "apparently dead," since lightning victims with vital signs generally survive without immediate medical attention, whereas those without vital signs can recover after prolonged resuscitation.

Electroconvulsive Therapy

Electroconvulsive therapy (ECT) may produce cardiac arrhythmias and electrocardiographic changes during the first few minutes after the shock. ECT produces brief, intense stimulation of the central nervous system. Cardiovascular complications may result from this stimulation or from the drugs used to modify the response. Patients with coronary artery disease should be pretreated with a beta blocker to blunt tachycardia and hypertension and reduce the frequency of ventricular ectopic beats. Patients with cardiac pacemakers can safely undergo ECT.

POISONS

Plants

Many plants contain poisonous substances, some of which produce cardiac effects similar to those of digitalis toxicity. Others can produce myocardial depression, cardiac arrhythmias, hypotension, circulatory collapse, and death. Plants with cardiac glycoside effects include foxglove, oleander, lily of the valley, Christmas rose, wallflower, and milkweed.

Snakes and Scorpions

Snake venoms affect the coagulation system, cellular components of the blood, endothelium, nervous system, and heart. Cardiac arrhythmias, severe hypotension, and cardiac arrest may occur. Multiple pulmonary emboli may be seen in patients who survive 12 h or longer. Scorpion venom may cause hypertension, myocardial infarction, arrhythmias, conduction disturbances, and myocarditis.

Marine Toxins

Scorpion fish cause envenomation that may result in rhythm disturbances and heart failure. Ingestion of pufferfish may cause severe bradycardia and cardiovascular collapse. Stingray venom contains phosphodiesterases and may rarely cause cardiac rhythm disturbances.

Halogenated Hydrocarbons

These substances are used in fire extinguishers, solvents, refrigerants, pesticides, plastics, paints, and glues. They can suppress myocardial contractility and produce arrhythmias and sudden death.

Carbon Monoxide

Carbon monoxide poisoning produces myocardial ischemia usually manifest by ST-segment and T-wave changes and atrial and ventricular arrhythmias. Extensive myocardial necrosis and cardiomyopathy can occur.

RADIATION

Radiation to the mediastinum can affect the pericardium, myocardium, endocardium, valves, and capillaries of the heart. Radiation may cause acute pericarditis, chronic pericarditis, and pericardial constriction. Pericardial involvement is most frequent 4 to 6 months after radiation therapy. Clinically important myocardial dysfunction related to radiation generally occurs in combination with pericardial disease. Myocardial fibrosis may result in diastolic dysfunction and, less commonly, systolic dysfunction. Radiation-induced valvular heart disease is rare but usually involves the aortic or mitral valves. Premature coronary artery disease may occur years after radiation therapy.

Radiation may result in fibrosis of the nodal and infranodal pathways. Right bundle branch block is especially common; complete heart block is rare.

SUGGESTED READING

Arsenian MA: Cardiovascular sequelae of therapeutic thoracic radiation. *Prog Cardiovasc Dis* 1991; 33:299–311.

Carleton SC: Cardiac problems associated with electrical injury. *Cardiol Clin* 1995; 13:263–277.

Devereux RB: Appetite suppressants and valvular heart disease (editorial). *N Engl J Med* 1998; 339:765–766.

Ellis MD: Poisonous plants, In: Ellis MD (ed): *Dangerous Plants, Snakes, Arthropods, and Marine Life*. Hamilton, IL: Hamilton Press; 1975, pp. 3–81.

Frishman WH, Sung HM, Yee HCM, et al: Cardiovascular toxicity with cancer chemotherapy. *Curr Probl Cardiol* 1996; 21:225–288.

Glassman AH, Roose SP, Bigger JT: The safety of tricyclic antidepressants in cardiac patients—risk benefit reconsidered. *JAMA* 1993; 269:2673–2675.

Kloner RA, Hale S, Alker Rezkalla S: The effects of acute and chronic cocaine use on the heart. *Circulation* 1992; 85:407–419.

Rosenberg L, Begaud B, Bergan U, et al: What are the risks of third generation oral contraceptives? *Hum Reprod* 1996; 11:687–693.

Shan K, Lincoff AM, Young JB: Anthracycline-induced cardiomyopathy. *Ann Intern Med* 1996; 125:47–58.

Smith AL, Schlant RC: Effect of noncardiac drugs, electricity, poisons, and radiation on the heart. In: Alexander RW, Schlant RC, Fuster V, O'Rourke RA, Roberts R, Sonnenberg EH (eds): *Hurst's The Heart*, 9th ed. New York, McGraw-Hill, 1998:2153–2166.

CHAPTER

32

NEOPLASTIC HEART DISEASE

Robert J. Hall

Tumors of the heart are uncommon and present in protean ways, challenging the acumen of the physician. Cardiac tumors may be primary in origin (see Table 32-1) or secondary from proximal or remote sites, and they are expressed in a limited number of ways. The general manifestations of these tumors are listed in Table 32-2, by site of cardiac involvement.

TABLE 32-1

MOST COMMON TUMORS AND CYSTS OF THE HEART AND PERICARDIUM

Benign	Percentage	Malignant	Percentage
Myxoma	24.4	Angiosarcoma	7.3
Pericardial cyst	15.4	Rhabdomyosarcoma	4.9
Lipoma	8.4	Mesothelioma	3.6
Papillary		Fibrosarcoma	2.6
fibroelastoma	7.9	Lymphoma	1.3
Rhabdomyoma	6.8	Others	3.7
Fibroma	3.2		
Hemangioma	2.8		
Teratoma	2.6		
Mesothelioma of the AV node	2.3		
Bronchogenic cyst	1.3		
Others	1.5		

SOURCE: McAllister HA Jr. Fenoglio JJ Jr: *Tumors of the Cardiovascular System.* Washington, DC, Armed Forces Institute of Pathology, 1978.

TABLE 32-2

MANIFESTATIONS OF CARDIAC TUMORS
(CLASSIFICATION BY SITE OF INVOLVEMENT)

Pericardial Involvement
  Pericarditis, pain
  Pericardial effusion
  Cardiac enlargement on chest x-ray
  Arrhythmias (predominantly atrial)
  Tamponade (usually with bloody fluid)
  Cardiac constriction
Myocardial Involvement
  Arrhythmias, both atrial and ventricular
  Electrocardiographic changes (usually nonspecific)
  Conduction defects and AV block
  Cardiac enlargement on chest x-ray
  Congestive heart failure
  Coronary involvement: angina, infarction
Intracavitary Involvement
  Cavity obliteration
  Valvular obstruction or insufficiency, or both
  Embolic phenomena: systemic, neurologic, coronary
  Constitutional symptoms: fever, weight loss

PRIMARY TUMORS OF THE HEART

Cardiac Myxomas

Intracardiac myxoma is the most common benign tumor of the heart. Approximately 75 percent of atrial myxomas are located within the left atrium and arise from the interatrial septum. Usually, they are pedunculated and prolapse through the mitral valve during diastole. Myxomas occur most commonly in women from 30 to 60 years of age. Patients generally present with symptoms from a triad of manifestations—constitutional, embolic, and obstructive. *Systemic manifestations*, which are noted in 90 percent of patients, consist of weight loss, fever, anemia, elevated sedimentation rate, and elevated immunoglobulin concentration (usually IgG).

In 50 percent of patients, *arterial emboli* occur, involving the brain (half of all emboli), heart, kidneys, extremities, and aortic bifurcation. Histologic examination of all surgically re-moved emboli can aid in diagnosing an unsuspected intra-cardiac myxoma. Emboli should arouse suspicion of myxoma in young patients, especially when they are in sinus rhythm. With left atrial myxomas, *obstruction* of either the mitral valve or the pulmonary veins may occur, producing pulmonary venous and arterial hypertension with secondary right-sided heart failure. Symptoms include dyspnea, orthopnea, manifes-tations of acute pulmonary edema, hemoptysis, dizziness, and syncope; occasionally, sudden death occurs. Change of the patient's position may also produce symptoms.

In the physical examination of a patient with a left atrial myxoma, auscultation reveals a loud S_1 and an accentuated S_2, followed by an early diastolic sound. This sound, the "tumor plop," is produced by prolapse of the tumor through the mitral valve. An apical diastolic or systolic murmur or both is usually present. The electrocardiographic and radiographic findings resemble those seen with mitral valve disease; however, sinus rhythm usually is present. The two-dimensional echocardio-gram is diagnostic and demonstrates the location, origin, and movement of the intracardiac mass. Because cardiac cathe-terization and angiography increase the risk of tumor embolization, they are indicated only for diagnosis of con-comitant cardiac or coronary disease. Treatment for left atrial myxoma consists of prompt surgical resection of the tumor, together with a generous portion of the atrial septum from which it arises.

Right atrial myxomas occur with one-fifth the frequency of left atrial myxomas and are characterized by manifestations of systemic venous hypertension: a prominent jugular *a* wave, hepatomegaly, ascites, and edema. There is a loud early sys-tolic sound at the lower sternal border, and systolic and diastolic murmurs at the tricuspid valve area may mimic a peri-cardial friction rub. As in cases of left atrial myxoma, echocardiography is helpful in the diagnosis, and prompt sur-gical removal is recommended.

Other Benign Primary Cardiac Tumors

Rhabdomyoma, which is the most common cardiac tumor in children, frequently accompanies tuberous sclerosis. Located in the myocardium, rhabdomyomas usually cause arrhythmias and obstructive manifestations. *Fibromas* may cause sudden death and, like *lipomas*, may grow to a large size.

Malignant Primary Tumors of the Heart

Almost all primary malignant tumors of the heart are *sarcomas*—most frequently *angiosarcomas*, which usually originate in the right atrium or pericardium. One-fourth of all angiosarcomas produce valvular obstruction and manifest right-sided heart failure and pericardial tamponade with hemorrhagic pericardial fluid. Tumor excision, radiation therapy, and chemotherapy may offer some relief of symptoms.

Tumors of the Pericardium

Since 75 percent of all *pericardial cysts* are asymptomatic, these "tumors" usually are found coincidentally on chest x-ray. They most often reside in the right cardiophrenic angle and are distinguished from solid tumors by echocardiography and by computed tomography (CT) scanning. *Mesothelioma*, a malignant tumor of the pericardium, may mimic several conditions, including pericarditis, constrictive pericardial disease, and venal caval obstruction. Aspiration and histologic examination of the usually bloody pericardial fluid is often diagnostic. The prognosis for patients with mesothelioma is poor. Surgical excision is rarely possible, and radiation and chemotherapy produce only temporary improvement.

SECONDARY TUMORS OF THE HEART

Metastatic tumors of the heart or pericardium occur 20 to 40 times more frequently than primary tumors. Tumor invasion of the heart may occur by direct contiguous growth from adjacent structures or along the vena cava or pulmonary veins. Metastatic invasion may also take place by hematogenous or lymphatic channels. The heart is the site of metastatic tumor in 2 to 20 percent of patients with malignant tumors; bronchogenic and breast carcinoma are the most common sources. Cardiac involvement is also common in patients with leukemia and the lymphomas and is seen with Kaposi's sarcoma in some patients with acquired immunodeficiency syndrome (AIDS) (Chap. 29).

Secondary tumors of the heart may involve the pericardium, myocardium, endocardium, valves, or coronary arteries. *Pericardial involvement* results in pleuritic pain and a pericardial friction rub. Cardiac enlargement, signs of tamponade, and reduced QRS amplitude in the electrocardiogram may be evident. Electrical alternans may be present with serious tamponade. The echocardiogram and CT scan will reveal pericardial effusion and aid in identifying intracavitary and pericardial masses. Diastolic collapse of the right atrium and ventricle is highly specific for significant tamponade. *Myocardial involvement* results in ST and T wave changes in the electrocardiogram and often arrhythmias, including atrial flutter and fibrillation, ventricular ectopic rhythms, conduction disturbances, and even complete heart block. Widespread myocardial involvement may produce congestive heart failure. Cardiac damage and myocardial failure may also result from the effects of radiotherapy and chemotherapy. *Coronary artery involvement* can result from tumor embolization or external compression as well as from the consequences of radiotherapy. *Intracavitary tumor* may be seen when there is extension of a primary tumor, such as renal cell carcinoma, hepatocellular carcinoma, or uterine leiomyoma, any of which may spread along the inferior vena cava and into the right atrium, or bronchogenic carcinoma, which may extend into the left atrium. Magnetic resonance imaging (MRI) and CT scanning provide a global view of the anatomy and reveal information about the location, extent, and attachment of a tumor. Successful surgical resection of intracavitary tumors has been reported.

DIAGNOSIS

Diagnosis of tumor involvement of the heart is fostered by a strong index of clinical suspicion. Cardiac enlargement, arrhythmias, chest pain, or features of congestive heart failure in any patient with a malignancy should arouse suspicion of cardiac metastases. Two-dimensional transthoracic and transesophageal echocardiography, CT scanning, and MRI provide highly diagnostic information. Cytologic examination of pericardial fluid and endomyocardial biopsy specimens may establish a histologic diagnosis.

TREATMENT

Depending on cytologic type, the treatment of choice for cardiac tumors is *radiation therapy*, with or without *systemic chemotherapy*. Pericardial effusion with tamponade requires urgent *pericardiocentesis*, either by needle aspiration or by limited subxiphoid surgical drainage. Recurrent malignant pericardial effusions respond to intrapericardial administration of various agents such as fluorouracil, radioactive gold, nitrogen mustard, and tetracycline. Persistent reaccumulation of pericardial fluid may require surgical creation of a *pericardial window*. An alternative nonsurgical approach consists of creation of a pericardial-pleural communication ("window") using a percutaneously introduced balloon catheter. Surgical removal of intracavitary obstructing secondary tumors may ameliorate symptoms and prolong life.

SUGGESTED READING

DeLoach JF, Haynes JW: Secondary tumors of the heart and pericardium: review of the subject and report of one hundred thirty-seven cases. *AMA Arch Intern Med* 1953; 91: 224–249.

Hall RJ, Cooley DA, McAllister HA Jr, et al: Neoplastic heart disease. In: Alexander RW, Schlant RC, Fuster V, O'Rourke RA, Roberts R, Sonnenblick EH (eds): *Hurst's The Heart*, 9th ed. New York, McGraw-Hill, 1998: 2295–2318.

Hurst JW, Hall RJ, Becker AE, et al: Neoplastic disease of the heart. In Hurst JW (ed): *Atlas of the Heart*. New York, McGraw-Hill, 1988: 13.1–13.14.

Kralstein J, Frishman W: Malignant pericardial diseases: Diagnosis and treatment. *Am Heart J* 1987; 113:785–790.

McAllister HA Jr: Primary tumors and cysts of the heart and pericardium. *Curr Probl Cardiol* 1979; 4:1–51.

McAllister HA Jr, Fenoglio JJ Jr: *Tumors of the Cardiovascular System*. Fascicle 15, Second Series, *Atlas of Tumor Pathology*. Washington, DC, Armed Forces Institute of Pathology, 1978.

McAllister HA Jr, Hall RJ, Cooley DA: Surgical pathology of tumors and cysts of the heart and pericardium. In Waller BF (ed): *Contemporary Issues in Surgical Pathology*, vol 12. New York, Churchill Livingstone, 1988.

Palacios IF, Tuzcu EM, Ziskind AA, et al: Percutaneous balloon pericardial window for patients with malignant pericardial effusion and tamponade. *Cathet Cardiovasc Diagn* 1991; 22:244–249.

Reeder GS, Khanderia BK, Seward JB, et al: Transesophageal echocardiography and cardiac masses. *Mayo Clin Proc* 1991; 66:1101–1109.

Salcedo EE, Cohen GI, White RD, et al: Cardiac tumors: diagnosis and management. *Curr Probl Cardiol* 1992; 17: 73–137.
Tsang TS, Freeman WK, Sinak LJ, et al: Echocardiographically guided pericardiocentesis: evolution and state-of-the-art technique. *Mayo Clin Proc* 1998; 73:647–652.

CHAPTER

33

HEART DISEASE DUE TO TRAUMA

Panagiotis N. Symbas

DEFINITIONS

The third leading cause of death in the United States and in most other industrialized nations is trauma. Injury to the heart or great vessels is the cause of death in a major number of trauma victims. The most common traumatic injuries to the heart are penetrating injuries from missiles and knives and blunt injuries from direct compressing or decelerating forces to the chest—usually suffered in vehicular accidents. Other types of cardiac injury include the iatrogenic injuries that occur during angioplasties, cardiac catheterization, cardiopulmonary resuscitation, and insertion of various catheters and pacemaker electrodes, and injuries due to ionizing radiation and electrical currents.

Penetrating wounds of the heart usually occur in the thorax and upper abdomen. These may cause injury to any of the cardiac structures—the cardiac wall, the atrial or ventricular septum, the cardiac valves, the coronary arteries, and/or the pericardium. Penetrating injury to the heart commonly causes cardiac tamponade or massive blood loss. In addition, such injury may lead to structural defects, such as the formation of aneurysms, septal defects, and fistulas between heart chambers and great vessels. Valvular and papillary muscle injury may result in acute valvular regurgitation; injury to coronary vessels may cause myocardial infarction, coronary arteriovenous fistula, or coronary artery aneurysm. Retained missiles or other foreign bodies may be embolized and other thrombotic events may occur; rhythm and conduction disturbances are not uncommon. Pericardial wounds may cause serofibrinous (usually), suppurative (occasionally), and constrictive (very rarely) pericarditis.

Nonpenetrating trauma also may cause injury to any of the cardiac structures, resulting in a variety of pathophysiologic changes. Contusion of the heart is the most common blunt injury. It can cause rhythm or conduction disturbances and can evolve to subsequent rupture of the free cardiac wall or septum or to aneurysm formation. Rupture of the cardiac valves, the free cardiac wall, or the cardiac septa may also result from non-penetrating trauma.

CLINICAL MANIFESTATIONS

The initial clinical presentation and course of cardiac injury depend on the type and site of injury and the degree of damage sustained. This may vary from no symptoms—particularly in patients sustaining blunt trauma—to immediate death, usually from cardiac tamponade or massive blood loss, from laceration or penetration of the free cardiac wall or coronary vessels, or from rhythm disturbances usually seen in blunt trauma. Delayed or residual sequelae—such as ventricular and atrial septal defects, shunts between cardiac chambers or great vessels, valvular regurgitation, endocarditis, and embolic phenomena—should be suspected and sought so that they can be recognized promptly and managed appropriately. Symptoms associated with arrhythmias and, rarely, symptoms of angina or congestive heart failure may occur. The physical examination varies, depending on the type of trauma and the time lapsed since the injury. Careful assessment of the vital signs should be made during the immediate postinjury period; this includes a search for evidence of hemodynamic instability and its causes. The cardiovascular examination should include a thorough search for evidence of cardiac tamponade, bleeding, or congestive failure resulting from myocardial damage, or valvular regurgitation. During the later postinjury period, the physical examination should be directed toward the detection of signs of residual damage from the cardiac injury—e.g., of valvular dysfunction, various shunts, or pericarditis.

DIAGNOSTIC EVALUATION

More than the usual diagnostic evaluation may be warranted. The extent of the evaluation depends on the patient's clinical condition and the time elapsed since the injury. An electrocardiogram is extremely useful in the evaluation of patients with suspected myocardial contusion. An echocardiogram is also very helpful in determining the presence of myocardial or valvular damage and in helping to document and qualify pericardial effusions. Measurements of cardiac isoenzymes should be obtained routinely in cases of suspected contusion. Occasionally, cardiac catheterization may be necessary for full assessment of the extent of damage before surgical repair can be undertaken. Patients exposed to major decelerating forces, such as those that occur in a motor vehicle accident, should be examined for aortic rupture. The blood pressure should be measured in both upper and lower extremities to help detect aortic injury. Chest x-rays are very helpful in suspected rupture of the aorta by demonstrating alteration of the mediastinal silhouette; aortography should be performed in all patients with such alteration.

NATURAL HISTORY AND PROGNOSIS

Unprecedented numbers of patients with traumatic injuries to the heart are now reaching the hospital alive, thanks to improved emergency facilities. About one-half of the victims of stab wounds to the heart survive long enough to reach the hospital, but the outlook for patients with projectile injuries is much more ominous. Advances in diagnostic, resuscitative, and surgical techniques now allow repair in a relatively large proportion of patients who reach the hospital alive. The long-term outlook for a given patient depends not only on the nature and degree of acute injury but also on the detection and appropriate management of residual or delayed sequelae of the trauma.

TREATMENT

The care of these patients must be individualized according to the nature of the injury. For penetrating wounds, prompt diagnosis of the clinical situation, with appropriate resuscitative and surgical intervention, is necessary. For myocardial contusion, the treatment is usually symptomatic and basically like that for myocardial infarction, with appropriate periods of bed rest and gradual ambulation. Monitoring for arrhythmias is mandatory in the early stages, with pharmacologic therapy given when necessary.

SUGGESTED READING

Baxter BT, Moore EE, Moore FA, et al: A plea for sensible management of myocardial contusion. *Am J Surg* 1989; 158:557–562.

Freshman SP, Wisner DH, Weber CJ: 2-D echocardiography: emergent use in the evaluation of penetrating trauma. *J Trauma* 1991; 31:902–906.

Rozycki GS, Feliciano DV, Schmidt JA, et al: The role of surgeon-performed ultrasound in patients with possible cardiac wounds. *Ann Surg* 1996; 224:1–8.

Symbas PN: *Cardiothoracic Trauma*. Philadelphia, Saunders, 1989.

Symbas PN: Traumatic heart disease. *Curr Probl Cardiol* 1991; 16:537–582.

Symbas PN: Traumatic heart disease. In: Alexander RW, Schlant RC, Fuster V, O'Rourke RA, Roberts R, Sonnenblick EH (eds): *Hurst's The Heart*, 9th ed. New York, McGraw-Hill, 1998:2319–2326.

34

HEART DISEASE
AND PREGNANCY

John H. McAnulty

An understanding of the cardiovascular changes of a normal pregnancy is important for optimal care. This is even more important if a woman has heart disease. With successful treatment of heart disease during childhood, usually with surgery, the number of women with heart disease is increasing. They now can survive to the age of childbearing and are able to conceive.

HEART ISSUES UNIQUE
TO PREGNANCY

Some basic issues must be considered when treating any woman with heart disease during pregnancy.

Health Priorities

Mother and child—the health of one importantly influences the other. The well-being of the fetus should be considered, but the safety of the mother is always the highest priority. Ideally, treatment of the mother with drugs, diagnostic studies, or surgery should be avoided, but if required for maternal safety, each should be used.

Maternal Fragility

Despite the advances in management of heart disease, pregnancy puts the mother at risk. The risk is so great with some cardiovascular abnormalities that a recommendation of avoidance or interruption of pregnancy is supportable (Table 34-1).

TABLE 34-1

CARDIOVASCULAR ABNORMALITIES PLACING
A MOTHER AND INFANT AT EXTREMELY HIGH RISK

Advise *avoidance* or *interruption of pregnancy*
    Pulmonary hypertension
    Dilated cardiomyopathy with congestive failure
    Marfan's syndrome with dilated aortic root
    Cyanotic congenital heart disease
Pregnancy counseling and close clinical follow-up
    required
    Prosthetic valve
    Coarctation of the aorta
    Marfan's syndrome
    Dilated cardiomyopathy in asymptomatic women
    Obstructive lesions

Fetal and Newborn Vulnerability

The fetus depends completely on its mother for a continuous
supply of oxygen and nutrients. While the maternal commitment to the fetus is exceptional, if the mother's safety is
threatened, blood is preferentially diverted away from the
uterus. In the woman with heart disease, where uterine blood
flow may already be compromised, the chance of inadequate
uterine perfusion increases. In addition, diagnostic studies,
drugs, or surgery required to protect the mother may increase
fetal loss, result in teratogenicity, or alter fetal growth. A newborn infant is also fragile—possibly due to previous marginal
uterine blood flow during pregnancy or to lingering effects of
medications used to treat the mother. Early infant nourishment
may be jeopardized if maternal heart disease is severe enough
to interfere with breast-feeding. If the mother is on medications, breast-feeding could further threaten the infant. Finally,
the infant is at risk of losing a parent, since life expectancy
with many forms of heart disease is significantly less than
normal.

Corrected Heart Disease

Many women with heart disease who become pregnant will
have had previous mechanical therapy, usually surgery but

increasingly with a catheter. It is best not to consider a previous lesion "corrected," because there is always some residual disease. In some cases, this disease—a shunt, ventricular dysfunction, or arrhythmia or persistent obstruction—may adversely affect the mother and, in turn, the fetus.

CARDIOVASCULAR ADJUSTMENTS DURING A NORMAL PREGNANCY

Maternal adaptation to pregnancy includes remarkable cardiovascular changes. This may result in symptoms and signs even in a normal pregnancy which are difficult to distinguish from those occurring with heart disease. This explains why some abnormalities are not well tolerated during pregnancy (Table 34-1).

Hemodynamic Changes

Resting cardiac output increases by over 40 percent during pregnancy. The increase begins early, with cardiac output reaching its highest levels by the 20th week. Body position influences cardiac output, particularly in the second half of the pregnancy, as the enlarged uterus diminishes venous return from the lower extremities. When compared to measurements made with the woman in the left lateral position, cardiac output falls by 0.6 L/min when she is supine and by 1.2 L/min when she assumes the upright position. In general, this results in few or no symptoms; but in some, maintenance of the supine position may result in symptomatic hypotension—the "supine hypotensive syndrome of pregnancy." This can be corrected by having the woman turn on to her side. Cardiac output increases still further at the time of labor and delivery. With each uterine contraction, its value rises by over 30 percent, resulting in a cardiac output that can be as great as 9 L/min. The cardiac output falls rapidly to near-normal nonpregnant values within days to weeks after delivery, although there is a slight elevation that can persist as long as lactation occurs.

Associated with and influencing these changes of cardiac output are changes in other hemodynamic parameters. The stroke volume increases dramatically early in pregnancy and then levels off or falls slightly as venous return is inhibited during the second half of pregnancy. Ventricular end-diastolic volume increases; the ejection fraction does not change. The heart rate increases steadily throughout pregnancy but rarely reaches a value greater than 100 beats per minute in the normal resting woman. Blood pressure falls slightly early in pregnancy and then returns to normal values for the rest of gestation. There is an early fall in systemic vascular resistance associated with the early rise in cardiac output, with a gradual return of vascular resistance to near normal levels by term. Oxygen consumption increases steadily during pregnancy, reaching a value that is 30 percent above nonpregnant levels by the time of delivery.

Distribution of Blood Flow

Where does it go? This is not fully understood, but the distribution is affected by changes in local vascular resistance. Renal blood flow increases by approximately 30 percent. Blood flow to the skin increases by 40 to 50 percent—a mechanism for heat dissipation. Not surprisingly, mammary blood flow increases—usually this is approximately 1 percent of the cardiac output, and it increases to 2 percent by term. In the nonpregnant woman, uterine blood flow is approximately 100 mL/min (2 percent of the cardiac output). This doubles by the 28th week of pregnancy and increases to approximately 1200 mL/min at term. During pregnancy, uterine blood vessels are maximally dilated; flow can increase, but it must result from increased maternal arterial pressure and flow. As mentioned earlier, the mother makes a remarkable commitment to a fetus; but if redistribution of total flow is required by the mother or if there is a fall in maternal blood pressure and cardiac output, uterine blood flow falls preferentially. Vasoconstriction caused by endogenous catecholamines, vasoconstrictive drugs, mechanical ventilation, and some anesthetics as well as that associated with preeclampsia and eclampsia can decrease perfusion of the uterus. Uterine blood flow can potentially be compromised even in a healthy woman. In the woman with heart disease, whose blood flow may already be compromised, the concern about diversion of flow from the uterus is greatest.

The mechanisms for the hemodynamic changes are also not fully understood. They may in part be due to volume changes. Total body water increases steadily throughout pregnancy by 6 to 8 L (most is extracellular). As early as 6 weeks after conception, plasma volume increases and, by the second trimester, approaches its maximum of $1\frac{1}{2}$ times normal, where it stays throughout the pregnancy. Red blood cell mass also increases, but not to the same degree; thus the hematocrit falls during pregnancy, though rarely to a value less than 30 percent. Vascular changes are importantly related to the hemodynamic changes of pregnancy. Arterial compliance increases, and there is an increase in venous vascular capacitance. These changes are advantageous in maintaining the hemodynamics of a normal pregnancy. There may be disadvantages as well. The arterial changes are associated with increased fragility and, when vascular accidents occur in women, they frequently do so in pregnancy. The increased level of circulating steroid hormones may be the major explanation for the vascular and myocardial changes.

DIAGNOSIS OF HEART DISEASE

Recognition and definition of heart disease may be difficult at any time; this is particularly true during pregnancy. Symptoms suggesting heart disease—fatigue, dyspnea, orthopnea, pedal edema, and chest discomfort—occur commonly in pregnant women with normal hearts. Still, they should at least alert the caregiver to the possibility of heart disease. The concern should increase if the dyspnea or orthopnea is progressive and limiting or if a woman develops hemoptysis, syncope with exertion, or chest pain clearly related to effort. Likewise, the examination can be confusing. Pedal edema, basilar pulmonary rales, a third heart sound, a systolic murmur, and visible neck vein pulsations are also part of a normal pregnancy. Cyanosis, clubbing, a loud systolic murmur ($>3/6$), cardiomegaly, a "fixed split" S_2 or evidence of pulmonary hypertension (a left parasternal lift and loud P_2) do not occur as part of a normal pregnancy and deserve attention. A diastolic murmur is unusual enough to indicate heart disease, although the murmurs can be confused with a venous hum or internal mammary flow sounds (souffle), which are normal findings during pregnancy.

Diagnostic Studies

It is preferable to evaluate the cardiovascular status with history and physical alone; but if the safety of the mother requires it, diagnostic studies should be performed. The electrocardiogram is safe and useful. Echocardiography with evaluation of flow by Doppler is so safe and is so diagnostically useful that overuse is the only significant concern. Expense and potential misinterpretations are reasons to consider its use only when required to answer a specific question. Transesophageal echocardiography has been used safely during pregnancy with no apparent increased risk.

All x-ray procedures should generally be avoided, particularly early in pregnancy, as they increase the risk of fetal organogenesis or of a subsequent malignancy in the child, particularly leukemia. If required, a study should be delayed to as late in pregnancy as possible and shielding of the fetus should be optimal. Every woman of childbearing age should be questioned about the possibility of pregnancy before any x-ray procedure. Chest x-rays are performed on occasion when pregnancy is not recognized (or intentionally, when it is). The exposure to a fetus is small (estimated between 10 and 400 µGy) and has not been associated with recognizable increase in problems—information that can be given to worried parents if a chest x-ray was performed. Magnetic resonance imaging been used for other purposes during pregnancy and without recognized adverse effects. Radionuclide studies are preferably avoided. Most radionuclide should attach to albumin and thus not reach the fetus, but separation can occur and fetal exposure is possible.

CARDIOVASCULAR DRUGS AND PREGNANCY

Nearly all cardiac drugs cross the placenta and are secreted in breast milk. Information about all of them is incomplete and it is best, when possible, to avoid their use. Again, if required for maternal safety, they should not be withheld.

Diuretics

These can and should be used for treatment of congestive heart failure uncontrolled by sodium restriction, and they are effective therapy for hypertension. Experience is greatest with the thiazide diuretics and with furosemide. Diuretics should not be used for prophylaxis against toxemia or for the treatment of pedal edema.

Inotropic Agents

Digoxin and digitoxin cross the placenta, and fetal serum levels are approximately those of the mother. Given the same digoxin dose, maternal serum levels during pregnancy are often lower than those of nonpregnancy. The measurement of these levels may be helpful, but the assay used should not be one that is affected by an immunoreactive substance found in pregnancy. Digitalis may shorten the duration of gestation and labor because of an effect on the myometrium that is similar to the effect on the myocardium. Intravenous inotropic and vaso-pressor agents (dopamine, dobutamine, and norepinephrine) should be used if necessary to protect the mother, but the fetus is jeopardized because of a reduction in uterine blood flow and the tendency to stimulate uterine contractions. Ephedrine is an appropriate initial vasopressor drug, as—at least in animal models—it does not adversely affect uterine blood flow. There is no available information about the use of phosphodiesterase inhibitors during pregnancy.

Adrenergic-Receptor Blocking Agents

There have been concerns about the use of beta-blocking drugs during pregnancy, but their safe use in large numbers of preg-nant women without adverse effects justifies their use for usual clinical indications. All available beta blockers cross the pla-centa and are present in human breast milk. If beta blockers are used, beta$_1$-selective agents are preferable. The newborn infant's heart rate, blood sugar, and respiratory status should be assessed immediately after delivery. Experience with the alpha-blocking agents is sparse. Clonidine, prazosin, and labetalol have been used for hypertension without clear detri-mental effects.

Calcium Channel–Blocking Drugs

Nifedipine, verapamil, diltiazem, nicardipine, isradipine, and amlodipine have been used to treat hypertension and arrhyth-mias without adverse effects on the fetus or newborn infant. The drugs may cause relaxation of the uterus; nifedipine has been used specifically for this purpose.

Antiarrhythmic Agents

Atrioventricular (AV) node blockade is occasionally required during pregnancy. This can be achieved with digoxin, beta blockers, and calcium-channel blockers, and early reports sug-gest that adenosine can be safely used as well.

When essential for recurrent arrhythmias or maternal safety, the standard antiarrhythmic agents should be used, but there is insufficient accumulated information to know whether or not they increase the risk to the fetus or child. If oral antiarrhythmic therapy is required, it may still be appropriate to begin with quinidine since, given its long-term availability, it has most frequently been used without clear adverse fetal effects. There is some information about procainamide, disopyramide, mexiletine, flecainide, and atenolol, but it is insufficient to recommend their use unless the drug is essential for the mother. The early available information concerning amiodarone would suggest an increased likelihood of fetal loss and deformity.

If intravenous drug therapy is required, lidocaine is reasonable first-line therapy, keeping maternal levels below 4 µg/L. There is no information about the effects of intravenous procainamide, amiodarone, ibutilide or bretylium during pregnancy.

Vasodilator Agents

When needed for emergency treatment of a hypertensive crisis or pulmonary edema, nitroprusside is the vasodilator drug of choice. This controversial recommendation is made despite a paucity of information about its use during pregnancy because the drug is so effective, works instantly, is easily titrated, and ceases to have any effect immediately after the drug is stopped. A concern about the use of this drug is that its metabolite, cyanide, can be detected in the fetus, but this has not been demonstrated to be a significant problem in humans. It is a reason to limit the duration of the use of this drug whenever possible. Intravenous hydralazine, nitroglycerin, and labetalol are options for parenteral therapy.

Chronic afterload therapy to treat hypertension, aortic or mitral regurgitation, or ventricular dysfunction during pregnancy has been achieved with the calcium blocking drugs, hydralazine, and methyldopa. The angiotensin-converting enzyme (ACE) inhibitors are contraindicated in pregnancy because they increase the risk of abnormalities in fetal renal development. No data are available on the angiotensin II blocking agents losartan and volsartan.

Antithrombotic Agents

Warfarin crosses the placenta, and fetal exposure during the first 3 months is associated with a 15 to 25 percent incidence of malformations that comprise the "warfarin embryopathy syndrome" (facial abnormalities, optic atrophy, digital abnormalities, epithelial changes, and mental impairment). This appears to be particularly true in women exposed to the drug during the seventh to twelfth weeks of gestation. Warfarin use at any time during pregnancy increases the risk of fetal bleeding or maternal uterine hemorrhage. In women who require anticoagulation during pregnancy, heparin is preferable. The drug does not cross the placenta. Self-administered subcutaneous high-dose heparin (16,000 to 24,000 units per day) has been proven feasible and efficacious, and accumulating data suggests that low-molecular-weight heparin may be as effective to use (once or twice daily without the need to follow serial blood tests), and as safe as standard heparin therapy. When anticoagulation is required, some have advocated using heparin for the first trimester, then warfarin for the next 5 months, and returning to heparin prior to labor and delivery. Successful pregnancy has been achieved with this approach, but the authors favor avoidance of warfarin during pregnancy.

Antiplatelet agents increase the chance of maternal bleeding, and they, too, cross the placenta. There are theoretical and demonstrated reasons to have concern about aspirin, but it has been used so frequently without problems and, in fact, has been recommended as prophylaxis against preeclampsia by some. As with any drug, it is probably preferable to avoid aspirin during pregnancy, but it can be used if necessary with reasonable safety. There are no data available on the effects of ticlopidine or clopidigril during pregnancy or on the effects of blockers of the glycoprotein IIb-IIIa receptor site blockers (see also Chap. 46).

MANAGEMENT OF CARDIOVASCULAR SYNDROMES

Cardiovascular complications can occur with any form of heart disease. Management must be individualized, but some recommendations are applicable in most cases.

Low-Cardiac-Output Syndrome

A low-cardiac-output syndrome is ominous in any patient, and this is particularly true in pregnancy. It results in signs of poor perfusion (mental obtundation, peripheral vascular constriction, low urine output, and often a low blood pressure). Potentially reversible causes such as tamponade or severe valvular stenosis should be considered, but in pregnancy it is most often due to intravascular volume depletion. While it is a concern in any pregnant woman, volume depletion is particularly dangerous in those with lesions that limit blood flow, such as pulmonary hypertension, aortic or pulmonic valve stenosis, hypertrophic cardiomyopathy, or mitral stenosis. The measures outlined in Table 34-2 are exceptionally important; they are necessary to prevent or treat a fall in central blood volume during pregnancy.

Congestive Heart Failure

Management of congestive heart failure during pregnancy should not differ greatly from that at other times. Standard therapy can and should be used, remembering that angiotensin-converting inhibitors are contraindicated during pregnancy. Congestive heart failure is one situation where maintaining a woman in the supine position may be beneficial by causing

---

TABLE 34-2

MEASURES TO PROTECT AGAINST A FALL IN CENTRAL BLOOD VOLUME

Position
    45–60° left lateral
    10° Trendelenburg
Full-leg stocking
Volume preloading for surgery and delivery
    1500 mL of glucose-free normal saline
Drugs
    Avoid vasodilator drugs
    Ephedrine for hypotension unresponsive to fluid
      replacement
    Anesthetics (if required)
      Regional: serial small boluses
      General: emphasis on benzodiazepines and
        narcotics, low-dose inhalation agents

preload reduction with reduction of return of blood from the inferior vena cava to the heart.

Thromboembolic Complications

The risk of venous thromboemboli increases fivefold during and immediately after pregnancy, and there is arguably an increase in arterial emboli as well. Both may be the result of a woman's hypercoagulable state during pregnancy, and venous thrombosis is enhanced by increased venous stasis. Prevention is optimal and prophylactic full-dose or low-molecular-weight heparin is indicated in those at particularly high risk—for example, those with a hypercoagulable state. If a thrombus or embolus is identified, 5 to 10 days of intravenous therapy followed by full-dose subcutaneous heparin is recommended. If a thromboembolus is life-threatening (for example, a massive pulmonary embolism or a thrombosed prosthetic valve), thrombolytic therapy can be used (see also Chap. 46).

Hypertension

Hypertension can be present before pregnancy (in 1 to 5 percent) and persist throughout pregnancy, or it can develop with pregnancy. When normotensive women become pregnant, 5 to 7 percent will develop hypertension. Because of the early fall in systemic vascular resistance, this often does not occur until the second half of pregnancy. It has been called *pregnancy-induced* or *gestational* hypertension or *toxemia*. When associated with proteinuria, pedal edema, central nervous system irritability, elevation of liver enzymes and coagulation disturbances, the hypertension syndrome is called *preeclampsia*. If convulsions occur, the diagnosis is *eclampsia*. It is not clear that hypertension alone puts the mother or fetus at risk during pregnancy, but preeclampsia increases maternal risk [a 1 to 2 percent chance of a central nervous system (CNS) bleed, convulsions, or severe systemic illness] and may cause fetal growth retardation (10 to 15 percent). This increases still more with eclampsia.

Treatment guidelines are evolving. Until more is known, an argument can be made for keeping the systolic pressure at least below 160 mmHg and the diastolic pressure below 100 mmHg. Nonpharmacologic therapy is preferable and possible, although it is not clearly defined. Strict bed rest is not generally recommended, although limitation of activity and reduction of stress is commonly advised. Unless the patient has previously demonstrated salt-sensitive hypertension, sodium restriction is generally inadvisable, since pregnant women have lower plasma values than normotensive women. If drug treatment is required, experience is greatest with methyldopa. This otherwise infrequently used antihypertensive agent has been demonstrated to promote fetal survival and to result in children with normal mental and physical development. It may be that other drugs will achieve the same goal. Initial therapy can also include a beta$_1$-selective blocker or diuretic. As mentioned earlier, ACE inhibitors should not be used.

Pulmonary Hypertension

No matter what its cause, pulmonary hypertension is associated with a maternal mortality rate ranging from 30 to 70 percent and a fetal loss exceeding 40 percent. Maternal death can occur at any time during pregnancy, but the woman is most vulnerable during the time of labor and delivery and in the first postpartum week. If pulmonary hypertension is recognized early, interruption of the pregnancy is advised. If this is declined or if pulmonary hypertension is recognized late in pregnancy, close follow-up is required. Intravascular volume depletion puts these patients at greatest risk, emphasizing again, the need to follow the measures outlined in Table 34-2. When pulmonary hypertension is due to increased pulmonary vascular resistance and associated with right-to-left shunting (Eisenmenger's syndrome), meticulous attention to avoidance of air or thrombus emboli from intravenous catheters is essential to avoid systemic emboli in these patients. At the time of labor and delivery, a central venous line allows adequate fluid administration, and a radial artery catheter makes blood pressure and oxygen saturation determinations easier. These lines should be used for 48 to 72 hours postdelivery.

Arrhythmias

The rules for treatment of arrhythmias should be the same as in the nonpregnant patient, with the possible exception that a rhythm causing hemodynamic instability should be treated somewhat more rapidly because of the concern about a diversion of blood flow away from the uterus. As always, if a potentially reversible cause can be identified, it should be corrected. If treatment is required, it should never be instituted without electrocardiographic documentation of the rhythm (see also Chap. 2).

Tachyarrhythmias are as frequent during pregnancy as at other times. Atrial and ventricular premature beats or sinus tachycardia are reasons to look for and correct the cause but not to initiate specific treatment.

Paroxysmal supraventricular tachycardia is the most common sustained abnormal rhythm of pregnancy. Initial treatment with vagal maneuvers is as appropriate as at other times. If medical treatment is required to convert the rhythm, intravenous adenosine and verapamil are effective. Cardioversion can be used, remembering that the rule "never cardiovert an awake patient" is just as applicable during pregnancy as at any other time. If recurrent episodes require day-to-day therapy, verapamil or beta blockers are optimal choices. Management of atrial fibrillation and flutter should be as in the nonpregnant woman. If they occur in a woman with mitral stenosis, severe left ventricular dysfunction, or a previous thromboembolic event, antithrombotic therapy with heparin is indicated.

Ventricular tachycardias should be treated as in a nonpregnant woman. Recurrent, sustained monomorphic ventricular tachycardia that has the appearance of originating in the right ventricle (i.e., a left bundle branch morphology) may be effectively treated with beta-blocker therapy.

Bradyarrhythmias may also occur during pregnancy. They, too, are a reason to look for a reversible cause. Treatment is generally not required unless the patient has clear hemodynamic compromise. Complete heart block, usually congenital in origin in this population, is consistent with a successful pregnancy; however, if necessary, a permanent pacemaker can be inserted.

Loss-of-Consciousness Spells

The supine hypotensive syndrome can result in loss-of-consciousness. Treatment, again, is to avoid the supine position. If a seizure cannot be excluded in the pregnant woman with loss of consciousness spells, appropriate evaluation with electroencephalography is indicated. If a seizure is unlikely, management should include a consideration of the usual causes of syncope.

Endocarditis

The clinical presentation of endocarditis is the same during pregnancy as at other times. *Streptococcus* is the most common cause. Intravenous drug abusers are more likely to have staphylococcal infections, and women with genitourinary tract infections are more likely to have gram-negative infections, most commonly *Escherichia coli*.

Optimal management includes prevention. The use of antibiotics as prophylaxis against endocarditis in women with structural heart disease (Table 21-6) is as appropriate during pregnancy as at other times. Most physicians caring for women with heart disease recommend antibiotic prophylaxis at the time of labor and delivery. If endocarditis does occur, it should be treated aggressively with medical therapy; the usual indications for surgery are appropriate during pregnancy.

Surgery

While surgery is not exactly a complication of pregnancy, pregnant women with heart disease have the same 0.5 to 2 percent chance of requiring surgery during pregnancy as those with normal hearts. If surgery is required, it is important to maintain venous return. Again, the rules in Table 34-2 apply. If possible, anesthesia is optimally achieved locally.

SPECIFIC FORMS OF HEART DISEASE

RHEUMATIC HEART DISEASE

Worldwide, rheumatic fever is common and virulent and probably the most common cause of heart disease during pregnancy. In the United States, clinically recognized rheumatic fever is uncommon. In a woman presenting with myocarditis, rheumatic fever as a cause should be considered, particularly if it is associated with fever, joint discomfort, subcutaneous nodules, erythema marginatum, or chorea and if there is evidence of a group A streptococcal infection.

Rheumatic fever is the cause of almost all mitral stenosis; some isolated mitral, aortic, or tricuspid regurgitation; and of some double and triple valve disease. Echocardiography will help to define valve morphology and thus the etiology. Recognition of rheumatic fever as the cause of heart disease is important because it identifies those who need antibiotic prophylaxis to prevent recurrence of the disease. Twice-daily penicillin is the treatment regimen of choice, and this should be continued throughout pregnancy (Table 13-1).

Valve Disease

MITRAL STENOSIS

Due almost exclusively to rheumatic fever, this lesion is more common in women than in men. The increased tachycardia and fluid retention of pregnancy may double the resting pressure gradient across a stenotic mitral valve. Symptoms associated with pulmonary vascular congestion occur in 25 percent of patients with mitral stenosis during pregnancy. This usually becomes apparent by about the 20th week and may be aggravated at the time of labor and delivery. While potentially at risk from the elevated left atrial pressure, the patient with mitral stenosis also depends on an adequate pressure to fill the left ventricle and maintain cardiac output. Preservation of an adequate intravascular volume is essential to prevent a dramatic fall in cardiac output (Table 34-2).

A woman with symptomatic mitral valve stenosis who is contemplating pregnancy should be treated with balloon dilation or valve surgery before conception. If mitral stenosis is first recognized during pregnancy and symptoms develop, standard medical therapy is appropriate. Balloon valvuloplasty can be performed (with appropriate radiation shielding of the fetus). Mitral valve surgery may also be performed. Atrial fibrillation is of particular concern during pregnancy—a reason to use daily digoxin on a prophylactic basis (recognizing some inadequacy of rate control with this drug).

MITRAL REGURGITATION

In a young woman, mitral regurgitation may also be due to rheumatic fever. Myxomatous changes of the valve, with prolapse, is the other common cause. Regurgitation is generally well tolerated during pregnancy. Recognized congestive heart failure should be treated as described earlier. Afterload reduction is an important component of this therapy, but the ACE inhibitors should not be used. While the examination findings of mitral valve prolapse may change during pregnancy, the possibly associated arrhythmias, endocarditis, cerebral emboli, and hemodynamically significant regurgitation are rare complications that are no more likely to occur in pregnancy than at other times. The physical examination is sufficient for diagnosis.

AORTIC STENOSIS

Almost always congenital in etiology, aortic stenosis is more common in males but does occur in women of childbearing age. Recent information suggests that pregnancy can succeed with little or no maternal mortality and with no clear increase in fetal loss. If severe stenosis is recognized before pregnancy, balloon valvotomy or surgical commissurotomy is recommended. If pregnancy does occur in the presence of severe aortic stenosis, measures to avoid hypovolemia are important (see Table 34-2). If severe symptoms persist, a balloon valvuloplasty or aortic valve surgery can be performed during pregnancy (see also Chap. 12).

AORTIC REGURGITATION

This lesion is generally well tolerated during pregnancy, but it is important to define the cause. Etiologies include rheumatic fever, endocarditis, dilation of the aortic root, or, more ominously, aortic dissection. An associated dilated root or dissection should raise the consideration of Marfan's syndrome (Chap. 26) as a cause. If congestive heart failure occurs, standard treatment is appropriate—once again, with a warning to avoid ACE inhibitors. If endocarditis occurs and the infection cannot be controlled, surgical therapy is indicated.

PULMONARY VALVE DISEASE

Much pulmonary valve disease will have been altered by previous surgery. The residual stenosis and the invariable regurgitation are potential concerns but in general do not adversely affect the outcome of pregnancy. In the rare patient presenting with pulmonic stenosis during pregnancy, it is again important to avoid intravascular volume depletion. If severe symptoms occur (recurrent syncope, uncontrolled dyspnea, or chest pain), balloon valvuloplasty can be performed.

TRICUSPID VALVE DISEASE

Tricuspid regurgitation (most often the result of previous intravenous drug use) usually requires no specific therapy during pregnancy. Tricuspid stenosis, of course, is rare. If encountered, avoiding intravascular volume depletion would seem to be important.

PROSTHETIC VALVE DISEASE

While many have benefited from prosthetic valves, all are left with "prosthetic heart valve disease," consisting of one or more of the major complications of thromboemboli, bleeding (from anticoagulation), endocarditis, valve dysfunction, reoperation, or death. This affects patients at a rate of 5 percent per year throughout their lives. Pregnancy increases the risk of each of these complications, and the prosthetic valve and its treatment can adversely affect the fetus as well. All these are reasons that a prosthetic valve is a relative contraindication to pregnancy. In women with mechanical valves, anticoagulation is required. Full-dose subcutaneous heparin (maintaining the partial thromboplastin time between 1.5 and 2.5 times normal) is the therapy of choice. Low-molecular-weight heparin is increasingly appearing to be a preferable alternative. A heterograft or homograft is an alternative to a mechanical prosthesis. The opportunity to avoid anticoagulation is a logical argument for using hetero- or homografts in young women. However, these valves do not completely eliminate the concern about thromboemboli, and the rate of heterograft degeneration is high in young women, resulting in the need for early valve replacement. The choice between insertion of a mechanical or a tissue valve in a woman of childbearing age is difficult. A young woman who is capable of using warfarin safely when not pregnant and heparin during pregnancy is best treated with a mechanical valve. If a woman's social situation or attention to her health are questionable in regard to the safety of anticoagulation therapy, a biological valve is more appropriate (see also Chap. 15).

CONGENITAL HEART DISEASE

Congenital heart disease is now the most common heart disease encountered in women of childbearing age in the United States. It has often been altered by surgery. Each abnormality is unique, but there are some issues that should be considered with all. First, some abnormalities significantly increase maternal morbidity and mortality (see Table 34-1). Second, there is an increased risk of fetal death, which increases with the severity of the maternal lesions. Third, the presence of a congenital cardiac lesion in either parent or sibling increases the risk of cardiac and congenital abnormalities in the fetus. Congenital heart disease is recognized in 0.8 percent of all live births in the United States. Its presence in a parent increases this risk to 2 to 15 percent. Actually, the risk goes up to 50 percent if the abnormality is transmitted as an autosomal dominant trait—for example, Marfan's syndrome, the congenital long-QT syndrome, or hypertrophic obstructive cardiomyopathy. Fourth, as a general rule, when maternal congenital heart disease is recognized, it should be corrected prior to pregnancy. Fifth, residual or inoperable lesions require careful understanding before pregnancy is undertaken. Finally, as with valve disease, antibiotic prophylaxis against endocarditis is as appropriate during pregnancy as at other times in patients with lesions that render them susceptible to this complication (see also Chap. 22).

Left-to-Right Shunts

Although left-to-right shunting increases the chance of pulmonary hypertension, right ventricular failure, arrhythmias, and emboli, it is not clear these are made more likely by pregnancy. The degree of shunting is affected by the relative resistances of the systemic and pulmonary vascular circuits; both fall to a similar degree during pregnancy, and, in general, there is no significant alteration in the degree of shunting. In the United States, most patients with left-to-right shunts will have undergone surgical correction prior to pregnancy. There is no clear increase in mortality in these patients with pregnancy as opposed to women with normal hearts. Correction does not influence the incidence of congenital heart disease in the offspring.

ATRIAL SEPTAL DEFECT

Since the symptoms and signs of an atrial septal defect can be subtle, this abnormality is occasionally encountered uncorrected during pregnancy. In women with an ostium secundum defect, pregnancy is generally well tolerated. Ostium primum defects are equally well tolerated unless they are associated with other congenital abnormalities. Surgical correction prior to pregnancy does not lower the 5 to 10 percent chance that offspring will have congenital heart disease (see Chap. 22).

VENTRICULAR SEPTAL DEFECT

Over half of ventricular septal defects close in childhood, and most of the remainder will have been corrected prior to the age of pregnancy. The woman with a ventricular septal defect, however, tolerates pregnancy well. Congestive heart failure or arrhythmias can be managed as previously discussed. Such a woman's baby has a 5 to 8 percent chance of being born with a cardiac defect. Again, this incidence is not altered by previous surgical correction (see Chap. 22).

PATENT DUCTUS ARTERIOSUS

This left-to-right shunt is also well tolerated during pregnancy. The occasional congestive heart failure can be treated in the standard fashion.

Right-to-Left Shunt ("Cyanotic" Heart Disease)

Right-to-left shunting can occur through an atrial or ventricular septal defect or a patent ductus arteriosus. This occurs when pulmonary vascular resistance exceeds systemic vascular resistance (Eisenmenger's syndrome) or when there is obstruction to right ventricular outflow with normal pulmonary vascular resistance. All are forms of "cyanotic" heart disease, and the presence of cyanosis, especially when it is sufficient to elevate hemoglobin levels, is associated with high maternal mortality, fetal loss, prematurity, and reduced infant birth weights. Eisenmenger's syndrome was discussed above under "Pulmonary Hypertension"—it is worth repeating that with this problem it is advisable to avoid or interrupt pregnancy. When the cyanosis is not due to elevated pulmonary vascular resistance but rather to right ventricular outflow obstruction, maternal mortality is less but cardiovascular complications are still increased.

TETRALOGY OF FALLOT

This is the most common form of right-to-left shunting resulting from obstruction to pulmonary flow with normal pulmonary vascular resistance. Successful pregnancy can be achieved if the lesion is uncorrected, but maternal mortality is high. After surgical correction, maternal mortality does not clearly exceed that of a women without heart disease. The offspring have a 5 to 10 percent chance of having congenital heart disease (see Chap. 22).

Obstructive Lesions

Two treatment recommendations apply in women with obstructive cardiac lesions. First, volume depletion should be avoided (Table 34-2), since it can result in a significant fall in cardiac output. Second, surgical or catheter treatment for obstructive lesion is recommended prior to pregnancy, not only to increase maternal safety but also because it reduces the chance of congenital heart disease in the offspring.

The two most common forms of obstruction to the right ventricle have already been mentioned—pulmonic valve stenosis and tetralogy of Fallot. Obstructions to the left side of the heart include aortic valve stenosis, also described above. Experience with isolated supravalvular or subvalvular bands is sparse; the approach recommended for aortic valve stenosis would seem applicable.

COARCTATION OF THE AORTA

This form of left ventricular obstruction is more common in men but may occur in women and may be associated with a bicuspid aortic valve. Maternal mortality can range from 3 to 8 percent. Surgical correction or balloon dilation reduces the risk of aortic dissection or rupture, and thus death, to less than 1 percent; its effects on the rate of rupture of associated intracranial aneurysms is not known. If pregnancy occurs in a woman with a coarctation, blood pressure control, as previously described, is recommended.

HYPERTROPHIC OBSTRUCTIVE CARDIOMYOPATHY

This is increasingly recognized as a left ventricular outflow tract obstructive lesion during pregnancy. Inherited as an autosomal dominant trait with variable penetrance, this abnormality has been shown to result in increased symptoms of dyspnea and chest discomfort and palpitations during pregnancy. It is not clear that pregnancy increases the approximately 1 to 3 percent chance per year of sudden death, although a death has been reported with this syndrome during pregnancy. This is still one more obstructive lesion where it is important to avoid hypovolemia (Table 34-2). Beta-blocker therapy is recommended at the time of labor and delivery (see also Chap. 47).

Complex Congenital Lesions

The predictability of outcome pregnancy is more difficult as complexity increases, but in general, maternal and fetal morbidity and mortality are high, particularly when the abnormality results in maternal cyanosis.

TRANSPOSITION OF THE GREAT VESSELS

While pregnancy has been reported in women with this syndrome, maternal and fetal outcomes are poor. Partial or complete correction of the lesion prior to pregnancy improves the outcome for the mother and the fetus. If "corrected" transposition is not complicated by cyanosis, ventricular dysfunction, or heart block, pregnancy should be well tolerated.

EBSTEIN'S ANOMALY OF THE TRICUSPID VALVE

This may be mild and unrecognized during pregnancy. Maternal risk increases in a woman with associated right ventricular dysfunction, right ventricular outflow tract obstruction, and right-to-left shunting with cyanosis. This last—the right-to-left shunting—is a reason to avoid pregnancy.

MARFAN'S SYNDROME

It may be difficult to diagnose Marfan's syndrome, but it is important to do so because pregnancy is particularly dangerous for affected women. First, the risk of death from aortic rupture or dissection is high, particularly if the aortic root is enlarged (one criterion has been greater than 40 mm by echocardiography). Second, the woman's life span is reduced, implying that her years of motherhood will be limited. Third, half of her offspring will be affected. These are reasons that women with Marfan's syndrome should be advised to avoid pregnancy. In the woman with a dilated aortic root, the risks are sufficient to recomment interruption if pregnancy has occurred. Should the parents elect to continue the pregnancy, the woman's activity should be restricted and hypertension should be prevented. The prophylactic use of beta blockers, while unproven, seems reasonable. This is the one cardiovascular syndrome where cesarean section is recommended because of the hemodynamic stresses of labor (see also Chap. 26).

MYOCARDIAL DISEASE

Hypertrophic Cardiomyopathy

The asymmetric form of hypertrophic cardiomyopathy has been discussed as an obstructive lesion. A concentric hypertrophic cardiomyopathy may be the result of aortic stenosis or hypertension. When the cardiomyopathy is *not* due to either of these, the cause, prognosis, and management are often unclear even unrelated to pregnancy. Again, hypovolemia should be avoided, and if congestive heart failure occurs, standard therapy is appropriate (see also Chap. 18).

Dilated Cardiomyopathy

This is a reason to recommend that pregnancy should be avoided. This recommendation is not supported by data from prospective trials, but it is given because myocardial dysfunction is the feature associated with increased maternal and fetal mortality in many forms of heart disease. On occasion, the cardiomyopathy may be caused by pregnancy—a *peripartum cardiomyopathy*. This occurs almost exclusively in the third trimester or in the first 6 weeks postpartum, suggesting that it is a unique entity. Myocarditis has been implicated as being an initial part of the syndrome, but data are limited. In the woman with a dilated cardiomyopathy during pregnancy, standard treatment for heart failure, thromboemboli, or arrhythmias is appropriate.

A subsequent pregnancy should be discouraged, even when heart function returns to normal, because the maternal mortality can approach 10 percent (see also Chap. 17).

Coronary Artery Disease

Chest discomfort is common during pregnancy but is due most often to gastroesophageal reflux or abdominal distention. Coronary artery disease, however, can cause angina and myocardial infarctions during pregnancy. The disease is rarely due to atherosclerosis. Other explanations have included dissection of the coronary artery, spasm, emboli, or vasculitis. Kawasaki's or Takayasu's disease has been implicated in some cases. When suspected or demonstrated, coronary artery disease should be treated with standard medical therapy. If symptoms are not relieved, angioplasty or surgery can be performed.

Pregnancy Following Cardiac Transplantation

Many cardiac transplant recipients are women of childbearing age. Successful pregnancies have been reported after transplantation. The potential hazards to the mother and fetus—which include maternal heart failure, immunosuppressive therapy, maternal infections, and serial diagnostic studies, as well as a potential for a shortened maternal life span—are reasons that a patient, at very best, should be counseled about the advisability of proceeding with pregnancy.

SUGGESTED READING

Haemostatis and Thrombosis Task Force: Guidelines on the prevention, investigation and management of thrombosis associated with pregnancy—maternal and neonatal haemostatis working papers of the haemostasis and thrombosis task force. *J Clin Pathol* 1993; 46:489–496.

McAnulty JH, Metcalfe J, Uleland K: Heart disease and pregnancy. In: Alexander RW, Schlant RC, Fuster V, O'Rourke RA, Roberts R, Sonnenblick EH (eds): *Hurst's The Heart*, 9th ed. New York, McGraw-Hill, 1998: 2389–2406.

National High Blood Pressure Education Program: Working Group report on high blood pressure in pregnancy. *Am J Obstet Gynecol* 1990; 163:1691–1712.

Presbytero P, Sommerville J, Stone S, et al: Pregnancy and cyanotic congenital heart disease: outcome of mother and fetus. *Circulation* 1994; 89:2673–2676.

Robson SC, Hunter S, Boys RJ, Dunlop W: Serial study of factors influencing changes in cardiac output during human pregnancy. *Am J Physiol* 1989; 256:H1060–H1065.

Whittemore R, Hobbins JCC, Engle MA. Pregnancy and its outcome on women with and without surgical treatment of congenital heart disease. *Am J Cardiol* 1982; 50:641–651.

CHAPTER

35

EXERCISE AND THE HEART

Peter M. Buttrick
James Scheuer

This chapter reviews the physiological responses that accompany acute exercise and the chronic cardiovascular adaptations that occur as a result of physical conditioning. In addition, the clinical features of the athlete's heart and the risks and benefits of conditioning as adjunctive therapy for several medical conditions are described.

ACUTE HEMODYNAMICS

In general, exercise is divided into two distinct types: *isotonic* (dynamic) and *isometric* (static). These differ in the physiological responses they evoke and in the demands they place on the heart, which, in turn, define the chronic adaptations that develop in conditioned athletes. The earliest hemodynamic response to dynamic exercise is a fall in systemic vascular resistance, which reflects vasodilation of the resistance vessels in the exercising muscle. This is prominent even at mild exercise intensity. Afterload falls and cardiac output is redistributed, so that, during maximal effort, more than 80 percent of the cardiac output may be directed to working muscle, versus about 20 percent at rest. Since the aerobic capacity of skeletal muscle is far greater than that of the splanchnic tissues (and since local factors may actually increase the capacity of skeletal muscle to extract oxygen), the net result of this redistribution of flow is an increase in systemic oxygen consumption. The primary cardiac response to dynamic exercise is an increase in heart rate. However, heart rate alone does not account for the increased cardiac output seen during exercise. There is an increase in venous return, probably mediated by vasoconstriction of the large veins as well as the mechanical effects of muscular contraction, which results in increased end-diastolic volume and augmented stroke volume (by the Frank-Starling mechanism). There is also an increase in neurohumoral sympathetic drive that augments cardiac contractility.

During isometric exercise, a discrete muscle group is enlisted and no external work is performed. The oxygen requirements necessary to sustain isometric exercise are proportional to the muscle mass involved and are generally modest. However, these demands cannot be met by an increase in blood flow, as local vasodilatation is limited by mechanical compression of the resistance vessels by the isometrically exercising muscle, and blood flow to the exercising muscle maintained may actually decrease. Muscle perfusion tends to be maintained via a rise in arterial pressure initiated by a reflex arc emanating from the contracting muscle, which results in increased systemic vascular resistance even in the face of only modest exertion. In concert with this, stroke volume may actually fall and the heart rate response to isometric exercise is exaggerated. Thus, in contrast to isotonic exercise, isometric exercise imposes a significant systolic, or pressure, load on the heart.

CHRONIC ADAPTATIONS

With conditioning induced by chronic bouts of dynamic exercise, a spectrum of cardiovascular adaptations develops that is reflected by a significant increase in maximal oxygen consumption. In addition to increasing maximal exercise capacity, these adaptations allow sustained submaximal work with an economy of effort. They affect both skeletal muscle and the heart. In skeletal muscle, capillary density increases, as do the number of mitochondria and the respiratory capacity of the muscle. Oxygen extraction by a conditioned skeletal muscle is increased at any given blood flow. The primary cardiac adaptations are a decrease in heart rate—both at rest and at any submaximal levels of exercise—and an increase in heart size, manifest mainly by greater right and left ventricular end-diastolic chamber dimensions. This is a result both of the lower heart rate and greater diastolic filling, which results in greater ventricular circumferences and eccentric cardiac hypertrophy. These changes allow an increase in stroke volume. In addition, the mechanical properties of the conditioned heart are altered, so that increased rates of systolic contraction and diastolic relaxation occur. Capillary density and coronary collaterals are also increased, but the clinical significance of these changes remains unclear.

THE ATHLETE'S HEART

Characteristic features in the clinical assessment of an athlete include a resting bradycardia, a slightly lateralized apex impulse, and both an S_3 and S_4 gallop (heard in up to 50 percent of athletes). Short systolic murmurs are also usual. Electrocardiographic (ECG) abnormalities are common and include sinus arrhythmias with pauses of up to 2.5 s. First-degree and Mobitz I second-degree AV blocks occur frequently, largely reflecting increased vagal tone. An increase both in the P wave and in the QRS voltages, associated with lateral T-wave inversions, may also be seen. QRS prolongation, axis deviations, and supraventricular tachycardias are not characteristic and may warrant further workup. Echocardiographic findings in a dynamically trained athlete include a slight symmetric increase in left ventricular wall thickness associated with an increase in end-diastolic dimension, and a normal (or even slightly diminished) end-systolic dimension. Concentric left ventricular hypertrophy may be seen in the isometrically conditioned athlete. Asymmetric septal hypertrophy is unusual.

EXERCISE AND SUDDEN DEATH

Cardiovascular death during exercise is extremely rare. Autopsy data overwhelmingly suggest that it is associated with recognized or occult cardiac diseases. In persons under age 35, common causes of sudden death during exercise include coronary anomalies and hypertrophic cardiomyopathy, and, less commonly, coronary artery disease, mitral valve prolapse, aortic rupture, and myocarditis. In persons over age 35, coronary artery disease is the predominant risk factor, though other entities, such as valvular heart disease and cardiomyopathies, also are seen. Recent reports suggest that unsupervised vigorous exercise in elderly or previously sedentary individuals may be associated with increased cardiovascular risk, so exercise cannot be recommended unconditionally in this population (see also Chap. 5).

A related question is whether or not exercise conditioning can prevent cardiovascular death; several large longitudinal studies suggest that it can. These include studies of Harvard alumni, San Francisco longshoremen, and British civil servants (among others), all of which showed that increased physical activity, independent of known coronary risk factors, was associated with a delay in the onset of symptomtic coronary disease and reduction in cardiovascular risk. The American Heart Association currently defines a sedentary lifestyle as an acquired risk factor for the development of coronary artery disease.

EXERCISE AND CONGESTIVE HEART FAILURE

Recently the role of exercise conditioning as adjunctive therapy for congestive heart failure (CHF) has been investigated. The rationale lies in the fact that CHF is characterized both by a decrease in blood flow to skeletal muscle as well as by reduced cardiac output, both of which might be favorably affected by conditioning. To date, clinical studies have suggested that substantial salutary effects can be derived from exercise training in patients with heart failure and that these effects reflect skeletal muscle and neural adaptations rather than direct cardiac effects. It is unknown whether or not chronic increases in venous return and in end-diastolic volume are potentially harmful in specific subsets of patients with heart failure.

SUGGESTED READING

Asmussen E: Similarities between static and dynamic exercise. *Circ Res* 1981; 48:I3–I10.

Buttrick PM, Sheuer J: Exercise and the heart: acute hemodynamics, conditioning training, the athlete's heart, and sudden death, in Alexander RW, Schlant RC, Fuster V, O'Rourke RA, Roberts R, Sonnenblick EH (eds): *Hurst's The Heart*, 9th ed. New York, McGraw Hill, 1998: 2425–2435.

Hanson P (ed): Exercise and the heart. *Cardiol Clin* 1987; 5:147–348.

Hambrecht R, Niebauer J, Fiehn E, et al: Physical training in patients with stable chronic heart failure: effects on cardiorespiratory fitness and ultrastructural abnormalities of leg muscles. *J Am Coll Cardiol* 1995; 25:1239–1249.

Huston TP, Puffer JC, Rodney WM: The athletic heart syndrome. *N Engl J Med* 1985; 313:24–32.

Maron BJ: Structural features of the athlete's heart as defined by echocardiography. *J Am Coll Cardiol* 1986; 7:190–203.

Paffenberger RS, Hyde RT, Wing AL, et al: The association of changes in physical activity level and other life style characteristics with mortality in men. *N Engl J Med* 1993; 328:538–545.

Pelliccia A, Maron BJ, Spataro A, et al: The upper limit of physiologic cardiac hypertrophy in highly trained athletes. *N Engl J Med* 1991; 324:295–301.

Schaible TF, Scheuer J: Cardiac adaptations to chronic exercise. *Prog Cardiovasc Dis* 1985; 27:297–324.

CHAPTER

36

CARDIOVASCULAR AGING AND ADAPTATION TO DISEASE

Steven P. Schulman

The perception that there is a general age-related decline in cardiovascular function is due to that fact that hypertension, coronary artery disease, and heart failure are so prevalent in the elderly. Indeed, in healthy individuals, resting left ventricular function is well maintained with increasing age. There are, however, several specific age-associated changes of the cardiovascular system that are found in animal models and healthy humans. Across species, there is prolonged duration of contraction and relaxation probably contributed by age-associated left ventricular hypertrophy and physical deconditioning. In healthy humans, noninvasive measurements of left ventricular systolic function do not change with increasing age. As in animal models, early diastolic left ventricular filling is decreased, consequent to prolonged relaxation, increases in afterload, mild left ventricular hypertrophy with increases in chamber stiffness, and an increase in regional heterogeneity in filling among left ventricular segments.

Studies of isolated cardiac muscle, ventricular myocytes, and intact animals show a striking decrease in the inotropic response to beta-adrenergic stimulation, likely due to a decrease in catecholamine-induced calcium released from the sarcoplasmic reticulum in senescent cardiac myocytes. In contrast to the inotropic response, catecholamine-enhanced relaxation is not impaired. Both older animals and humans also have a decrease in the chronotropic response to catecholamines as compared with their younger counterparts. A third age-associated cardiovascular change is an increase in arterial stiffness due to changes in the composition and distribution of arterial elastin and collagen. These changes result in an increase in both arterial characteristic impedance (pulsatile load) and peripheral vascular resistance (nonpulsatile load) in older animals and humans. As a result of this increase in arterial stiffness, predictable age-related cardiovascular changes include an increase in systolic blood pressure, a widening of pulse pressure, and an increase in left ventricular mass. Because of an increase in pulse-wave velocity during left ventricular ejection into a stiffened aorta, arterial pressure waves are reflected back to the heart in systole. In an animal model, greater ischemic dysfunction occurs with transient coronary occlusion when the left ventricle ejects into a stiffened conduit compared to a compliant conduit due to a greater reliance of coronary blood flow on systolic blood pressure. In addition to these changes in the vasculature with age, both the large and small artery vasodilating response to catecholamines is impaired in older animals and humans. This impaired vasodilatation may limit the cardiovascular response to exercise in the elderly.

CARDIOVASCULAR RESPONSE TO EXERCISE

Peak exercise capacity declines linearly with increasing age in healthy humans. During maximal exercise, cardiac index and ejection fraction decline across the age span. There are three major age-associated differences in the exercise response in healthy men and women. First, older subjects have a lower heart rate response to exercise due to the decrease in beta-adrenergic responsiveness. Second, young individuals have a decrease in end-systolic volume from rest to peak exercise, reflecting the increase in inotropic state and arterial vasodilation with a reduction in afterload; both are beta-adrenergic—mediated responses. End-systolic volume rises progressively with increasing age reflecting both the decrease in inotropic response and decrease in arterial vasodilator response. Third, end-diastolic volume does not change in young subjects from rest to peak exertion, but it progressively increases in older subjects, resulting from the use of the Frank-Starling mechanism to augment stroke volume and cardiac output during exercise. These age differences in the exercise response are abolished when young subjects are pretreated with beta-adrenergic blockade, suggesting that the differences in catecholamine responsiveness across the age span result in differences in the exercise response. Importantly, these age changes in cardiac function can be partially allayed by aerobic conditioning, which results in significant increases in peak exercise capacity, peak ejection fraction, and cardiac index and a decrease in end-systolic volume index. Although beta-adrenergic responsiveness does not change with exercise training, arterial stiffness in senior athletes is reduced compared with age-matched sedentary controls. Reduction in afterload with conditioning may be one mechanism whereby peak-exercise left ventricular function improves.

HYPERTENSION IN THE ELDERLY

Systolic blood pressure increases progressively with increasing age, as does the prevalence of isolated systolic hypertension in the population. Systolic blood pressure is a powerful predictor of cardiovascular morbidity. The prevalence of left ventricular hypertrophy also increases with age and is also an independent predictor for cardiovascular morbidity. Therapy for the treatment of both diastolic and isolated systolic hypertension in the elderly reduces cardiovascular mortality and stroke compared with placebo.

ISCHEMIC HEART DISEASE IN THE ELDERLY

Older patients often present with atypical symptoms of acute myocardial ischemia and infarction as compared with younger patients, and a myocardial infarct may be completely unrecognized. Age is a powerful independent predictor of both short- and long-term mortality in patients with an acute myocardial infarction, even in those thought to be at relatively low risk and those with a first myocardial infarct and eligible for thrombolytic therapy. Elderly patients with acute infarction also suffer a higher frequency of heart failure and cardiogenic shock even with smaller indices of infarct size. Elderly subjects who are eligible for thrombolytic therapy have a mortality benefit, and age per se is not a contraindication for thrombolytic therapy. On the other hand, age is an important predictor for both hemorrhagic and nonhemorrhagic stroke following thrombolytic therapy. Primary angioplasty in centers with a highly trained staff may have beneficial effects on mortality in thrombolytic-eligible patients with acute myocardial infarction. Beta-blocker therapy and aspirin are important therapeutic agents in elderly patients with acute myocardial infarction; both agents reduce mortality yet both are underused in this age group. Finally, angiotensin-converting enzyme inhibitors save lives in elderly patients with acute myocardial infarction and left ventricular dysfunction.

Revascularization procedures are used with increasing frequency in the elderly for both chronic and acute coronary syndromes. There have been reductions in morbidity and mortality in the elderly following both coronary artery bypass grafting and angioplasty techniques over the last 10 years, likely reflecting the increased use of the internal mammary grafts and stents. Although no randomized trial comparing medical management with bypass surgery has included the elderly, there have been significant improvements in the quality of life of elderly patients with symptomatic coronary disease.

CONGESTIVE HEART FAILURE IN THE ELDERLY

The prevalence of congestive heart failure in the elderly has risen dramatically over the last 10 years, reflecting both the increased survival of patients with coronary disease and hypertension. Many of these patients develop heart failure as they live to older ages, and the interaction of disease with age-associated cardiovascular changes may exacerbate heart failure in this age group. Increasing arterial stiffness and impaired left ventricular diastolic filling will increase afterload and left ventricular filling pressures, respectively. The age-associated decrease in beta-adrenergic responsiveness also limits the older patient with heart failure. Vasodilator therapy in older patients with heart failure from left ventricular systolic dysfunction reduces mortality and improves symptoms. Careful clinical follow-up of the older patients with heart failure may result in less frequent readmissions to the hospital and improved quality of life. About one-third of elderly patients with heart failure have normal left ventricular systolic function on noninvasive testing. This group of patients likely are symptomatic from left ventricular diastolic dysfunction, which results in an abnormal diastolic pressure-volume relationship and elevated diastolic pressures. These patients often improve symptomatically with beta-blocker therapy or calcium channel blockers.

ELECTROPHYSIOLOGY IN THE ELDERLY

Due to age-associated loss of conducting cells, elderly subjects are more prone to develop sick sinus syndrome. Atrial fibrillation is common in the elderly, being present in 5 to 6 percent of subjects ≥65 years of age and often associated with underlying heart disease. Anticoagulation lessens the risk of cardioembolic strokes in the elderly, although this age group also has an increased risk of intracranial hemorrhage; careful monitoring of the International Normalized Ratio is important to minimize the bleeding risk (see Chap. 46).

VALVULAR HEART DISEASE IN THE ELDERLY

Calcific trileaflet aortic stenosis is unique to the elderly. Asymptomatic elderly patients with aortic stenosis need to be followed carefully, since clinically significant disease can develop over a short time period. In symptomatic patients with aortic stenosis, aortic valve replacement often results in marked improvement in quality of life. Mitral regurgitation in this age group is often related to coronary artery disease. For elderly patients requiring valve replacement, the choice of a mechanical valve with lifelong anticoagulation must be balanced against the choice of a bioprosthetic valve and an increased risk of structural deterioration. Additional factors include the risks of anticoagulation, other requirements for anticoagulation, and the valve position, with bioprosthetic aortic valves having a greater longevity than valves in the mitral position (see also Chap. 15).

SUGGESTED READING

Aronow WS, Tresch DD (eds): Clinics in geriatric medicine. *Coronary Artery Disease in the Elderly.* Philadelphia, Saunders, 1996.

Lakatta EG: Cardiovascular regulatory mechanisms in advanced age. *Physiol Rev* 1993; 73:413–467.

Maggioni AP, Maseri A, Fresco C, et al: Age-related increase in mortality among patients with first myocardial infarctions treated with thrombolysis. *N Engl J Med* 1993; 329: 1442–1448.

Schulman SP, Weisfeldt ML: Cardiovascular aging and adaptation to disease, in Alexander RW, Schlant RC, Fuster V, O'Rourke RA, Roberts R, Sonnenblick EH (eds): *Hurst's The Heart*, 9th ed. New York, McGraw-Hill, 1998: 2437–2449.

Wei JY: Age and the cardiovascular system. *N Engl J Med* 1992; 327:1735–1739.

CEREBROVASCULAR DISEASE AND NEUROLOGIC MANIFESTATIONS OF HEART DISEASE

Louis R. Caplan

Vascular diseases often affect both heart and brain. Heart diseases cause secondary brain lesions; central nervous system diseases affect the heart and its functions.

BRAIN COMPLICATIONS OF HEART DISEASE

- Heart pumps foreign matter into the circulation (embolism).
- Pump function fails—brain hypoperfusion.
- Drugs given for heart disease have neurologic side effects.
- Encephalopathies due to associated organ dysfunction appear.

Cardiogenic Brain Embolism (Table 37-1)

SOURCES
- Cardiac wall abnormalities—cardiomyopathies, akinetic regions, ventricular aneurysms.
- Valve disease, including mitral annulus calcification, fibrofibrous endocarditis.
- Arrythmias—atrial fibrillation, sick sinus syndrome
- Myxomas and chamber thrombi
- Septal defects, including patent foramen ovale
- Aortic plaques, especially protruding atheromas

RECIPIENT ARTERY LOCATIONS
- Anterior circulation, mainly the middle cerebral artery (MCA).

- Posterior circulation, mainly the intracranial vertebral (ICVA), distal basilar (BA), and posterior cerebral arteries (PCAs).
- Some 80 percent of emboli go to the anterior circulation, 20 percent to the posterior.

FINDINGS IN PATIENTS WITH EMBOLI IN RECIPIENT ARTERY LOCATIONS

- MCA upper division: contralateral hemiparesis, hemisensory loss in arm, face, and/or leg; left, aphasia; right, neglect of left space, unawareness of deficit, motor impersistence.
- MCA lower division: contralateral upper quadrantanopia; left, Wernicke aphasia; right, agitation, left neglect, poor drawing and copying.
- MCA mainstem: mix of findings in upper and lower divisions.
- ICVA, cerebellar infarct: ataxia, vomiting, occipital headache.

TABLE 37-1

EMBOLIC MATERIALS

Cardiac	Intraarterial
Red fibrin-dependent thrombi	Red fibrin-dependent thrombi
White platelet-fibrin nidi	White platelet-fibrin nidi
Material from marantic endocarditis	Combined fibrin-platelet and fibrin-dependent clots
Bacteria from vegetations	Cholesterol crystals
Calcium from valves and mitral annulus calcification	Atheromatous plaque debris
Myxoma cells and debris	Calcium from vascular calcifications
	Air
	Mucin from tumors
	Talc or microcrystalline cellulose from injected drugs

- Distal BA: hypersomnolence, bilateral visual loss, poor upgaze, amnesia.
- PCA: contralateral hemianopia, hemisensory loss.

ONSET AND COURSE
In about 80 percent of patients with embolism, neurologic signs are maximal at onset; in 20 percent, there is later progression, often in a step, in the first 24 to 48 h due to distal passage of embolus.

DIAGNOSTIC TESTING
- Computed tomography (CT)
- Computed tomography angiography (CTA)
- Magnetic resonance imaging (MRI)
- Magnetic resonance angiography (MRA)
- Transesophageal echocardiography (TEE)
- Transthoracic echocardiography (TTE)
- CT and MRI favor emboli if there are
  Superficial or superficial and deep, wedge-shaped infarcts.
  Lower division MCA, cerebellum, PCA.
  Hemorrhagic infarct.
  Hyperdense MCA without contrast.
  Multiple superficial or superficial and deep infarcts in different vascular territories.
- MRI and/or CT for hemorrhagic changes and posterior circulation.
- TEE and/or TTE for potential cardiac sources.
- Extracranial ultrasound and CTA and MRA are useful for detecting arterial sources of emboli.
- Transcranial Doppler (TCD) is useful for monitoring emboli.

TREATMENT
Trials show low-dose warfarin (Coumadin) [International Normalized Ratio (INR 2 to 3)] to be effective in preventing brain embolism in nonvalvular atrial fibrillation.[1–3] Recent congestive heart failure (CHF), history of hypertension, and large left atrium predict a high rate of embolism in patients with atrial fibrillation.[4]

Brain Hypoperfusion Due to Cardiac Pump Failure (Usually Arrest)

VULNERABLE LOCATIONS AND CLINICAL FINDINGS[5]
- Cerebral cortex
  Diffuse and severe—coma followed by persistent vegetative state.
  Laminar necrosis—multifocal seizures and myoclonus.
  Watershed ischemia—bilateral arm and thigh weakness ("man in a barrel"), bilateral visual loss
  Temporal lobe—amnesia, Korsakoff-like syndrome
- Cerebellum—ataxia and intention myoclonus
- Brainstem—death or persistent deep coma; no brainstem reflexes

PROGNOSTIC SIGNS
- Alertness, response to environment and stimuli
- Brainstem reflexes—corneal, caloric, and "doll's eye" pupils

DIAGNOSTIC TESTS
- Imaging not helpful; electroencephalogram (EEG) useful—mostly clinical signs.
- TCD useful for brain death determination and clinical.

Complications of Cardiac Surgery[6]

- Embolism, the most important, may be delayed. Emboli arise from aortic atheromas after clamping, arrythmias, and preexistent cardiac lesions. Before coronary surgery, advise TEE to assess aorta and potential embolic sources.
- Occlusive carotid and vertebral lesions are often also present but rarely cause intraoperative or postoperative stroke.
- Encephalopathy is common in the postoperative period—caused by hypoxia-ischemia, metabolic disorder (lung, liver, kidney, etc.), and drugs, especially haloperidol (Haldo), pain medications, sedatives. These should be avoided or minimized.
- Intracranial hemorrhage occurs occasionally in children who have rapid correction of congenital defects.

CARDIAC EFFECTS
OF BRAIN LESIONS

Brain lesions in various loci, especially the lateral brainstem and insula, have been shown to cause:

- Cardiac lesions—myocytolysis, subendocardial hemorrhages
- Arrhythmias
- Electrocardiographic (ECG) and enzyme changes
- Pulmonary edema
- Sudden death

COEXISTENT OCCLUSIVE
VASCULAR DISEASE[5]

Common Locations and Symptoms
and Signs (Fig. 37-1 and Table 37-2)

TREATMENT[5]
- Carotid endarterectomy (CEN) is effective in symptomatic patients with >70 percent stenosis if done by surgeons with mortality/morbidity not greater than 2 to 4 percent (Chap. 39).
- Vertebral artery surgery feasible in some patients.
- In patients with TIAs or strokes due to intraarterial emboli from arteries with minor to moderate stenosis, use platelet antiaggregants (aspirin, ticlopidine) to prevent white cell–platelet-fibrin thromboemboli.
- Patients with severe stenosis–warfarin to prevent red fibrin clots: INR 2 to 3. Monitor lesions noninvasively by ultrasound CTA, or MRA.
- Patients with complete occlusion who have minor or moderate strokes, short-term heparin, then warfarin while clot becomes organized and adherent.
- Thrombolytic treatment and angioplasty are of growing importance.

Guidelines for the use of antiplatelets and anticoagulants are given in Table 37-3.

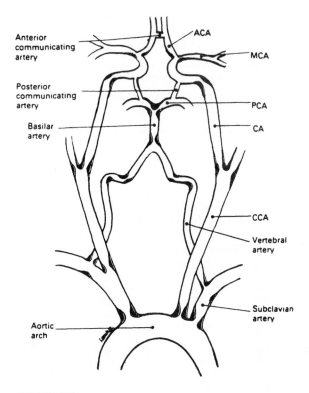

FIGURE 37-1

Sites of predilection for atherosclerotic narrowing: black areas represent plaques. (From Caplan LR, Stein RW: *Stroke: A Clinical Approach.* Boston, Butterworth, 1986, with permission.)

TABLE 37-2

COMMON SIGNS IN CEREBROVASCULAR OCCLUSIVE DISEASE AT VARIOUS SITES

ICA origin	Ipsilateral transient monocular blindness; MCA and ACA signs
ICA siphon (proximal to ophthalmic artery)	Same as ICA origin
ICA siphon (distal to ophthalmic artery)	MCA and ACA signs
ACA	Contralateral weakness of the lower limb and shoulder shrug
MCA	Contralateral motor, sensory, and visual loss Left: Aphasia Right: Neglect of left space, lack of awareness of deficit, apathy, impersistence
AChA	Contralateral motor, sensory, and visual loss, usually without cognitive changes
Subclavian artery (proximal to VA)	Lack of arm stamina, cool hand, transient dizziness, veering, diplopia
VA origin	Same as subclavian but no ipsilateral arm or hand findings
VA intracranially	Lateral medullary syndrome; staggering and veering (cerebellar infarction)
BA	Bilateral motor weakness; ophthalmoplegia and diplopia
PCA	Contralateral hemianopia and hemisensory loss Left: Alexia with agraphia Right: Neglect of left visual space

Key: ICA = internal carotid artery; ACA = anterior cerebral artery; MCA = middle cerebral artery; AChA = anterior choroidal artery; VA = vertebral artery; BA = basilar artery; PCA = posterior cerebral artery.

TABLE 37-3

SUGGESTED USE OF ANTICOAGULANTS
AND PLATELET ANTIAGGREGANTS

Heparin (Standard Dose)

Short-term, 2–4 weeks. Usually given by intravenous infusion
keeping APTT between 60 and 100 s (1.5–2 × control APTT).
Immediate therapy of definite cardiac origin cerebral embolism (large cerebral infarct, hypertension, bacterial endocarditis, or sepsis would delay or contraindicate this use).
Patients with severe stenosis or occlusion of the ICA origin,
ICA siphon, MCA, vertebral artery, or basilar artery with less
than a clinical deficit. Subsequent treatment could be warfarin or surgery.

Heparin (Subcutaneous Minidose)

For prophylaxis of deep vein occlusion in patients immobilized by
stroke (unless contraindicated) (see Chap. 23).

Warfarin

Usually overlapped with heparin; keeping prothrombin time
around INR of 2.0–3.0 (approximately 1.3–1.5 × control).
Long-term (>3 months)
  Patients with cardiogenic cerebral embolization and
  rheumatic heart disease, atrial fibrillation with large atria or
  prior cerebral embolism, prosthetic valves, and some hypercoagulable states.
  Patients with severe stenosis of the ICA origin, ICA siphon,
  MCA stem, vertebral artery, and basilar artery. Used until
  studies show artery has been occluded for at least 3 weeks.
Short term (3–6 weeks)
  Patients with recent occlusion of the ICA, MCA, vertebral, or
  basilar arteries.

Platelet Antiaggregants (Aspirin, Ticlopidine)

Patients with plaque disease of the extracranial and intracranial
arteries without severe stenosis.
Patients with polycythemia or thrombocytosis and related
ischemic attacks.

Key: APTT = activated partial thromboplastin time; ICA = internal
carotid artery; MCA = middle cerebral artery; INR = International
Normalized Ratio.

SURGERY FOR COEXISTENT CORONARY AND OCCLUSIVE VASCULAR DISEASE

Many patients have both coronary and cerebrovascular occlusive disease. In candidates for both CEN and CABS, controversy exists over timing—whether both should be done together or whether they should be staged.

Studies suggest that patients with active coronary artery disease and asymptomatic carotid disease should have CABS only. If they have symptomatic carotid disease and have no (or stable) coronary symptoms, CEN is done first. If both coronary and carotid symptoms are active, a combined procedure is best.[7,8]

REFERENCES

1. Petersen P, Boysen G, Godtfredsen J, et al: Placebo-controlled, randomized trial of warfarin and aspirin for prevention of thromboembolic complications in chronic atrial fibrillation: Copenhagen AFASAK study. *Lancet* 1989; 1:175–179.
2. Boston Area Anticoagulation trial for atrial fibrillation investigators: the effect of low-dose warfarin on the risk of stroke in patients with nonrheumatic atrial fibrillation. *N Engl J Med* 1990; 323:1505–1511.
3. Stroke prevention in atrial fibrillation investigators: The stroke prevention in atrial fibrillation trial: final results. *Circulation* 1991; 84:527–539.
4. Stroke prevention in atrial fibrillation investigators: Clinical and echocardiographic features of risk. *Ann Intern Med* 1992; 116:1–5, 6–12.
5. Caplan LR: *Stroke: A Clinical Approach*, 2d ed. Boston, Butterworth, 1993.
6. Barbut D, Caplan LR: Brain complications of cardiac surgery. *Curr Probl Cardiol* 1997; 22:455–476.
7. Hertzer NR, Loop FD, Beven EG: Management of coexistent carotid and coronary artery disease. In: Furlan A (ed): *The Heart and Stroke*. London, Springer-Verlag, 1987: 305–318.
8. Pettigrew LC: Surgical considerations. In: Rolak L, Rokey R (eds): *Coronary and Cerebral Vascular Disease*. Mt Kisco, NY, Futura, 1990:349–377.

SUGGESTED READING

Caplan LR: Cerebrovascular disease and neurologic manifestations of heart disease. In: Alexander RW, Schlant RC, Fuster V, O'Rourke RA, Roberts R, Sonnenblick EH (eds): *Hurst's The Heart*, 9th ed. New York, McGraw-Hill, 1998:2483–2504.

Matchar DB, McCrory DC, Barnett HJM, et al: Medical treatment for stroke prevention. *Ann Intern Med* 1994; 121:41–55.

CHAPTER

38

DISEASES OF THE AORTA

Joseph Lindsay, Jr.

DEFINITION

The aorta is affected by diverse disease processes, but its uncomplicated structure and function limit the number of clinical syndromes that result. Weakening of the aortic wall by congenital or acquired disease may lead to aneurysm, rupture, or dissection. When the process affects the proximal aorta, aortic valvular incompetence may result from dilatation of the aortic ring or involvement of the valve leaflets themselves. Disease may also narrow the aorta or the origin of one of its branches.

ETIOLOGY

Atherosclerosis of the aorta accompanies aging in most individuals in the western world but varies in severity from subject to subject. It is accelerated by diabetes, hypercholesterolemia, hypertension, and smoking. Aortic atherosclerosis is clinically manifest most often by embolization of plaque material from its luminal surface to more distal arterial segments. Obstruction of the infrarenal aorta by atherosclerosis may produce ischemic symptoms in the lower extremities. Infrarenal abdominal aneurysms have, in the past, been attributed to atherosclerosis, but currently most authorities believe that such aneurysms are the consequence of medial weakening. Penetration of an atherosclerotic plaque into the aortic media may expose the media to the pulsatile force of the bloodstream. Although the aortic media is very resistant to radial stress, axial stress from this event may produce splitting of the medial fibers—that is, an aortic dissection.

As with atherosclerosis, *medial degeneration* accompanies aging. As a consequence, the aorta becomes elongated, tortuous, and inelastic in older individuals. Severe and premature

551

medial degeneration is the cardiovascular hallmark of Marfan's syndrome. The gene that controls the production of fibrillin, a matrix protein of the aorta and other connective tissue, has been identified. Aneurysmal dilatation of the proximal aorta, including the sinuses of Valsalva, is characteristic of this heritable disease. The resulting bulb-shaped aortic root has been termed *anuloaortic ectasia*. In addition, longitudinal splitting of the media (aortic dissection) may occur. While such aneurysms are characteristic of Marfan's syndrome, most patients who have them show no other manifestations of that genetic disorder. Medial degeneration may be severe enough to produce clinical disease in individuals with coarctation of the aorta, bicuspid aortic valve, polycystic kidneys, and Turner's syndrome. Its association with these congenital/heritable diseases adds credence to the idea that additional genetically determined medial defects await identification.

Infectious aortitis may result from extension of aortic valve endocarditis to the adjacent wall and, less often, from spread of infection in periaortic soft tissue. Although the intact aortic intima is resistant to blood-borne infection, areas damaged by atherosclerosis or other disease may be invaded by pathogens from the bloodstream. The late effects of syphilitic aortitis (aneurysm and aortic regurgitation) are still encountered.

Nonspecific aortitis, often associated with evidence of an "autoimmune" process, has been termed *Takayasu's aortitis*. Moreover, a nonspecific aortitis of the proximal aorta often accompanies giant cell arteritis, ankylosing spondylitis, and Reiter's syndrome.

Common congenital anomalies encountered after early childhood include aortic coarctation and aneurysm of the sinuses of Valsalva.

Trauma that is not immediately fatal may result in the formation of a false aneurysm at the site of injury.

CLINICAL MANIFESTATIONS

Most aortic aneurysms are asymptomatic and are first detected in the course of a routine physical examination or an imaging study directed at another organ. An expanding aneurysm may impinge on or erode into an adjacent structure, thereby producing pain or other symptoms. For example, an aneurysm of the aortic arch may produce hoarseness or dysphagia. It is wise to regard any symptom referable to an aneurysm as evidence of expansion and impending rupture.

Rupture of an aneurysm is life-threatening and demands immediate operative treatment. The sudden onset of abdominal discomfort in a patient with an abdominal aneurysm or of chest pain in one with aneurysm of the thoracic aorta heralds such events. Evidence of blood loss often accompanies the pain and alerts the examiner to the vascular nature of the illness.

Aortic regurgitation is a common manifestation of proximal aortic disease. Severe valvular incompetence most often results from anuloaortic ectasia, with its characteristic dilatation of the aortic sinuses, ascending aorta, and aortic valve ring. When a regurgitant valve is the initial manifestation of aortic root disease, as is often the case, the underlying nature of the disease may not be suspected until echocardiography is performed.

Aortic dissection, transaxial splitting of the medial fibers, produces a dramatic clinical syndrome characterized by the abrupt onset of severe midline chest, back, or abdominal pain or, less commonly, by sudden syncope. The aortic valve is often rendered incompetent when the ascending aorta is involved. Moreover, a major branch vessel off the aorta may be obstructed, producing absent arterial pulses and evidence of obstruction to flow to the brain, spinal cord, kidneys, viscera, or limbs.

Obstruction of one or more of the branches of the aortic arch, as a consequence of aortitis or atherosclerosis, may produce ischemic symptoms of the central nervous system or an upper extremity. Atherosclerotic obstruction of the infrarenal aorta may result in effort-produced discomfort of the low back, buttocks, or thighs (the Leriche syndrome). Abrupt occlusion of the aortic bifurcation by a saddle embolus, aortic dissection, or in situ thrombosis results in limb-threatening lower extremity ischemia accompanied by weakness or frank paralysis.

Atherosclerotic aortic segments are often the source of *emboli* to the central nervous system, the viscera, or the extremities.

Physical Examination

Detection of the murmur of aortic regurgitation, of a supraclavicular or abdominal bruit, or of diminution of one or more of the carotid, radial, or femoral pulses may direct the examiner's attention to aortic disease or may support a suspicion of that possibility raised by other information. Moreover, every abdominal examination should include a search for the expansile mass of an abdominal aneurysm. The presence of a tracheal tug or of a lift of either sternoclavicular joint may reflect aneurysm or dissection of the aortic arch.

One should be particularly alert for evidence of aortic disease in patients whose body habitus suggests either Marfan's or Turner's syndrome.

Usual X-ray Studies

The aortic silhouette on the plain chest film is usually distorted in the presence of disease of the thoracic aorta. Among the exceptions to this generalization are processes limited to the aortic root (aortitis and anuloaortic ectasia). In that case, any aortic deformity may lie entirely within the cardiac shadow. Notching of the underside of the ribs is virtually pathognomonic of aortic coarctation.

Abdominal x-rays, particularly when proper techniques are employed, frequently disclose the calcified outline of an infrarenal aneurysm. A "cross-table lateral" may be rewarding.

DIFFERENTIAL DIAGNOSIS

The possibility of aortic disease should be considered in a variety of clinical situations. It may be responsible for newly discovered aortic regurgitation. It may manifest as pain, suggesting myocardial infarction or an acute abdominal illness. At times the manifestations of an infectious or an autoimmune process may seem to dominate. The mass of an aneurysm may lead to the suspicion of a neoplasm.

SPECIAL DIAGNOSTIC EVALUATION

The aorta may be imaged in exquisite detail by several techniques. The "gold standard" of a decade ago, contrast aortography, can now be reserved for special needs. Computed tomography (CT) is perhaps the most readily available screening tool for aneurysm or dissection. Ultrasound study offers advantages over both CT and aortography in that it is noninvasive and does not require the injection of contrast material. Abdominal ultrasound, long a valuable tool, has been joined by transesophageal echocardiography (TEE) as a means of providing aortic images of great value. The latter may be the diagnostic method of choice in acutely ill patients suspected of having dissection. Magnetic resonance imaging (MRI) provides remarkably detailed images and may be an important tool in selected cases.

NATURAL HISTORY AND PROGNOSIS

Rupture or dissection of the aorta is an acute, highly lethal emergency. Survival for more than a few days is exceptional unless surgical repair can be effected. An exception to this generalization is dissection located distal to the aortic arch. Such patients may survive with effective medical management.

Aortic aneurysms tend to enlarge gradually. The greater their size, the more likely they are to rupture. An abdominal aneurysm of 5 cm or more presents a sufficient threat to the patient to justify operation. In the thorax, 6 cm is the critical dimension usually cited. Smaller aneurysms also rupture, but less often. One should bear in mind that advanced age or associated illness may be more threatening to the patient's well-being than the aneurysm.

Aortic regurgitation and aortic branch obstruction also tend to be progressive.

TREATMENT

Surgical repair of aortic lesions is now usually possible in suitable patients and is the only definitive approach in most cases. For rupture of an aneurysm or for dissection of the proximal aorta, surgical treatment is the lifesaving option. *Medical treatment* of dissection or unruptured aneurysm is generally supportive or aimed at preventing progression of a demonstrated lesion. For example, the patient with medial degeneration and anuloaortic ectasia may be given beta-blocking agents and other antihypertensive medications to reduce wall stress and the severity of aortic regurgitation.

Aggressive reduction of blood pressure has a special role in many cases of acute aortic dissection. Systolic arterial pressure should be lowered promptly to 100 to 120 mmHg using a drug regimen such as those suggested in Table 38-1. This antihypertensive program should be maintained through the preoperative period. In selected patients whose dissection begins beyond the aortic arch, chronic antihypertensive therapy may be preferable to surgery.

TABLE 38-1

TREATMENT OF AORTIC DISSECTION
ANTIHYPERTENSIVE AGENTS

Agent	Dose	Comments
Nitroprusside	0.5–10.0 µg/kg/min IV	Also use beta blocker
Esmolol		
Loading dose	0.5 mg/kg IV for 1 min	
Infusion	0.05 mg/kg/min, increase by 0.05 mg/kg/min q 4 min	Until pulse slows No more than 0.3 mg/kg/min
Labetalol	20 mg over 2 min IV	Additional 40–80 mg*
Trimethaphan	1 to 2 mg/min IV	Tachyphylaxis common
Propranolol	0.5 mg/5 min	Until pulse slows or dose reaches 1.5 mg/10 kg

*At 5-min intervals until effect or a total of 300 mg.

556

SUGGESTED READING

Dolmatch BL, Gray RJ, Horton KM, Rundback JH: Diagnostic imaging in the evaluation of aortic disease. In: Lindsay J Jr (ed): *Diseases of the Aorta*. Philadelphia, Lea & Febiger, 1994:197–250.

Ernst CB: Abdominal aortic aneurysm. *N Engl J Med* 1993; 328:1167–1172.

Goldstein SA: Ultrasound in the diagnosis of aortic disease. In: Lindsay J Jr (ed): *Diseases of the Aorta*. Philadelphia, Lea & Febiger, 1994:251–284.

Halloran BG, Baxter BT: Pathogenesis of aneurysms. *Semin Vasc Surg* 1995; 8:85–92.

Kronzon I, Tunick P: Atheromatous disease of the thoracic aorta: pathologic and clinical implications. *Ann Intern Med* 1997; 126:629–637.

Lindsay J Jr, Beall AC Jr, DeBakey ME: Diagnosis and treatment of diseases of the aorta. In: Alexander RW, Schlant RC, Fuster V, O'Rourke RA, Roberts R, Sonnenblick EH (eds): *Hurst's The Heart*, 9th ed. New York, McGraw-Hill, 1998:2461–2482.

Lindsay J Jr: Diagnosis and treatment of diseases of the aorta. *Curr Probl Cardiol* 1997; 22:488–542.

Roman MJ, Devereux RB: Heritable aortic disease. In: Lindsay J Jr (ed): *Diseases of the Aorta*. Philadelphia, Lea & Febiger, 1994:55–74.

DISEASES OF THE PERIPHERAL ARTERIES AND VEINS

Robert C. Allen
Mark J. Kulbaski
Robert B. Smith III

LOWER EXTREMITY PERIPHERAL ARTERIAL DISEASE

Definition

Peripheral arterial diseases are abnormalities that impair the ability of the circulatory system to deliver blood to the lower extremities or result in aneurysmal dilatation of the artery due to a weakening of the vessel wall.

Etiology

Atherosclerosis is the most common cause of arterial disease of the lower extremities. Risk factors for the development and progression of atherosclerotic disease include smoking, lipid abnormalities, diabetes mellitus, hypertension, and homocysteinemia. The arteritides (i.e., giant-cell arteritis, Takayasu's disease, Buerger's disease) can be contributory and are due to an inflammatory and/or autoimmune process, ultimately resulting in either arterial occlusive disease or aneurysmal degeneration. Other important etiologies of compromised blood flow to the lower extremities are emboli, dissection, trauma, vasoconstrictive medications, and profound systemic shock.

Clinical Manifestations

HISTORY

Intermittent claudication is the classic symptom of chronic lower extremity arterial insufficiency. This usually presents as calf fatigue and cramping with exercise that is relieved promptly by rest. Progression of disease with involvement of multiple levels and inadequate collateral circulation can result in pain with minimal exertion or at rest. Rest pain presents as a severe burning discomfort in the foot and may progress to actual tissue necrosis. Acute ischemia produces excruciating pain with progressive sensory and motor deficits.

PHYSICAL EXAMINATION

The hallmark of peripheral occlusive disease is diminished or absent pulses. Chronic severe arterial insufficiency may also result in dependent rubor, elevation pallor, thickened nails, absent hair, and atrophic skin. The leg is often cool and mottled and may have tissue loss or distal gangrene. Bruits may be present, representing turbulent flow in vessels due to stenotic or aneurysmal changes. The acutely ischemic limb will manifest "the six Ps": pulselessness, pain, pallor, paresthesia, poikilothermia, and paralysis.

Differential Diagnosis

The variety of vascular diagnoses must also be differentiated from the nonvascular causes of lower extremity pain. These include predominantly neurogenic or muscular disorders and collectively may be termed *pseudoclaudication*.

Diagnostic Evaluation

The noninvasive vascular laboratory is an essential tool in the diagnosis of lower extremity arterial insufficiency. Plethysmographic pulse volume recordings and Doppler ultrasound can very accurately detect the presence of lower extremity arterial insufficiency, along with the anatomic level and severity. Contrast arteriography remains the "gold standard" to define the presence and severity of arterial disease and is reserved to define anatomy prior to operative intervention.

Natural History/Prognosis

Chronic lower extremity arterial insufficiency is typically characterized by a slow progression in the distribution and severity of disease. Over their lifetime only 25 percent of patients with complaints of calf claudication will require surgical intervention, with only 5 to 10 percent eventually requiring amputation. In contrast, without revascularization, patients with rest pain or

gangrene face a very high rate of limb loss. Acute limb ischemia is even more ominous, owing to inadequate collateral
circulation, and requires emergent intervention for limb sal-
vage. Peripheral aneurysms in the femoral or popliteal arteries
often result in leg ischemia due to thrombosis or distal embolization, but rupture rarely causes acute limb threat.

Treatment

MEDICAL

The single most important factor in the care of patients with
atherosclerotic occlusive disease is the cessation of smoking.
An exercise program that stresses daily activity is important to
maximize the development of collateral pathways and the effi-
ciency of muscle metabolism. Lipid abnormalities and
homocysteinemia should be aggressively treated with dietary
alterations and possibly medication. Diabetics require both
strict control of their blood glucose and preventive foot care.
Pharmacologic intervention with the rheologic agent pent-
oxifylline may benefit some patients with claudication (dose:
400 mg PO tid). A trial of vasodilators is indicated in patients
with vasospastic disorders; glucocorticoids may be beneficial
in the inflammatory arteriopathies.

SURGICAL/INTERVENTIONAL

Stable claudication that does not interfere with patient lifestyle
is not an indication for intervention. However, progressive
short-distance claudication and severe ischemia with rest pain,
tissue loss, or gangrene should be treated promptly. Open surgical techniques include removal of the arterial obstruction
(endarterectomy) or revascularization using a bypass graft.
Catheter-based endovascular techniques are becoming increasingly popular and are appropriate in selected patients with
limited occlusive disease. Angioplasty and stenting of the common iliac artery has the best durability; the long-term patency
rates after angioplasty of smaller vessels are lower. Thrombolytic therapy, atherectomy, and stent grafting are other
techniques that must be individualized. Aneurysmal disease is
usually treated by aneurysmorrhaphy or exclusion of the
aneurysm and bypass. The repair of aortic and peripheral artery
aneurysms with endoluminal stent grafts is now being investigated in clinical trials. Patients with vasospastic diseases or
with nonreconstructable vasculature and shallow ulceration
may benefit from sympathectomy. Life-threatening sepsis,

advanced gangrene, and nonreconstructable occlusive anatomy are indications for primary amputation.

CAROTID ARTERY CEREBROVASCULAR DISEASE

Definition

Carotid artery disease specifically refers to those pathologic processes afflicting the extracranial carotid arteries that supply the anterior cerebral circulation and the ipsilateral eye.

Etilogy

Atherosclerosis is the dominant disease entity with a strong predilection for the lesion to involve the carotid bifurcation and the origin of the internal carotid artery. The majority of transient ischemic attacks (TIAs) and fixed neurologic deficits are due to embolization of material from ulcerated plaques, but reduced flow related to severe stenosis may also account for some ischemic deficits. Alternative disease processes include fibromuscular dysplasia (FMD), aneurysmal disease, inflammatory arteritis, radiation therapy injury, and carotid body tumors (see also Chap. 37).

Clinical Manifestations

HISTORY

Many patients with carotid atherosclerotic occlusive disease are asymptomatic. Symptoms may be specific to the hemispheric distribution or may be vague and nonspecific owing to global hypoperfusion. The neurologic complaints may be temporary (TIAs), fluctuating/progressive (stroke in evolution), or fixed (completed stroke). Temporary monocular blindness (amaurosis fugax) results from microembolization of the ipsilateral ophthalmic artery and is described as a "shade coming down over the eye."

PHYSICAL EXAMINATION

An asymptomatic carotid bruit may be the only indication of significant cerebrovascular disease. A pulsatile neck mass may be indicative of a carotid aneurysm or carotid body tumor. Fundoscopic examination in patients with visual changes can detect embolic cholesterol plaque in the retinal artery. Neurologic findings are variable and reflect the extent of the cerebral ischemia (see Chap. 37).

Differential Diagnosis

The presence of atherosclerotic carotid disease should be presumed until formally excluded by objective testing. Primary central nervous system processes should be considered, including neoplasms, abscess, and epidural/subdural hematoma. Other important mimicking conditions are carotid kinks, dissection, cardiogenic embolization, and subclavian steal.

Diagnostic Evaluation

Duplex ultrasonography is an excellent noninvasive test to evaluate the extent and severity of carotid disease. Computed tomography or magnetic resonance imaging is used to exclude or localize cerebral infarct or hemorrhage. Echocardiography, especially transesophageal (TEE), is helpful in excluding a cardiac source for embolism. Although cerebral angiography has traditionally been used to define carotid lesions and anatomy, carotid endarterectomy may safety be performed based on information from duplex or magnetic resonance angiography if the clinical and imaging data are congruous.

Natural History/Prognosis

Knowledge of the long-term outcome of patients with carotid artery cerebrovascular disease is essential because of the potential for neurologic sequelae. However, the underlying generalized atherosclerosis associated with coronary disease makes cardiac disease the most important cause of mortality in this patient population. Asymptomatic lesions with a greater than 50 percent stenosis carry a risk of stroke of approximately 4 percent per year. Asymptomatic high-grade (greater than 80 percent) carotid lesions become symptomatic or occlude in 35 percent of patients within 2 years. Patients with symptomatic high-grade lesions have a stroke rate at 18 months of 24 percent with medical management versus 7 percent with surgical intervention.

Treatment

MEDICAL

Medical management involves antiplatelet drugs, with aspirin the preferred medication. The optimal dose of aspirin is currently under investigation. Newer antiplatelet drugs such as ticlopidine may be more efficacious. Patients should be counseled with regard to cessation of smoking. Blood pressure as well as lipid and homocyteine abnormalities should also be controlled.

SURGICAL/INTERVENTIONAL

Carotid endarterectomy is the procedure of choice for extracranial carotid artery occlusive disease. Operative indications are (1) appropriate neurological symptoms with a significant atherosclerotic lesion; (2) asymptomatic stenosis with a high-grade obstruction (greater than 75 to 80 percent); and (3) neurologic symptoms with a complex ulcerated plaque. When carotid endarterectomy is performed by experienced surgeons, the combined operative mortality and morbidity in patients with symptomatic disease is 5 percent and in patients with asymptomatic disease, less than 2 percent. Carotid stenting for the treatment of carotid bifurcation disease is now being investigated within clinical trials. However, in early studies, stroke rates with carotid stenting have been higher than with open endarterectomy.

LOWER EXTREMITY DEEP VENOUS THROMBOSIS

Definition

Deep venous thrombosis (DVT) is clotting that develops in the deep veins of the calf, thigh, or pelvis.

Etiology

Virchow's triad of stasis of blood, vessel wall injury, and increased coagulability concisely describes the primary etiologic factors that precipitate venous thrombosis. Stasis of blood results from reduced venous return, as occurs in prolonged bed rest, limb paralysis, surgical procedures, and venous valvular insufficiency. Vessel wall injury may be due to surgical injury, iatrogenic catheterization, or blunt/penetrating trauma. Hypercoagulability may be a primary or secondary state: its causes include protein deficiencies (protein C, protein S, antithrombin III), malignancy, thrombocytosis, the lupus anticoagulant, and the factor V Leiden mutation.

Clinical Manifestations

HISTORY

A majority of cases of DVT are silent and require a high index of suspicion for early diagnosis. The patient may have only vague, nonspecific complaints. Local symptoms may include pain, swelling, or tenderness in the involved extremity. Chest pain and shortness of breath with a "feeling of doom" is an infrequent but not rare presentation due to an acute pulmonary embolus (PE) secondary to the DVT.

PHYSICAL EXAMINATION

Clinical findings are usually similarly unimpressive, but localized limb swelling and tenderness can be present. An extremity with marked physical findings is indicative of an extensive underlying iliofemoral thrombotic process (phlegmasia cerulea dolens). Palpable cords are present in superficial thrombophlebitis but are uncommon in DVT. Homans' sign (pain or resistance on passive dorsiflexion of the ankle) may be present and raises clinical suspicison but, contrary to past dogma, is a nonspecific finding. The venous occlusion may become severe and, in rare cases, may result in a pale leg with compromised arterial perfusion (phlegmasia alba dolens).

Differential Diagnosis

DVT must be accurately distinguished from other causes of a painful, swollen lower extremity. The important disorders in this list are malignancy, musculoskeletal injury, infections, lympedema, congestive heart failure, and a ruptured Baker's cyst. Bilateral leg edema as a side effect of medication with calcium channel blockers is commonly seen.

Diagnostic Evaluation

Objective evaluation is essential, since clinical criteria are inaccurate in more than 50 percent of cases. A bedside examination with a simple Doppler ultrasound device can be highly accurate in the diagnosis of DVT. Impedance and strain-gauge plethysmography are simple, noninvasive tests that are available in most noninvasive vascular laboratories, with an accuracy above the knee of greater than 90 percent. Venous duplex scanning, by combining ultrasound imaging and Doppler analysis, is extremely accurate in the diagnosis of DVT and should be considered the noninvasive test of choice. If these tests are inconclusive, contrast venography remains the "gold standard" for the diagnosis and definition of the venous anatomy. However, venography requires contrast injection and may cause renal toxicity or chemical phlebitis in a small percentage of patients.

Natural History/Prognosis

The origin of most lower extremity venous thrombi is the calf, with subsequent propagation into the thigh. The incidence of proximal propagation is approximately 20 to 30 percent in an untreated patient with a calf DVT. Possible serious sequelae of untreated leg DVT are chronic venous insufficiency with the postphlebitic syndrome and pulmonary embolus. The postphlebitic syndrome is characterized by lower extremity pain, edema, fibrosis, and ulceration.

Treatment

MEDICAL

Traditionally, patients with a lower extremity DVT have been treated with strict bed rest, leg elevation, and systemic anticoagulation with intravenous heparin, which is then switched to oral warfarin for 3 to 6 months. Recently, outpatient therapy with self-administered subcutaneous injections of low-molecular-weight heparin has been shown to be safe and effective. Lower extremity support stockings are an essential adjunct to minimize leg edema. The use of thrombolytic therapy to minimize venous valvular injury due to the thrombosis is currently under investigation (see also Chap. 23).

SURGICAL/INTERVENTIONAL

The need for surgical therapy for DVT is rare. Open surgical thrombectomy should be performed if thrombolytic therapy for phlegmasia cerulea dolens has failed or is contraindicated and there is impending venous gangrene. Inferior vena cava filter insertion is indicated in the following conditons: (1) recurrent or progressive thromboembolic disease despite therapeutic anticoagulation; (2) contraindication to or complication of anticoagulation; and (3) prophylaxis in a patient at high risk for a pulmonary embolus. Thrombolytic therapy is also an option in the treatment of a PE provided that no contraindications exist. Cardiopulmonary bypass with pulmonary artery embolectomy or suction embolectomy should be considered in patients with a documented PE who fail to respond to maximal medical management.

SUGGESTED READING

Callow AD, Ernst CB (eds): *Vascular Surgery, Theory and Practice*. Stamford, CT, Appleton & Lange, 1995.

Dodson TF, Smith RB III: Surgical treatment of peripheral vascular disease. In: Alexander RW, Schlant RC, Fuster V, O'Rourke RA, Roberts R, Sonnenblick EH (eds): *Hurst's The Heart*, 9th ed, New York, McGraw-Hill, 1998:2529–2539.

Joyce JW, Rooke TW: Diagnosis and management of diseases of the peripheral arteries and veins. In: Alexander RW, Schlant RC, Fuster V, O'Rourke RA, Roberts R, Sonnenblick EH (eds): *Hurst's The Heart*, 9th ed, New York, McGraw-Hill, 1998:2505–2527.

Moore WS (ed): *Vascular Surgery, A Comprehensive Review*, 5th ed. Philadelphia, Saunders, 1998.

Rutherford RR (ed): *Vascular Surgery*, 4th ed. Philadelphia, Saunders, 1995.

PERIOPERATIVE EVALUATION AND MANAGEMENT OF PATIENTS WITH KNOWN OR SUSPECTED CARDIOVASCULAR DISEASE WHO UNDERGO NONCARDIAC SURGERY

Robert C. Schlant
Kim A. Eagle

PATIENTS WITH KNOWN CARDIAC DISEASE

Coronary Artery Disease

Patients with known coronary artery disease (CAD) have an increased risk for perioperative cardiac morbidity (PCM), including myocardial infarction (MI), unstable angina pectoris, congestive heart failure (CHF), serious arrhythmia, and cardiac death. Patients with a documented previous MI have a perioperative risk of about 6 percent of a recurrent MI. The risk is greater in patients operated upon less than 3 to 6 months post-MI.

The mortality of a perioperative MI ranges from about 26 to 70 percent, with an average of about 50 percent. Most perioperative MIs occur within the first 4 postoperative days, with a peak incidence on the second day. Often they are clinically silent, although frequently associated with ST-segment changes in the preceding 3 to 6 h.

In general, medication regimens for patients with CAD should be continued up to, during, and following surgery. A frequent exception is aspirin, which is often discontinued 5 to 7 days prior to surgery to lessen blood loss. Patients with known or suspected CAD who are not taking a beta blocker should be considered for such therapy if there is no contraindication.

Although it is common practice, there are few data to support the intraoperative administration of intravenous nitroglycerin in patients with known or suspected CAD.

Systemic Arterial Hypertension

Untreated or poorly controlled systemic arterial hypertension (SAH) is associated with an increased incidence of perioperative myocardial ischemia, arrhythmias, and transient neurologic symptoms. Whenever possible, SAH should be controlled prior to surgery and antihypertensive medications continued up to, during, and following surgery. Special care should be taken to avoid rebound hypertension and tachycardia, which can occur following discontinuation of oral clonidine or, occasionally, a beta blocker.

Valvular Heart Disease

Patients with valvular heart disease should have antibiotic prophylaxis to prevent endocarditis before either lithotripsy or surgery likely to be associated with bacteremia (see Chap. 21). Patients who are on chronic warfarin (Coumadin) therapy for a prosthetic heart valve or atrial fibrillation should have the warfarin discontinued 5 to 7 days prior to surgery. Whenever possible, they should then be hospitalized and given intravenous heparin to maintain the activated partial thromboplastin time at 1.5 to 2.0 times the control value. The effect of heparin should be reversed by intravenous protamine sulfate immediately before surgery. Postoperatively, both intravenous heparin and warfarin are restarted when there is no evidence or significant risk of bleeding. The heparin is discontinued after the prothrombin time is within therapeutic range (see also Chap. 46). In emergencies, the effect of warfarin can be reversed by intravenous vitamin K_1, 10 to 25 mg, or fresh frozen plasma. Administration of vitamin K_1 often will delay the anticoagulant effect of warfarin when it is restarted (see also Chaps. 15 and 46).

Patients with moderate to severe mitral stenosis (Chap. 13), aortic stenosis (Chap. 12), or hypertrophic cardiomyopathy (Chap. 18) are very sensitive to changes in venous return. In such patients, the monitoring of pulmonary artery pressures by a percutaneous balloon catheter frequently is very helpful, both during surgery and postoperatively.

Congestive Heart Failure

Patients with diabetes mellitus or a history of MI or dysrhythmias have an increased risk of postoperative CHF.

The presence of CHF is associated with an increased risk of PCM. Most patients with CHF should be treated with triple therapy [diuretic, angiotensin-converting enzyme (ACE) inhibitor, and digoxin] to optimize cardiac performance and minimize the symptoms and signs of CHF prior to elective surgery (see also Chap. 1). Special care should be taken to avoid hypokalemia, hypomagnesemia, excess depletion of blood volume, and excess digitalis.

Patients are more difficult to manage if they do not have known CHF and have not been on therapy but are discovered to have moderate or marked cardiomegaly on a chest film and/or significantly decreased left ventricular function by echocardiography. In many such patients, it is advisable to initiate therapy with an ACE inhibitor preoperatively. Diuretics are used only if there is evidence of edema on physical examination or the chest film. Digoxin is used only if the left ventricular ejection fraction is less than about 40 percent and if there is cardiomegaly, an S3 gallop, and increased jugular venous pressure. Whenever possible, therapy with digoxin should be initiated several days before surgery (see also Chap. 43).

Inotropic support for severe heart failure during or following surgery can best be provided by intravenous dobutamine and/or milrinone (see Chap. 43).

Congenital Heart Disease

Most patients with congenital heart disease should receive antibiotic prophylaxis against endocarditis when undergoing surgery likely to be associated with bacteremia (see Chap. 21). Patients with cyanotic congenital heart disease who have an hematocrit greater than 65 percent should be considered for cautious preoperative phlebotomy with replacement of the blood volume to lessen the change of vascular thrombosis and postoperative bleeding.

Arrhythmia and Conduction Disorders

Patients with a history of a cardiac arrhythmia have an increased risk of postoperative CHF.

About 60 percent of all patients have perioperative arrhythmias; most of them require no therapy. Preoperatively, patients with very frequent or symptomatic atrial or junctional premature beats can be treated with a beta blocker or digoxin. Patients with CAD who have more than 5 premature ventricular contractions (PVCs) per minute, particularly if they are multifocal or occur in runs, should be considered for intravenous lidocaine (2 to 3 g in 1 L of 5 percent glucose in water), initiated at a rate of 1 mL/min and continued during surgery (see Chap. 2).

Adult patients undergoing emergency surgery who have supraventricular tachyarrhythmias should be considered for electrical cardioversion or rate control with intravenous beta blocker, verapamil, diltiazem, or digoxin (see Chap. 2).

Patients with a history of Stokes-Adams attacks, complete heart block, high-degree (Möbitz II) atrioventricular block, or prolonged sinoatrial pause or block should be considered for a prophylactic temporary right heart pacemaker. On the other hand, patients with asymptomatic bifascicular block (left or right bundle branch block with left anterior or posterior fascicular block), with or without prolongation of the PR interval, usually do not require a prophylactic pacemaker prior to surgery, although a pacing catheter should be readily available (see also Chaps. 2 and 4).

The intraoperative use of electrocautery can rarely interfere with the function of a cardiac pacemaker or an implantable cardioverter-defibrillator (ICD). During electrocautery surgery, the electrosurgical tip and the ground plate should be placed as far away from the pacemaker or ICD as possible and electrosurgery limited to 2- to 3-s periods if there is any evidence of pacemaker suppression (see also Chap. 4).

Pulmonary Embolism

Patients at increased risk of pulmonary embolism should be considered for graduated-pressure compression stockings and for prophylactic low-dose heparin (5000 to 10,000 units every 8 to 12 h), preferably beginning 10 to 24 h before surgery. Heparin should be avoided in patients undergoing brain or eye surgery or spinal anesthesia. Patients with a history of pulmonary embolism should be started on external pneumatic compression of the lower extremities at least 24 h preoperatively in the absence of contraindications (see also Chaps. 23 and 46).

Diabetes Mellitus

Patients with diabetes mellitus have an increased risk of asymptomatic myocardial ischemia and postoperative CHF. Perioperative therapy of diabetes should attempt to maintain blood sugar in a range of 80 to 180 mg/dL.

PREOPERATIVE EVALUATION OF PATIENTS WITH KNOWN OR SUSPECTED CARDIAC DISEASE

The two major cardiovascular conditions related to PCM are CAD and left ventricular dysfunction. Accordingly, a careful preoperative evaluation of patients for subjective and objective evidence of these conditions comprise important components of several multifactorial cardiac risk indices, such as the Goldman Multifactorial Cardiac Risk Index, which separated patients into four classes of risk, and the Detsky Modified Multifactorial Cardiac Risk Index, which considered coronary artery disease, especially angina pectoris, more than the original Cardiac Risk Index. These risk indices provided useful guidelines for the identification of patients at increased risk during noncardiac surgery.

The ACC/AHA Guidelines for the perioperative cardiovascular evaluation for noncardiac surgery provide a stepwise algorithm for clinicians that emphasizes clinical markers of risk, functional status, surgery-specific considerations, and selected use of noninvasive and invasive tests (Fig. 40-1). The clinical predictors of increased perioperative cardiovascular risk are shown in Table 40-1 and the cardiac risk of various noncardiac surgical procedures is shown in Table 40-2. The greatest risk is found in patients with recent unstable coronary syndromes, advanced or poorly controlled CHF, or symptomatic arrhythmias. Immediate-risk predictors include diabetes mellitus, mild stable angina, prior MI, and compensated or prior stable CHF.

The steps and questions corresponding to the algorithm in Fig. 40-1 appear on pages 579–580.

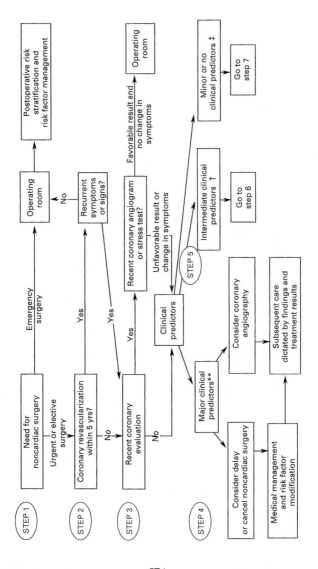

STEP 1

Need for noncardiac surgery

Urgent or elective surgery → Emergency surgery → Operating room → Postoperative risk stratification and risk factor management

STEP 2

Coronary revascularization within 5 yrs? — Yes → Recurrent symptoms or signs? — No → Operating room

STEP 3

Recent coronary evaluation — Yes → Recent coronary angiogram or stress test? — Favorable result and no change in symptoms → Operating room

No ↓

Unfavorable result or change in symptoms → Clinical predictors

STEP 4

Clinical predictors

Major clinical predictors** → Consider delay or cancel noncardiac surgery → Medical management and risk factor modification → Subsequent care dictated by findings and treatment results

Major clinical predictors** → Consider coronary angiography → Subsequent care dictated by findings and treatment results

STEP 5

Clinical predictors

Intermediate clinical predictors † → Go to step 6

Minor or no clinical predictors ‡ → Go to step 7

574

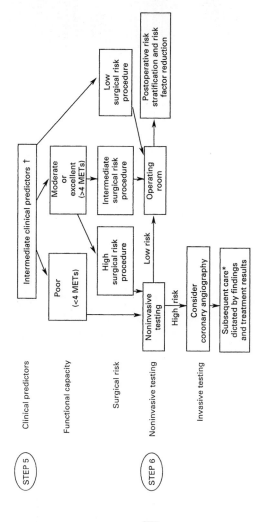

STEP 5

Clinical predictors

Functional capacity

Surgical risk

STEP 6

Noninvasive testing

Invasive testing

Intermediate clinical predictors †

Poor (<4 METs)

Moderate or excellent (>4 METs)

High surgical risk procedure

Intermediate surgical risk procedure

Low surgical risk procedure

Noninvasive testing

Low risk

High risk

Consider coronary angiography

Subsequent care* dictated by findings and treatment results

Operating room

Postoperative risk stratification and risk factor reduction

(continued on pages 586–87)

575

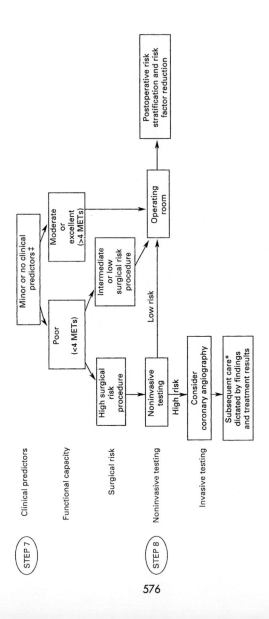

FIGURE 40-1

Stepwise approach to preoperative cardiac assessment. Steps are discussed in the text (page 11). [Report of the American College of Cardiology/American Heart Association Task Force on Practice Guidelines (Committee on Perioperative Cardiovascular Evaluation for Noncardiac Surgery): Guidelines for perioperative cardiovascular evaluation for noncardiac surgery. *J Am Coll Cardiol* 1996; 27:910–948 and *Circulation* 1996; 93:1278–1317.]

*Subsequent care may include cancellation or delay of surgery, coronary revascularization followed by noncardiac surgery, or intensified care.

**Major clinical predictors: unstable coronary syndromes, decompensated CHF, significant arrhythmias, severe valvular disease.

†Intermediate clinical predictors: mild angina pectoris, prior MI, compensated or prior CHF, diabetes mellitus.

‡Minor clinical predictors: advanced age, abnormal ECG, rhythm other than sinus, low functional capacity, history of stroke, uncontrolled systemic hypertension. (From the ACC/ACA Guidelines for Perioperative Cardiovascular Evaluation for Noncardiac Surgery.[21] Reproduced with permission from the publisher.)

TABLE 40-1

CLINICAL PREDICTORS OF INCREASED PERIOPERATIVE
CARDIOVASCULAR RISK (MYOCARDIAL INFARCTION,
CONGESTIVE HEART FAILURE, DEATH)

Major
  Unstable coronary syndromes
    Recent myocaridial infarction[a] with evidence of
      important ischemic risk by clinical symptoms
      or noninvasive study
    Unstable or severe[b] angina (Canadian class III
      or IV)[c]
  Decompensated congestive heart failure
  Significant arrhythmias
    High-grade atrioventricular block
    Symptomatic ventricular arrhythmias in the
      presence of underlying heart diease
    Supraventricular arrhythmias with uncontrolled
      ventricular rate
  Severe valvular disease
Intermediate
  Mild angina pectoris (Canadian class I or II)
  Prior myocardial infarction by history or pathologic
    Q waves
  Compensated or prior congetive heart failure
  Diabetes mellitus
Minor
  Advanced age
  Abnormal electrocardiogram (left ventricular hyper-
    trophy, left bundle branch block, ST-T
    abnormalities)
  Rhythm other than sinus (e.g., atrial fibrillation)
  Low functional capacity (e.g., inability to climb one
    flight of stairs with a bag of groceries)
  History of stroke
  Uncontrolled systemic hypertension

[a] The American College of Cardiology National Database Library
defines *recent MI* as greater than 7 days but less than or equal to
1 month (30 days).

[b] May include "stable" angina in patients who are unusually sedentary.

[c] Campeau L, Grading of angina pectoris. *Circulation* 1976; 54:
522–523.

TABLE 40-2

CARDIAC RISK^a STRATIFICATION
FOR NONCARDIAC SURGICAL PROCEDURES

High (reported cardiac risk often >5%)
  Emergent major operations, particularly in the
    elderly
  Aortic and other major vascular
  Peripheral vascular
  Anticipated prolonged surgical procedures associ-
    ated with large fluid shifts and/or blood loss
Intermediate (reported cardiac risk generally <5%)
  Carotid endarterectomy
  Head and neck
  Intraperitoneal and intrathoracic
  Orthopedic
  Prostate
Low^b (reported cardiac risk generally <1%)
  Endoscopic procedures
  Superficial procedure
  Cataract
  Breast

^{a} Combined incidence of cardiac death and nonfatal myocardial infarction.
^{b} Do not generally require further preoperative cardiac testing.
SOURCE: ACC/AHA Guidelines for Perioperative Cardiovascular Evaluation for Noncardiac Surgery. Reproduced with permission from the
publisher.

Step 1. What is the urgency of noncardiac surgery?
Step 2. Has the patient undergone coronary revascularization
         in the past 5 years?
Step 3. Has the patient had a coronary evaluation in the past
         2 years?
Step 4. Does the patient have an unstable coronary syndrome
         or a major clinical predictor of risk?
Step 5. Does the patient have intermediate clinical predictors
         of risk?
Step 6. Patients without major but with intermediate pre-
         dictors of clinical risk and moderate or excellent

functional capacity can generally undergo intermediate-risk surgery with little likelihood of perioperative MI or death. Conversely, additional noninvasive testing is often considered for patients with poor functional capacity prior to moderate-risk elective and especially for patients with two or more intermediate predictors.

Step 7. Noncardiac surgery is generally safe for patients with neither major nor intermediate predictors of clinical risk and moderate or excellent functional capacity [four metabolic equivalents (METS) or greater]. Further testing may be indicated on an individual basis for patients without clinical indicators but with poor functional capacity who are facing higher-risk operations, especially patients with several minor predictors of risk who are to undergo vascular surgery.

Step 8. The results of noninvasive testing can be used to determine further perioperative management.

In most patients, preoperative risk stratification should include specific tests only if the results are likely to influence patient management. For example, most patients with lower extremity ischemic pain at rest from peripheral vascular disease and most patients with possible carcinoma in whom the chance of surgical cure would be decreased by delay do not require testing for coronary artery disease. *Preoperative testing should be limited to circumstances in which the results will affect patient management. Coronary revascularization before noncardiac surgery to enable the patient to "get through" the noncardiac procedure is appropriate for only a small subset of patients at very high risk.*

Coronary arteriography should be considered in most patients with either unstable angina or angina pectoris that is refractory to medical therapy and in most high-risk patients who either have high-risk results on noninvasive testing or are scheduled for elective high-risk noncardiac surgery and have nondiagnostic noninvasive test results.

Age

Perioperative MI is the leading cause of postoperative death in elderly patients undergoing noncardiac surgery.

Twelve-Lead Electrocardiogram (ECG)

This should be obtained routinely in patients over age 40 and in younger patients who have increased risk factors for cardiovascular disease.

Chest Roentgenogram

A routine preoperative chest film in patients over age 40 is useful primarily as a baseline for comparison with postoperative films when the patient may have such complications as pneumonia, atelectasis, or pulmonary embolism. In addition, it may detect cardiomegaly, which can be the first sign of decreased left ventricular function, or the presence of a tortuous or calcified aorta.

Exercise Stress Testing

The cost effectiveness of routine preoperative exercise stress testing is controversial. On the other hand, in selected patients, it can be very useful (see below).

Radionuclide Angiography (RNA)

The left ventricular ejection fraction measured by RNA at rest or during exercise is a relatively insensitive and nonspecific predictor of perioperative complications.

Radionuclide Scintigraphy

Radionuclide scintigraphy employing thallium has good sensitivity and specificity in the identification of patients at increased risk of PCM. It can be performed utilizing treadmill exercise or, if patients are unable to exercise, the infusion of dipyridamole (Persantine). In general, thallium scintigraphy has been found to have greater predictive value than either exercise ECG testing or the estimation of left ventricular function at rest or during exercise by gated blood pool scanning. Most but not all studies have found radionuclide scintigraphy to be of particular value in patients with stable angina pectoris, a history of myocardial infarction, or diabetes mellitus.

Ambulatory (Continuous or Holter) Electrocardiography

Most studies have found that patients with evidence of myocardial ischemia on ambulatory electrocardiographic (AECG) (Holter) recordings have an increase in perioperative cardiac events. At present, however, the relatively low specificity and sensitivity of AECG and its relatively high cost limit its usefulness. On the other hand, in the postoperative period careful, continuous multiple-lead ECG monitoring should be more widely employed to detect asymptomatic myocardial ischemia.

Transthoracic Echocardiography

The use of routine transthoracic echocardiography (TTE) has not been well evaluated in preoperative evaluation of patients. Although TTE can identify significant left ventricular dysfunction and/or segmental wall motion abnormalities, it is unlikely to be cost-effective.

Transesophageal Echocardiography

Transesophageal echocardiography (TEE) is very useful intraoperatively during cardiac surgery in selected patients for the detection of myocardial ischemia or the evaluation of a procedure.

Exercise and Stress Echocardiography

Exercise and pharmacologic [dobutamine (Dobutrex)] stress echocardiography appear to be safe and useful methods to identify selected patients at high or low risk of PCM. The diagnostic accuracy, sensitivity, and specificity of the test should be determined for each institution or laboratory performing the test because of the special skills necessary for proper interpretation.

Coronary Arteriography

A considerable number of patients undergoing vascular surgery and of older patients undergoing nonvascular surgery are found to have significant, surgically correctable coronary artery disease on routine coronary arteriography. In carefully selected patients scheduled for elective, nonurgent surgery, it may be appropriate to perform coronary arteriography and coronary artery revascularization by surgery or coronary angioplasty before the elective surgery. Noncardiac surgery can usually be performed several days after coronary angioplasty. On the other hand, if the noncardiac surgery is emergent, urgent, or necessary to prevent spread of a malignancy, or if the patient is not a candidate for coronary artery surgery for other reasons, the results of coronary arteriography would not change patient management and the procedure is not justified. In general, the indications for revascularization by surgery or angioplasty should be the same as if the patient were not undergoing the noncardiac surgery.

POSTOPERATIVE PREDICTION OF PERIOPERATIVE CARDIAC MORBIDITY

Postoperative Ischemia

Postoperatively, ECG evidence of myocardial ischemia on a two-channel ECG without clinical symptoms is relatively frequent in men. Although the majority of such patients do not progress to an ischemic event, those with ECG evidence of ischemia are much more likely to have an ischemic event than are those who do not have ECG changes. Some patients who progress to an event development dyspnea and/or chest discomfort, whereas others have no symptoms. The major preoperative predictors of postoperative myocardial ischemia are a history of coronary artery disease, systemic arterial hypertension, diabetes mellitus or the use of digoxin, and the finding of left ventricular hypertrophy on the ECG.

Most patients with known CAD or at significant risk of having CAD should have continuous postoperative ECG monitoring using at least 2 leads, daily 12-lead ECGs, and, in selected high risk patients or patients who develop clinical, hemodynamic, or ECG evidence of ischemia or dysfunction, determinations of creatine kinase-MB every 8 h for several days following major noncardiac surgery. Postoperative tachycardia and hypertension are treated with a beta blocker in the absence of contraindications (see Chap. 45). Postoperative supraventricular tachycardias are often related to hypoxia, infection, electrolyte abnormalities, or medications. Atrial fibrillation with a rapid rate can usually be managed by intravenous diltiazem, a short-acting beta blocker (Chap. 45), or cardioversion (see also Chap. 2).

SUGGESTED READING

Abraham SA, Coles NA, Coley CM, et al: Coronary risk of noncardiac surgery. *Prog Cardiovas Dis* 1991; 34: 205–234.

Coley CM, Eagle KA: Preoperative assessment and perioperative management of cardiac ischemic risk in noncardiac surgery, *Curr Prob Cardiol* 1996; 21:291–382.

Eagle KA, Brundage BH, Chaitman BR, et al: Guidelines for perioperative cardiovascular evaluation for noncardiac surgery: Report of the American College of Cardiology/American Heart Association Task Force on Practice Guidelines (Committee on Perioperative Cardiovascular Evaluation for Noncardiac Surgery). *J Am Coll Cardiol* 1996; 27:910–948, and *Circulation* 1996; 93:1276–1317.

Eagle KA, Rihal CS, Mickel MC, et al: Cardiac risk of noncardiac surgery: influence of coronary disease and type of surgery in 3,368 operations. *Circulation* 1997; 96: 1882–1887.

Hollenberg M, Mangano DT, Browner WS, et al: Predictors of postoperative myocardial ischemia in patients undergoing noncardiac surgery. *JAMA* 1992; 268:205–209.

Mangano DT, Goldman L: Preoperative assessment of patients with known or suspected coronary disease. *N Engl J Med* 1995; 333:1750–1756.

Massie BM, Mangano DT: Assessment of preoperative risk: Have we put the cart before the horse? (editorial). *J Am Coll Cardiol* 1993; 21:1353–1356.

Schlant RC, Eagle KA: Perioperative evaluation and management of patients with known or suspected cardiovascular disease who undergo noncardiac surgery. In: Alexander RW, Schlant RC, Fuster V, O'Rourke RA, Roberts R, Sonnenblick EH (eds): *Hurst's The Heart*, 9th ed. New York, McGraw-Hill, 1998:2243–2255.

Shaw LJ, Eagle KA, Gersh BJ, et al: Meta-analysis of intravenous dipyridamole-thallium-201 imaging (1985 to 1994) and dobutamine echocardiography (1991 to 1994) for risk stratification before vascular surgery. *J Am Coll Cardiol* 1996; 27:787–798.

41

MANAGEMENT OF THE PATIENT AFTER CARDIAC SURGERY

Douglas C. Morris

The main justification for specialized postoperative care following cardiac surgery is to allow recovery of physiologic systems disrupted by cardiopulmonary bypass. A number of organ systems are subject to insult during cardiopulmonary bypass, much of which can be attributed to the generalized inflammatory response caused by the contact of blood with the synthetic surfaces of bypass equipment. Upon the patient's arrival in the intensive care unit (ICU), these disrupted physiologic systems will pose multiple and changing management problems. The patient's condition on arrival varies depending on preoperative status, length of time on cardiopulmonary bypass, success of the operative procedure, and intraoperative complications.

IMMEDIATE POSTOPERATIVE CARE

On arrival in the ICU, the patient is still feeling the effects of anesthesia, hypothermia, and generally one or more pharmacologic agents affecting the systemic circulation; he or she is also being mechanically ventilated. Once the patient is satisfactorily connected to the bedside monitors and ventilator, all hemodynamic measurements should be recorded, a portable supine chest x-ray should be acquired, and a 12-lead electrocardiogram (ECG) obtained.

Improvements in technology coupled with changes in anesthesia and perfusion techniques now frequently allow extubation within several hours of surgery. This specific intraoperative management, resulting in an abbreviated need for cardiopulmonary support after surgery and earlier tracheal extubation, has been termed "fast tracking."

The criteria for early extubation would include the following: the patient awake, neurologically intact and cooperative;

stable and satisfactory hemodynamics; normothermia and satisfactory gas exchange off the ventilator.

EARLY POSTOPERATIVE MANAGEMENT

The basic pathophysiology during the initial postoperative hours revolves around the following variables: warming from hypothermia, transient left ventricular dysfunction, mediastinal bleeding, emergence from anesthesia, and capillary leak. Hypothermia predisposes the patient to ventricular arrhythmias; increases systemic vascular resistance; precipitates shivering, which increases oxygen consumption; and produces a generalized impairment of the coagulation cascade.

Clinical evidence of left ventricular dysfunction during the first 1 to 24 h postoperatively with a gradual recovery to preoperative levels occurs in the vast majority of patients. Myocardial depression in the early postoperative period has been suggested by some studies to be a sequela of inadequate myocardial protection or the effects of cold cardioplegia. A release of oxygen free radicals in response to reperfusion injury is now generally accepted as the explanation for the transient postoperative ventricular dysfunction. Ventricular function is generally depressed by 2 h and is at its worst at 250 to 300 min. Significant recovery of function has usually occurred by 8 to 10 h and full recovery has been reached by 24 to 48 h. Systemic vascular resistance, while not rising significantly immediately postoperatively, does increase as ventricular function worsens.

The capillary leak syndrome may last from a few hours to 24 h depending to a large degree on the length of the pump run. When the capillary leak ceases, additional fluid does not need to be administered because fluid tends to remain in the intravascular space and is also mobilized from the interstitial tissues. At this time, diuretics are beneficial to eliminate excessive fluid. Hypothermia with the patient's core temperature below 35°C (95°F) is frequently present despite what appears to be adequate rewarming to 37°C (98.6°F) on cardiopulmonary bypass. Sodium nitroprusside or glycerol trinitrate likely were begun in the operating room to optimize whole-body heat exchange during rewarming. If the patient is still hypothermic, these agents should be continued and titrated to the maximum rate that still allows for an adequate arterial perfusion pressure. Monitoring of body sites other than the blood and brain can guide more complete rewarming. Hypothermia causes peripheral vasoconstriction and contributes to the hypertension frequently seen after cardiac surgery. Furthermore, hypothermia causes a decrease in cardiac output by producing bradycardia.

Hypercarbia will cause catecholamine release, tachycardia, and pulmonary hypertension. If the patient cannot increase the cardiac output, mixed venous desaturation and metabolic acidosis can result. Arterial blood gases should be analyzed frequently to indicate the increased carbon dioxide production and guide increases in ventilation. These consequences of rewarming are most profound when the patient shivers; they can be reduced by muscle paralysis. As the temperature increases, the vasoconstriction and hypertension associated with hypothermia are replaced by vasodilatation, tachycardia, and hypotension. Volume loading during the rewarming process helps reduce the rapid swings in blood pressure.

The commonly reported frequency of severe postoperative bleeding (more than 10 U of blood transfused) following cardiac surgery is 3 to 5 percent. While approximately one-half of the patients who undergo reoperation for excessive bleeding exhibit incomplete surgical hemostasis, the remainder bleed because of various acquired hemostatic defects, most often related to acquired platelet dysfunction. While the exact mechanism responsible for the transient platelet dysfunction remains undefined, it appears to be related to contact of platelets with the synthetic surfaces of the extracorporeal oxygenator and to the hypothermia.

CARDIOVASCULAR MANAGEMENT

Low-Cardiac-Output Syndrome

A satisfactory cardiac output following cardiac surgery is a cardiac index of greater than 2.5 L/min/m^2 with a heart rate below 100 beats per minute. A marginal output is present with a cardiac index between 2.5 and 2.0 L/min/m^2. A cardiac index below 2.0 L/min/m^2 is unacceptably low, and therapeutic intervention is usually indicated.

ETIOLOGIES

The most common causes of low cardiac output postoperatively are related to a decreased left ventricular preload. Common causes of reduced left ventricular preload include hypovolemia due to bleeding or vasodilation as a consequence of warming, vasodilators, or drugs; cardiac tamponade; or right ventricular dysfunction. Decreased contractility due to a preexisting low ejection fraction or to intraoperative or postoperative ischemia or infarction may be the explanation for the low output. Tachy- or bradyarrhythmias may be contributing to the problem by reducing cardiac filling time, causing the loss of atrial contraction and AV synchrony, or reducing the number of effective ventricular contractions per minute. Marked alterations in the systemic vascular resistance by either vasoconstriction from hypothermia or vasodilatation from sepsis or anaphylaxis could also be causative.

MANAGEMENT
The multiple parameters constantly being monitored usually provide sufficient clues as to the cause of the low output. If there is no obvious noncardiac cause, the first step is to optimize the preload [pulmonary capillary wedge pressure (PCWP) of 15 to 18 mmHg]. The next step is to optimize the heart rate by either cardiac pacing or antiarrhythmic drugs. Postoperative myocardial performance is usually best at a rate of 90 to 100/min. If these measures prove unsuccessful, pharmacologic intervention with either inotropic agents, vasodilators, vasopressors, or a combination of these agents must be considered (Table 41-1). The exception to the use of these pharmacologic agents would be the suspicion of anaphylaxis, which should be treated accordingly, or the presence of elevated left- and right-sided filling pressures and a recent absence of mediastinal drainage, which suggest tamponade. Transesophageal echo has been very helpful in clarifying these situations. The final therapeutic step if the above measures have proven inadequate is the use of aortic counterpulsation.

HYPERTENSION
Systemic arterial hypertension in the presence of a high left ventricular filling pressure and marginal cardiac output is most appropriately treated by an arterial vasodilator. The available choices include nitroprusside, nicardipine, nitroglycerin, and hydralazine (Table 41-2).

ARRHYTHMIAS
The most common rhythm disturbance after cardiac surgery is sinus tachycardia. This condition is appropriately treated by searching for and treating the underlying cause (pain, anxiety, low cardiac output, fever, or beta-blocker withdrawal). The second most common arrhythmia is ventricular ectopy. Again,

TABLE 41-1

MEDICATIONS USED IN LOW-CARDIAC-OUTPUT SYNDROME

Medication	Hemodynamic Properties	Dosage Range
Dopamine	Low dose, dopaminergic effect Moderate dose, inotropic effect High dose, vasopressor effect	2–20 µg/kg/min
Dobutamine	Positive inotropic agent	2–20 µg/kg/min
Epinephrine	Positive inotropic agent	1–4 µg/min
Amrinone	Positive inotropic agent	10–15 µg/kg/min
Isoproterenol	Potent inotropic agent, pronounced chronotropic effect	0.5–10 µg/min
Norepinephrine	Potent vasopressor effect, inotropic effect	1–100 µg/min
Phenylephrine	Potent vasopressor agent	10–500 µg/min

TABLE 41-2
INTRAVENOUS ANTIHYPERTENSIVE AGENTS

Drug	Peak Effect	Duration	Dosage
Nitroprusside	Immediate	2–5 min	0.3–1.0 µg/kg/min
Nitroglycerin	Immediate	2–5 min	5–100 µg/kg/min infusion
Nicardipine	5–60 min	20–40 min	2.5 mg over 5 min; may repeat ×4 at 10-min intervals; infusion 2–7.5 mg/h
Esmolol	2–5 min	8–10 min	1-min loading infusion of 0.25–0.5 mg/kg, sustained infusion of 50–200 µg/kg/min
Enalaprilat	15–30 min	6 h or more	0.625–1.25 mg slowly over 5 min every 6 h
Hydralazine	15–20 min	3–4 h	5- to 10-mg bolus may be repeated every 15 min up to a total of 40 mg
Diltiazem	3–30 min		20- to 25-mg bolus, may repeat; infusion of 10–15 mg/h
Verapamil	2–3 min	20–40 min	5- to 10-mg bolus, may repeat in 10 min
Labetalol	5–15 min	2–6 h	20-mg bolus over 2 min, then 40- to 80-mg boluses every 15 min until effect achieved (to total dose of 300 mg)

an underlying cause such as myocardial ischemia, hypo-kalemia, hypoxia, or administration of an inotropic agent must be sought and corrected. It is also important to review the patient's preoperative record to learn whether or not he or she had preexisting ectopy. Patients with chronic ventricular ectopy will frequently have their ectopy exaggerated postoper-atively. In the presence of active myocardial ischemia, pharmacologic suppression is advisable for complex ventricu-lar ectopy. If the ectopy does not respond to lidocaine, the options are to not use an antiarrhythmic agent unless ventricu-lar tachycardia occurs or to use intravenous amiodarone.

After cardiac surgery, a few patients will develop sustained ventricular tachycardia (either monomorphic or polymorphic) or ventricular fibrillation. The ventricular tachycardia in these patients rarely responds to lidocaine and usually requires amio-darone. In some instances, a combination of amiodarone and beta blockers is required; in a rare circumstance use of aortic counterpulsation has seemed to be of benefit. Every encounter with a wide, complex tachycardia requires careful considera-tion as to the possibility of supraventricular tachycardia with aberrant conduction. In the presence of atrial fibrillation with a rapid ventricular response, right bundle branch aberrant con-duction often mimics ventricular tachycardia. Care must be given to avoid lidocaine in these situations, since it may result in a more rapid rate and even more aberrant conduction.

With the exception of sinus tachycardia, the more common supraventricular arrhythmias are atrial fibrillation and atrial flutter. These rhythm disturbances occur in 10 to 28.4 percent of patients following cardiac surgery. The most dominant fac-tor promoting the development of atrial fibrillation is the age of the patient.

The prophylactic use of beta blockers has a protective effect against the development of atrial fibrillation or atrial flutter. Intravenous infusions of either esmolol or diltiazem can be used to control the ventricular rate with atrial fibrillation or flutter and may occasionally convert the rhythm to sinus. The atrial epicardial placing wires provide the means of atrial pac-ing to convert some cases of atrial flutter to sinus rhythm. Short bursts (15 to 30 s) of atrial pacing at rates of 300 to 600 per min may be effective in converting atrial flutter. If hemo-dynamic compromise is present and aggravated by a supra-ventricular tachyarrhythmia, electrical cardioversion or the intravenous administration of 1 mg of ibutilide over 10 min may be used earlier rather than later. The prevalence of intra-ventricular conduction abnormalities after coronary bypass surgery is reported to be from 1 to 45 percent, with approximately 10 percent seeming most reasonable. The most common conduction defect is right bundle branch block. This may be due to selective sensivity of the right bundle to the effects of hypothermia and the extracorporeal circulation process. Only about 5 percent of the patients are left with a permanent conduction abnormality and the prognosis for these patients is no worse than that of comparable patients with no conduction defect. The development of second- or third-degree atrioventricular block is an indication for temporary pacing via the epicardial pacing wires.

RESPIRATORY MANAGEMENT

Expected Respiratory Changes after Cardiac Surgery

Pulmonary problems are the most significant cause of morbidity following cardiopulmonary bypass. The sternotomy or thoracotomy has a deleterious effect on the functioning of the muscular action of the chest wall. Phrenic nerve damage can result in diaphragm dysfunction. Furthermore, the diaphragm is passively displaced cephalad by the abdominal contents in the anesthetized, paralyzed patient on mechanical ventilation. Pain due to the surgery and the presence of chest tubes may interfere with normal respiratory function. Elevated left-sided cardiac pressures may cause alveolar edema and in some patients increased capillary permeability may exist.

Atelectasis is the most common pulmonary complication, occurring in about 70 percent of patients. During cardiopulmonary bypass, the lungs are not perfused and are usually allowed to collapse. Once the lungs are reexpanded, a variable amount of atelectasis remains. The preponderance of atelectasis occurs in the left lower lobe because of its compression by the heart during cardiopulmonary bypass, the tendency to suction the more direct-arising right mainstem bronchus during blind bronchial suctioning, and the greater likelihood to open the left pleural space in isolating the left internal mammary artery.

Management of Mechanical Ventilation

The goals of mechanical ventilation are the maintenance of satisfactory arterial oxygenation and alveolar ventilation. The

Pa_{O_2} is generally used to assess the adequacy of oxygenation. While monitoring the Pa_{O_2}, the relation between this value (the amount of oxygen dissolved in solution) and the oxygen saturation (amount of oxygen bound to hemoglobin) must be kept in mind. The oxygen-homoglobin dissociation curve demonstrates that a Pa_{O_2} below 65 mmHg will result in a precipitous fall in the oxygen saturation. With hypothermia or with profound respiratory alkalosis, the curve will shift to the left, resulting in more avid binding of oxygen to hemoglobin and less release to the tissues. Pulse oximetry is the best means of continuously measuring Sa_{O_2}. Hypocarbia is best treated by reducing the ventilator rate. Hypercarbia in the immediate postoperative period usually indicates that minute ventilation is inadequate. The problem can be rectified by increasing the ventilator rate or the tidal volume. Severe hypercarbia should raise a concern about mechanical problems such as ventilator malfunction, endotracheal tube malposition, or a pneumothorax.

VENTILATORY WEANING AND EXTUBATION

Controlled ventilation should be discontinued when the cardiovascular system has become stable and the arterial oxygen tension is satisfactory (Pa_{O_2} above 70) on an Fi_{O_2} of 0.5 and a peak end-expiratory pressure (PEEP) of 5 cm. The patient should also be alert, normothermic, and with no active bleeding.

BRONCHOSPAMS

Severe bronchospasm during cardiopulmonary bypass is an unusual event, but it can occur. A few patients cannot have the chest cavity closed at the end of surgery because of hyperinflated lungs. The most likely cause of this fulminant bronchospasm is activation of human C5a anaphylatoxin by the extracorporeal circulation. Other causes of bronchospasm in the postoperative period include bronchospasm secondary to cardiogenic pulmonary edema; a simple exacerbation of preexisting bronchopastic disease triggered by instrumentation, secretions, or cold anesthetic gas; beta-adrenergic blockade–induced bronchospasm in susceptible individuals; and allergic reaction to protamine.

The initial therapy of bronchospasm in the postoperative patient, once a diagnosis of heart failure is excluded, should be inhaled beta$_2$-receptor agonists (terbutaline, metaproterenol, albuterol) or inhaled cholinergic agents (ipratropium bromide or glycopyrrolate). In the inhaled form, these rather potent bronchodilators have minimal cardiovascular effects. In addition to their bronchodilator effect, these agents may augment mucociliary transport and aid in clearing secretions. A combination of beta$_2$-receptor agonists and cholinergic agents should be tried in the patient refractory to a single agent. Even more refractory bronchospasm requires either a short course of systemic steroids or intravenous aminophylline.

POSTOPERATIVE OLIGURIA AND RENAL INSUFFICIENCY

Etiology

Following cardiopulmonary bypass, there is a substantial incidence of postoperative renal dysfunction (up to 30 percent) but a relatively low incidence of severe renal impairment requiring dialysis (1 to 5 percent). Renal blood flow and glomerular filtration rate during bypass are reduced by 25 to 75 percent, with partial but not complete recovery in the first day after bypass. This reduction in renal function is attributed to renal artery vasoconstriction, hypothermia, and the loss of pulsatile perfusion. Angiotensin II levels are higher with nonpulsatile flow. While renal dysfunction cannot be consistently related to the systemic blood pressure and pump flow rate during nonpulsatile bypass, there is a definite relationship between the incidence of postbypass renal dysfunction and the duration of bypass. In addition to the duration of bypass, the risk of developing postbypass renal failure seems to be a function of the patient's underlying renal function (also affected by age) and the perioperative circulatory status.

Management

There are three agents ("renoprotective" drugs) that might be used during or immediately after bypass to prevent an ischemic insult to the kidneys. Mannitol given during bypass may moderate ischemic insult, probably by volume expansion and hemodilution. It may also initiate an osmotic diuresis to prevent tubular obstruction and also serve as a free radical scavenger. Furosemide also appears to improve renal blood flow when given during bypass. Postoperatively, low-dose dopamine may maintain renal blood flow and urine output. Once renal failure has developed, none of these drugs is likely to offer any beneficial effect. A megadose of furosemide (200 to 300 g) may be tried; but if there is no response, it should not be repeated. Similarly, a single dose of mannitol (12.5 to 25 g), either with or without furosemide, may be tried but should not be repeated if there is no effect.

POSTOPERATIVE GASTROINTESTINAL DYSFUNCTION

Gastrointestinal Consequences of Cardiopulmonary Bypass

The gastrointestinal consequences of bypass appear to be minimal. Reviews of the subject report a 1 percent prevalence. Most patients are eating within 24 to 48 h after an uncomplicated elective procedure. Transient elevations in liver function tests and hyperamylasemia occur after cardiac surgery. The risk factors for such include long bypass times, multiple transfusions, and multiple valve replacements. The appearance of jaundice portends a poor prognosis. Severe gastrointestinal complications are usually ischemic in nature and are often associated with a low-output syndrome.

POSTOPERATIVE METABOLIC DISORDERS

Potassium Imbalance

In the patient following cardiopulmonary bypass, there are multiple factors at play that can produce large and rapid shifts in the serum potassium levels: the high-potassium cardioplegia solution the patient received during the surgery; some degree of renal dysfunction that is probably present, with associated oliguria and decreased clearance of potassium; low-cardiac-output states accompanied by oliguria and acidosis; release of potassium by hemolyzing red cells; and the loss of potassium by diuresis. The therapy of severe hyperkalemia should include counteracting the toxic cardiac effects of the elevated potassium with intravenous calcium gluconate or calcium chloride and lowering the serum level of potassium with sodium bicarbonate and/or regular insulin and glucose. Hypokalemia is treated with the intravenous administration of potassium chloride (KCl) at a rate of no more than 10 to 15 meq/h. The serum potassium rises approximately 0.1 meq/L for each 2 meq of KCl administered.

Hypomagnesemia

Hypomagnesemia is common following cardiac surgery using cardiopulmonary bypass. Its cause is unknown, but it is probably multifactorial. Many patients will be hypomagnesemic preoperatively. The loop diuretics, thiazides, digoxin, alcohol, and type I diabetes mellitus can all contribute to reduced magnesium. In the postoperative period, magnesium can be administered as magnesium sulfate (2 g in 100-mL solution) to raise serum levels to 2 meq/L.

Hyperglycemia

During cardiopulmonary bypass, there is an increase in blood glucose levels. The rise in blood glucose is modest during hypothermia and becomes more marked during rewarming. This rise in glucose is due in part to increased glucose mobilization related to marked rises in cortisol, catecholamine, and growth hormone levels during cardiopulmonary bypass. Also, there is an apparent failure of glucose secretion, particularly during hypothermia. The insulin response is blunted for the first 24 h after surgery. These changes are exaggerated in the diabetic patient.

POSTOPERATIVE FEVER

Etiology

Fever is a likely occurrence in the postoperative patient. It generally is a consequence of pleuropericarditis, atelectasis, or phlebitis. The fact that some 70 percent of patients after cardiac surgery have atelectasis makes this the most likely etiology of postoperative fever. The appropriate therapeutic approach is an intensified effort at incentive spirometry.

Sternal wound infections occur in 0.4 to 5 percent of patients after sternotomy. In the various studies that have addressed this issue, multiple and differing factors have been identified as increasing the risk to develop sternal infection. These include pneumonia; prolonged ventilation, especially with tracheostomy; emergency operations; postoperative hemorrhage with mediastinal hematoma; early reexploration; obesity; diabetes; and the use of bilateral internal mammary grafts.

Approximately 1 percent of coronary bypass surgery patients experience leg wound infections that necessitate extra care. Leg infections seem to occur more frequently in obese women, especially if the thigh veins are harvested.

NEUROLOGIC AND NEUROPSYCHOLOGICAL DYSFUNCTION

Mechanism

The mechanisms thought to account for most cerebral injury during cardiac surgery are macroembolization of air, debris from aortic atheroma or left ventricular thrombus; microembolization of aggregates of granulocytes, platelets, and fibrin; and cerebral hypoperfusion. Death or disabling stroke occurs in about 2 percent of patients, with another 3 percent experiencing transient or minor functional disability secondary to cerebral infarction. Alteration of mental status (encephalopathy and delirium) will be seen in approximately 30 percent of the patients following cardiopulmonary bypass. While the appearance of these encephalopathic symptoms likely reflects cerebral injury, other causes—including sepsis, fever, hypoxia, medications, ethanol withdrawal, renal failure, and hyperosmolar state—must be excluded.

Recognition of this entity is important because the family can be assured that the patient's mental status is very likely to be normal by the time of hospital discharge. Agitation and acute psychosis in these patients usually responds to intravenous haloperidol 2 to 10 mg repeated as needed to produce adequate sedation.

The third serious neurologic complication of coronary bypass surgery is brachial plexopathy. This neurologic dysfunction involving C8 and T1 usually results from mechanical trauma secondary to sternal retraction, but it may be due to penetration by a posterior fractured segment of the first rib of injury during internal jugular cannulation. There is no specific therapy for this condition; recovery could possibly take 6 months, with a few cases being permanent.

SUGGESTED READING

Aps C: Fast-tracking in cardiac surgery. *Br J Hosp Med* 1995; 54:139–142.

Aglio LS, Standord GG, Maddi R, et al: Hypomagnesemia is common following cardiac surgery. *J Cardiothorac Anesth* 1991; 5:201–208.

Baerman JM, Kirsch MM, de Buitleir M, et al: Natural history and determinates of conduction defects following coronary artery bypass surgery. *Ann Thorac Surg* 1987; 44:150–153.

Bolli R: Oxygen derived free radical and postischemic myocardial dysfunciton. *J Am Coll Cardiol* 1988; 12: 239–249.

Bojar RM: *Manual of Perioperative Care in Cardiac and Thoracic Surgery* 2d ed. Boston, Blackwell, 1994.

Breuer AC, Furlan AJ, Hanson MR, et al: Central nervous system complications of coronary artery bypass graft surgery: prospective analysis of 421 patients. *Stroke* 1983; 14:82–87.

Cameron D: Initiation of white cell activation during cardiopulmonary bypass: cytokines and receptors. *J Cardiovasc Pharmacol* 1996; 27(Suppl 1):S1–S5.

Chong JL, Pillai R, Fisher A, et al: Cardiac surgery, moving away from intensive care. *Br Heart J* 1992; 68:430–433.

Collins JD, Bassendine MF, Ferner R, et al: Incidence and prognostic importance of jaundice after cardiopulmonary bypass surgery. *Lancet* 1983; 1:1119–1123.

De Laria GA, Hunter JA, Goldin MD, et al: Leg wound complications associated with coronary revascularization. *J Thorac Cardiovasc Surg* 1981; 81:403–407.

Frater RW, Oka Y, Kadish A, et al: Diabetes and coronary artery surgery. *Mt Sinai J Med* 1982; 49:237–240.

Fuller JA, Adams GG, Buxton B: Atrial fibrillation after coronary artery bypass grafting: is it a disorder of the elderly? *J Thorac Cardiovasc Surg* 1989; 97:821–825.

Hanks JB, Curtis SE, Hanks BB, et al: Gastrointestinal complications after cardiopulmonary bypass. *Surgery* 1982; 92:394–400.

Harker L, Malpass TW, Branson HE, et al: Mechanism of abnormal bleeding in patients undergoing cardiopulmonary bypass: acquired transient platelet dysfunction associated with selective alpha-granule release. *Blood* 1980; 56:824–834.

Hazelrigg SR, Wellons HA, Schneider JA, Kolm P: Wound complications after median sternotomy: relationship to internal mammary grafting. *J Thorac Cardiovasc Surg* 1989; 98:1096–1099.

Jindosi A, Aps C, Neville E, et al: Postoperative cardiac surgical care: an alternative approach. *Br Heart J* 1993; 69:59–64.

Leith JW, Thomson D, Baird DK, Harris PJ: The importance of age as a predictor of atrial fibrillation and flutter after coronary artery bypass grafting. *J Thorac Cardiovasc Surg* 1990; 100:338–342.

Levy JH, Salemenpera MT, Bailey JM, Ramsey JG: Postoperative circulatory control. In: Kaplan JA (ed): *Cardiac Anesthesia*, 3d ed. Philadelphia, Saunders, 1993:1168–1193.

Przyklenk K, Kloner RA: "Reperfusion injury" by oxygen derived free radicals? *Circ Res* 1989; 64:86–96.

Ralley FE, Wynando JE, Rams JG, et al: The effects of shivering on oxygen consumption and carbon dioxide production in patients rewarming from hypothermic cardiopulmonary bypass. *Can J Anaesth* 1988; 35:332–337.

Ramsey J: The respiratory, renal and hepatic systems: effects of cardiac surgery and cardiopulmonary bypass. In: Mora CT (ed): *Cardiopulmonary Bypass*. New York, Springer-Verlag, 1995:147–168.

Shaw PJ, Bates D, Cartlidge NE, et al: Early neurological complications of coronary artery bypass surgery. *Br Med J* 1985; 91:1384–1387.

Sladden RN, Berkowitz DE: Cardiopulmonary bypass and the lung. In: Gravlee GP, Davis RF, Utley IR (eds): *Cardiopulmonary Bypass*, Philadelphia, Williams & Wilkins, 1993:468–487.

Smith LW, Dimsdale JE: Postcardiotomy delirium: conclusions after 25 years? *Am J Psychiatry* 1989; 146: 452–458.

Swanson DK, Myerowitz PD: Effect of reperfusion temperature and pressure on the functional and metabolic recovery of preserved hearts. *J Thorac Cardiovasc Surg* 1983; 86:242–251.

Topol EJ, Lerman BB, Baughman KL, et al: De novo refractory ventricular tachyarrhythmias after coronary revascularization. *Am J Cardiol* 1986; 57:57–59.

Tuzcu EM, Emre A, Goormastic M, Loop FD: Incidence and prognostic significance of intraventricular conduction abnormalities after coronary bypass surgery. *J Am Coll Cardiol* 1990; 16:607–610.

Ulicny KS, Hiradzka SF: The risk factors of median sternotomy infection: a current review. *J Cardiac Surg* 1991; 6:338–351.

Welling RE, Rath R, Albers JE, Glaser RS: Gastrointestinal complications after cardiac surgery. *Arch Surg* 1986; 121:1178–1180.

Woodman RC, Harker LA: Bleeding complications associated with cardiopulmonary bypass. *Blood* 1990; 76:1680–1697.

CHAPTER

42

DIURETICS

Juha P. Kokko

Diuretics increase excretion of salt and water. While the primary clinical indication for their use is to decrease extracellular fluid volume, they have other uses not strictly related to their diuretic properties. This chapter begins with a figure and a summary of the sites and molecular mechanisms of diuretic action; the second part outlines the use and complications of the five major specific diuretic families. Finally, the use of diuretics in congestive heart failure (CHF) is reviewed.

SITES AND MOLECULAR MECHANISM OF DIURETIC ACTION

Figure 42-1 shows schematically the various segments of the mammalian nephron and summarizes the principal transport mechanism existing across these sites. The figure also contains an inset of the various families of diuretics with associated numbers indicating those nephron segments where they exert their major effect.

Proximal Convoluted Tubule

The proximal convoluted tubule reabsorbs some two-thirds of the glomerular filtrate. However, it should be appreciated that the inhibition of proximal fluid tubule reabsorption has a relatively small effect on net fluid homeostasis, since nephron segments distal to the proximal tubule have tremendous unused capacity to reabsorb more salt and water if greater amounts are delivered to them. As a result, diuretics that have their major effect across the proximal tubule are weak diuret-

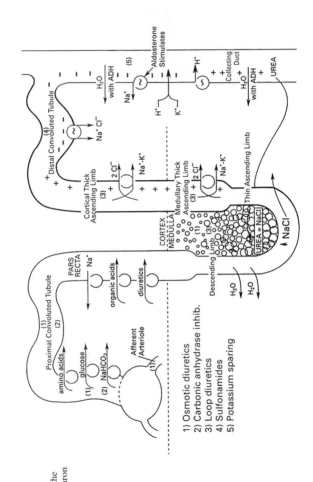

FIGURE 42-1
Major transport processes along the mammalian nephron and the sites of action of various diuretics

ics. However, since one of the primary active transport processes across the proximal convoluted tubule is that of reabsorption of sodium bicarbonate by carbonic anhydrase–dependent mechanisms, it stands to reason that carbonic anhydrase inhibitors increase urinary pH by increasing urinary excretion of bicarbonate.

The Pars Recta

The pars recta is the straight component of the proximal tubule, which is functionally important with respect to the mechanism of diuretic action, since it secretes many organic anions and cations, including diuretics. The loop diuretics and thiazide diuretics are not filtered to a significant degree but rather gain access to the luminal side by virtue of being secreted by the pars recta. This is important, since these diuretics work only from the luminal and not the blood side of the epithelium. Since organic anions and cations compete respectively with diuretics that are organic acids or organic bases for secretory sites, it is not surprising that, in such clinical circumstances as uremia and organic acidosis, there is a lower concentration of organic acid diuretic in the urine for any given concentration of diuretic in blood. It is therefore important that the pars recta be intact and able to secrete diuretics to the urinary side of the tubule, where they may exert their diuretic effect in more distal nephron segments.

Loop of Henle

The loop of Henle is composed of three functionally distinct segments: descending limb of Henle; thin ascending limb of Henle; and thick ascending limb of Henle. Of these segments, only the thick ascending limb of Henle has an important role in increasing salt excretion in response to diuretic function, since neither the descending nor the ascending limb of Henle has active transport processes. In the thick ascending limb of Henle, the loop diuretics act to inhibit a Na-K-Cl cotransporter on the luminal side. Since the cortical thick ascending limb forms dilute urine while the function of the medullary thick ascending limb is to generate surroundings allowing the formation of concentrated urine, it stands to reason that diuretics affecting the thick ascending limb of Henle inhibit the capacity to generate a maximally dilute and concentrated urine.

Also, significant amounts of calcium and magnesium are re-absorbed across the segments; therefore, loop diuretics that inhibit salt reabsorption across this segment also inhibit calcium and magnesium reabsorption and increase excretion of these divalent cations.

Distal Convoluted Tubule

The primary importance of the distal convoluted tubule in diuretic action is that it is a diluting segment (i.e., makes urine that has a low salt concentration) by virtue of having a luminal surface electroneutral sodium chloride transport mechanism. This mechanism can be inhibited by thiazides. Thus, thiazides may induce hyponatremia, since inhibition of this segment would decrease the renal ability to excrete dilute urine.

The Collecting Duct

The cortical collecting duct segments absorb sodium and secrete potassium and hydrogen by mechanisms that are sensitive to circulating levels of aldosterone. Thus, diuretics that have their main action in the collecting duct increase sodium excretion while decreasing the secretion of potassium and hydrogen. While these diuretics do not have a major effect on net fluid excretion, they nevertheless do have important effects on sodium, potassium, and hydrogen homeostasis.

DIURETIC FAMILIES

Tables 42-1 to 42-5 summarize in tabular form the mechanisms of action, complications, advantages, disadvantages, and uses of the five families of diuretics.

While specific complications are listed in these tables, a few general comments are in order. First, it should be recognized that the metabolic complications of diuretics are more common with long-acting diuretics than with short-acting diuretics. The reason for this is that if a diuretic does not suppress the nephron segment for a significant time, these nephron segments can compensate for the abnormalities caused during the time that the diuretics exert their action. For instance, thiazide diuretics cause a higher incidence of potassium de-

TABLE 42-1

THIAZIDE AND NONTHIAZIDE SULFONAMIDES

Mechanism:
    ↑ Active chloride reabsorption in distal convoluted
    tabule (↑ diluting but not concentrating capacity)
    Varying degrees of carbonic anhydrase inhibition →
    not important to diuretic effect.
    Onset 1–2 h   Duration 6–24 h
Complications
    Unpredictable hypersensitivity type: vasculitis,
      interstitial pneumonitis, dermatitis, pancreatitis,
      thrombocytopenia
    Predictable but varying metabolic consequences:
      hyperuricemia, hyperglycemia, hypercalcemia,
      hypokalemia, hyponatremia
      ↓ GFR in moderate renal failure
Advantages:
    Inexpensive
    Effective orally
    Mild diuresis
Disadvantages:
    Less or not useful in advanced edematous states
    Ineffective and possibly contraindicated in moderate
      to severe renal insufficiency
Use:
    Hypertension
    Mild CHF
    Hypercalciuria
    Diabetes insipidus

ficiency than do loop diuretics. Thus, thiazide-induced hypo-kalemia is more often associated with a number of untoward clinically significant side effects, including decreased insulin release, increased incidence of arrhythmias, and potentiation of rhabdomyolysis. While 20 to 30 percent of patients receiving a normal dose of thiazides become hypokalemic, only approximately 10 percent of patients develop hypokalemia with loop diuretics. The second major point with respect to complications can be appreciated by recognizing that metabolic com-

TABLE 42-2

LOOP DIURETICS

Mechanism
  $\uparrow$ Active chloride reabsorption in medullary and
    cortical thick ascending limb of Henle ($\downarrow CH_2O$
    and T^cH_2O under appropriate circumstances)
  Redistribution of renal blood flow
  Onset 2–10 min IV   1 h PO   Duration 2–3 h IV
  6 h PO
Complications:
  Too rapid diuresis
  Ototoxicity
  Hyperuricemia
  Hypokalemia (alkalosis)
  Hyperglycemia ($<$ thiazides)
  Magnesium depletion
Advantages:
  Potent and rapid diuresis
  Effective under all electrolyte disturbances
Disadvantages:
  None, except for rare complications
Use:
  Advanced edematous states—CHF, cirrhosis,
    nephrotic syndrome
  CHF with acute MI
  Hypertension—especially in chronic renal failure
  To "prevent" ATN
  Acute treatment of symptomatic hypercalcemia

plications of diuretic use are unique and specific, depending on
the site of action of the specific diuretic. For example, proximal tubule diuretics are associated with metabolic acidosis,
loop diuretics may be associated with increased calcium and
magnesium excretion, and collecting duct diuretics may be
associated with hyperkalemic acidosis.

Since loop diuretics must be secreted by the pars recta to
have their effect on the luminal side, and since this secretory
process is inhibited by uremic organic acids, it is evident that
the dose of loop diuretic in patients with chronic renal insuffi-

TABLE 42-3

POTASSIUM-SPARING AGENTS

Mechanism:
  Spironolactone antagonizes aldosterone in the
    collecting duct
  Triamterene and amiloride inhibit active sodium
    reabsorption in collecting duct by decreasing
    conductivity of the sodium channels
  All ↓ distal H^+ and K^+ secretion
Complications:
  Hyperkalemia
  Gynecomastia, amenorrhea—aldactone
  Nausea, vomiting—triamterene
Advantages:
  ↓ K^+ excretion
  Good adjuncts to other diuretics
Disadvantages:
  Weak diuretics by themselves
Use:
  Adjunctive Rx in refractory edema—advanced CHF
    or cirrhosis with ascites
  Adjunctive Rx in preventing clinically significant
    hypokalemia
  Diagnostic and therapeutic tool in mineralocorticoid
    excess states—aldactone

ciency must be increased to have a similar urinary concentration and thus an effective diuresis. Table 42-6 summarizes the starting and ceiling doses of two common loop
diuretics in patients with different degrees of renal failure. No
clinically significant extra diuresis can be anticipated if the
ceiling doses of diuretics are exceeded.

Most of the diuretics within specific families are quite similar in their effect at the respective equivalent dose levels.
Table 42-7 lists the common oral diuretics with respect to their
generic name, trade name, usual daily dose, onset of action,
peak effect, and duration of action.

TABLE 42-4

CARBONIC ANHYDRASE INHIBITORS

Mechanism:
  Inhibit carbonic anhydrase in proximal tubule
  $\downarrow$ Proximal reabsorption of $NaHCO_3$, $NaCl$, H_2O
  $\uparrow$ Distal delivery of "nonreabsorbable" anion—HCO_3
  Onset—30 min (PO)   Duration 8–12 h
Complications:
  Drowsiness, parasthesias with >1 g/24 h
  Exacerbation of hepatic encephalopathy
Advantages:
  Alkalize urine
Disadvantages:
  Weak diuretic
  Ineffective in seriously edematous patients
  Self-limiting diuresis secondary to development of
    metabolic acidosis
Use:
  $\uparrow$ excretion of uric acid and various drugs that are
    more soluble in alkaline urine
  Glaucoma— $\downarrow$ rate of aqueous humor formation

TABLE 42-5

OSMOTIC DIURETICS

Mechanism:
  $\uparrow$ Renal plasma flow and GFR
  $\downarrow$ Proximal reabsorption
  Medullary "wash out"
  Onset—immediate
  Duration—until excreted
Complications:
  Overexpansion of intravascular compartment if not
    excreted
Advantages:
  Probably useful in "pre-ATN" phase to prevent its
    development
Disadvantages:
  Must be given IV
  Require renal excretion
Uses:
  "Pre-ATN"
  Catharsis (PO administration) in various overdoses
  Acute Rx cerebral edema

Key: GFR, glomerular filtration rate; ATN, acute tubular necrosis.

TABLE 42-6

TITRATION TO DEFINE A SINGLE EFFECTIVE
DOSE OF LOOP DIURETIC IN PATIENTS WITH
CHRONIC RENAL INSUFFICIENCY

Creatinine clearance (mL/min)	20–50	<20
Starting dose		
Furosemide	40 mg IV	80 mg IV
Furosemide	80 mg PO	160 mg PO
Bumetanide	1 mg IV or PO	4 mg IV or PO
Ceiling dose		
Furosemide	120–160 mg IV	160–200 mg IV
Furosemide	240–320 mg PO	320–400 mg PO
Bumetanide	4–6 mg IV or PO	8–10 mg IV or PO

Source: Brater, 1988, with permission.

USE OF DIURETICS IN CONGESTIVE HEART FAILURE

In acute congestive heart failure, the sympathetic nervous system is activated mainly to maintain hemodynamic stability, while the renin-angiotensin-aldosterone is activated mainly to decrease renal blood flow and increase reabsorption of salt and water. Although some of the antinatriuresis in congestive heart failure may be mediated by activation of the alpha receptors, the main therapeutic approaches have not included use of alpha inhibitors but rather use of diuretics to increase salt excretion; use of angiotensin-converting enzyme (ACE) inhibitors to reduce afterload and increase renal blood flow; reduction of salt and water intake; and use of inotropic agents to increase cardiac output.

Generally speaking, the loop diuretics are most effective in moderate or severe congestive heart failure. They have the advantage of being potent natriuretic agents and are effective in patients with a wide variety of electrolyte abnormalities. Satisfactory diuresis can usually be obtained in patients with normal renal function with intravenous doses of furosemide, 40 mg twice a day; ethacrynic acid, 50 mg twice a day; or

TABLE 42-7
ORAL DIURETIC AGENTS

Generic Name	Trade Name	Usual Daily Dosage, mg	Onset of Action, h	Peak Effect, h	Duration of Action, h
Thiazide and related diuretics					
Bendroflumethiazide	Naturetin	2.5–1.5	2	4	6–12
Benzthiazide	Aquatag, Exna, Marazide	50–200	2	4–6	6–12
Chlorothiazide	Diuril, Diachlor, Diurigen	500–2000	1–2	4	6–12
Chlorthalidone	Hygroton, Hylidone	25–100	2	2–6	24–72
Cyclothiazide	Anhydron	2	Within 6	7–12	18–24
Hydrochlorothiazide	Esidrix, HydroDIURIL	25–100	2	4–6	6–12
Hydroflumethiazide	Diucardin, Saluron	50–200	2	4	6–12
Indapamide	Lozol	2.5–5	1–2	Within 2	Up to 36
Methylclothiazide	Enduron, Ethon	2.5–5	2	4–6	24
Metolazone	Diulo, Zaroxolyn	2.5–5	1	2	12–24

610

Polythiazide	Renese	2–4	2	6	24–48
Quinethazone	Hydromox	50–100	2	6	18–24
Trichlormethiazide	Metahydrin, Naqua	2–4	2	6	24
Loop diuretics					
Bumetanide	Bumex	0.5–2	0.5–1	1–2	4–6
Ethacrynic acid	Edecrin	50–100	Within 0.5	2	6–8
Furosemide	Lasix	29–80	Within 1	1–2	6–8
Potassium-sparing diuretics					
Amiloride	Midamor	5–20	2	6–10	24
Spironolactone	Aldactone	25–200	24–48	48–72	48–72
Triamterene	Dyrenium	200–300	2–4	6–8	12–16
Combination diuretics					
Amiloride and hydrochlorothiazide	Moduretic	5–10(A) 50–100(H)	See individual agents above	See individual agents above	See individual agents above
Spironolactone and hydrochlorothiazide	Aldactazide, Alazide	25–200(S) 25–200(H)	See individual agents above	See individual agents above	See individual agents above
Triamterene and hydrochlorothiazide	Maxide, Dyazide	37.5–100(T) 25–50(H)	See individual agents above	See individual agents above	See individual agents above

SOURCE: Compiled by Clyde Buchanan from *Facts and Comparisons Drug Information*, 1992.

bumetanide, 1 mg twice a day. If renal disease is present, the maximum necessary intravenous dose should be increased as summarized in Table 42-6. If adequate diuresis is not achieved by the loop diuretics, a more proximally acting diuretic such as metolazone, 5 mg once a day, or more distally acting diuretics such as potassium-sparing diuretics can be added. Since a combination of metolazone and loop diuretics can lead to life-threatening hypokalemia, it is important to monitor serum potassium concentrations carefully.

One of the recent advances in the treatment of congestive heart failure is the appreciation of the importance of inhibiting the renin-angiotensin system. Beneficial effects, both acutely and chronically, have been noted. When ACE inhibitors are used in congestive heart failure and in a setting of presumed high angiotensin levels, it is prudent to make sure that an initial low dose is used first so as to avoid an untoward hypotensive episode does not occur.

As noted in Chap. 1, a key component of the treatment of congestive heart failure is the restriction of sodium intake. Indeed, it has been shown that the loop diuretics given once a day fail to achieve negative sodium balance if salt intake is not limited. Thus, it is essential to limit sodium intake to ensure negative sodium balances. Balance studies on normal individuals have demonstrated that significant negative sodium balance can be predictably obtained with loop diuretics if sodium intake is limited to 20 mEq/day (460 mg/day).

SUGGESTED READING

Bleich M, Greger R: Mechanism of action of diuretics. *Kidney Int Suppl* 1997; 59:S11–5.

Brater DC: Use of diuretics in chronic renal insufficiency and nephrotic syndrome. *Semin Nephrol* 1988; 8:333–341.

Brater DC: Diuretic therapy. *N Engl J Med* 1998; 339: 387–395.

Capasso G, Pica A, Saviano C, et al: Clinical complications of diuretic therapy. *Kidney Int* 1997; 51:S16–S20.

Captopril Multicenter Research Group: A placebo-controlled trial of captopril in refractory congestive heart failure. *J Am Coll Cardiol* 1983; 2:7550–7630.

Dyckner T, Wester PO: Ventricular extrasystoles and intracellular electrolytes before and after potassium and magnesium in patients in diuretic treatment. *Am Heart J* 1979; 97:12–18.

Dzau VJ, Colucci WS, Hollenberg NK, et al: Relation of the renin-angiotensin-aldosterone system to clinical state in congestive heart failure. *Circulation* 1981; 63:645–651.

Knauf H, Mutschler E: Sequential nephron blockade breaks resistance to diuretics in edematous states. *J Cardiovasc Pharmacol* 1997; 29:367–372.

Kokko JP: Diuretics. In: Alexander RW, Schlant RC, Fuster V, O'Rourke RA, Roberts R, Sonnenblick EH (eds): *Hurst's The Heart*, 9th ed. New York, McGraw-Hill, 1998: 783–798.

Morgan DB, Davidson C: Hypokalemia and diuretics: an analysis of publications. *BMJ* 1980; 280:905–908.

Moser M: Diuretics in the prevention and treatment of congestive heart failure. *Cardiovasc Drugs Ther* 1997; 11: 273–277.

CHAPTER

43

DIGITALIS AND NONGLYCOSIDIC CARDIOTONIC AGENTS

Edmund H. Sonnenblick
Thierry H. LeJemtel

Cardiotonic agents improve contractility by increasing systolic Ca^{2+} in the myocyte, either by Na^+, K^+-ATPase inhibition (e.g., digitalis glycosides) or by increasing adenylate cyclase (e.g., catecholamines) or decreasing its breakdown [e.g., phosphodiesterase (PDE) inhibitors—milrinone, amrinone]. Their use in systolic ventricular failure is to increase myocardial contractility so as to increase cardiac output and reduce ventricular filling pressures and hence improve circulatory performance. Other actions on heart rate, conduction, and reflex activity are noted below.

DIGITALIS

The most commonly used digitalis glycoside is digoxin (Lanoxin), which is excreted by the kidneys. Given orally, a loading dose of 1.0 to 2.0 mg is required, followed by 0.125 to 0.25 mg oral maintenance; given intravenously, the dose is approximately one-half. Peak action occurs in 6 to 8 h for the oral dose and in 1 to 4 h for that given intravenously; plasma half-life is 32 to 48 h, which is extended with renal insufficiency. Digitoxin, which is metabolized by the liver, has a half-life of several days. In the elderly, loading and maintenance dose must be reduced primarily because of decreased muscle mass and the slower renal clearance of digoxin.

Pharmacologic Actions

Digitalis increases myocardial contractility in both the normal and failing heart by inhibiting Na^+, K^+-ATPase, which results in increased intracellular calcium. In the nonfailing circulation, digoxin can produce arterial and venous vasoconstriction that is blocked by calcium antagonists but not alpha blockade; of note, in the failing circulation, where vasoconstriction is present, digoxin can produce vasodilatation due to withdrawal of increased sympathetic tone.

Digitalis has an important effect on the central nervous system, increasing vagal and decreasing sympathetic efferent activity. Indeed, these effects may be the major beneficial action of digitalis glycosides. This occurs through enhancement of arterial baroreceptor activity as well as a central action to increase vagal output. The enhancement of parasympathetic activity and the decrease in efferent sympathetic activity contribute importantly to the efficacy of digitalis in the treatment of supraventricular arrhythmias as well as congestive failure.

Electrophysiologically, digitalis slows the rate of the sinus node rate owing increased vagal tone as well as reduced sympathetic tone. In heart failure, improved ventricular performance may also contribute to this. With a "sick sinus syndrome," digitalis may increase sinus pauses and thus should be used with caution. Vagal enhancement shortens the atrial refractory period and may accelerate atrial conduction. At the same time, digitalis prolongs conduction and increases the refractory period of the AV node, primarily because of increased vagal and reduced sympathetic tone.

The refractory periods for antegrade conduction in fast and slow AV nodal fibers are increased. The antegrade effective refractory period in the Wolff-Parkinson-White syndrome accessory pathways are affected variably, while their retrograde refractory period is unaffected.

Clinical Uses of Digitalis Glycosides

Digitalis is used to treat systolic ventricular dysfunction (heart failure) and supraventricular tachycardias.

TREATMENT OF CONGESTIVE HEART FAILURE (SEE CHAP. 1)

Digitalis may be helpful in relieving symptoms in congestive heart failure, characterized by systolic ventricular dysfunction with a reduced ejection fraction and evidence of peripheral congestion with pulmonary rales, elevated venous pressures, and peripheral edema. It is generally used along with diuretics and vasodilators such as angiotensin converting enzyme

(ACE) inhibitors. Even lower doses of digitalis may be beneficial by augmenting vagal tone and reducing sympathetic tone, thus decreasing sinus tachycardia as well as enhancing vascular dilatation. Inotropic effects are modest but help to reduce ventricular filling pressures and increase cardiac output. Studies support the action of digoxin to reduce worsening heart failure and hospitalization without necessarily prolonging life.

With diastolic ventricular dysfunction characterized by elevated filling pressures with a normal ejection fraction, digitalis may not be helpful except in reducing heart rate and thus permitting more time for ventricular filling. With ventricular dysfunction without congestive failure (i.e., normal peripheral resistance), digitalis may even be detrimental in augmenting peripheral resistance, in contrast to the benefits seen in late congestive failure, when peripheral resistance is reduced by digitalis.

TREATMENT OF ARRHYTHMIAS (SEE CHAP. 2)

Paroxysmal Supraventricular Tachycardia (PSVT) Since most PSVT is due to reentry in the AV node, digitalis may prevent PSVT by increasing both the effective and functional refractory period of the AV node, primarily via the antegrade fibers. This effect may also terminate an attack of PSVT. Beta blockers or digitalis or their combination, as may be needed in about 25 percent of cases, may be used for prevention of PSVT. Calcium-channel blockers such as diltiazem and verapamil given intravenously may be useful in treating the acute episode (see also Chap. 2).

Atrial Fibrillation (AF) and Flutter In the presence of AF, digitalis is very useful in reducing ventricular rate through two mechanisms. First, it directly increases the refractory period of the AV node. Second, by shortening the refractory period of the atrium, it speeds atrial rate, which sends more atrial pulses to the AV node. This leads to greater concealed conduction, which results in greater depolarization and thus a slower heart rate. The effect of slowing heart rate can be augmented with beta-blocking agents or calcium-channel blockers such as diltiazem or verapamil. These agents are often used in combination if the control of the heart rate during exercise is inadequate. Digoxin commonly does not prevent recurrence of AF when the patient is converted to sinus rhythm, nor is it useful for conversion to sinus rhythm except to the degree that it decreases heart failure.

Digitalis is poorly effective in controlling atrial flutter except in very high doses, which would be necessary to markedly block AV conduction. Often flutter converts to AF when the patient is digitalized. Thus, in flutter, beta blockers or calcium-channel blockers are much more effective in controlling ventricular rate (see also Chap. 2).

Digoxin Administration

The digoxin loading dose ranges from 1.0 to 2.0 mg in divided doses over 24 h with a maintenance of 0.125 to 0.5 mg/day, generally 0.25 mg. The intravenous dose is about 70 percent of the oral dose. The dosage is about half of this in the elderly or patients with renal insufficiency. With sinus rhythm, a serum level should approximate 1.0 mg/mL, but the range is wide. With atrial fibrillation, heart rate is the best measure of adequate dose.

Digitalis Toxicity

Symptoms of toxicity include nausea, appetite loss, and visual changes. Digitalis arrhythmias reflect increased ventricular ectopy and vagal tone. Enhanced vagal effects include sinus bradycardia and Wenckebach AV nodal block. Ectopy includes atrial extrasystoles, junctional tachycardia, and ventricular ectopy including bigeminy, tachycardia, and eventually ventricular fibrillation. Atrial tachycardia with AV block reflects vagal and ectopy effects. Hypokalemia amplifies effects of toxicity.

Safe therapeutic serum levels of digoxin range from 1.0 to 1.5 ng/mL; toxicity is generally associated with levels greater than 2.5 ng/nL. A diagnosis of toxicity is suspected clinically and confirmed by drug levels. Pharmacokinetic interactions with digoxin are listed in Table 43-1.

Treatment of digitalis toxicity reflects the severity of concurrent arrhythmias. Digitalis is discontinued and hypokalemia corrected. Severe bradycardia or complete heart block may require a temporary pacemaker. Ectopy may respond to lidocaine or procainamide. Amiodarone and intravenous magnesium have been used in unresponsive severe arrhythmias. With life-threatening arrhythmias, Fab fragments, which bind digoxin for its subsequent renal excretion, have been very successfully utilized. The only limitation is expense and potential anaphylaxis should they be needed again.

NONGLYCOSIDIC CARDIOTONIC AGENTS

Inotropic stimulation of the heart by a nonglycosidic cardiotonic agent may be indicated in two clinical situations, which differ greatly pathophysiologically. In acute heart failure complicating myocardial infarction or coronary bypass surgery, residual noninfarcted or stunned myocardium and the peripheral circulation are normally responsive to hormonal and metabolic alterations. The major aims of therapy are to increase cardiac output and reduce left ventricular filling pressure without tachycardia and exacerbation of myocardial ischemia. In chronic end-stage heart failure, myocardial inotropic reserve is limited; peripheral arterial resistance is increased and, to a certain extent, fixed. The profound abnormalities of the peripheral circulation are responsible for the symptoms of congestive heart failure and contribute to further reduction in left ventricular performance. Thus, the aim of therapy for chronic congestive heart failure is to reverse the abnormalities of the peripheral circulation without increasing pathologic damage or inducing further ventricular arrhythmias.

Nonglycosidic cardiotonic agents can be classified in two major groups. The first includes the catecholamines (norepinephrine, dopamine) and their derivatives (dobutamine, isoproterenol), which are available only for parenteral use. These agents activate specific receptors (Table 43-2) that increase the adenylate-cyclase system with augmentation of intracellular Ca^{2+} to increase contraction while enhancing its removal for better relaxation. The second group are the new, specific type III phosphodiesterase inhibitors, which decrease the breakdown of cyclic AMP. Some of these agents may, in addition, increase the affinity of the regulatory site of troponin C for Ca^{2+}, which may contribute to their positive inotropic effects. These newer inotropic agents, which are also orally active but are available only for parenteral use, also produce direct peripheral arterial and venous vasodilatation, which, in certain instances, may be the predominant factor in improving cardiac performance. Thus, specific phosphodiesterase inhibition, which leads to increased cyclic AMP levels in the cells, enhances contractility of the cardiac myocyte and relaxes the vascular smooth muscle. Amrinone was the first new inotropic agent approved by the FDA. It has now been largely supplanted by milrinone, a second-generation specific type III phosphodiesterase inhibitor available only for parenteral administration.

TABLE 43-1

PHARMACOKINETIC INTERACTIONS WITH DIGOXIN

Interfering drugs	Mean magnitude of interaction, %[a]	Mechanism of interaction	Suggested dose adjustments
Drugs that alter absorption or bioavailability of digoxin			
Antacid, bran	−25	Unknown; ? adsorption of digoxin	Give digoxin 1–2 h before antacid or bran
Kaolin-pectin (Kaopectate)	20 to 30	Adsorption of digoxin	Give digoxin 1–2 h before kaolin-pectin
Cholestyramine (Questran)	−30	Physical binding to resin	Avoid by dosing cholestyramine bid, 8 h from digoxin administration
Metoclopramide (Reglan)	−25	↓ Bioavailability by intestinal motility	↑ (?) Administer digoxin as elixir or as Lanoxicaps
Propantheline (Pro-Banthine)	+25	↑ Bioavailability by intestinal motility	↓ (?) Administer digoxin as elixir or as Lanoxicaps
Sulfasalazine (Azulfidine)	−20	↓ Bioavailability	
Erythromycin or tetracycline[b]	+43 to +150	↑ Bioavailability due to inactivation of gut flora	Measure serum digoxin level during coadministration of these antibiotics with digoxin

620

		Decrease	
Neomycin	−28	↓ Bioavailability	Measure serum digoxin levels
Cancer chemotherapy	−50	↓ Bioavailability due to damage to intestinal mucos	Measure serum digoxin levels

Drugs that interfere with elimination of digoxin

		Decrease	
Amiodarone (Cordarone)	+100	In renal and total body clearance of digoxin	Reduce digoxin dose by one-half
Cyclosporine (Sandimmune)	+50	Vol of distribution and plasma clearance	Reduce digoxin dose by one-half
Indomethacin (Indocin)[c]	+50	Glomerular filtration	Reduce digoxin dose by one-fourth to one-half
Propafenone (Rhythmol)	+20 to +80	Nonrenal clearance of digoxin	Reduce digoxin dose by one-fourth to one-half
Quinidine	+100	(?) Absorption Vol of distribution Renal and total body clearance of digoxin	Reduce digoxin dose by one-half
Quinine	+75	Renal clearance of digoxin	Recuce digoxin dose by one-half
Verapamil	+75	Renal and total body clearance of digoxin	Reduce digoxin dose by one-half

[a] For single-dose studies, the magnitude of the anticipated change in serum digoxin concentration was estimated from pharmacokinetic data, particularly the change in total body clearance.

[b] Expected to occur in 10% of patients—those who have substantial conversion of digoxin to dihydro-derivatives in the gut.

[c] Interaction only in premature infants.

621

TABLE 43-2

RECEPTOR ACTIVITY OF SYMPATHOMIMETIC AMINES

	α_1	β_1	β_2	Dopaminergic
Norepinephrine	4+	4+	0	0
Epinephrine	4+	4+	2+	0
Dopamine	4+	2+	+	4+
Isoproterenol	0	4+	4+	0
Dobutamine	3+	4+	2+	0

Catecholamines and their Derivatives

Dopamine (Intropin), a naturally occurring precursor of norepinephrine, and *norepinephrine* itself may be used to increase myocardial contractility. However, both dopamine in high doses and norepinephrine constrict the peripheral arterioles; this limits both drugs' usefulness. Of greater therapeutic importance in managing patients with acute and decompensated chronic heart failure, dopamine at doses below 2.0 μg/kg/min, the "renal dose," selectively increases renal blood flow, which in turn potentiates the effects of diuretics. At doses greater than 3 μg/kg/min, dopamine stimulates alpha$_1$ receptors and tends to increase arterial pressure. This latter effect becomes predominant at doses greater than 10 μg/kg/min, the "pressor dose." In the presence of acidosis, the ability of dopamine to increase arterial pressure is markedly limited and norepinephrine is generally required.

Dobutamine (Dobutrex), a synthesized catecholamine derivative without vasoconstrictor activity, has become the most widely used parenteral inotropic agent. Endogenous catecholamines and dobutamine have different physiologic actions that derive from their relative specificities for alpha and beta andrenoreceptors (Table 43-2). Alpha receptors include alpha$_1$ receptors, which are postsynaptic and are located in the vascular smooth muscle or the myocardium. The smooth muscle alpha$_1$ receptors are responsible for vasoconstriction. The alpha$_2$ receptors are mostly presynaptic and are responsible for reducing norepinephrine release in the peripheral nerve terminals and reducing sympathetic outflow via the central nervous system.

Beta-adrenergic receptors consist of $beta_1$ receptors, which are located in the myocardium and are responsible for positive inotropic, chronotropic, and dromotropic responses; and $beta_2$ receptors, which are located in the smooth muscles and mediate vasodilatation. $Beta_2$ receptors may also be located in the sinoatrial node and may be responsible for positive chronotrophy. In addition, there are specific dopaminergic receptors in the renal and mesenteric vascular beds; these are responsible for arterial vasodilatation.

DOBUTAMINE (DOBUTREX)

The synthesis of dobutamine resulted from systematic modification of the chemical structure of isoproterenol. Dobutamine produces a potent positive inotropic action mediated through direct stimulation of $beta_1$-adrenergic receptors in the myocardium, which in turn increase cyclic AMP. Unlike dopamine, dobutamine does not stimulate the heart indirectly by releasing norepinephrine from the nerve endings. Dobutamine's relative lack of positive chronotropic effect is not well understood; it may result from specific stimulation of $beta_1$ receptors and/or $alpha_1$ myocardial receptors. Whatever its exact mechanisms of action, dobutamine is the cardiotonic agent that exerts the most potent inotropic action while producing limited effects on heart rate and blood pressure. With acute myocardial infarction and left ventricular dysfunction, dobutamine administration is safe and most often improves cardiac performance without obviously exacerbating myocardial ischemia. The rate of infusion of dobutamine should start at 2 μg/kg/min and then, if these measurements are available, be titrated upward to obtain maximal cardiac output while reducing left ventricular filling pressure. If not, heart rate and blood pressure should be closely monitored to prevent tachycardia and excessive reduction in blood pressure. The most serious side effect of dobutamine is the precipitation of ventricular tachycardia, which may necessitate dose reduction or discontinuation. In chronic congestive heart failure, administration of dobutamine is useful either during acute decompensation or for intermittent inotropic support; the latter requires short-term hospitalization or use of a small, portable infusions pump on an outpatient basis. Although patients with severe congestive heart failure have decreased density and affinity of myocardial $beta_1$-adrenergic receptors, most experience a hemodynamic and clinical improvement after administration of dobutamine. The fate of the beta receptors and their coupling with adenylate cyclase during long-term administration of dobutamine is unknown. The mechanism of the prolonged clinical benefits of dobutamine after discontinuation of the infusion are also unknown.

Phosphodiesterase Inhibitors

MILRINONE (PRIMACOR)

Currently, the parenteral form of milrinone is the only nonglycosidic, noncatecholamine cardiotonic agent marketed in the United States for use in congestive heart failure. Besides its positive inotropic action, which is at least partly mediated by specific type III myocardial phosphodiesterase inhibition, milrinone has direct arteriolar and venous vasodilating properties. The peripheral vasodilation induced by milrinone is cyclic AMP–mediated. The exact mechanisms by which rising levels of cyclic AMP promote the vascular smooth muscle relaxation are incompletely understood. The arterial dilatation appears to occur preferentially in the limb vasculature and, to a lesser extent, in the renal vasculature.

Intravenous milrinone increases cardiac output and reduces left ventricular filling pressure in patients with chronic congestive heart failure without producing excessive changes in heart rate or systemic blood pressure. The changes in myocardial contractility, as evidenced by the rate of increase in left ventricular pressure, vary from patient to patient. Consequently, in patients experiencing minimal changes in myocardial contractility, arteriolar vasodilatation induced by milrinone contributes substantially to the improvement in left ventricular performance; thus, these hemodynamic changes are often accompanied by a decrease in myocardial oxygen requirements. Milrinone, as compared with dobutamine, tends to produce greater reduction in pulmonary capillary wedge pressure, while the increase in cardiac output is similar.

Therapy with milrinone should be initiated at a loading dose of 50 μg/kg, which is administered slowly over 10 min and followed by a continuous infusion maintenance dose ranging from 0.375 to 0.75 μg/kg/min.

Milrinone is primarily excreted by the kidneys. The plasma elimination half-life of milrinone in normal subjects is 50 min, while it is approximately 100 min in patients with congestive heart failure. In patients with severe renal insufficiency, the rate of infusion should be reduced to prevent toxic plasma levels of milrinone and potential adverse reactions, such as ventricular tachycardia and hypotension.

As is the case with amrinone, an important therapeutic application of intravenous milrinone is its concomitant use with dobutamine. When dobutamine and milrinone are administered simultaneously, the initial dose of dobutamine should not be greater than 5 µg/kg/min and the loading dose of milrinone should be halved—i.e., 25 µg/kg over 10 min; the maintenance dose of milrinone should be initiated at 0.375 µg/kg/min. When the combination of dobutamine and milrinone is well tolerated without precipitating ventricular arrhythmia or excessive reduction in systemic arterial pressure, the dose of these drugs can be gradually increased. The combined use of dobutamine and milrinone may allow administration of dobutamine at a lower dose than when dobutamine is administered alone, thus reducing the metabolic cost to the myocardium of increasing left ventricular performance. The beneficial effects of milrinone can also complement those of vasodilators. For example, ACE inhibitors preferentially benefit the renal circulation, while specific phosphodiesterase inhibition predominantly increases blood flow to the limbs. This argues for combined use of these agents.

Short-term intravenous administration of milrinone is most often associated with hemodynamic and clinical benefits in patients with acute and chronic congestive heart failure. Thrombocytopenia is extremely rare with milrinone, unlike the case with amrinone. Thus, milrinone can be safely administered for longer periods than amrinone without daily monitoring of platelet count.

SUGGESTED READING

Anderson JL, Baim D, Fein SA, et al: Efficacy and safety of sustained (48 hour) intravenous infusions of milrinone in patients with severe congestive heart failure: a multicenter study. *J Am Coll Cardiol* 1987; 9:711–722.

Antman EM, Smith TW: Current concepts in the use of digitalis. *Adv Intern Med* 1989; 34:425–454.

Colucci WS: Positive inotropic/vasodilator agents. *Cardiol Clin* 1989; 7:131–144.

Colucci WS, Wright RF, Braunwald E: New positive inotropic agents in the treatment of congestive heart failure: Mechanisms of action and recent clinical developments. *N Engl J Med* 1986; 314:290–299, 349–358.

Leier CV: Current status of non-digitalis positive inotropic drugs. *Am J Cardiol* 1992; 69(suppl):120G–129G.

LeJemtel TH, Sonnenblick EH, Frishman WH: Diagnosis and management of heart failure: In: Alexander RW, Schlant RC, Fuster V, O'Rourke RA, Roberts R, Sonnenblick EH (eds): *Hurst's The Heart*, 9th ed. New York, McGraw-Hill, 1998:745–781.

Packer M, Gheorghiade M, Young JB, et al: Withdrawal of digoxin from patients with chronic heart failure treated with angiotensin-converting-enzyme inhibitors. *N Engl J Med* 1993; 329:1–7.

Smith TW: Digitalis: mechanisms of action and clinical use. *N Engl J Med* 1988; 318:358–365.

Sonnenblick EH, Frishman WH, LeJemtel TH: Dobutamine: a new synthetic cardioactive sympathetic amine. *N Engl J Med* 1979; 300:17–22.

44

DRUGS USED TO CONTROL VASCULAR RESISTANCE AND CAPACITANCE

Robert C. Schlant

Vasoactive drugs that alter vascular resistance and capacitance are of importance in the management of patients with cardiovascular disorders—particularly heart failure, coronary artery disease, systemic arterial hypertension, hypotension, and pulmonary hypertension. Some of these agents are listed in Table 44-1. The antihypertensive drugs, including diuretics, adrenergic inhibitors, vasodilators, and angiotensin converting enzyme (ACE) inhibitors, are discussed in Chap. 11 (see Table 11-1); the calcium antagonists are discussed in Chap. 48 (see Tables 48-1 and 48-2), as well as in Chap 11 (Table 11-1). Phosphodiesterase inhibitors, which have vasodilatory properties in addition to their inotropic effects, are discussed in Chap 43, together with other cardiotonic agents.

DIRECT-ACTING VASODILATOR DRUGS

Nitrates

Organic nitrates and nitroprusside are "nitrovasodilators," which provide a source of nitric acid (NO), which activates the soluble isoform of guanylyl cyclase to increase the synthesis of cyclic guanosine monophosphate (GMP) from guanosine triphosphate (GTP). In vascular muscle, this leads to a cascade of events, including cyclic GMP-dependent protein activation, phosphorylation of many smooth muscle proteins, decreased activity of phospholipase C, decreased phosphoinositide turnover, decreased cytosolic calcium, decreased phosphorylation of myosin light chain, and vascular relaxation. Prolonged exposure to nitrates results in vascular tolerance, thought to be

TABLE 44-1
DIRECT-ACTING VASODILATORS

Drugs	Route of administration	Dosage	Onset of effect	Side effects
Nitroglycerin* (Tridil, Nitro-Bid)	Intravenous	5–200μg/min	Immediate	Headache, hypotension, withdrawal symptoms, methemoglobinemia, tolerance
Nitroglycerin*; (Nitrostat)	Sublingual	0.15–0.6 mg prn	30 s	Headache, hypotension, tolerance
Nitroglycerin*; (Nitro-Bid)	Oral	2.5–9.0 mg bid or tid	1 h	Headache, hypotension, tolerance
Nitroglycerin* (Nitrol ointment 2%, Nitro-Dur, Transderm-Nitro, Nitrodisc, Deponit)	Transdermal	1.25–5.0 cm q 8–24 h 2.5–15 mg patch/day	1 h	As above
Nitrolingual spray*	Oral spray	0.4 mg/spray dose prn	30 s	As above

Isosorbide dinitrate* (Isordil, Sorbitrate) Dilatrate-SR	Sublingual or oral	SL: 2.5–20 mg q 6 h PO: 5–80 mg q 8–12 h	5–30 min	As above
Isosorbide mononitrate (ISMO)	Oral	10–40 mg bid at 7-h intervals	30 min	Headache, dizziness, hypotension
Isosorbide mononitrate; (extended release; Imdur)	Oral	30–240 mg d in single dose	30 min–1 h	Headache, dizziness, hypotension
Sodium nitroprusside; (Nipride)	Intravenous	0.2–10 µg/kg/min	Immediate	Hypotension, metabolic acidosis, accumulation of thiocyanate
Hydralazine (Apresoline)	Oral	10–75 mg q 6 h	30 min	Headache, tachycardia, autoimmune lupus-like reaction, increased toxicity in slow acetylators, angina, edema, tolerance
Minoxidil (Loniten)	Oral	5–40 mg q d in single or divided doses	30 min	Tachycardia, water retention, pericardial effusion, hirsutism
Diazoxide; (Hyperstat)	Intravenous	1–3 mg/kg q 10–15 min	Immediate	Hyperglycemia, hypotension, tachycardia, hyperuricemia

KEY: SL = sublingual; PO = by mouth.

*Chronic nitrate therapy should include a daily 8- to 12-h period without.

due to the oxidation of sulfhydryl (SH) groups necessary for the formation of NO. Accordingly, nitrate-free periods of approximately 10 to 12 h per 24 h are now advocated. Nitrates have a greater vasodilator effect on veins than arterioles. Nitrates are effective in a variety of forms (Table 44-1).

Sodium Nitroprusside

Intravenous sodium nitroprusside is an effective, balanced arterial and venous vasodilator that can be very useful in the management of patients with systemic arterial hypertension or acute heart failure. Nitroprusside is converted to cyanmethemoglobin and free cyanide in red blood cells, and the free cyanide is converted by the hepatic enzyme rhodanese to thiocyanate. Thiocyanate is cleared by the kidneys, with a half-life of 7 days in patients with normal renal function. With prolonged administration of nitroprusside, thiocyanate can accumulate in association with a metabolic (lactic) acidosis.

If infusions exceed 48 to 72 h, monitoring for nitroprusside toxicity includes monitoring blood acid-base determinations, thiocyanate blood level (toxic symptoms occur with plasma thiocyanate levels over 5 to 10 mg/dL) and, if available, measurements of mixed venous oxygen tension and arteriovenous oxygen difference. Thiocyanate levels can be within normal limits in patients with very definite acidosis and an increase in mixed venous oxygen tension from cyanide toxicity. In some other patients with lactic acidosis due to poor perfusion, however, nitroprusside can be beneficial. Solutions of sodium nitroprusside should be freshly made, administered with an infusion pump, and shielded from light during infusion. Because nitroprusside has a strong, very rapid effect, patients receiving it require close monitoring. In patients with acute aortic dissection, sodium nitroprusside should not be administered without concurrent administration of a beta-adrenergic blocker. The initial rate of infusion (0.2 to 1.0 μg/kg/min or 10 μg/min) is increased by 0.2 μg/kg/min or 10 μg/min about every 5 to 10 min, up to an average dose of 3 μg/kg/min (maximal 10 μg/kg/min) or between 15 and 200 μg/min, with a maximum dose of about 800 μg/min. Thiocyanate toxicity is manifest by fatigue, nausea, disorientation, psychotic behavior, muscle spasms, convulsions, a skin rash, and bone marrow depression. If nitroprusside-cyanide-thiocyanate toxicity is suspected, the sodium nitroprusside should be discontinued and a bolus of sodium thiosulfate 150 mg/kg (or 12.5 g) given intravenously to provide a sulfate donor for the conversion of cyanide to thiocyanate. Additional treatment includes the infusion of 5 mg/kg of sodium nitrite in 20 mL water over 3 to 4 min and possibly the inhalation of amyl nitrite. The nitrite converts hemoglobin to methemoglobin, which combines with cyanide to form nontoxic cyanmethemoglobin. Pressor agents may also be necessary.

Hydralazine

Hydralazine is predominantly an arteriolar dilator with a very slight positive inotropic effect. It is used primarily in the treatment of patients with heart failure who cannot tolerate treatment with an angiotensin converting enzyme (ACE) inhibitor or who need additional arteriolar vasodilatation. In such patients it is frequently administered in combination with nitrates, such as isosorbide dinitrate, which produce venous vasodilatation. It is also used in the management of patients with systemic arterial hypertension.

Side effects include headache, nausea, reflect tachycardia, fluid retention, and a syndrome similar to lupus erythematosus. The latter syndrome occurs in about 10 to 20 percent of patients receiving 400 mg/day or more hydralazine for more than 6 months. It occurs in about 10 percent of patients receiving 200 mg/day and in about 5 percent of those receiving 100 mg daily. The syndrome, which is usually manifest by arthralgia or arthritis, is usually reversible on stopping the drug. It occurs almost exclusively in slow acetylators. Fatal outcome and renal involvement are rare. The antinuclear antibody or factor (ANF) is positive in about 98 percent of patients with hydralazine-induced lupus.

The usual oral dosage is 50 to 75 mg every 6 h, with a maximal dose of 800 mg/day.

Minoxidil (Loniten)

Minoxidil (Loniten) produces arteriolar vasodilatation by opening potassium channels. Side effects, including hirsutism and pericardial disease, limit its use to patients with resistant hypertension. Diuretics are usually required because of fluid retention, and beta-blocker therapy often is required to reduce reflex tachycardia. In general, it is used only rarely to treat patients with systemic arterial hypertension who are resistant to multidrug regimens; it is particularly effective when renal failure is present (see Chap. 11).

The initial dosage is 5 mg given as a single daily dose. Daily dosage can be increased to 10, 20, and then to 40 mg in single or two divided doses. The usual daily dose is 10 to 40 mg/day. The maximum recommended dosage is 100 mg/day.

Diazoxide

Diazoxide (Hyperstat) is useful intravenously for the emergency reduction of blood pressure. The dosage is the minibolus administration of 1 to 3 mg/kg repeated at intervals of 5 to 15 min up to a maximum of 150 mg. Repeat injections at 4-to-24-h intervals usually will maintain blood pressure reduction until oral agents become effective. Diazoxide requires very close monitoring of the patient's blood pressure at frequent intervals.

ALPHA$_1$-ADRENERGIC BLOCKERS

The competitive alpha$_1$-selective agents in this group include prazosin (Minipress), terazosin (Hytrin), and doxazosin (Cardura); the use of these agents in the treatment of patients with systemic arterial hypertension is discussed in Chap. 11 (Table 11-1). The nonselective agents in this group include phentolamine (Regitine) and phenoxybenzamine (Dibenzyline).

Phentolamine Mesylate

Phentolamine mesylate (Regitine) is predominantly a short-acting competitive, nonselective alpha-adrenergic blocking drug. It also has some vasodilator effects on vascular smooth muscle and mild inotropic and chronotropic effects on cardiac muscle. It is used intravenously primarily to prevent or control blood pressure in patients with pheochromocytoma. Preoperatively, 5 mg is injected IV or IM 1 to 2 h before surgery and repeated as necessary. During surgery, 5 mg is administered intravenously to help control or prevent paroxysms of hypertension, tachycardia, respiratory depression, or convulsions. Phentolamine also has been used as a test for pheochromocytoma, with a positive response being a reduction in blood pressure of more than 35 mmHg systolic and 25 mmHg diastolic. The maximal effect usually is seen within 2 min. Phentolamine is also useful to prevent tissue necrosis and sloughing following extravasation of norepinephrine. In this situation 5 to 10 mg of phentolamine in 10 mL saline is injected into the area of extravasation within 12 h.

Phenoxybenzaline

Phenoxybenzaline (Dibenzaline) is a long-acting, noncompetitive, nonselective alpha-receptor blocking agent that produces a chemical sympathectomy. It is used primarily to control hypertension and sweating in patients with pheochromocytoma. The initial dose of 10 mg twice a day is increased every other day up to 20 to 40 mg two or three times a day until adequate blood pressure control is obtained.

CENTRAL AND PERIPHERAL SYMPATHOLYTICS

Centrally Acting Alpha$_2$-Adrenergic Agonists

These agents include methyldopa (Aldomet), clonidine (Catapres), clonidine TTS (patch; Catapres TTS), guanabenz (Wytensin), and guanfacine (Tenex). Their use in the treatment of patients with systemic arterial hypertension is discussed in Chap. 11 (Table 11-1).

Peripherally Acting Adrenergic Antagonists

These agents include the rauwolfia alkaloids reserpine (Serpasil), guanethidine (Ismelin), and guanadrel (Hylorel). Their use in the treatment of patients with systemic arterial hypertension is discussed in Chap. 11 (Table 11-1).

Ganglion-Blocking Agents

Trimethaphan camsylate (Arfonad) is a ganglionic blocking agent that is diluted and administered intravenously to produce controlled hypotension during surgery or to control blood pressure in hypertensive emergencies. There is a marked variation in the response of different patients and frequent blood pressure determinations are essential. Infusion rates vary from 0.3 mg to 6 mg/min. Because of its many side effects and adverse reactions, it is currently used only infrequently.

ANGIOTENSIN CONVERTING ENZYME (ACE) INHIBITORS

These agents inhibit the angiotensin-converting enzyme (ACE) that cleaves off two amino acids from the biologically inactive decapeptide angiotensin I to form the octapeptide angiotensin II, which is a powerful vasoconstrictor and stimulator of aldosterone secretion. Angiotensin II decreases the "opening" of receptor-operated calcium channels and decreases the permissive effect of angiotensin II on norepinephrine (NE) release from terminal neurons. The same converting enzyme also inactivates the vasodilatory enzyme bradykinin. While the beneficial effects of ACE inhibitors are partly due to the suppression of angiotensin II and partly to the potentiation of bradykinin, the allergic or toxic reactions are specific for individual ACE inhibitors and are unlikely to recur with other ACE inhibitors.

Adverse effects common to most ACE inhibitors include cough, which occurs in about 4 to 16 percent of patients (especially in older female white patients); functional renal insufficiency due to changes in intrarenal hemodynamics in patients with stenosis of the main artery to a solitary kidney, bilateral renal artery stenoses, or severe generalized atherosclerosis with pre-existing chronic renal failure; and, rarely, angioedema. It is thought that the cough and angioedema are due to bradykinin. These adverse effects are reversible after discontinuation of the ACE inhibitor.

Allergic or toxic reactions include taste alterations, pruritus, skin rash, and, rarely, blood dyscrasias that may range from granulopenia to pancytopenia. These reactions are more likely to occur in patients with altered immune states or severe renal insufficiency receiving high doses of an ACE inhibitor and are usually reversible with switching to a different ACE inhibitor.

ACE inhibitors are very useful in the treatment of patients with heart failure (Chap. 1), with left ventricular dysfunction following myocardial infarction (Chap. 10), and with systemic arterial hypertension (Chap 11).

Intravenous enalaprilat can be used for the treatment of severe hypertension.

Angiotensin II Receptor Blockers (ARBs)

These agents (losartan, or Cozaar; valsartan, or Diovan; and irbesartan, or Avapro) predominantly block the angiotensin II subtype 1 (AT$_1$) receptor and thereby block the effects of angiotensin II but do not block the conversion of bradykinin. Presumably, this is why their use is not associated with cough. Like ACE inhibitors, they are also contraindicated with pregnancy. They are currently approved for the treatment of patients with systemic arterial hypertension in whom an ACE inhibitor may be indicated but is not well tolerated.

ALPHA$_1$-ADRENERGIC AGONISTS

Methoxamine (Vasoxyl)

Methoxamine is an alpha-receptor stimulant that produces a rise in blood pressure due to direct peripheral vasoconstriction without inotropic or chronotropic effect. It is used primarily for maintaining blood pressure during operations under spinal anesthesia. The usual intravenous dose is 3 to 5 mg, injected slowly. The usual intramuscular dose is 10 to 15 mg given shortly before or at the time of spinal anesthesia.

Phenylephrine (Neo-synephrine)

Phenylephrine directly stimulates alpha$_1$-receptors to produce vasoconstriction. It is injected subcutaneously, intramuscularly, or intravenously. For hypotensive emergencies during spinal anesthesia, an initial intravenous dose is 0.2 mg. For the prophylaxis of hypotension during spinal anesthesia, 2 to 3 mg is injected subcutaneously or intramuscularly 3 or 4 min before injection of the spinal anesthetic.

Mephentermine (Wyamine)

This is used to prevent hypotension during spinal anesthesia. It causes the release of norepinephrine in addition to a direct effect.

Metaraminol (Aramine)

This increases blood pressure by both a direct effect and by the release of norepinephrine.

Dopamine-1 Receptor Agonists

Fenoldopam can be given intravenously for the treatment of severe systemic arterial hypertension. It can cause reflex tachycardia or an increase in intraocular pressure.

SUGGESTED READING

Ewy GA, Bressler R: *Cardiovascular Drugs and the Management of Heart Disease*, 2d ed. New York, Raven Press, 1992:1–496.

Frishman WH: *Cardiovascular Drugs*, 2d ed. Philadelphia, Current Medicine, 1995:1–317.

Frishman WH, Sonnenblick EH: Beta-adrenergic blocking drugs and calcium channel blockers. In: Alexander RW, Schlant RC, Fuster V, O'Rourke RA, Roberts R, Sonnenblick EH (eds): *Hurst's The Heart*, 9th ed. New York, McGraw-Hill, 1998:1583–1618.

Frishman WH, Sonnenblick EH: *Cardiovascular Pharmacotherapeutics*. New York, McGraw-Hill, 1997:1–1722.

Gifford RW Jr: Treatment of patients with systemic arterial hypertension. In: Alexander RW, Schlant RC, Fuster V, O'Rourke RA, Roberts R, Sonnenblick EH (eds): *Hurst's The Heart*, 9th ed. New York, McGraw-Hill, 1998: 1673–1696.

Hardman JG, Limbird LE, Molinoff PB, Ruddon RW (eds): *Goodman and Gilman's The Pharmacological Basis of Therapeutics*, 9th ed. New York, McGraw-Hill, 1996.

Messerli FH: *Cardiovascular Drug Therapy*. Philadelphia, WB Saunders, 1990:1–1709.

Opie LH: *Drugs for the Heart*, 4th ed. Philadelphia, WB Saunders, 1997.

CHAPTER

45

DRUGS USED TO TREAT CARDIAC ARRHYTHMIAS

Raymond L. Woosley

The antiarrhythmic drugs are among the most controversial drugs in cardiology. Our understanding of whether and how to treat arrhythmias has changed in recent years. For example, there is disagreement over the significance of "warning arrhythmias," and the risk-benefit ratio of lidocaine therapy for patients with acute myocardial infarction (MI) is now in question. A metanalysis of the controlled trials with lidocaine in acute MI found that lidocaine did not improve mortality. Another analysis of trials in acute MI patients found that the incidence of ventricular fibrillation has fallen to <0.35 percent in 1990, a level at which drug efficacy would be difficult to detect. This observation and the results of the Cardiac Arrhythmia Suppression Trial (CAST) conducted in patients with ventricular arrhythmias after recovery from MI have led physicians to reserve antiarrhythmic drugs to those patients who are highly symptomatic or have arrhythmias that are clearly life-threatening—i.e., ventricular tachycardia or fibrillation. Because of their toxicity, these drugs should be prescribed only when benefit is likely. The clinician should carefully review the indications for use of these drugs before prescribing any of them.

Table 45-1 lists the actions of antiarrhythmic drugs. Until recently, the Vaughan-Williams classification was the primary framework for organizing an increasing list of available drugs. However, its limitations and potential for being clinically misleading prompted cardiologists to seek another approach. The *Sicilian gambit* has been proposed as a replacement for the outdated Vaughan-Williams classification (see Suggested Reading). It emphasizes that drugs have many actions and the clinical outcome may depend on more than one of these actions interacting with a complex arrhythmia substrate. Most drugs have multiple actions (e.g., amiodarone) and cannot be classi-

TABLE 45-1
ANTIARRHYTHMIC DRUG ACTIONS*

Vaughn-Williams class	Drug	Channels			Receptors				Clinical effects					ECG changes
		Na	Ca	K	α	β	ACh	Ado	Pro-Arrhy	LV	PX	Heart rate	Extra-cardiac	
I A	Quinidine	◨		●	○				●				◨	A
	Procainamide	◨		●			◨		◨				●	B
	Disopyramide (Norpace)	◨		◨			◨		○	⇊			◨	C
I B	Lidocaine (Xylocaine)	○							○				○	
	Mexiletine (Mexitil)	○							○				○	
I C	Propafenone (Rythmol)	●				◨			◨	⇊	↓		◨	
	Flecainide (Tambocor)	●							◨	⇊	⇊		●	
IV	Verapamil (Calan, Isoptin)		●						○	⇊	↓		○	
	Diltiazem (Cardizem)		◨						○				○	

Class	Drug														ECG
III	Bretylium (Bretylol)					○ ▲			●	○	→		○		⋰ (dotted)
	Sotalol (Betapace)			○ △					●	○			○		
	Amiodarone (Cordarone)	◐	◐			◐	◐		●	●	→→→		●		
	Ibutilide (Corvert)								●	○	→		●		
II	β-adrenergic antagonists					●			●	○		⇉	○		◠
Misc	Adenosine (Adenocard)							△	●	○	→		○		◠

Antagonist relative potency ○ Low ◐ Moderate ● High   △= Agonist ▲= Agonist/Antagonist

*This table is a modification of the Sicilian Gambit drug classification system and includes designation using the Vaughan-Williams system. The sodium-channel blockers are subdivided into A, B, and C subgroups based upon differing effects of drugs on the QRS and QT. The targets of antiarrhythmic drugs are listed across the top of the table. The channels are sodium, calcium, and potassium; the receptors are alpha-adrenergic, beta-adrenergic, cholinergic (ACh), and adenosinergic (Ado). The next columns compare the drugs' clinical actions. These include proarrhythmic potential (ProArrhy), effect on left ventricular function (LVFX), effects on heart rate (Heart rate), and potential for extracardiac side effects (Extracardiac). The ECG tracings indicate the portion that is altered by the drug: PR interval (⌒), QRS interval (▓), and QT interval (⋰). The drugs are listed with their brand names shown in parentheses. The symbols in the table indicate the drugs' relative potency as agonists or antagonists. The solid triangle indicates the biphasic effects of bretylium to release norepinephrine and act as an agonist and then block further release and subsequently act as an antagonist of adrenergic stimulation. The number and direction of the arrows indicate the magnitude and direction of effect of the drugs on heart rate and left ventricular function—i.e., inotropy.

fied simply. The Sicilian gambit is a tabulation of actions and Table 45-1 is a modification of the original tabulation that incorporates the Vaughan-Williams nomenclature and electro-cardiographic (ECG) changes induced by the drugs. The drug actions are in two major groups: in vitro actions and clinical actions. The in vitro actions are subdivided into the effects on channels and receptors, the major determinants of cardiac electrophysiology. The actions of drugs on the sodium, calcium, and potassium (I_K) channels are indicated. Sodium channel blockade is subdivided into three groups of actions characterized by their effect on the ECG intervals. Drug interaction with receptors (alpha- and beta-adrenergic), muscarinic acetylcholine (ACh) and adenosine (Ado) are indicated. Circles indicate antagonist actions; unfilled triangles indicate direct or indirect acting agonists or stimulators. The darkness of the symbol increases with the intensity of the action. Solid triangles for bretylium indicate its biphasic action initially to stimulate alpha- and beta-receptors by release of norepinephrine, followed by subsequent block of norepinephrine release and indirect antagonism of these receptors.

The likelihood of clinically effective doses of the drugs to induce arrhythmia (ProArr), to alter left ventricular function (LVFX), sinus heart rate, and electrocardiographic intervals (PR, QRS, or JT) is indicated. Hatched circles are used to indicate the relative potency for extracardiac side effects (Extracard), such as hepatitis or other organ toxicity.

QUINIDINE

Quinidine is one of the oldest antiarrhythmic agents, but much is still to be learned about its actions. It is effective for both ventricular and supraventricular arrhythmias and is one of the most effective drugs for preventing recurrence of atrial fibrillation or flutter. Because its vagolytic action enhances AV nodal conduction, patients in atrial fibrillation or flutter should be given digoxin prior to starting quinidine. As with other sodium channel–blocking antiarrhythmic agents, it has only limited efficacy in preventing recurrence of ventricular tachycardia. Approximately 20 to 25 percent of patients respond when programmed ventricular stimulation is used to guide therapy. The use of lower doses in combination with mexiletine is often necessary to increase effectiveness and reduce the incidence of side effects. Because of quinidine's effects to slow conduction and increase refractoriness in accessory pathways, it may be effective in preventing rapid ventricular response during atrial fibrillation in patients with the Wolff-Parkinson-White (WPW) syndrome. Digoxin may be deleterious in this setting, as it may counteract a beneficial effect in the bypass tract tissue.

Quinidine is very poorly tolerated because of side effects that characteristically occur both early and late. The most common early side effect is diarrhea, which has been reported in 10 to 40 percent of patients during the first 4 weeks of therapy. During chronic therapy, rash, fever, photosensitivity of the skin, and thrombocytopenia are occasionally seen. Excessive dosages lead to a classic toxicity called *cinchonism* (tinnitus, blurred vision, and headache). Another classic toxicity of serious concern is the syndrome of torsades de pointes; it has been estimated to occur in 1 to 4 percent of patients and usually is self-limiting, but can be lethal if unrecognized. It is far more common in women and generally occurs in one of three settings during quinidine therapy: (1) within the first few days of therapy at usual or low dosages; (2) later, during stable therapy, when a patient becomes hypokalemic or bradycardiac; and (3) at high dosages. It is facilitated by any factors that prolong repolarization, slow heart rate, or reduced potassium levels. Examples are diuretic therapy, congenital long-QT syndrome, hypothyroidism, sick-sinus syndrome, and diarrhea. It is preceded by marked prolongation of the QT interval, usually with plasma quinidine concentrations that are in the low or subtherapeutic range. It is treated best by (1) increasing the heart rate (pacing or isoproterenol infusion), (2) administration of magnesium sulfate rapid infusion (2 g over 10 to 20 min and 3 to 20 mg/min), and (3) normalizing the potassium and magnesium levels. The use of other antiarrhythmic drugs can be dangerous in this setting and has led to death in some patients.

The usual initial dose of quinidine sulfate is 200 mg every 6 h. The elimination half-life is highly variable and ranges from 4 to 20 h, so steady state is reached in 1 to 4 days. Since the metabolites are active, it is best to wait 2 to 3 days between dose escalations when possible. The QT interval and QRS duration should be monitored regularly; continuous ECG monitoring is advisable. Many cardiologists recommend hospitalization and ECG monitoring during initiation of any antiarrhythmic drug, especially quinidine. Measurement of trough levels is helpful; concentrations between 2 and 5 μg/mL are usually associated with arrhythmia control. Side effects may occur at any concentration; the measurement of peak levels is of little value. Many patients require dosages of quinidine sulfate of 300 or 400 mg every 6 h. Patients who are receiving drugs that induce hepatic metabolism may require 400 to 600 mg every 6 h to achieve a therapeutic plasma concentration.

The sustained-release forms of quinidine have slower absorption and often have trough levels below the usual therapeutic range. The quinidine gluconate formulation is said to be better tolerated, probably because it contains less quinidine base—60 percent, versus 80 percent for quinidine sulfate. The gluconate salt is also available for intravenous infusion but should be given at rates ≤20 mg/min to prevent serious hypotension. Because quinidine is an alpha-adrenergic blocker and reduces afterload, it only rarely worsens heart failure. However, it may produce excessive orthostatic hypotension if the patient is taking vasodilators, especially nitrates or dipyridamole. Digoxin or digitoxin dosages should be reduced and/or plasma concentrations monitored to prevent accumulation to toxic levels when they are used with quinidine. Quinidine blocks the metabolism of many drugs; interactions should be watched for with metoprolol, timolol, flecainide, propafenone, many of the tricyclic antidepressants, and some protease inhibitors.

PROCAINAMIDE (PRONESTYL, PRONESTYL-SR, PROCAN-SR)

Procainamide is widely thought to be similar to quinidine. However, they are very different drugs with few similarities. Procainamide has similar potency for preventing recurrence of ventricular tachycardia (about 20 percent of patients), but it may be effective in patients refractory to quinidine—and vice versa. It is said to be effective in preventing recurrence of atrial fibrillation, but no controlled studies are available to prove this, and it is generally considered less effective than quinidine. In general, it is not a good drug for chronic oral therapy because of a very high incidence of side effects (see next paragraph). Intravenous therapy is often effective for ventricular arrhythmias but requires approximately 20 to 30 min to reach effective levels safely. Rapid infusion (>50 mg/min) can lead to hypotension because of ganglionic blocking actions seen at high levels and/or negative inotropic actions. A relatively safe regimen is 20 mg/min for 30 to 40 min. This yields concentrations in the effective range (4 to 8 µg/mL) in most patients, with little effect on blood pressure or intraventricular conduction. Some patients may require a total loading dose of 1 to

2 g, but this produces levels that are usually not tolerated during chronic therapy. Blood pressure and the QRS interval should be monitored during the infusion. If the patient has recently been given other cardioactive drugs, lower initial dosages of procainamide should be used. Response to intravenous procainamide does not reliably predict response to oral therapy, because of the varied production of procainamide's potentially active metabolite, *N*-acetylprocainamide (NAPA). Because NAPA has mainly potassium channel blocking antiarrhythmic action and little, if any, potency as a sodium channel blocker, it is very difficult to predict clinical response to oral procainamide therapy. Plasma level monitoring is of limited value, and there is no validity in the often recommended practice of using the sum of the procainamide and NAPA concentrations to monitor therapy.

During chronic oral therapy, approximately 50 percent of patients withdraw because of intolerable side effects. Nausea and vomiting are common early problems. Rash and drug fever occur sporadically during therapy, and approximately 20 percent of patients per year develop a lupus-like syndrome. Antinuclear antibodies (ANA) are seen in almost all patients after 9 to 12 months, but only those with high or rapidly rising titers develop symptoms that include a malar rash, arthralgias, and fever. If this lupus-like syndrome goes undiagnosed and the drug is continued, the syndrome can progress to a more generalized rash, arthritis, and/or pleuritis. Several cases of pericarditis have been described; some have progressed to tamponade and death. The symptoms usually resolve with a time course similar to that of their onset. However, ANA may remain intermittently positive for years. The current estimates are that 1 patient out of 500 will develop agranulocytosis due to procainamide. Regular monitoring of white blood cells (WBCs) is recommended by the manufacturers every 2 weeks for the first 3 months.

Oral dosages are variable and highly influenced by renal function. Because of its short half-life, the usual starting dose (500 mg) is administered every 3 to 4 h. The slow-release formulations allow dosing every 6 to 8 h. Because NAPA accumulates more slowly and reaches higher levels than procainamide in rapid acetylators (55 percent of Caucasians and 90 percent of Asians), one should wait at least 2 days to allow achievement of steady state (>4 days in renal failure). Many patients require total daily dosages of 6 to 8 g/day (750 mg every 3 h or, for the sustained-release formulation, 1.5 to 2.0 g every 6 h).

DISOPYRAMIDE (NORPACE)

Disopyramide is of limited use for patients with ventricular arrhythmias because of its negative inotropic actions but it is useful in patients with preserved ventricular function or with supraventricular arrhythmias and WPW syndrome. Dose-related anticholinergic side effects limit therapy in many patients by causing urinary retention, dry mouth, and blurred vision. The usual dosage is 150 mg every 6 h. Some patients require dosages of 200 to 250 mg every 8 h. The sustained-release form allows dosing every 12 h, and is usually initiated at 200 mg every 12 h.

Plasma concentration monitoring is of little value because of variable and saturable binding of the drug to plasma proteins. A doubling of total concentration in blood may actually be a sixfold increase in the *free* drug level, potentially causing toxicity. For this reason, loading doses or rapid increases in dosage are very dangerous and not recommended. There is little organ toxicity associated with chronic therapy; if the drug is initially tolerated, chronic therapy is usually without problems. Disopyramide is metabolized by the liver. Agents that induce hepatic metabolism (phenytoin, rifampin, and barbiturates) can cause levels to fall, with loss of clinical benefit.

TOCAINIDE (TONOCARD)

Tocainide is an orally active lidocaine congener that should be used only for life-threatening ventricular arrhythmias that have failed to respond to other agents, including mexiletine. Only 10 to 15 percent of patients evaluated respond to therapy. Approximately 1 patient in 500 develops agranulocytosis or pulmonary fibrosis, making tocainide a dangerous agent for chronic therapy. Troublesome side effects, such as tremor, lightheadedness, pruritus, and nausea, occur in 30 to 40 percent of patients, limiting therapy. Usual doses are from 400 to 600 mg q 8 h. Lower doses must be used in renal patients.

MEXILETINE (MEXITIL)

Mexiletine is a lidocaine congener for oral therapy. It has little serious organ toxicity and therefore need not always be reserved for refractory patients. Only 10 to 15 percent of patients with ventricular tachycardia respond, however, and it is best used in combination with quinidine or sotalol to increase efficacy and reduce side effects.

The most common side effects are seen early in therapy; 30 to 35 percent of patients may develop nausea, vomiting, tremor, blurred vision, or pruritus. Rare cases of hepatitis have been described. Most side effects are dose-related and may be controlled by lowering the size of each dose and/or giving it with food or aluminum antacids. Monitoring of plasma concentrations is of little value. Dosage usually begins with 150 mg orally every 8 h, and may be increased every 2 to 3 days to 200 mg q 8, 250 mg q 8, or 25 mg q 6 h.

FLECAINIDE (TAMBOCOR)

Flecainide (see below) is a potent drug that prolongs the QRS and PR intervals at dosages usually required to suppress arrhythmias. It has been found to increase mortality in patients with prior myocardial infarction and asymptomatic ventricular arrhythmias. It is indicated for life-threatening or highly symptomatic ventricular arrhythmias, but in these patients it has a low response rate (about 20 percent) and a very high incidence of aggravation of the arrhythmia (5 to 10 percent). Obviously, therapy in these patients should be initiated under monitored conditions by physicians experienced in the care of such patients. In lower dosages (50 mg every 8 to 12 h) it can be safe and effective for patients with supraventricular arrhythmias, particularly those with WPW syndrome. However, it is not recommended for patients with structural heart disease because of the potential for life-threatening proarrhythmia and death.

LIDOCAINE (XYLOCAINE)

Lidocaine is one of the most effective and most often misused drugs in cardiology. It reduces the incidence of ventricular fibrillation but has not been found to improve mortality in met-analyses of the numerous conflicting studies. However, it is rapidly effective in suppressing ventricular arrhythmias, especially in the setting of acute ischemia. Once reversible causes have been excluded, it should be given as a series of loading and maintenance infusions. In patients *without* heart failure, the initial injection of 1 mg/kg should be given over 2 min and a maintenance infusion of 2 mg/min started as soon as possible. After 8, 16, and 24 min, additional loading injections of

0.5 mg/kg should be given to maintain effective levels. If the patient develops side effects at any time during the loading, subsequent loading infusions should not be given. If serious side effects such as heart block or hypotension are seen, the drug should be discontinued entirely and, if necessary, pacing used or volume expanders given to correct the side effects. Most side effects are transient and require nothing more than observation. The sizes of the initial injection and the three serial loading injections should be halved in patients with heart failure. For those with severe heart failure or liver disease, the maintenance infusion, too, should be halved. Plasma concentration monitoring is essential to adjust dosage to prevent excessive accumulation or subtherapeutic levels.

VERAPAMIL (ISOPTIN, CALAN, VERELAN)

Verapamil is a reasonably effective agent for converting paroxysmal supraventricular tachycardia. Adenosine, however, is preferred in most patients. Diagnosis of the arrhythmia is essential, because inadvertent administration of verapamil to patients in ventricular tachycardia almost uniformly produces hemodynamic decompensation, shock, or even death. Verapamil is contraindicated in patients with WPW syndrome and atrial fibrillation, who have occasionally been reported to develop acceleration of ventricular rhythm after administration of verapamil. In contrast, patients with WPW syndrome and narrow-complex supraventricular tachycardia respond well to verapamil. It is usually given as an infusion of 5 to 10 mg over 2 to 5 min. Oral therapy is in the range of 80 to 160 mg q 8 h.

PROPRANOLOL (INDERAL)

Propranolol, like many other beta blockers, is an effective antiarrhythmic agent. By slowing conduction in the AV node, it can convert or suppress supraventricular arrhythmias and, when used with digoxin, it can be used to slow ventricular response in atrial fibrillation. Patients with exercise-induced tachyarrhythmias often can be controlled with oral therapy. Propranolol is often effective for acute termination of supraventricular tachycardia and occasionally effective for ventricular tachycardia. However, severe hypotension is frequently seen after intravenous administration to patients in ventricular tachycardia.

Intravenous dosages of 0.1 mg/kg at a rate of 1 mg/min can be administered for acute therapy of arrhythmias. A maintenance infusion of 0.1 mg/min will maintain an effective level of propranolol in patients with relatively normal hepatic and cardiac function. Oral therapy requires titration from a usual starting dose of 40 mg q 6–8 h. The dosage can be increased every 1 to 2 days. There is little benefit above 120 mg q 6 h. Neither propranolol nor any other beta blocker is generally effective as a single agent for prevention of recurrent sustained ventricular tachycardia, unless the tachycardia is reproducibly induced by exercise. Propranolol is often used in combination with other antiarrhythmic drugs (e.g., quinidine).

ESMOLOL (BREVIBLOC)

Esmolol is an effective cardioselective beta blocker for intravenous use that was specifically developed to be ultra–short-acting. It can be useful for any arrhythmias responsive to beta-blockers, and has the advantage of short duration of action in case adverse effects occur. Therapy should be initiated with a loading infusion of 500 μg/kg over 1 min, followed by 50 μg/kg/min. If inadequate efficacy is obtained, additional loading doses of 500 μg/kg should be given and the maintenance dose increased by 50 μg/kg/min. The usually effective maintenance dose is 100 μg/kg/min; few patients require over 200 μg/kg/min. Hypotension is often a complication.

BRETYLIUM (BRETYLOL)

Bretylium is a reasonably effective antiarrhythmic drug for malignant ventricular arrhythmias, available only for intravenous use. Because of its complex pharmacologic actions, it usually is reserved for patients who have failed to respond to lidocaine and, in some cases, procainamide. For patients with recurrent ventricular fibrillation or sustained ventricular tachycardia, it should be given as a rapid intravenous injection into a central line. If the patient is awake, rapid injection will often cause nausea and vomiting; the drug should be given at a rate of 8 to 10 mg/min in such cases.

Bretylium is cleared entirely by the kidneys. The dosage should be reduced in patients with renal insufficiency. The usual initial dose is 5 mg/kg as a rapid injection or 8 to 10 mg/min as a slow infusion. A second dose of 5 to 10 mg/kg is usually given if the patient fails to respond to the first loading dose. A maximum loading dose of 20 mg/kg is recommended, though no dose-related toxicity other than nausea has been described. Maintenance doses of 1 to 2 mg/min are usually recommended. Very low doses of bretylium block adrenergic neuronal function, and blood pressure is not reduced further by higher dosages. Supine blood pressure should be regulated by volume expansion to maintain CNS perfusion and urine production.

AMIODARONE (CORDARONE)

Amiodarone is available for oral therapy of life-threatening ventricular tachycardia or fibrillation; but because of severe and sometimes lethal toxicity, many recommend that it be used in low dosages or be reserved for patients in whom all other available drugs have failed. The toxic effects are generally dose-related but have been seen in some patients with low dosages within 2 to 3 months of beginning therapy. During initial therapy, loading doses of 300 to 600 mg bid are given for 10 to 14 days. Variable regimens have been used in the past; there are no studies allowing comparison of different dosing methods. Onset of action is delayed unless loading dosages are used. Abrupt change in dosage can lead to recurrence of the arrhythmia, so tapering of the dose in the first few weeks to the lowest effective dosage is recommended. Most patients with serious ventricular arrhythmias require 400 mg/day as a single dose, but a few can be controlled on 200 mg/day. Some require 600 mg/day. Most patients with less serious arrhythmias should be treated with 200 mg/day or less.

During the loading period, the patient should be observed for evidence of amiodarone's effects on conduction (heart block, bundle branch block, severe bradycardia). Worsening of the arrhythmia, including torsades de pointes, has been reported, but is highly unusual and may simply reflect inadequate therapy in some cases. Before or very soon after the beginning of therapy, baseline tests to screen for toxicity should be obtained. These tests include chest radiograph, pulmonary function tests in patients with preexisting lung disease (including CO diffusion capacity), thyroid function, and serum chemistry (including liver function tests). These should be repeated whenever toxicity is suspected; the chest radiograph and serum chemistry should be repeated every 3 to 6 months. Approximately 30 percent of patients treated with 400 mg/day develop elevated levels of serum transaminase that usually resolve with dosage reduction. The most serious side effect is pulmonary toxicity, which can occur acutely or develop slowly and insidiously. It is estimated to occur in 5 to 10 percent of patients treated for ventricular tachycardia/fibrillation each year; 10 percent of cases are lethal. Other serious toxic effects in these patients include hypo- and hyperthyroidism, peripheral neuropathy, and incapacitating nausea and vomiting. There are many other side effects and potentially serious drug interactions. Lower dosages (100 to 200 mg/day) have been found to have a much lower incidence of the serious adverse effects. Numerous mortality trials in patients with non–life-threatening arrhythmias have either been neutral (no benefit) or have been found to have only a small benefit. A reasonable conclusion is that is does not increase mortality when used in lower dosages.

ADENOSINE (ADENOCARD)

Adenosine is very effective for the acute conversion of paroxysmal supraventricular tachycardia (PSVT) due to reentry involving the AV node. Sixty percent of patients respond at a dose of 6 mg and an additional 32 percent respond when given a higher dose of 12 mg. Some patients with PSVT associated with the WPW syndrome also have been found to respond to adenosine. Because of the fleeting and previously assumed selective action of adenosine on the AV node, some cardiologists suggest that it be used as a diagnostic tool in patients with narrow and wide complex tachycardia. However, adenosine-sensitive ventricular tadycardia has been observed, and it would seem much more reasonable to use the appropriate procedures to make the correct diagnosis. Adenosine has a direct effect of slowing AV nodal conduction, which can result in transient AV block. Adenosine usually has no effect on anterograde or retrograde accessory pathway conduction. However, pathways that demonstrate decremental conduction often respond to adenosine, probably because they are partially depolarized and can be hyperpolarized by adenosine. Slow injections into a peripheral line often produce no clinical benefit or changes in blood pressure or heart rate. Maximal pharmacologic effects are seen within 30 s after injection into a peripheral intravenous line but occur within 10 to 20 s when given into a central line.

Adenosine should be injected intravenously into a proximal tubing site and flushed quickly with saline. For adults, the recommended initial dose is 6 mg injected over 1 to 2 s. If the arrhythmia persists, a 12-mg dose can be injected 1 to 2 min later. This can be repeated, but doses larger than 12 mg are not recommended by the manufacturer. A dosage regimen based on body weight has been proposed: an initial dose of 50 μg/kg, incremented by 50 μg/kg until PSVT is terminated or side effects become intolerable. Higher doses may be required for patients who have received caffeine or theophylline because of their antagonistic effects at A_1 receptors. Lower doses are recommended if the patients are receiving dipyridamole or carbamazepine. Although the pharmacokinetics of adenosine is unlikely to be altered in patients with renal or hepatic disease, these patients often have electrolyte imbalances that could alter the clinical response. The manufacturer notes that it has been used without obvious problems in patients who have been treated with beta-receptor antagonists, calcium-entry blockers, digitalis glycosides, quinidine, and angiotensin converting enzyme inhibitors.

Adenosine is contraindicated in sick sinus syndrome or second- or third-degree heart block unless the patient has a functioning artificial pacemaker. Because of the rapid clearance of adenosine, side effects usually last less than 60 s. The most common side effects of adenosine are facial flushing, dyspnea, or chest pressure. Although intrapulmonary administration of adenosine has precipitated bronchospasm in asthmatic patients, this has not been reported with intravenous administration. The manufacturer recommends caution in treating patients with asthma. Other less frequent side effects were nausea, lightheadedness, headache, sweating, palpitations, hypotension, and blurred vision. Intravenous theophylline has been recommended to reverse the effects of adenosine.

SOTALOL (BETAPACE)

Sotalol is unlike other beta-adrenergic antagonists in that it prolongs the action potential, producing a parallel increase in refractoriness of cardiac tissues. The unique combination of properties makes sotalol effective in a variety of supraventricular and ventricular arrhythmias. It has been found to be effective in ~30 percent of patients with sustained ventricular tachycardia evaluated by programmed ventricular stimulation. Sotalol slows heart rate, decreases AV nodal conduction, and increases refractoriness of atrial, ventricular, AV nodal, and

AV accessory pathways in both the anterograde and retrograde directions. When given in dosages between 160 and 640 mg/day, there is a 40- to 80-ms increase in QT_c.

Sotalol has a desirable pharmacokinetic profile. Oral bioavailability is greater than 90 percent and peak concentrations are seen 2.5 to 4 h after a dose. It is not bound to plasma proteins and is eliminated by the kidneys unchanged, with an elimination half-life of approximately 12 h. Because of the relatively long half-life and twice-daily dosing regimen, it is recommended that testing for efficacy be conducted near the end of the dosing interval at steady state.

The recommended initial dose of sotalol is 80 mg q 12 h. In patients with relatively normal renal function, steady state will occur in 2 to 3 days. If evaluation on this dosage indicates lack of response without evidence of excessive effects on repolarization (QT < 550 ms), the dosage may be increased to 160 mg bid and, if necessary, to 240 mg bid. A more rapid dose escalation has been used by some investigators. Some patients with life-threatening arrhythmias have required dosages of 640 mg/day. For patients with a creatinine clearance >60 mL/min, the usual dosing interval is q 12 h. However, if the creatinine clearance is between 30 and 60 mL/min, the recommended interval between doses is 24 h. For patients with Cl_{CR} between 10 and 30 mL/min, the interval should be every 36 to 48 h or the usual dose halved and given every 24 h. Because of the increased risk of proarrhythmia and congestive heart failure, patients with reduced cardiac output should be given lower doses and monitored carefully.

A major concern with sotalol treatment has been the occurrence of torsades de pointes (TdP). Reports of this syndrome predominantly have been cases of suicidal overdose or in patients, especially women, who were receiving concomitant diuretics and inadequate potassium replacement. Clearly, female gender, hypokalemia, and bradycardia are predisposing factors for the development of this arrhythmia. The manufacturer observed an overall incidence of TdP of 2 percent (patients with sustained ventricular tachycardia, 4 percent, and supraventricular arrhythmias, 1.5 percent). It is more common in patients with congestive heart failure and a history of sustained ventricular tachycardia (7 percent). The incidence of TdP can be held to a minimum by careful screening for predisposing factors (bradycardia, baseline prolongation of the QT interval, and electrolyte disturbances, especially hypokalemia), careful dose escalation beginning at 160 mg/day, and limiting the maximum QT interval prolongation to less than 550 ms.

The incidence of new or worsened congestive heart failure is only about 3 percent. This is somewhat lower than expected for a drug with potent beta-blocking actions but may be attenuated because of the increased inotropy produced by its action to prolong repolarization. Other side effects typical of beta blockers are to be expected (bronchospasm in asthmatic patients, masking the signs and symptoms of hypoglycemia in diabetics, and catecholamine-hypersensitivity withdrawal syndrome).

PROPAFENONE (RYTHMOL)

Propafenone is similar to other sodium channel–blocking agents in overall efficacy and patient tolerance, but it has been found effective in some patients refractory to other agents and has a role in treatment of many types of arrhythmias, including supraventricular arrhythmias.

It has been described as having potent ability to slow conduction velocity with little change in action potential duration. The time constants for onset and recovery of sodium-channel blockade are intermediate between quinidine and flecainide. Therefore, compared to flecainide, slightly less prolongation of the QRS is seen at effective dosages. It has structural similarity to propranolol and produces clinically significant beta-adrenergic blockade in many but not all patients. Seven percent of Caucasians are deficient in activity of the cytochrome P450 responsible for the metabolism of propafenone and fail to form measurable quantities of the potentially active metabolite, 5-hydroxypropafenone. The accumulation of high concentrations of propafenone leads to significant beta-receptor antagonism at both low and high dosages in poor metabolizers but only at high dosages in extensive metabolizers of propafenone. Although metabolic phenotype does not seem to dramatically influence the antiarrhythmic response to propafenone, it clearly influences the degree of beta blockade occurring during therapy.

Effective dosages range from 300 to 900 mg/day in two to four divided doses. In order to prevent unexpected accumulation of pharmacologic action, propafenone dosage should not be changed more frequently than every 3 days.

Patients receiving propafenone should be carefully monitored for deterioration in ventricular function, which may result from beta-adrenergic receptor antagonism seen in some patients and the direct negative inotropic actions seen in some patients with reduced ventricular function. Dosage should be reduced by 20 to 30 percent in patients with hepatic dysfunction.

It is very likely that there will be drug interactions between propafenone and other agents that utilize or inhibit cytochrome P450 for their metabolism. Such an interaction has been documented already between propafenone and metoprolol and should be expected with timolol, tricyclic antidepressants, fluoxetine, and perhaps other agents. Quinidine, which inhibits this cytochrome, inhibits the formation of 5-hydroxy-propafenone in extensive metabolizers. One would expect greater beta blockade to occur after combining quinidine with propafenone therapy because of the higher propafenone concentrations.

IBUTILIDE (CORVERT)

Ibutilide is the first drug given FDA approval for the rapid conversion of recent-onset atrial fibrillation or flutter. It has not been tested in other arrhythmias or in patients with atrial fibrillation or flutter of long duration (>90 days). It should not be given to patients who have hypokalemia, hypomagnesemia, or QTc prolongation at baseline >440 ms. In pacebo-controlled studies summarized in the manufacturer's labeling, the placebo conversion rate for atrial fibrillation or flutter was approximately 2 percent. Ibutilide terminated the arrhythmia in approximately 44 percent of patients treated with 1 mg followed by either 0.5 or 1 mg. Approximately 20 percent of patients responded to the first infusion and approximately 25 percent of those not responding to the first infusion responded to the second infusion. Response usually occurred at 20 to 30 min, ranging from 5 to 88 min after infusion.

Ibutilide is a remarkably potent methanesulfonamide analog of sotalol that prolongs cardiac action potential duration and has class III action to prolong cardiac refractoriness and action potential duration.

Ibutilide is available at this time only for intravenous administration. When given over 10 min, it distributes rapidly in a multiexponential fashion with the relevant component, having a half life from 2 to 12 h (mean, 6 h). The plasma concentration and pharmacokinetics are highly variable and dosing is recommended on the basis of weight. The drug is mainly eliminated by oxidative hepatic metabolism and systemic clearance is rapid (about 29 mL/min/kg).

Ibutilide is given undiluted or diluted in saline as an infusion over 10 min. The recommended dose for a patient over 60 kg is 1 mg and under 60 kg, 0.01 mg/kg. For patients whose arrhythmias have not converted by 10 min after completion of the first dose, a second dose of equal size can be administered. Since conversion of the arrhythmias is usually associated with peak levels, slower infusion rates are not likely to be as effective.

It is essential that patients receiving ibutilide be treated in a carefully monitored environment during and at least 4 h subsequent to treatment. The FDA-approved labeling recommends that skilled personnel, facilities, and medication for defibrillation or resuscitation must be readily available.

Although specific studies with heart failure and with renal or hepatic disease have not been conducted, current information does not indicate that any dosage adjustments should be necessary in these conditions. However, patients with severe LV dysfunction have a higher risk of developing ventricular arrhythmias, including torsades de pointes. Since the duration of drug effect is determined by distribution, it is very possible that patients with severe congestive heart failure will have decreased volumes of distribution as well as exaggerated and prolonged duration of effect.

The most serious adverse reaction is torsades de pointes which has been seen in 1.7 percent of patients. However, in only 586 patients participating in trials, patients with a $QT_c > 440$ ms or potassium <4 mEq/L were excluded. In spite of these precautions, the incidence of sustained polymorphic ventricular tachycardia requiring cardioversion was 1.7 percent. Another 2.7 percent developed nonsustained polymorphic VT, 4.9 percent nonsustained monomorphic VT, 1.5 percent AV block, and 1.9 percent bundle branch block. The risk of polymorphic VT was highest in patients who were female and/or had evidence of reduced ventricular performance. The incidence of these adverse effects may well be higher in general clinical use where electrolyte disorders and concomitant therapies may be more common.

No specific drug interaction studies have been performed. Concomitant beta-receptor or calcium-channel antagonists apparently do not interact, although data are limited.

SUGGESTED READING

CAST Investigators: Preliminary report: effect of encainide and flecainide on mortality in a randomized trial of arrhythmia suppression after myocardial infarction. *N Engl J Med* 1989; 321:406–412.

ESVEM Investigators: Determinants of predicted efficacy of antiarrhythmic drugs in the Electrophysiologic Study versus Electrocardiographic Monitoring Trial. *Circulation* 1993; 87:323–329.

Katzung BG: New concepts in antiarrhythmic drug action. In: Yu PN, Goodwin JF (eds): *Progress in Cardiology*, vol 15. Philadelphia, Lea & Febiger, 1987:5–18.

Task Force of the Working Group on Arrhythmias of the European Society of Cardiology: The Sicilian gambit: a new approach to the classification of antiarrhythmic drugs based on their actions on arrhythmogenic mechanisms. *Circulation* 1991; 84:1831–1851.

Woosley RL: Antiarrhythmic drugs. In: Alexander RW, Schlant RC, Fuster V, O'Rourke RA, Roberts R, Sonnenblick EH (eds): *Hurst's The Heart*, 9th ed. New York, McGraw-Hill, 1998:969–994.

46

ANTICOAGULANT, ANTIPLATELET, AND THROMBOLYTIC AGENTS

Adam Schussheim
Valentin Fuster

While the vast majority of cardiovascular morbidity and mortality is rooted in thrombosis, the initiation of antithrombotic therapy depends on an understanding of the immediate and long-term clinical risk to the patient and also on the particular therapeutic agent. This chapter briefly reviews the major antithrombotic therapies and then formulates an approach to risk stratification that will guide the use of anticoagulation.

ANTITHROMBOTICS

Platelet Inhibitor Therapy

Platelet activation is increasingly appreciated as a pivotal initial step in cardiovascular pathophysiology. Vessel injury exposes subendothelial elements that allow adherence by otherwise quiescent circulating platelets. Collagen and fibronectin in the vessel wall are bridged to platelets through von Willebrand factor and platelet surface glycoprotein receptors Ia/IIa and Ic/IIa. Initial adhesion results in a platelet conformation change, release of granular contents, and activation of further platelets and the coagulation system (Fig. 46-1). Thus, platelet granules release multiple vasoactive and thrombotic substances including ADP, serotonin, beta-thromboglobulin, platelet factor 4, and platelet growth factor. Phospholipases provide arachidonic acid to the cyclooxygenase pathway, producing thromboxane A_2 (TxA_2), a potent vasoconstrictor and platelet activator, and also leukotrienes, leading to leukocyte recruitment and activation. Surface glycoprotein IIb/IIIa under-

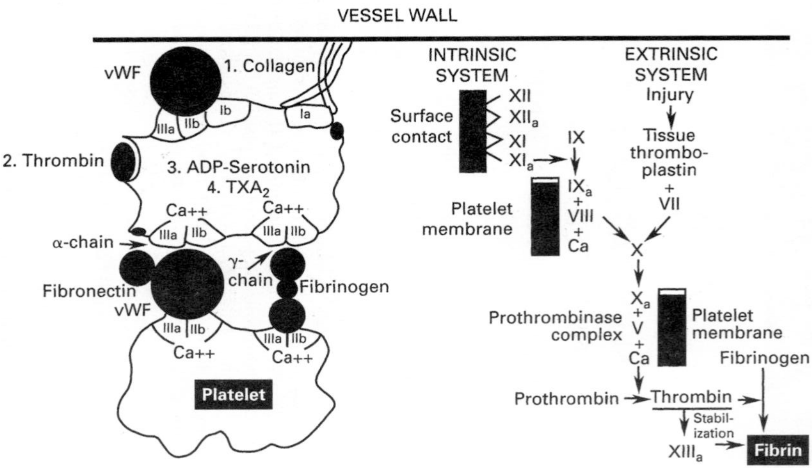

FIGURE 46-1
Interactions among platelet membrane receptors (glycoproteins Ia, Ib, and IIb/IIIa), adhesive macromolecules, and the disrupted vessel wall (*left panel*) and a flowchart of the intrinsic and extrinsic systems of the coagulation cascade (*right panel*). In the left panel, arabic numerals indicate the pathways of platelet activation that are dependent on (1) collagen, (2) thrombin, (3) ADP and serotonin, and (4) thromboxane A_2 (TXA$_2$); there are also some reports suggesting the binding of von Willebrand factor (vWF) (polymeric protein) to collagen or heparin. Note the interaction of the right panel between clotting factors (XII, XIIa, XI, XIa, IX, IXa, VII, VIII, X, Xa, V, and XIIIa) and the platelet membrane. (From Fuster V, Badimon L, Badimon JJ. The pathogenesis of coronary artery disease and the acute coronary syndromes. *N Engl J Med* 1992; 326:315. Copyright Massachusetts Medical Society.)

goes a conformation change, exposing it to the circulation and allowing platelet aggregation through cross-linking with plasma fibrinogen, fibronectin, and endothelial thrombospondin. Currently available antiplatelet agents therefore target one or more steps in the cycle of platelet activation.

ASPIRIN

Aspirin, an antiplatelet drug in clinical use for over 100 years, selectively inhibits TxA_2 production by irreversible acylation of cyclooxygenase. It does not, however, affect initial platelet adherence to atherosclerotic plaques and only partially inhibits platelet activation mediated by ADP, collagen, and thrombin. The ideal dose of aspirin for primary or secondary prevention of cardiovascular events is not well established. Although doses between 324 and 1300 mg daily seem to induce similar reductions in cardiovascular complications, complete inhibition of cyclooxygenase-dependent platelet activation is observed at doses between 1 and 2 mg/kg. Increasingly, it appears that the antithrombotic effect of aspirin is not completely accounted for through this cyclooxygenase-dependent pathway. The anti-inflammatory properties of aspirin are probably quite important, as the multiple triggers of plaque destabilization and rupture are becoming better understood. Recent studies, for example, have demonstrated that patients with coronary disease and markers of chronic inflammation derive greater benefit from aspirin; this is an area of active investigations. In addition, the identification of two cyclooxygenase enzymes (COX-1 and COX-2) may further elucidate the mechanisms of aspirin's effects.

Major side effects of aspirin include gastritis and possible hemorrhage. These complications appear to be dose-related. In addition, caution should be used in patients with impaired renal function or states of "effective hypovolemia," such as congestive heart failure, where impairment of prostaglandin-dependent renal artery control may cause worsening of renal function as well as fluid retention.

DIPYRIDAMOLE

The mechanism of dipyridamole's antiplatelet activity is unclear but probably involves the inhibition of phosphodiesterase, resulting in an increase in platelet cAMP levels and the inhibition of adenosine reuptake, causing elevated levels of extracellular adenosine; this, in turn, inhibits platelet activity and causes vessel vasodilation and perhaps the blocking of platelet TxA_2 generation. Dipyridamole by itself has little or no antiplatelet activity. In addition, it appears not to add much antiplatelet activity to adequately dosed aspirin in the secondary prevention of myocardial infarction. When combined with warfarin, dipyridamole (100 mg four times daily) has been shown to be effective in decreasing thromboembolic events in patients with prosthetic valves. The main side effects are nausea and abdominal discomfort, which are dose-related. Overall—given its cost, variable absorption, and equivocal clinical utility—dipyridamole's use as an antiplatelet agent is waning.

TICLOPIDINE AND CLOPIDOGREL

These two thienopyridine derivatives require activation in the liver and are noncompetitive and selective antagonists of ADP-dependent platelet aggregation, inhibiting glycoprotein (Gp) IIb/IIIa activation. In animal models, both of these compounds have been found to be more efficacious than aspirin or dipyridamole. For ticlopidine, adverse effects include gastrointestinal (diarrhea in about 20 percent of treated patients), dermatologic (rash and urticaria), and hemorrhagic disorders (epistaxis and ecchymosis); the most potentially dangerous, requiring close supervision, is bone marrow suppression and leukopenia. Although this side effect is rare, occurring in less than 2 percent of recipients, patients begun on ticlopidine require early (within 2 weeks) blood tests and should be alerted to seek medical attention during febrile illnesses. Clopidogrel has recently been approved by the FDA and it appears to lack the bone marrow suppression side effect.

These agents do not achieve platelet inhibition immediately, requiring up to 48 h, and are therefore not used in acute clinical situations. Thus far, ticlopidine is frequently used for short-course therapy to prevent stent occlusion and may also be used in patients intolerant of aspirin or who are at high risk with cerebrovascular or peripheral vascular disease. Recent trials examining clopidogrel have demonstrated its superiority to aspirin in reducing cardiovascular and cerebrovascular events in high-risk individuals. Given its favorable side-effects profile, other issues such as cost and future studies will help guide its use.

MONOCLONAL ANTIBODIES AGAINST GP IIB/IIIA RECEPTORS

Exposure of Gp IIb/IIIa receptors, members of the integrin family of membrane-bound receptors, is the final common pathway leading to platelet activation and therefore a logical point of pharmacologic inhibition. Abciximab (Reopro or 7e3), a fusion of a mouse-derived monoclonal antibody with a human IgG Fc region, is the first clinically available agent capable of selectively blocking Gp IIb/IIIa. Abciximab is available only for intravenous use, causes a dramatic prolongation in bleeding time, has a half-life of approximately 3 days, and has no available antidote. Its effects can be reversed only by extensive platelet transfusion. Currently, this agent is approved for use in high-risk angioplasty, but recent trials examining its efficacy in acute coronary syndromes are in progress. The increased bleeding associated with its use may be mitigated by concomitant reductions in heparin therapy, which should be carefully monitored. Thrombocytopenia, sometimes severe, is also an encountered side effect. A recent trial has demonstrated that abciximab may have a beneficial effect at a 30-day combined end point of all-cause death, myocardial infarction, and urgent revascularization when it is started in patients with unstable coronary syndromes destined for intervention. The 60-day end point, however, showed no difference in outcome, and bleeding complications were increased.

Several newer synthetic compounds for both oral and intravenous use that are also capable of selectively inhibiting Gp IIb/IIIa are under development. Eptifibatide (Integrilin) is a peptide inhibitor of Gp IIb/IIIa for use in patients undergoing coronary intervention and in acute coronary syndromes. It shares the bleeding side effects of abciximab, but its comparative efficacy is unclear. Tirofiban (Aggrastat), a synthetic agent with a shorter half-life than abciximab, has also been studied for intravenous use in higher-risk patients with coronary syndromes and has recently been approved for use in unstable angina. In addition, drug development is under way for inhibitors of the RGD sequence common to several integrins. Interestingly, abciximab inhibits the Gp IIb/IIIa receptor, the vitronectin receptor, and possibly other integrin receptors. Its potential benefit in coronary intervention may partially derive from antagonism at these other sites, which may influence multiple growth-signaling pathways in the vessel wall.

Heparin and Inhibitors of Thrombin

Platelet aggregates provide a microenvironment that can greatly accelerate thrombin generation. The platelet has specific binding sites for coagulation factors and its phospholipid membrane provides a surface upon which the coagulation system's enzymatic cascade can generate thrombin from its precursor (zymogen) form. This occurs through the similar activation, usually by limited proteolysis, of several other precursor proteins. Traditionally, the coagulation system has been conceptualized as two separate pathways, which differ according to whether activation occurs from factors intrinsic to the circulating blood (the "intrinsic" pathway) or from other elements (the "extrinsic" pathway) (Fig. 46-2). Recently, extensive overlap between the two has been demonstrated.

Activated factor X is at the convergence of the intrinsic and extrinsic pathways and can be activated by either the tenase complex or by the tissue factor–VIIa complex. Thrombin represents the culmination of the coagulation cascade and exerts multiple effects. These include catalyzing its own generation; activating the inhibitory proteins C and S; increasing endothe-

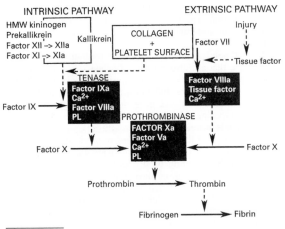

FIGURE 46-2

Clotting factor interactions. Coagulation is initiated by either an intrinsic or extrinsic pathway. In the intrinsic pathway, negatively charged surfaces initiate the contact activation and the phospholipid (PL) is furnished by platelets. In the extrinsic system, the phospholipid portion of tissue thromboplastin functions in conjunction with factor VIIa on the activation of factor X. From factor Xa on, both pathways converge upon a common path. Omitted from the diagram are inhibitors of the various steps, the augmentation of action of each pathway by activated factors, and the interaction between the intrinsic and extrinsic systems.

lial production of prostacyclin, NO, ADP, and plasminogen activator inhibitor (PAI-1); and activating platelets. Thrombin also cleaves fibrinogen, yielding insoluble fibrin and causing clot formation.

Inhibitors of blood coagulation work by interrupting one or more points of the self-amplifying clotting cascade. Agents in clinical use today may potentiate the actions of the body's natural inhibitors, modify naturally occurring antagonists, or interfere with the natural synthesis of coagulation proteins. In addition, fibrinolytic agents degrade fibrin in already formed clots.

UNFRACTIONATED HEPARIN

Heparin comprises a heterogeneous group of mucopolysaccharide chains of varying length and composition and, by itself, possesses no anticoagulant activity. Heparin accelerates the anticoagulant activity of two endogenous inhibitors—antithrombin III (an inhibitor of thrombin and activated factors X, IX, XI, and XII) and heparin cofactor II (inhibiting only thrombin). A unique pentasaccharide, present in only one-third of heparin molecules, is responsible for high-affinity binding to antithrombin III but does not account for heparin's activity with heparin cofactor II, which occurs only at very high doses. Bound components of the clotting system, whether to the platelet surface or to fibrin in clots, are less accessible to the inhibitory effects of the heparin-antithrombin III complex. This, in part, accounts for why more heparin is required to retard clot propagation than to inhibit its initiation. In addition, multiple circulating and bound inhibitors of heparin's actions exist and result in highly variable bioavailability and efficacy. Thus, frequent monitoring of the aPTT is required (even after a therapeutic level is achieved). The most common side effect of heparin is bleeding; this risk is higher when heparin is given intermittently and increases with advanced age and duration of therapy, occurring in up to 5 percent of patients who receive heparin for 5 to 7 days. Protamine sulfate may be given to reverse the effects of heparin at a dose of 1 mg for 100 U of heparin, given by slow intravenous infusion to reduce the risk of hypotension.

Heparin-induced thrombocytopenia, which may present with extensive arterial and venous thrombosis, occurs in its severe form in up to 2.4 percent of patients receiving therapeutic heparin, necessitating vigilant examinations of platelet counts. This complication classically occurs beginning 5 to 15 days after heparin therapy is started but may be seen earlier in patients with prior exposure. Osteoporosis, alopecia, urticaria, transient hepatitis, and aldosterone suppression are additional complications that may be encountered. Heparin dosing is best carried out using a weight-based nomogram such as that in Table 46-1. Adjustments of heparin dose can then be made accordingly.

LOW-MOLECULAR-WEIGHT HEPARINS

Low-molecular-weight heparin (LMWH) exerts its anticoagulant activity primarily through antithrombin III–dependent inhibition of factor Xa. By emphasizing smaller mucopolysaccharide chains, LMWH limits interactions with extraneous proteins, minimizing side effects and maximizing dosing stability and thereby allowing for once- or twice-daily subcutaneous injections without the need for laboratory monitoring (see Table 46-2 and Fig. 46-3). It is believed that concentrated inhibition earlier in the clotting cascade may

TABLE 46-1

WEIGHT-BASED HEPARIN DOSING NOMOGRAM[a]

aPTT Result	Action[b]
aPTT < 35 s (< 1.2 × control)	80 U/kg bolus, then increase infusion by 4 Ug/kg/h
aPTT 35–45 s (1.2 to 1.5 × control)	40 U/kg bolus, then increase by 2 Ug/kg/h
aPTT 46–70 s (1.5 to 2.3 × control)	No change
aPTT 71–90 s (2.3 to 3 × control)	Decrease infusion rate by 2 Ug/kg/h
aPTT > 90 s (> 3 × control)	Hold infusion 1 h, then decrease rate by 3 U/kg/h

[a] Initial dose: 80 Ug/kg bolus, then 18 Ug/kg hourly infusion.
[b] After initial bolus and after any subsequent change, stat aPTT should be checked at 4 to 6 h. In addition, daily aPTT and platelet counts should be checked in all patients on heparin.
SOURCE: From Raschke RA et al: The weight-based heparin dosing nomogram compared with a "standard care" nomogram. *Ann Intern Med* 1993; 119:874–881.

exert greater anticoagulant effect by preventing future amplification of steps. LMWH can be used in most instances where heparin is indicated, guided by issues of cost and ease and reliability of dosing. Recent studies have even suggested the superiority of LMWH in some clinical situations, in part due to its lower incidence of side effects (bleeding and thrombocytopenia) and also related to its more potent and specific efficacy. Enoxaparine is an available LMWH and is in clinical use for the prevention of deep venous thrombosis (DVT) in high-risk patients (e.g., those undergoing hip or knee replacement) and may perhaps be useful in patients who develop thrombocytopenia with heparin. Recent studies examining the role of LMWH in the treatment of unstable angina and of DVT have had promising results. The reliable dose response of subcutaneous injections of LMWH may even enable outpatient treatment. Mild thrombocytopenia can still occur with LMWH, although it is less common.

SPECIFIC THROMBIN INHIBITORS

Antithrombin III–independent thrombin antagonism can be accomplished with the use of hirudin and hirulog, neither of which is generally available for clinical use (see Table 46-2). These agents are modified proteins that after intravenous administration can specifically and directly inhibit thrombin and are active against clot-bound thrombin, which is less accessible to conventional heparin therapy. Several large-scale clinical trials have suggested the potential superiority of hirudin and hirulog to conventional heparin, although the results are not conclusive. Future studies with these drugs and with additional direct thrombin inhibitors hold promise to allow more specific inhibition of clot-bound thrombin. In addition, these drugs may be available for oral use and may combine specific inhibition of various integrin motifs.

Warfarin and Related Anticoagulants

Warfarin sodium (Coumadin) and its derivatives are presently the only anticoagulants available for oral use (Fig. 46-4). These compounds depress hepatic synthesis of the four vitamin K–dependent procoagulants (factors II, VII, IX, and X) and of two inhibitors (proteins C and S). Since factor VII and the anticoagulant protein C have the shortest half-lives, they are depleted first with the initiation of warfarin therapy. This may lead to an early imbalance (especially in patients with

TABLE 46-2

COMPARISON OF UNFRACTIONATED HEPARIN, LOW-MOLECULAR-WEIGHT HEPARIN (LMWH), AND THE DIRECT ANTITHROMBIN HIRUDIN

Unfractionated Heparin	LMWH	Hirudin
Inhibits to the same extent thrombin and factor VII, much less IXa and XIa	Inhibits mainly factor Xa, thrombin to some extent	Specific and potent inhibitor of thrombin
Antithrombin III–dependent Neutralized by heparinase, several plasma proteins, platelet factor 4, and endothelium	Antithrombin III–dependent Neutralized by heparinase, weak endothelium binding	Antithrombin III–independent Not neutralized by heparinase, endothelium, macrophages, fibrin monomer, and plasma proteins

Does not inactivate clot-bound thrombin and factor VII	Does not inactivate clot-bound thrombin and factor VII	Inactivates clot-bound thrombin
Inhibits platelet function	Inhibits platelet function	Prevents thrombin-induced aggregation but not other platelet agonists
Induced thrombocytopenia is not rare	Can induce thrombocytopenia	Does not induce thrombocytopenia
Bioavailability after SC injection, 30%	Bioavailability after SC injection, <90%	Good bioavailability after SC injection, ~ 85%
Poor dose-effect response	Fair dose-effect response	Fair dose-effect response
Not immunogenic	Not immunogenic	Not or barely immunogenic
Transient increase of liver enzymes is common	Transient increase of liver enzymes possible	No liver toxicity
Increases vascular permeability	No increase of vascular permeability	No increase of vascular permeability

Pentasaccharide containing
fractions with less than
18 saccharide units
50–75%

Pentasaccharide containing
fractions with18 or more
saccharide units
25–50%

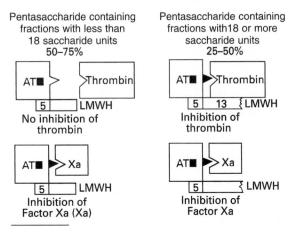

No inhibition of
thrombin

Inhibition of
thrombin

Inhibition of
Factor Xa (Xa)

Inhibition of
Factor Xa

FIGURE 46-3

Low-molecular-weight heparin (LMWH) activity. Approximately 25 to 50 percent of the LMWH molecules of different commercial preparations contain at least 18 saccharide units; these molecules inhibit both thrombin and factor Xa. The remaining 50 to 75 percent of LMWH molecules contain fewer than 18 saccharide units and inhibit only factor Xa.

deficiencies of protein C), tending to produce thrombosis of skin capillaries and venules, with cutaneous necrosis. In addition, the prolonged prothrombin time early in treatment with warfarin may reflect only depressed factor VII and not a meaningful interference with the intrinsic and common coagulation pathways. Thus, heparin anticoagulation should be continued for at least 1 day after a therapeutic prothrombin time is achieved.

Patients receiving warfarin therapy should have frequent measurements of anticoagulation, and this should be performed using the International Normalized Ratio (INR), which accounts for the varying potencies of laboratory thromboplastin reagents. Intercurrent illness, altered hepatic metabolism, and the initiation of new medications may all vary the efficacy of warfarin, making frequent monitoring essential during these periods. Drugs known to *increase* the effects of an unchanged warfarin dose include digoxin, amiodarone, acetaminophen, and many antibiotics. Drugs that may *decrease* the anticoagulant effect of warfarin include cholestyramine, rifampin,

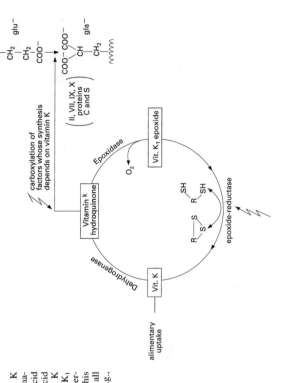

FIGURE 46-4
Vitamin K in its reduced form (vitamin K hydroquinone) is essential for the gamma-carboxylation reaction of glutamic acid (Glu)- to gamma-carboxyglutamic acid (Gla-). In this carboxylation vitamin K hydroquinone is converted to vitamin K_1 epoxide and an epoxide-reductase regenerates the active vitamin K hydroquinone. It is this regeneration step which is blocked by all coumarin-type anticoagulant drugs (e.g., warfarin).

phenytoin, and barbiturates. The major side effect of warfarin is bleeding, commonly associated with trauma or excessive anticoagulation. Warfarin's effects can be *temporarily* reversed with the administration of fresh frozen plasma. Vitamin K administration (2 to 5 mg) will reverse anticoagulation through renewed hepatic synthesis but may make reanticoagulation difficult. Very low doses of vitamin K (0.5 to 1.0 mg) should be used if moderate lowering of the INR is required (e.g., INR 6.0 to 10.0 without clinical bleeding). Recent studies suggest that oral administration of 2.5 mg of vitamin K will reverse mildly excessive anticoagulation without impairing the reinitiation of warfarin. Warfarin is also teratogenic and should generally be avoided during pregnancy, especially during the first trimester.

When warfarin therapy is begun, a 1- to 2-day delay is usual before the INR starts to rise, owing to the prolonged half-lives of most of the factors. Thus, adjustments in warfarin dosing require attention to the INR of the previous several days, the doses given on those days, and the rate of change in the INR. For example, a patient with an INR of 2.0 may not require any warfarin if the previous day's INR was 1.2 but *may* require an increase in dose if it was 3.0. The average daily dose of warfarin to maintain an INR of 2.0 to 3.0 is 4 to 5 mg. When an early onset of therapeutic anticoagulation is required, 10 mg can be given, and this can be changed as needed based on daily INR levels. If there is no urgency, one can begin with an anticipated therapeutic dose and expect steady-state anticoagulation in 5 to 7 days. This approach may minimize the likelihood of "overshooting" anticoagulation in frail or poorly nourished patients.

Thrombolytic Agents

All thrombolytic agents, whether modified forms of human endogenous proteins or derived from bacterial sources, are activators of plasminogen. The major clinical use of thrombolytics is in the setting of acute myocardial infarction, although hemodynamically significant pulmonary embolism, peripheral artery occlusion, and nonhemorrhagic stroke are also potential clinical indications. Hemorrhage, including bleeding at venous and arterial puncture sites and intracerebral hemorrhage, is the feared complication of thrombolytic use, mandating a careful understanding of risks and potential benefits. The use of thrombolytic agents as well as absolute and relative contraindications to it are outlined in Table 46-3.

TABLE 46-3

ANTITHROMBOTIC AND THROMBOLYTIC RECOMMENDATIONS FOR ACUTE MYOCARDIAL INFARCTION[a]

All patients should receive aspirin, 160–325 mg chewable, unless there is active bleeding or aspirin allergy, and daily thereafter.

Thrombolysis

   t-PA regimen: Accelerated t-PA should be given to patients presenting within 4 h of onset of chest pain, age under 75, and large or anterior wall MI.

     Accelerate t-PA (total dose ≤ 100 mg over 90 min).

      15-mg bolus, then 0.75 mg/kg over 30 min (dose not to exceed 50 mg), then 0.50 mg/kg over 60 min (dose not to exceed 35 mg).

      Heparin 5000-U bolus and an infusion of 1000 U/h immediately to maintain PTT at 1.5–2.0 control (or use other weight-based nomogram).

   Reteplase can be given in place of t-PA with two administrations of 10 mg IV separated by 30 min.

   Streptokinase 1.5 million U IV over 1 h (if prior SK exposure, use t-PA). Heparin is not advised in first 24–48 h unless patient has additional risk for thromboembolism (e.g., anterior MI with LV thrombus, atrial fibrillation, etc.). If heparin is given, SC dosing (12,500 U bid) started 4 h after SK is completed is equivalent to intravenous.

Patients not eligible for thrombolytics should be treated with heparin, aspirin, and other conventional therapies.

Heparin therapy is usually continued for 3–5 days. Warfarin anticoagulation for at least 3 months should be considered for those with large anterior infarctions (see text and Chap. 10).

[a] Contraindications to thrombolysis are as follows:

Absolute: active internal bleeding (not including menses), any hemorrhagic stroke (at any time), or other stroke or CNS event within 12 months. CNS structural damage (i.e., neoplasm, aneurysm, intracranial surgery).

Relative: Uncontrolled severe hypertension. Current warfarin use with INR > 2.0, or known bleeding diathesis. History of peptic ulcer disease, especially within 6 months. Serious advanced illness such as malignancy. Major surgery or trauma or CPR involving extensive (>10 min) chest compression. Known pericarditis. Pregnancy.

STREPTOKINASE

Streptokinase (SK) is a nonenzyme protein derived from beta-hemolytic streptococci that indirectly activates plasminogen. Most individuals have low levels of antibodies directed against this bacterial protein, but this can be overcome with proper dosing. After therapy, however, or following streptococcal infection, these antibodies remain elevated for 4 to 6 months. The conventional dose of streptokinase for acute myocardial infarction is 1.5 million units, given intravenously over 1 h. Heparin is usually not given concurrently with SK. The major side effect is bleeding, at venous access sites or internally. The stroke risk with SK of about 1 to 1.5 percent, however, is lower than that with t-PA. Minor side effects are usually related to antibody-mediated responses and include transient hypotension, fever, rash, and sometimes bronchospasm.

ANISOYLATED PLASMINOGEN-STREPTOKINASE ACTIVATOR COMPLEX (APSAC, ANISTREPLASE)

APSAC was constructed with the aim of controlling the enzymatic activity of the SK-plasminogen complex, which normally forms during SK administration, by reversible acylation of its catalytic site. Deacylation uncovers this site and allows conversion of plasminogen to plasmin. This occurs, however, both in the circulation and at the fibrin surface, resulting in only marginal fibrin specificity of thrombolysis with APSAC. Preexisting antibodies against streptokinase will inhibit and react with APSAC.

TISSUE-TYPE PLASMINOGEN ACTIVATOR (t-PA)

Recombinant t-PA (Activase) is a single-chain form of native t-PA, a serine protease, which is converted by plasmin into a two-chain form. t-PA has several protein domains, including two kringle-loop structures, that account for its binding to fibrin and its fibrin-specific plasminogen activation. Fibrin markedly enhances the activity of t-PA by two or three orders of magnitude and also protects the generated plasmin from inactivation by α_2-antiplasmin.

t-PA–DERIVED VARIANTS (RETEPLASE, rt-PA-TNK)

Reteplase (Retevase) is a deletion variant of t-PA and has a longer half-life of almost 20 min, compared with 2 min for t-PA. This allows more convenient dosing and appears to provide equivalent results. rt-PA-TNK is a mutant of t-PA that has a prolonged half-life and resistance to plasminogen activator inhibitor-1 (PAI-1). Clinical trials are under way to examine its efficacy.

CLINICAL APPROACH TO
ANTITHROMBOTIC THERAPY

The use of antithrombotic therapy in cardiovascular disease depends on an understanding of the clinical risk of various thrombotic syndromes. We have formulated an approach based on risk stratification in which a high risk is defined as a greater than 6 percent event rate, justifying more aggressive therapy; medium risk, between 2 to 6 percent; and low risk, less than 2 percent. With this formulation, thrombotic syndromes can further be separated as occurring in large arteries and the coronary arteries, the cardiac chambers, and in association with prosthetic valves. This division supports a conceptual approach to selecting therapy based on patholophysiologic principles and clinical risk. Table 46-4 outlines this approach.

High-Risk Coronary Syndromes and Revascularization

The fissuring of an atherosclerotic plaque and the subsequent formation of obstructive thrombus constitutes the unstable nidus of an acute coronary syndrome. Plaque rupture exposes highly thrombogenic subendothelial surfaces, macrophages, and a lipid pool, which rapidly cause platelet adherence and thrombus formation. Various local and systemic factors may influence the severity and extent of thrombus formation (Table 46-5). The degree and duration of disturbance in coronary flow dictates the clinical syndrome. In patients with unstable angina, for example, the thrombus formation may be transient and produce chest pain lasting 10 to 20 min. In non–Q-wave myocardial infarction, flow limitation is more severe and persistent. When coronary flow is completely suspended for more than 1 h, transmural cellular necrosis and Q-wave myocardial infarction occurs. Since both platelet aggregation and activation of the coagulation cascade are involved in this pathogenesis, therapy must be directed aggressively at both. In most cases, when the thrombotic obstruction spontaneously lyses, the residual denuded vessel is highly thrombogenic and carries a significant risk of rethrombosis.

Acute Myocardial Infarction

Transmural myocardial infarction is caused by an occlusive thrombus formed on a severely ruptured atherosclerotic plaque; it requires prompt intervention to restore patency and

TABLE 46-4

RISK STRATIFICATION IN CARDIOVASCULAR DISEASE AND THE ROLE OF ASPIRIN AND ANTICOAGULANTS[a]

	Thromboembolic Risk		
	High Risk, > 6% per year	Medium Risk, 2–6% per year	Low Risk, <2% per year
Coronary arteries Fibrin and platelets *(Rx A/C and/or PI)*	Acute coronary syndromes Vein grafts (1–3 months old) Coronary interventions *(A/C + PI)*	Stable CAD, s/p MI Vein grafts < 1 year old *(A/C or PI)*	Primary prevention of CAD *(PI?)*
Cardiac chambers Fibrin *(Rx A/C)*	Mitral stenosis and Afib Afib with previous thromboembolism *(INR, 2.5–3.5)*	Nonvalvular Afib Anterior MI LV dysfunction *(INR, 2.0–3.0)*	Mitral stenosis and NSR Lone Afib Chronic LV aneurysm *(No therapy)*
Prosthetic heart valves Fibrin > platelets *(Rx A/C > PI)*	Old prosthetic valve Prosthetic valve and previous thromboembolism *(INR, 3.0–4.5 or INR, 2.5–3.5 + ASA)*	Prosthetic valve Bioprosthetic valve early Postoperatively or with Afib (Mechanical: *INR, 2.5–3.5* Bioprosthesis: *INR, 2.0–3.0*)	Bioprosthetic valve and NSR *(No therapy)*

KEY: CAD, coronary artery disease; A/C, anticoagulant; PI, platelet inhibitor; Afib, atrial fibrillation; ASA, aspirin; MI, myocardial infarction; NSR, normal sinus rhythm; Rx, suggested therapy; s/p, status post, history of.
[a]Italics suggest pathogenesis as primarily rooted in fibrin deposition and/or platelet activation and indicate target INR levels of anticoagulation.

674

TABLE 46-5

ACUTE PLAQUE THROMBOSIS: SYSTEMIC AND LOCAL THROMBOGENIC RISK FACTORS

Local factors
    Degree of plaque disruption
    Degree of stenosis
    Plaque composition (i.e., lipid-rich plaque, tissue
       factor expression)
    Surface of residual thrombus
    Vasoconstriction
Systemic factors
    Catecholamines (i.e., smoking, stress, cocaine)
    Cholesterol, lipoprotein (a), homocysteinemia,
       hypertension, diabetes
    Fibrinogen, impaired fibrinolysis (i.e., plasminogen
       activator inhibitor-1)
    Inflammatory mediators (i.e., cytokines)

adequate blood flow to stabilize the surface of the ruptured vessel and to prevent reocclusion (see also Chap. 10).

ANTIPLATELET THERAPY

Multiple large, randomized trials including the landmark Second International Study of Infarct Survival (ISIS-2) have established the critical role of aspirin in the management of acute myocardial infarction. In the ISIS-2 trial, in which a 2×2 design tested the effects of aspirin (160 mg/day) either alone or in combination with SK, there was a 23 percent mortality reduction in patients treated with aspirin alone as well as 40 and 46 percent reductions in nonfatal reinfarction and nonfatal stroke, respectively. The benefits of aspirin and SK were additive. Based on this trial and several others, all patients with an acute myocardial infarction and without absolute contraindications, such as active bleeding or allergy, should promptly receive chewable aspirin 160 to 325 mg and should thereafter be treated with daily aspirin for at least a year if not for life.

THROMBOLYTIC THERAPY

Several large-scale clinical trials have established the efficacy of thrombolytic therapy in reducing mortality in acute myocardial infarction. Table 46-3 lists the major dosing recommendations for SK and t-PA. The greatest benefit with the use of thrombolytics is achieved when they are administered early. This mandates a rapid clinical evaluation in order to minimize the "door-to-needle time" for thrombolytic administration.

Choice of the appropriate thrombolytic therapy is controversial. The GUSTO (Global Utilization of Streptokinase and Tissue Plasminogen Activator for Occluded Coronary Arteries) trial examined whether a regimen of accelerated t-PA dosing, intravenous heparin, and aspirin—designed to produce rapid and sustained infarct-vessel recanalization—was associated with improved survival when compared with conventional administration of SK with heparin given either subcutaneously or intravenously. The relative reduction in mortality with accelerated t-PA versus the SK group was 14 percent, and the absolute mortality reduction was 1 percent, which was significant. However, there was a significant increase in hemorrhagic and total stroke among those treated with t-PA as compared with those treated with SK. As mentioned earlier, the incidence of stroke risk increases with age, hypertension, and a history of a previous cerebrovascular event. A composite outcome of death or nonfatal disabling stroke was decreased from 7.8 percent with SK to 6.9 percent with t-PA. Therefore, for every 1000 patients treated with t-PA rather than SK within 6 h of onset of acute myocardial infarction, there would be 10 fewer deaths or 9 fewer occurrences of the composite outcome of death or nonfatal disabling stroke (see also Chap. 10).

HEPARIN THERAPY

The use of heparin as adjunctive therapy with thrombolytics has been studied in several investigations, but few statistically significant results were shown. The GUSTO trial, mentioned above, for example, used heparin in all four arms, but it can only be interpreted as to demonstrate the lack of benefit of intravenous over subcutaneous heparin with SK. Smaller studies using angiographic end points suggest that heparin used with t-PA results in better coronary patency. Currently, therefore, heparin therapy is recommended with t-PA, but no study has unequivocally demonstrated a benefit of heparin with streptokinase when aspirin is adequately dosed.

RECOMMENDATIONS

Given the above information and the greater cost of t-PA, accelerated t-PA appears to confer the greatest advantage over SK for patients under the age of 75 with large infarctions who present within 4 h of onset. Aspirin's efficacy has been proven and it should be given to all patients without contraindications. Heparin should be used with t-PA and in all patients who would otherwise benefit from anticoagulation for the prevention of systemic emboli, such as those with large anterior infarctions, congestive heart failure, or atrial fibrillation. After they are discharged from the hospital, we recommend that patients at higher risk be considered for aspirin (80 to 160 mg) plus warfarin (INR, 2 to 3) therapy for a period of 4 to 6 weeks following infarction. However, the utility of combination warfarin and aspirin therapy is being actively investigated. Otherwise, it is clear that long-term antiplatelet therapy should be given to all patients following infarction (Chap. 10).

Unstable Angina

Unstable angina represents the clinical manifestations of a highly thrombotic lesion with high clinical risk, in which both fibrin deposition and platelet activation participate. Therapy, therefore, must be directed at both arms to prevent thrombotic recurrence and its consequences (see also Chap. 9).

HEPARIN THERAPY ALONE AND WITH ASPIRIN

Several trials have demonstrated the efficacy of heparin in reducing the risk of death, myocardial infarction, and refractory angina in patients with unstable angina. The additive benefit of aspirin when heparin is given has been suggested by several studies. It appears clear, however, that long-term antiplatelet therapy should be given to patients with unstable angina. In addition, there is a well-documented rebound phenomenon when heparin is discontinued, and aspirin may mitigate this risk. In fact, recent trials have examined, in addition to aspirin, the use of longer durations of anticoagulation, using either low-molecular weight heparin or warfarin. This may allow more adequate "healing" of the highly thrombogenic, disrupted plaque. Finally, newer antithrombotics including oral inhibitors of Gp IIb/IIIa may offer added advantages.

RECOMMENDATIONS

All patients with unstable angina are at high clinical risk and should therefore be treated aggressively. At the present time, this includes prompt administration of 325-mg chewable aspirin and intravenous heparin to raise the aPTT to approximately 45 to 70 s. Ticlopidine or clopidrogel can be used for patients intolerant of aspirin. At the present time, thrombolytic agents are not recommended for patients with unstable angina. In patients not undergoing intervention, heparin therapy should probably be continued for 3 to 5 days. A complete evaluation, including a search for potential extracardiac precipitants of angina (e.g., anemia, infection, thyroid disease), should be undertaken. Finally, after discharge from the hospital, the use of aspirin (80 to 160 mg) and warfarin (INR, 2 to 3) or possibly LMWH should be considered for the high-risk individual for 4 to 6 weeks. Otherwise, long-term antiplatelet therapy should be given to all patients (see Chap. 9).

Nonsurgical Coronary Revascularization

Percutaneous transluminal coronary angioplasty (PTCA) and other catheter-based interventional techniques cause disruption of the atherosclerotic plaque, exposing denuded endothelial surfaces and causing deep vessel injury. It follows, therefore, that local activation of platelets and deposition of fibrin will pose an acute risk of thrombosis and restenosis and may, in fact, play a role in the longer-term response of the vessel wall to injury. Currently, pretreatment with aspirin and adequate heparinization throughout the procedure are recommended. In the EPIC and EPILOG trials, however, the use of Gp IIb/IIIa antagonists conferred a significant benefit in reducing early occlusion and possibly late restenosis in high-risk lesions. Abciximab (Reopro) is currently indicated for this use and is usually given as an intravenous bolus in the catheterization lab, followed by an infusion over several hours. When it is used, heparin dosing must be adjusted and platelet counts monitored closely.

The optimal antithrombotic regimen for the first few weeks following the placement of coronary stents is under investigation. As of this writing, the combination of ticlopidine (250 mg bid) and aspirin (325 mg daily) appears superior. Treatment with ticlopidine is usually given for 2 weeks, during which the patients must be alerted to the rare but potentially fatal complication of neutropenia. A complete blood count is usually checked by day 10 of therapy; while late neutropenia has been reported, it is exceedingly rare. The role of clopidogrel, an agent related to ticlopidine but without many of its side effects, has not yet been examined in the prevention of stent occlusion and restenosis.

CORONARY REVASCULARIZATION—SAPHENOUS VEIN BYPASS GRAFTS

Without antithrombotic therapy, occlusion rates of aortocoronary saphenous vein grafts are 8 to 18 percent per anastomosis at 1 month following surgery; this reaches 16 to 26 percent at 1 year. Antiplatelet agents are the cornerstone of therapy. Aspirin therapy, 325 mg/day, should be started early following surgery (within 48 h). Preoperative aspirin use conferred no additional benefit but was, rather, associated with increased bleeding in one Veterans Administration study. Ticlopidine is an acceptable alternative for those intolerant of aspirin within 1 to 3 months. The problem is that despite aspirin use, the incidence of early graft occlusion is high; therefore, the combination of aspirin and warfarin, LMWH, or ticlopidine/clopidogrel needs to be investigated. After at least 1 year following operation, the results of the Post Coronary Artery Bypass Graft Trial have demonstrated that while the addition to aspirin of low-dose warfarin (target INR, 2.0) was not helpful, aggressive cholesterol lowering showed a significant reduction in vein atherosclerosis and in the need for repeat revascularization.

High Risk of Thromboembolism of Cardiac Chambers

While most conditions predisposing to cardiac chamber thromboembolism fall within our medium-risk classification, there are two notable exceptions—atrial fibrillation with mitral stenosis and a history of previous embolism.

ATRIAL FIBRILLATION WITH MITRAL STENOSIS OR WITH A PREVIOUS EMBOLISM

Atrial fibrillation is a common arrhythmia in mitral stenosis and can be quite troublesome because it often occurs in young patients and may, in fact, be the first clinical manifestation of stenotic disease. Although no randomized trials have been undertaken, the high event rates and published reports of protective events are convincing. Therefore, all patients with mitral stenosis in whom atrial fibrillation has been documented should be anticoagulated with warfarin with a target INR of 2.5 to 3.5. Debate arises over the use of anticoagulation in patients with mitral stenosis and sinus rhythm. In an effort to anticipate the onset of atrial fibrillation and potentially avert a thromboembolism, some advocate anticoagulation when the diagnosis of mitral stenosis is made, while others rely on a particular valve area.

ATRIAL FIBRILLATION WITH PREVIOUS STROKE

Previous stroke of cardiac origin is a significant risk factor for subsequent embolism in patients with atrial fibrillation, with a risk approaching 10 percent per year. Therefore, patients with atrial fibrillation with a history of a previous cerebrovascular event require warfarin anticoagulation with a target INR of 2.5 to 3.5. Therapy should begin within 1 to 2 weeks of stroke (depending on the size of the stroke and the risk of hemorrhage). Heparin anticoagulation is often begun earlier in the case of smaller infarcts when there is no sign of hemorrhage on imaging tests. Those unable to receive anticoagulation should be treated with aspirin, which confers some benefit in secondary prevention.

High Risk of Thromboembolism of Prosthetic Valves

OLD MECHANICAL VALVES

Multiple factors—such as better design of prosthetic valves, earlier operation, better patient selection, and improved surgical technique resulting in preserved ventricular function—have contributed to an overall decreasing risk of thromboembolism after valve replacement. Older mechanical valves (usually higher-profile valves more than 10 to 15 years old) pose a medium to high risk of embolic events that is cumulative and persists for the life of the valve. In addition, all new prosthetic valves exhibit a medium to high risk of embolism within the first 30 postoperative days. Warfarin anticoagulation with a target INR of 3.0 to 4.5 is indicated for old mechanical valves (and may be considered for the first month after valve replacement). The combination of lower-intensity warfarin (INR, 2.5 to 3.5) combined with either low-dose aspirin (100 mg daily, particularly in patients with known or suspected coronary disease) or dipyridamole (100 mg four times daily), may also be considered (see below).

PREVIOUS EMBOLISM WITH A MECHANICAL VALVE

As in patients with atrial fibrillation, a previous thromboembolic event in patients with a prosthetic valve is associated with a very high risk of recurrent embolism, greater than 20 percent in one Mayo Clinic series. This risk appears to be concentrated within the first 2 weeks, with approximately a 1 percent risk per day. Immediate anticoagulation may lead to hemorrhagic transformation, however. Patients with previous embolism and a mechanical valve should receive warfarin anticoagulation with a target INR of 3.0 to 4.5. Anticoagulation may be restarted within 48 h with smaller infarcts provided that there is no evidence of hemorrhage on computed tomography (CT) and no hypertension. Patients with extensive infarction should probably have anticoagulation delayed for 5 to 7 days, pending a repeat CT scan. Lower-intensity anticoagulation in combination with antiplatelet therapy as described above may be considered (see also Chap. 15).

Medium-Risk Coronary Syndromes

POST–MYOCARDIAL INFARCTION

Secondary prevention therapy following myocardial infarction aims to reduce events from a baseline risk of approximately 2 to 6 percent per year. Antiplatelet therapy, therefore, is quite popular, given its ease of use and reduced need for close follow-up. Despite multiple trials examining the efficacy of aspirin in secondary prevention, no single trial has demonstrated conclusive results. A metanalysis of trials pooling results from 18,000 patients showed that antiplatelet regiments reduced mortality by 13 percent, nonfatal reinfarction by 31 percent, and nonfatal stroke by 42 percent, with an overall risk reduction of 25 percent. In this analysis, medium-dose aspirin (75 to 325 mg) was as effective as higher doses, and the addition of dipyridamole was not helpful. A recent study of secondary prevention has suggested that clopidogrel may be as effective as or even slightly superior to aspirin therapy.

The benefit of warfarin therapy in patients following myocardial infarction appears to be similar to that of aspirin, although there is little information of a direct comparison. The combination of aspirin with very low fixed-dose warfarin (INR, < 1.5) was not superior to aspirin alone in the CARS (Coronary Artery Reinfarction Study) and is being examined with relatively low-intensity warfarin (INR, 1.5 to 2.5) in current trials.

STABLE CORONARY DISEASE

Aspirin in patients with stable angina has been shown to reduce new angiographic lesion formation and also to reduce the incidence of myocardial infarction. Again, older trials employing warfarin suggest benefit as well.

Recommendation Because of its lower cost and ease of administration, aspirin (75 to 325 mg daily) is the preferred therapy for secondary prevention in patients following myocardial infarction and in those with stable coronary artery disease. In those patients intolerant to aspirin, ticlopidine or clopidogrel and warfarin can be considered. Warfarin is not only of benefit in the prevention of coronary thrombosis, but, unlike aspirin, should also be considered in patients at risk for embolism from cardiac chambers (e.g., those with mural thrombi or early after a large anterior infarction, those with significant left ventricular dysfunction, or those with atrial fibrillation). The benefits of combination therapies are under investigation.

Medium Risk of Thromboembolism of Cardiac Chambers

Thrombus formation in cardiac chambers is primarily driven by fibrin deposition through the clotting cascade; therefore anticoagulation, as opposed to platelet inhibition, is usually the preferred therapy. The clinical decision lies in identifying those patients in whom anticoagulation should be considered and those in whom it is not necessary.

NONVALVULAR ATRIAL FIBRILLATION

Atrial fibrillation is the most common sustained cardiac arrhythmia and raises stroke risk sixfold in the elderly to about 5 percent per year. This risk may be higher if subclinical episodes of brain ischemia are included. Several large, randomized clinical trials have demonstrated that systemic warfarin anticoagulation (INR, 1.8 to 4.2) reduces this risk by more than two-thirds. Aspirin probably confers some benefit but is significantly less effective than warfarin. In these trials, bleeding risk was less than 1 percent per year. The elderly, however, in whom atrial fibrillation is most prevalent, are more prone to both stroke and bleeding complications. It remains to be established whether the net balance of risk and benefit may lie with lower-intensity anticoagulation in elderly patients.

It is clear that the risk of stroke is heterogeneously distributed among patients with atrial fibrillation. The ability to risk stratify, therefore, would allow tailoring of the intensity and mode of anticoagulation. The Stroke Prevention in Atrial Fibrillation Study I (SPAF) and the Atrial Fibrillation Study Investigators identified several risk factors associated with increased risk of systemic embolization (Table 46-6). The risk

TABLE 46-6

RISK STRATIFICATION IN ATRIAL FIBRILLATION: INDEPENDENT PREDICATORS OF THROMBOEMBOLIC RISK

	SPAF I Placebo Patients	AFI Pooled Analysis
High-risk variables	History of hypertension Prior stroke/TIA Diabetes Recent heart failure	History of hypertension Prior stroke/TIA Diabetes Age > 65 years
Thromboembolic rate (95% CI)		
Low risk	1.4% per year (0.5–3.7)	1.0% per year (0.3–3.1)
High risk	>7% per year	>5% per year
Percentage of cohort "low-risk"	38%	15%

NOTE: These figures are based on large, prospectively acquired data sets analyzed by multivariate techniques. The Stroke Prevention in Atrial Fibrillation Study I (SPAF I) placebo data set was included in the pooled analysis of clinical trials by the Atrial Fibrillation Investigators (AFI).

from chronic or intermittent atrial fibrillation is similar. In the future, imaging or biochemical tests may further refine our ability to tailor therapy. Transesophageal echocardiography appears to be very promising, as recent studies have identified left atrial appendage size, blood flow, and the presence of spontaneous echogenic densities ("smoke") as predictors of future events. Contributions from other comorbid conditions that increase the risk of cerebrovascular events, such as aortic atherosclerotic plaques, may also be useful.

Recommendations Based on the above evidence, all patients with atrial fibrillation not at low risk for thromboembolism (Table 46-6) should receive warfarin anticoagulation with a target INR of 2.0 to 3.0. Patients at high risk, as discussed above, should have higher INR targets. Lone atrial fibrillation is described in a subsequent section and is defined as atrial fibrillation without structural heart disease and age below 60. Elderly patients over 75 years of age should receive anticoagulation with a target INR close to 2.0. Patients unable to receive systemic anticoagulation or those at low risk should be treated with aspirin.

THROMBOEMBOLISM IN DILATED CARDIOMYOPATHY AND POST–MYOCARDIAL INFARCTION

The combination of tissue injury exposing subendothelial thromogenic surfaces, regional hypocontractility and relative blood stasis, and a systemic hypercoaguable state results in conditions favoring the formation of intracavitary thrombosis in patients suffering large, usually anterior myocardial infarctions. This may occur in up to 30 percent of patients with anterior infarctions but in less than 5 percent of those with infarctions in other areas. The potential for embolism is probably related both to the morphology of the thrombus (sessile versus mobile and protuberant) and to its exposure to blood flow. For example, in chronic aneurysms, thrombus is ostensibly excluded from dynamic circulatory forces, thus limiting its embolic potential. Dilated cardiomyopathy, however, in which mural thrombi are exposed to blood flow, may represent a higher risk.

A recent study of over 4000 patients has identified the presence of atrial fibrillation, advanced age (>70), a history of hypertension, prior infarction or stroke, and anterior location as potent risk factors for a cerebrovascular event following myocardial infarction. Transthoracic echocardiography has a reported 90 percent specificity and 75 to 90 percent sensitivity for thrombus detection. The identification of a left ventricular thrombus by echocardiography, especially one that is mobile or protruding, requires anticoagulation, with heparin and then warfarin, with periodic follow-up.

No prospective randomized trial of systemic anticoagulation is dilated ischemic or nonischemic cardiomyopathy is available, although some prospective information is evolving. The annual incidence of clinically apparent thromboembolism in dilated cardiomopathy has been reported to be between 1 to 12 per 100 patients. Recent subset analyses of heart failure trials seem to suggest that anticoagulation with warfarin or antiplatelet therapy may reduce this risk. In addition, past studies suggest that pulmonary embolism occurs with a frequency of from 5 to 11 percent among those not treated with antithrombotic therapy and is associated with significant morbidity and mortality.

Recommendations The patient suffering a large anterior myocardial infarction should receive warfarin anticoagulation (INR, 2.0 to 3.0) following heparin therapy for 3 months. The value of chronic anticoagulation or platelet inhibitor therapy in patients with LV dysfunction, whether due to ischemic disease or idiopathic dilated cardiomyopathy, has not been proven, but prospective trials and registries may offer some answers shortly. Until such information is available, a reasonable approach would be to utilize anticoagulants in patients with LV dysfunction (i.e., EF <35 percent) who are at higher risk, including those with atrial fibrillation, the presence of a thrombus or echocardiography, or a history of thromboembolism. Spontaneous echocardiographic contrast in the left ventricle may also be a useful marker of higher risk. The use of anticoagulation in patients with old ventricular aneurysms is generally not indicated because of the low clinical risk.

Moderate Risk of Thromboembolism of Prosthetic Valves

As discussed above, the current annual risk of thromboembolism with anticoagulation for prosthetic heart valves is about 2.5 percent for the Starr-Edwards and Omniscience prostheses, 2 percent for the Medtronic Hall prosthesis, 1.5 percent for the St. Jude Medical prosthesis, and 1 percent for the pericardial and porcine bioprostheses. Without anticoagulation, the risks tend to double. While this risk is cumulative and persistent, it is elevated during the first 30 postoperative days for both

mechanical and bioprosthetic valves. Patients with older valves
and previous embolism are also at high risk, as discussed. The
risk of valve thrombosis, which may be as high as 5 to 6 percent without anticoagulation, is reduced to between 0.2 to
1.8 percent per year with therapy (see Chap. 15).

The optimal level of anticoagulation for patients with prosthetic heart valves, balancing the risk of thromboembolism
with the risk of bleeding, is yet to be determined, but current
evidence suggests that the previous targets may have been too
high (Fig. 46-5). In addition, several trials have addressed the
benefit of combined therapy using antiplatelet agents. Turpie et
al. examined the addition of low-dose aspirin (100 mg daily)
with warfarin (INR, 3.0 to 4.5) and found a significant reduction in annualized event rates for embolism and death (1.9

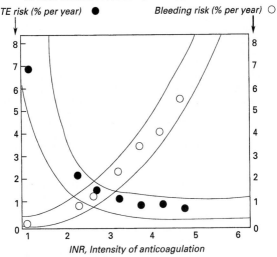

St. Jude Medical Prosthesis (n = 435) (Av 122 months)

TE risk (% per year) ●   Bleeding risk (% per year) ○

INR, Intensity of anticoagulation

FIGURE 46-5

Variations in thromboembolic (•) and bleeding (○) risk among 435
patients with St. Jude valve replacements across a range of anticoagulation intensity. INR, international normalized ratio; TE, thromboembolic. (Modified from Piper C, Schulte HD, Horstkotte D: Opti-
mization of oral anticoagulation for patients with mechanical heart
valve prosthesis. *J Heart Valve Dis* 1995; 4:127. Used with
permission.)

versus 8.5 percent). Although the rate of hemorrhage was greater in the combined therapy group, the overall rate of major hemorrhagic events was not statistically different. The achieved mean INR was 3.0, perhaps accounting for the relatively high event rates. In addition, it is not clear if the benefit of aspirin was concentrated in a reduction in coronary events. Recent trials with ticlopidine found similar results.

Recommendations (See Table 46-7)   Patients with *mechanical prosthetic valves* should receive anticoagulation—ideally, although not yet fully accepted, beginning with heparin following chest tube removal and continued until oral anticoagulants are therapeutic. Warfarin therapy can usually be started within the first 48 h of surgery. Current evidence suggests a target INR of 2.5 to 3.5. It is essential to stress that INR levels must be vigilantly checked every 3 to 4 weeks and additionally if drugs are changed or during intercurrent illnesses.

TABLE 46-7
ANTITHROMBOTIC THERAPY—PROSTHETIC HEART VALVES

Mechanical valve	
Routine	Medium-dose warfarin (INR, 2.5–3.5)
Old prosthesis, previous thromboembolism, or severe atherosclerosis	High-dose warfarin (INR, 3.0–4.5) or Medium-dose warfarin + aspirin or dipyridamole
Anticoagulation problems	Low-dose warfarin (INR, 2.0–3.0) Low-dose warfarin + dipyridamole or aspirin
Recurrent embolism	High-dose warfarin + consider other causes
Bioprosthetic valve	
Routine (normal sinus rhythm)	Low-dose warfarin for 3 months (then aspirin?)
Atrial fibrillation, thrombus or previous thromboembolism	Medium-dose warfarin for 3 months, then low-dose warfarin

Higher-risk patients, as discussed previously, include those with prior embolism and valves implanted before the mid-1970s. The addition of antiplatelet therapy may obviate the need for higher-intensity warfarin therapy. Patients who suffer hemorrhagic complications should receive lower-intensity anticoagulation (INR, about 2.0) in combination with aspirin or dipyridamole and a medical evaluation of the source of bleeding. Patients with *bioprosthetic valves* should receive anticoagulation (INR, 2.0 to 3.0) postoperatively for 3 months. Those at higher risk for embolism—including patients with atrial fibrillation, severe left ventricular dysfunction, or previous embolism—should receive oral anticoagulation indefinitely (see also Chap. 15).

Low Risk of Coronary Thrombosis

Two studies, the United States Physicians' Health Study and the British Doctors' Trial, have examined the use of aspirin in the primary prevention of myocardial infarction. The Physicians' Health Study—a blinded, randomized study of 22,000 male physicians which was terminated prematurely—demonstrated a 44 percent relative reduction in acute myocardial infarction (from approximately 0.4 to 0.2 percent per year) among those receiving aspirin (325 mg every other day) over 5 years. This benefit was limited to those older than 50 years. There was no significant difference in cardiovascular death, and those receiving aspirin had an increase in gastrointestinal bleeding and a nonsignificant increase in hemorrhagic stroke. The British Doctors' Trial was unable to demonstrate a difference in the rate of myocardial infarction among male physicians receiving 500 mg of aspirin daily, but this trial was smaller, not blinded, and not controlled. Again, however, there was a slight increase in the risk of disabling stroke. Both trials examined populations at fairly low risk and therefore demonstrated an absolute reduction of less than 2 events per 1000 patient-years. The recent Thrombosis Prevention Trial examined the utility of daily aspirin (75 mg), low-dose warfarin (target INR, 1.5), and their combination in primary prevention among male patients deemed by their physicians to be at high risk on the basis of family history, the presence of hypertension, smoking, obesity, high cholesterol, and high fibrinogen. Low-dose aspirin significantly reduced the incidence of nonfatal myocardial infarction but also resulted in a slight increase in strokes and bleeding. Low-dose warfarin demonstrated a significant reduction in fatal myocardial infarction, and the combination of the two provided added protection—but with an increased incidence of bleeding. Again, because of the low overall event rate, while the relative reductions were similar to the Physicians' Health Study, the absolute reduction was small. Of note, the use of aspirin for primary prevention in women has not been studied, and we await the U.S. Women's Health Study to examine this question.

Recommendations   While aspirin therapy at a dose of 75 to 325 mg daily is indicated in those with established coronary artery disease (secondary prevention), its use in primary prevention is unclear. Aspirin use should be considered in men with poorly controlled risk factors who are over the age of 50 and at relatively higher risk for coronary events. It should be used cautiously, if at all, in those with uncontrolled hypertension. Overall, it should be used as an adjunct and not as a substitute for aggressive risk factor reduction.

Low Risk of Thromboembolism of Cardiac Chambers

LONE ATRIAL FIBRILLATION

As discussed above, patients with atrial fibrillation who have no evidence of structural heart disease and are less than 60 years old are considered to have "lone" atrial fibrillation and suffer thromboembolic events at rates below 2 percent per year. These patients may require no therapy at all, but those with diastolic hypertension or risk factors for coronary artery disease may benefit from aspirin therapy.

VALVULAR DISEASE AND SINUS RHYTHM

The risk of thromboembolism in patients with mitral regurgitation or aortic valve disease and sinus rhythm is low and requires no therapy. *Mitral valve prolapse*, a highly prevalent condition affecting more than 10 percent of adult women, is most frequently well tolerated and asymptomatic. A small percentage of patients, however, may suffer cerebrovascular events, perhaps resulting from fibrin deposition on highly redundant valve leaflets. Routine antithrombotic prophylaxis is not indicated. Following a clinical event, as a rigorous medical evaluation is conducted, aspirin may be considered.

Low Risk of Thromboembolism of Prosthetic Valves

BIOPROSTHETIC VALVES

Patients with bioprosthetic valves and normal sinus rhythm are at low risk for thromboembolism and do not require therapy. All patients, however, should receive anticoagulation (INR 2.0 to 3.0) beginning as soon as possible after surgery and continuing for 3 months. Again, those with atrial fibrillation or severe left ventricular dysfunction should receive therapy indefinitely (see Chap. 15).

SUGGESTED READING

Antiplatelet Trialists' Collaboration: Collaborative overview of randomized trials of antiplatelet therapy: Parts, I, II, III. *BMJ* 1994; 308:159–235.

Cairns JA, Fuster V, Kennedy JW: Coronary thrombolysis. *Chest* 1995; 108(4 suppl):401S–423S.

Cairns JA, Lewis HD, Meade TW, et al: Antithrombotic agents in coronary artery disease. *Chest* 1995; 108(4 suppl): 380S–400S.

CAPRIE Steering Committee: A randomized, blinded trial of clopidogrel versus aspirin in patients at risk of ischemic events (CAPRIE). *Lancet* 1996; 348:1329–1339.

Cohn M, Demers C, Gurfinkey EP, Turpie AGG, et al: A comparison of low-molecular-weight heparin with unfractionated heparin for unstable coronary artery disease. *N Engl J Med* 1997; 337:447–452.

Fuster V: Coronary thrombolysis: a perspective for the practicing physician (editorial). *N Engl J Med* 1993; 329: 723–725.

Fuster V, Badimon L, Badimon J, Chesebro J: The pathogenesis of coronary artery disease and the acute coronary syndromes. *N Engl J Med* 1992; 326:242–250, 310–318.

Fuster V, Dyken ML, Vokonas PS, Hennekens C: Aspirin as a therapeutic agent in cardiovascular disease. *Circulation* 1993; 87:659–675.

Hirsh J, Fuster V: AHA Medical/Scientific statement guide to anticoagulant therapy. Part 1: Heparin. *Circulation* 1994; 89:1449–1468.

Hirsh J, Fuster V: Guide to anticoagulant therapy: Part 2. Oral anticoagulants. *Circulation* 1994; 89:1469–1480.

Hirsh J, Levine MN: Low molecular weight heparin. *Blood* 1992; 79:1–17.

Laupacis A, Albers G, Dalen J, Dunn M, et al: Antithrombotic therapy in atrial fibrillation. *Chest* 1995: 108(suppl): 3525–3595.

Noble S, Goa K: Ticlopidine: a review of its pharmacology, clinical efficacy and tolerability in the prevention of cerebral ischaemia and stroke. *Drugs Aging* 1996; 8:214–232.

Peto R, Gary R, Collins R, et al: A randomized trial of the effects of prophylactic daily aspirin among male British doctors. *BMJ* 1988; 216:313–316.

Popma JJ, Coller BS, Ohman EM, et al: Antithrombotic therapy in patients undergoing coronary angioplasty. *Chest* 1995; 108(suppl):486S–501S.

Steering Committee of the Physician's Health Study Research Group: Final report on the aspirin component of the ongoing Physician's Health Study. *N Engl J Med* 1989; 321: 129–135.

Stroke Prevention in Atrial Fibrillation Investigators: Warfarin versus aspirin for prevention of thromboembolism in atrial fibrillation: stroke prevention in atrial fibrillation II study. *Lancet* 1994; 343:687–691.

The Capture Investigators: Randomized placebo-controlled trial of abciximab before and during coronary intervention in refractory unstable angina: the CAPTURE study. *Lancet* 1997; 349:1429–1435.

The EPILOG Investigators: Platelet glycoprotein IIb/IIIa receptor blockade and low-dose heparin during percutaneous coronary revascularization. *N Engl J Med* 1997; 336:1689–1696.

The GUSTO Investigators: An international randomized trial comparing four thrombolytic strategies for acute myocardial infarction. *N Engl J Med* 1993; 329:678–682.

The IMPACT Investigators: Randomized placebo-controlled trial of effect of eptifibatide on complications of percutaneous coronary intervention: IMPACT-II *Lancet* 1997; 349:1422–1428.

Turpie AGG, Gent M, Laupacis A, et al: A comparison of aspirin with placebo in patients treated with warfarin after heart valve replacement. *N Engl J Med* 1993; 329: 524–529.

Verstraete M, Fuster V: Thrombogenesis and antithrombotic therapy. In: Alexander RW, Schlant RC, Fuster V, O'Rourke RA, Roberts A, Sonnenblick EH (eds): *Hurst's The Heart*, 9th ed. New York, McGraw-Hill, 1998: 1501–1551.

47

BETA-ADRENERGIC BLOCKING DRUGS

William H. Frishman

The beta-adrenergic blocking drugs (beta blockers) were used originally for angina pectoris and arrhythmias. They have also gained widespread acceptance for treatment of many other conditions, both cardiac and noncardiac, including hypertrophic cardiomyopathy, systemic hypertension, congestive heart failure, dissecting aneurysm, mitral valve prolapse, migraine, glaucoma, and essential tremor. The drugs also are indicated, in oral and parenteral forms, for reducing the risk of death and nonfatal reinfarction in survivors of an acute myocardial infarction.

BASIC PHARMACOLOGIC DIFFERENCES

Fifteen beta blockers are currently marketed in the United States (Table 47-1): propranolol, atenolol, nadolol, timolol, sotalol, carvedilol, carteolol, metoprolol, penbutolol, pindolol, acebutolol, esmolol, labetalol, betaxolol, and bisoprolol. Others are in clinical trials or are about to be approved.

Most marketed beta blockers are racemic mixtures of optical isomers (D, L). More beta-blocking activity is found in the L or levorotatory form. This difference in activity of the two isomers allows for differentiation of the pharmacologic properties of membrane stabilizing activity and beta blockade.

Membrane-Stabilizing Activity

In high concentrations, some beta blockers have a quinidine-like effect on the cardiac action potential. This is unrelated to

TABLE 47-1

PHARMACODYNAMIC PROPERTIES AND CARDIAC EFFECTS OF β-ADRENOCEPTOR BLOCKERS

Drug	Trade Name	Relative β₁ Selectivity*	PAA	MSA	R-HR	E-HR	R-Myocardial Contractility	R-BP	E-BP	R-AV Conduction	Antiarrhythmic Effect
Acebutolol	Sectral	+	+	+	$\updownarrow\rightarrow$	→	→	→	→	→	+
Atenolol	Tenormin	++	0	0	→	→	→	→	→	→	+
Betaxolol	Kerlone	++	0	+	→	→	→	→	→	→	+
Bisoprolol	Zebeta	++	0	0	→	→	→	→	→	→	+
Bucindolol**		0	+	0	$\updownarrow\rightarrow$	→	$\updownarrow\rightarrow$	→	→	$\updownarrow\rightarrow$	+
Carteolol	Cartol	0	0	+	$\updownarrow\rightarrow$	→	$\updownarrow\rightarrow$	→	→	→	+
Carvedilol†	Coreg	0	0	0	$\updownarrow\rightarrow$	→	$\updownarrow\rightarrow$	→	→	$\updownarrow\rightarrow$	+
Esmolol	Brevibloc	++	0	0	→	→	→	→	→	→	+
Labetalol‡	Normodyne, Trandate	0	+	0	$\updownarrow\rightarrow$	→	$\updownarrow\rightarrow$	→	$\updownarrow\rightarrow$	$\updownarrow\rightarrow$	+

694

		β₁ Selectivity*	PAA	MSA							
Metoprolol	Lopressor	++	0	0	↓	↓	↓	↓	↓	↓	+
Nadolol	Corgard	0	0	0	↓	↓	↓	↓	↓	↓	+
Penbutolol	Levatol	0	+	0	↓	↓	↔↓	↔↓	↓	↓	+
Pindolol	Visken	0	++	0	↓	↓	↔↓	↔↓	↓	↓	+
Propranolol	Inderal	0	0	++	↓	↓	↓	↓	↓	↓	+
Sotalol	Betapace	0	0	0	↓	↓	↓	↓	↓	↓	+
Timolol	Blocadren	0	0	0	↓	↓	↓	↓	↓	↓	+
Isomer											
D-propranolol		0	0	++	↔↓¶	↔	↔	↔↓¶	↔	↔↓¶	+¶

*β₁ selectivity is seen only with low therapeutic drug concentrations. With higher concentrations, β₁ selectivity is not seen.

**Bucindolol has direct peripheral vasodilating activity.

†Carvedilol has peripheral vasodilating activity and additional α₁-adrenergic blocking activity.

‡Labetalol has additional α₁-adrenergic blocking properties and direct β₂-adrenergic vasodilator activity.

¶Effects of D-propranolol with doses in human beings well above the therapeutic level. The isomer also lacks β-blocking activity.

Key: PAA, partial agonist activity; MSA, membrane stabilizing activity; R-, resting; E-, exercise; BP, blood pressure; AV, atrioventricular. ++, strong effect; +, modest effect; 0, absent effect; ↓, reduction; ↔, no change.

SOURCES: From Frishman, 1984, and Frishman and Sonnenblick, 1998, with permission.

beta blockade; when present, it is exhibited equally by the two stereoisomers of each drug.

Beta₁ Selectivity

Beta blockers can be classified as selective or nonselective, based on their ability to block the action of endogenous cate-cholamines in some tissues at lower drug concentrations than required in other tissues. In relatively low therapeutic doses, selective agents such as atenolol, metoprolol, acebutolol, and esmolol block cardiac beta₁-receptors more than beta₂-receptors found in bronchial or vascular smooth muscle. In higher doses, selective agents block both types of beta receptors; thus, they should not be used in patients with absolute contraindications to beta blockade, e.g., bronchospasm.

Intrinsic Sympathomimetic Activity

Some beta blockers exhibit intrinsic sympathomimetic activity or partial agonist activity (PAA) at the beta receptor. These agents cause less slowing of the heart rate than do beta blockers without this property; exercise-induced elevations in heart rate are similarly blunted. Agents with PAA reduce peripheral vascular resistance and theoretically might be better tolerated in patients with peripheral vascular disease. These agents also may have less effect on serum lipids than nonselective beta blockers without PAA.

Alpha-Adrenergic Blocking Ability

Several beta blockers, such as labetalol and carvedilol, have the additional property of alpha-adrenergic blockade. Labetalol has been shown to be 4 to 16 times less potent at alpha receptors than at beta blockers. Theoretically the property of alpha blockade should produce a decrease in peripheral vascular resistance, which in turn should act to preserve cardiac output. Whether or not this is clinically significant is unclear. One advantage of these agents is their efficacy in certain populations in whom conventional beta blockers are less effective (e.g., blacks). Labetalol is available in parenteral form for treatment of hypertensive emergencies and urgencies. Labetalol also has a neutral effect on plasma lipids.

Compared to labetalol, carvedilol has a ratio of alpha₁ to beta blocker of 1:10. On a milligram-to-milligram basis, carvedilol is about two to four times more potent than propran-olol as a beta blocker. In addition, carvedilol has antioxidant and antiproliferative activities. Carvedilol has been used for the treatment of hypertension and angina pectoris and was recently approved as a treatment for patients having symptom-atic heart failure

Direct Vasodilator Activity

Bucindolol is a nonselective beta blocker, which, in addition, has direct peripheral vasodilatory activity. It is currently under-going clinical evaluation as a treatment for symptomatic congestive heart failure.

CLINICAL EFFECTS AND THERAPEUTIC APPLICATIONS (TABLE 47-2)

Hypertension

Beta blockers have been shown to be effective in reducing blood pressure in patients with systemic hypertension, includ-ing those with isolated systolic hypertension, and they are considered a first-line therapy. The mechanism by which they reduce blood pressure is not entirely clear; it is likely that a number of drug actions account for this, including negative chronotropic and inotropic effects and effects to lower periph-eral vascular resistance (beta blockers with PAA and $alpha_1$ antagonism) and plasma renin.

Angina Pectoris

Beta blockers act primarily on the demand side of the myocar-dial oxygen supply-demand equation. The three main factors contributing to myocardial oxygen demand are heart rate, sys-tolic blood pressure, and left ventricular wall tension. The first two appear to be most important. Beta blockers slow resting heart rate, which has two favorable effects: blood pressure is decreased, and the longer diastolic filling time that accompa-nies a slower heart rate also may allow for increased coronary perfusion time. Beta blockade also decreases the increments in heart rate and blood pressure that accompany exercise. Most available beta blockers cause an increase in work capacity in patients with stable angina pectoris. Beta blockers also have

TABLE 47-2

REPORTED CARDIOVASCULAR INDICATIONS FOR β-ADRENOCEPTOR BLOCKING DRUGS

Hypertension* (systolic and diastolic)
Isolated systolic hypertension in the elderly
Angina pectoris*
"Silent" myocardial ischemia
Supraventricular arrhythmias*
Ventricular arrhythmias*
Reducing the risk of mortality and reinfarction in survivors of acute myocardial infarction*
Hyperacute phase of myocardial infarction*
Dissection of aorta
Hypertrophic cardiomyopathy*
Reversing left ventricular hypertrophy
Digitalis intoxication
Mitral valve prolapse
QT interval prolongation syndrome
Tetralogy of Fallot
Mitral stenosis
Congestive cardiomyopathy (class II–III)
Fetal tachycardia
Neurocirculatory asthenia

*Indications formally approved by the FDA.

been shown to decrease nitroglycerin consumption and the frequency of chest pain. They have been successfully used in combination with nitrates and with some of the calcium antagonists. In the latter case, one must be concerned with potentiation of negative inotropic effects and impairment of cardiac conduction.

Arrhythmias

Beta blockers function as antiarrhythmic drugs predominantly by competitive inhibition of adrenergic stimulation. This decreases the slope of phase 4 depolarization as well as the spontaneous firing rate of sinus or ectopic pacemakers. Some of these agents also may have membrane-stabilizing activity, also known as the "quinidine-like" or "local anesthetic" effect. Sotalol has additional type III antiarrhythmic activity, which might provide greater efficacy and toxicity. The drugs have been used to treat various supraventricular and ventricular tachyarrhythmias.

Survivors of Acute Myocardial Infarction

Several studies have demonstrated the efficacy of beta-adrenergic blockade for reducing mortality and nonfatal reinfarction in survivors of an acute myocardial infarction. Propranolol and timolol have been approved for reducing the risk of mortality in infarct survivors when initiated 5 to 28 days after myocardial infarction. Metoprolol and atenolol are similarly approved, and can be given intravenously during the initial stages of an infarction, with and without thrombolysis.

Congestive Cardiomyopathy

Controlled trials over the last 20 years with several different beta blockers in patients with both ischemic and nonischemic cardiomyopathy have shown that these drugs, as demonstrated primarily with metoprolol and carvedilol, could improve symptoms, ventricular function, and functional capacity while reducing the need for hospitalization. The combined results of a series of placebo-controlled clinical trials with the alpha-beta blocker carvedilol showed mortality benefit in patients with New York Heart Association class II–III heart failure when the drug was used in addition to diuretics, angiotensin-converting enzyme inhibitors, and digoxin. Additional placebo-controlled mortality studies are currently in progress in patients with congestive heart failure, evaluating immediate and sustained-release metoprolol, bisoprolol, and the vasodilator-beta blockers bucindolol and carvedilol.

The mechanisms of benefit from beta-blocker use are not currently known. Possible mechanisms for beta-blocker benefit in chronic heart failure include an upregulation of impaired beta-receptor expression in the heart and an improvement in impaired baroreceptor functioning that can inhibit excess sympathetic outflow.

Other Cardiovascular Applications

Beta blockers are used to treat patients with hypertrophic cardiomyopathy. Propranolol, for example, controls the symptoms of dyspnea and chest pain seen with this condition, and may lower the intraventricular pressure gradient. Beta blockers also have been used to treat asymptomatic myocardial ischemia with reductions in the amount and degree of ECG ST depression, as determined by 24-hour ambulatory electro-cardiographic (Holter) recording. In patients with mitral valve prolapse, these drugs also relieve chest pain and palpitations. In the initial management of dissecting aortic aneurysm, they have been used in combination with arterial vasodilators. Other clinical conditions in which these agents have been used include tetralogy of Fallot, QT prolongation syndrome, and thyrotoxicosis. For more detailed discussion of these subjects, we refer the reader to "Suggested Reading" at the end of the chapter.

Other Noncardiovascular Uses

Some beta-adrenergic blockers are now approved in topical form for reducing intraocular pressure in patients with open-angle glaucoma, and in oral form for treating patients with benign essential tremor and for prophylaxis against migraine headache syndrome.

ADVERSE EFFECTS

Adverse effects from beta blockers fall into two categories: those related to beta-adrenergic blockade and those related to other reactions. The former group includes asthma, heart failure, bradycardia and heart block, intermittent claudication, Raynaud's phenomenon, glycemic changes, and central nervous system effects. There may be some difference in the frequency of these side effects, depending on the particular agent used. It is important to remember that the property of $beta_1$ specificity is relative; thus, in patients with absolute contraindications to $beta_2$ blockade, the use of a $beta_1$-selective agent may still be unacceptable—e.g., bronchospasm.

A number of rare side effects unrelated to beta blockade have been reported. These include uveitis and sclerosing peritonitis, purpura, and agranulocytosis.

Overdosage

Suicidal and accidental ingestions with beta blockers have been reported. The competitive pharmacologic properties of these drugs permit antidotal treatment with parenteral infusions of beta-agonist agents such as isoproterenol or dobutamine. Glucagon, amrinone, and milrinone have also been used for this purpose. Close monitoring of cardiopulmonary status is especially important in patients with known or suspected beta-blocker overdose.

Beta-Blocker Withdrawal

Exacerbation of angina pectoris, myocardial infarction, and arrhythmias—and in some instances, death—have been reported in patients in whom beta-blocker therapy was abruptly halted. Recently, exacerbation of silent myocardial ischemia was observed after abrupt cessation of beta-blocker therapy. Double-blind controlled trials have also confirmed the existence of a true beta-blocker withdrawal reaction. Thus, patients on chronic beta-blocker therapy should be tapered off this therapy slowly and observed closely.

CLINICAL USE AND DRUG INTERACTIONS

Those situations where a particular beta blocker or beta-blocker subclass should be used are shown in Table 47-3. The common drug interactions with beta blockers are shown in Table 47-4.

TABLE 47-3

CLINICAL SITUATIONS THAT WOULD INFLUENCE THE CHOICE OF A BETA-BLOCKING DRUG

Condition	Choice of Beta Blocker
Asthma, chronic bronchitis with bronchospasm	Avoid all beta blockers if possible, but small doses of beta$_1$-selective blockers can be used; beta$_1$ selectivity is lost with higher doses; drugs with partial agonist activity and labetalol with α-adrenergic blocking properties can also be used.
Congestive heart failure	Drugs with partial agonist activity and vasodilatory activity (carvedilol, labetalol) may have an advantage, although all beta blockers should be used with caution.
Angina	In patients with angina at low heart rates, drugs with partial agonist activity are probably contraindicated; patients who have angina at high heart rates but who have resting bradycardia may benefit from a drug with partial agonist activity; in vasospastic angina, labetalol may be useful; other beta blockers should be used with caution.
Atrioventricular conduction defects	Beta blockers are generally contraindicated, but drugs with partial agonist activity and labetalol can be tried with caution.

Bradycardia	Beta blockers with partial agonist activity and labetalol have less of a pulse-slowing effect and are preferable.
Raynaud's phenomenon, intermittent claudication, cold extremities	$Beta_1$-selective blocking agents, labetalol, and agents with partial agonist activity may have an advantage.
Depression	Avoid propranolol; substitute a beta blocker with partial agonist activity.
Diabetes mellitus	$Beta_1$-selective agents and partial agonist drugs are preferable.
Thyrotoxicosis	All agents will control symptoms, but agents without partial agonist activity are preferred.
Pheochromocytoma	Avoid all beta blockers unless an alpha blocker is given; labetalol may be used as a treatment of choice.
Renal failure	Use reduced doses of compounds largely eliminated by renal mechanisms (nadolol, sotalol, atenolol) and drugs whose bioavailability is increased in uremia (propranolol); also consider possible accumulation of active metabolites (propranolol).
Insulin and sulfonylurea use	There is a danger of hypoglycemia; possibly less using drugs with $beta_1$ selectivity.
Clonidine	Avoid nonselective beta blockers; there is a severe rebound effect with clonidine withdrawal.
Oculomucocutaneous syndrome	Stop drug; substitute any beta blocker.
Hyperlipidemia	Avoid nonselective beta blockers; use aqents with partial agonism or beta, selectivity, or alpha-blocking activity.

SOURCES: From Frishman, 1984, and Frishman and Sonnenblick, 1998, with permission.

703

TABLE 47-4

DRUG INTERACTIONS THAT MAY OCCUR WITH BETA-ADRENOCEPTOR BLOCKING DRUGS

Drug	Pharmacokinetic Interactions	Pharmacodynamic Interactions	Precautions
Alcohol	Enhanced first-pass hepatic degradation	None	May need increased doses of lipid-soluble agents
α-Adrenergic blockers		Increased risk for first-dose hypotension	Use with caution
Aluminum hydroxide gel	Decreased β-blocker absorption	None	Clinical efficacy rarely altered
Amiodarone	None	Enhanced negative chronotropic activity	Monitor response
Aminophylline	Mutual inhibition		Observe patient's response
Ampicillin	Impaired GI absorption leading to decreased β-blocker bioavailability		May need to increase β-blocker dose
Angiotensin II receptor blockers (losartan)	None	Enhanced blood pressure effects and bronchospasm	Monitor response

704

Antidiabetics	Both enhanced and blunted responses seen	None	Monitor for altered diabetic response
ACE inhibitors	None	Enhanced blood pressure effects and bronchospasm	Monitor response
Calcium	Decreases β-blocker absorption		May need to increase β-blocker dose
Calcium channel inhibitors	Decreased hepatic clearance of lipid- and water-soluble β blockers; decreased clearance of calcium blockers	Potentiation of AV nodal, negative inotropic and hypotensive responses	Avoid use if possible, although few patients show ill effects
Cimetidine	Decreased hepatic clearance of lipid-soluble β blockers	None	Combination should be used with caution
Clonidine	None	Nonselective agents exacerbate clonidine withdrawal phenomenon	Use only β₁-selective agents or labetalol
Diazepam	Diazepam metabolism reduced		Observe patient's response
Digitalis glycosides	None	Potentiation of brady-cardiac and AV blocks	Observe patient's response; interactions may benefit angina patients with abnormal ventricular function

(continued)

TABLE 47-4

DRUG INTERACTIONS THAT MAY OCCUR WITH BETA-ADRENOCEPTOR BLOCKING DRUGS (*continued*)

Drug	Pharmacokinetic Interactions	Pharmcodynamic Interactions	Precautions
Epinephrine	None	Severe hypertension and bradycardia	Administer epinephrine cautiously; cardioselective β blocker may be safer
Ergot alkaloids	None	Severe hypertension and peripheral artery hyperperfusion have been seen, though β blockers are commonly coadministered	Observe patient's response; few patients show ill effects
Fluvoxamine	Decreased hepatic clearance of propranolol		Use with caution
Glucagon	Enhanced clearance lipid-soluble β blockers	None	Monitor for reduced response
Halofenate			Observe for impaired response to β blockade

Hydralazine	Decreased hepatic clearance of lipid-soluble β blockers	Enhanced hypotensive response	Observe patient's response
Indomethacin and ibuprofen	None	Reduced efficacy in treatment of hypertension	Avoid concurrent use or choose selective β₁ blocker
Isoproterenol	None	Cancels pharmacologic effect	Monitor for altered response; interaction may have favorable results
Levodopa		Antagonism of hypotensive and positive inotropic effects of levodopa	
Lidocaine	Decreased hepatic clearance of lidocaine by lipid soluble β blockers	Enhanced lidocaine toxicity	Combination should be used with caution; use lower doses of lidocaine
Methyldopa	Uncertain	Hypertension during stress	Monitor for hypertensive episodes
Monoamine oxidase inhibitors		Enhanced hypotension	Manufacturer of propranolol considered concurrent use contraindicated
Nitrates	None	Enhanced hypotension	Monitor response
Omeprazole	None	None	None
Phenobarbital	Increased hepatic metabolism of β blockers		May need to increase lipid-soluble β-blocker dose

(continued)

TABLE 47-4

DRUG INTERACTIONS THAT MAY OCCUR WITH BETA-ADRENOCEPTOR BLOCKING DRUGS (continued)

Drug	Pharmacokinetic Interactions	Pharmacodynamic Interactions	Precautions
Phenothiazines	Increased phenothiazine and β-blocker blood levels	Additive hypotensive response	Monitor for altered response, especially with high doses of phenothiazine
Phenylpropanolamine		Severe hypertensive reaction	Avoid use, especially in hypertension controlled by both methyldopa and β blockers
Phenytoin		Additive ventricular depressive effects	Use with caution
Reserpine		Depression, possible enhanced sensitivity to β-adrenergic blockade	Monitor closely
Ranitidine	Not marked	None	Observe response
Smoking	Enhanced first-pass metabolism	None	May need to increase dose of lipid-soluble β blockers

Sulindac and naproxen	None	None	
Tricyclic antidepressants	None	Inhibits negative inotropic and chronotropic effects; enhanced hypotension	Use with caution with sotalol because of additive effects on ECG QT interval
Tubucuraine		Enhanced neuromuscular blockade	Observe response in surgical patients, especially after high doses of propranolol
Type I antiarrhythmics	Propafenone and quinidine decrease clearance of lipid-soluble β blockers	Disopyramide is a potent negative inotropic and chronotropic agent	Cautious coprescription; use with sotalol can be dangerous because of additive effects on ECG QT interval
Warfarin	Decreased clearance of warfarin	None	Monitoor response

SUGGESTED READING

Frishman WH: β-Adrenergic blockers. In: Izzo JL Jr, Black HR (eds): *Hypertension Primer*. American Heart Association, TX, 1998. In press.

Frishman WH: *Clinical Pharmacology of the Beta-Adreno-ceptor Blocking Drugs*, 2d ed. Norwalk, CT, Appleton-Century-Crofts, 1984.

Frishman WH: Alpha- and beta-adrenergic blocking drugs. In: Frishman WH, Sonnenblick EH (eds): *Cardiovascular Pharmacotherapeutics*. New York, McGraw-Hill, 1997: 59–94.

Frishman WH, Skolnick AE: Secondary prevention post-infarction: the role of β-adrenergic blockers, calcium-channel blockers and aspirin. In: Gersh BJ, Rahimtoola SH (eds): *Acute Myocardial Infarction*, 2d ed. New York, Chapman & Hall, 1996:766–796.

Frishman WH, Sonnenblick EH: Beta-adrenergic blocking drugs and calcium channel blockers. In: Alexander RW, Schlant RC, Fuster V, O'Rourke RA, Roberts R, Sonnenblick EH (eds): *Hurst's The Heart*, 9th ed. New York, McGraw-Hill, 1998:1583–1618.

Frishman WH: Alpha- and beta-adrenergic blocking drugs. In: Frishman WH, Sonnenblick EH (eds): *Cardiovascular Pharmacotherapeutics Companion Handbook*. New York, McGraw-Hill, 1998:23–64.

Moser M, Frishman WH: Results of therapy with carvedilol, a β-blocker vasodilator with antioxidant properties, in hypertensive patients. *Am J Hypertens* 1998; 11:15s–22s.

The Sixth Report of the Joint National Committee on Prevention, Detection, Evaluation and Treatment of High Blood Pressure (JNC VI). *Arch Intern Med* 1997; 157:2413–2446.

Opie LH, Sonnenblick EH, Frishman WH, Thadani U: β-blocking drugs. In: Opie LH, et al (eds): *Drugs for the Heart*, 4th ed. Philadelphia, Saunders, 1995:1–30.

48

CALCIUM CHANNEL BLOCKERS

William H. Frishman

The calcium antagonists are a heterogeneous group of drugs with widely variable effects on heart muscle, sinus node function, atrioventricular conduction, peripheral blood vessels, and the coronary circulation. Eleven of these drugs–nifedipine, nicardipine, nimodipine, felodipine, isradipine, amlodipine, nisoldipine, verapamil, diltiazem, bepridil, and mibefradil—are approved in the United States for clinical use. Their predominant effect is on the slow channels of the cell membrane, although the exact mechanism(s) of action vary considerably. In addition to their membrane effects, calcium channel blockers can inhibit the availability of calcium ions for excitation coupling at intracellular sites.

CARDIOVASCULAR EFFECTS
(TABLES 48-1, 48-2)

Calcium channel blockers inhibit contraction of vascular smooth muscle and, to a lesser extent, cardiac and skeletal muscle. This differential effect is due to the greater reliance of vascular smooth muscle on external calcium entry. Cardiac and skeletal muscle rely on recirculating internal calcium stores. This allows for dilatation of arterial smooth muscle with less of an effect on myocardial contractility and virtually no effect on skeletal muscle. The calcium antagonists are less active on venous smooth muscle, and at therapeutic doses are ineffective for decreasing venous capacitance.

All calcium antagonists exert some degree of negative inotropic effect, which appears to be dose-dependent. Since excitation-contraction coupling in smooth muscle is 3 to 10 times more sensitive to the action of calcium antagonists than that in myocardial fibers, the usual therapeutic doses may not

TABLE 48-1

PHARMACOLOGIC EFFECTS OF THE CALCIUM CHANNEL BLOCKERS

Drug	Trade Name	Heart Rate		Conduction		Myocardial Contractility	Peripheral Vasodilator	CO	Coronary Blood Flow	Myocardial O₂ Demand
		Acute	Chronic	SA Node	AV Node					
Diltiazem	Cardizem	↑	↓	↓	↓	↓	↑	V	↑	↓
Bepridil	Vascor	↑	↓	↓	↓	V	–	V	↑	↓
Verapamil	Calan, Isoptin, Verelan	↓	↓	↓	↓	↓	↑	V	↑	↓
Amlodipine	Norvasc	↑	↑	–	–	↓	↑↑	↑	↑	↓
Felodipine	Plendil	↑	↑	–	–	–	↑↑	–	↑	↓
Isradipine	DynaCirc	↑	↑	–	–	–	↑↑	–	↑	↓
Nicardipine	Cardene	↑	↑	–	–	–	↑↑	–	↑	↓
Nifedipine	Adalat	↑	↑	–	–	↓	↑↑	↑	↑	↓
Nimodipine	Nimotop	↑	↑	–	–	–	V	–	↑	↓
Nisoldipine	Sular	↕	↕	↕	↕	↕	↑	↕	↑	↓

KEY: ↑, increase; ↓, decrease; –, no change; V, variable; SA, sinoatrial; AV, atrioventricular; CO, cardiac output.

SOURCES: From Frishman WH, Stroh JA, Greenberg SM, Suarez T, Karp A, Peled HB: Calcium channel blockers in systemic hypertension. *Med Clin North Am* 1988; 72:449–499; and Frishman WH: Calcium-channel blockers. In: Schlant RC, Alexander RW, O'Rourke RA, Roberts R, Sonnenblick EH (eds): *Hurst's The Heart*, 8th ed. New York. McGraw-Hill, 1993:1291–1308.

produce significant negative inotropic effects. Additionally, the negative inotropic effect may be attenuated by the reflex augmentation of beta-adrenergic tone that accompanies arterial vasodilation. However, this mechanism is unlikely to benefit patients with significantly impaired left ventricular function in whom the baroreceptor reflex is markedly attenuated.

Electrophysiologic Effects

The calcium antagonists differ somewhat in their effects on the cardiac conduction system. Verapamil and diltiazem, which are L-channel blockers, both prolong conduction and refractoriness in the atrioventricular node. Sinus node discharge rates are depressed by all calcium channel blockers. This effect may be compensated for clinically by baroreceptor reflexes, however, particularly with the dihydropyridines. Bepridil has, in addition, sodium ion channel inhibitory activity and activity against ventricular arrhythmias.

CLINICAL APPLICATIONS
(TABLE 48-3)

The calcium antagonists available in the United States are approved for various cardiovascular and cerebrovascular disorders.

The antianginal action of these agents appears to be multifactorial. These drugs exert vasodilator effects on the coronary and peripheral vessels, in addition to their mild effects on contractility. They are particularly effective in coronary spasm. Calcium antagonists seem to be as effective as beta blockers in chronic stable angina pectoris, and combined angina pectoris/hypertension when used as monotherapy. However, there may be some limitations when using short-acting dihydropyridines because of the reflex tachycardia that sometimes occurs. Combination therapy with nitrates and/or beta-adrenergic blockers may be more efficacious than monotherapy. Because adverse effects can occur with combination therapy, patients must be carefully selected and monitored.

Both verapamil and diltiazem have been widely used for atrial fibrillation/flutter and paroxysmal supraventricular tachycardia. In atrial fibrillation, the predominant effect is to decrease the ventricular response rather than conversion to normal sinus rhythm. One caution should be remembered when dealing with wide complex tachyarrhythmias: *If there is any*

(*Text continues on page 718.*)

TABLE 48-2

CLINICAL USE OF CALCIUM CHANNEL BLOCKERS

Agent	Dosage		Onset of Action		Therapeutic PC	Site of Metabolism	Active Metabolites	Excretion, Percent
	Oral	IV	Oral	IV				
Diltiazem	30–90 mg q6–8h	Bolus: 0.25 mg/kg (20 mg) Infusion: 5–15 mg/h	<30 min	<10 min	50–200 ng/mL	Deacetylation N-deacetylation O-demethylation Major hepatic first-pass effect	Yes	60 (fecal) 2–4 (unchanged in urine)
Diltiazem SR	60–120 mg q12h		30–60 min		50–200 ng/mL		Yes	
Diltiazem CD	180–360 mg q24h		30–60 min		50–200 ng/mL		Yes	
Dilacor XR	180–540 mg q24h		30–60 min		40–200 ng/mL		Yes	
Tiazac	120–360 mg q24h		1–2 h		40–200 ng/mL		Yes	
Verapamil	80–120 mg q6–12h	75–150 µg/kg (5–10 mg)	<30 min	<5 min	>100 ng/mL	N-dealkylation O-demethylation Major hepatic first-pass effect	Yes	15 (fecal) 70 (renal) 3–4 (unchanged in urine)

714

	Dose				Metabolites		Elimination (%)
Verapamil SR	240 mg q12 or 24h or 480 mg q24h	<30 min		>50 ng/mL		Yes	15 (fecal) 70 (renal) 3–4 (unchanged in urine)
Verelan	120–480 mg q24h			>50 ng/mL		Yes	16 (fecal) 70 (renal) 3–4 (unchanged in urine)
Verapamil Coer 24	180–240 mg q24h		4 h			Yes	16 (fecal) 70 (renal) 3–4 (unchanged in urine)
Nifedipine	10–30 mg q6–8h	<20 min	3 min SL	25–100 ng/mL	A hydroxycarbolic acid and a lactone with no known activity	No	20–40 (fecal) 50–80 (renal) <0.1 (unchanged in urine)
Nifedipine GITS	30–120 mg q24h	2 h			Major hepatic first-pass effect	No	

(*continued*)

TABLE 48-2
CLINICAL USE OF CALCIUM CHANNEL BLOCKERS (*continued*)

Agent	Dosage		Onset of Action		Therapeutic PC	Site of Metabolism	Active Metabolites	Excretion, Percent
	Oral	IV	Oral	IV				
Nicardipine	20–30 mg tid	1.15 mg/h	<20 min	<5 min	28–100 ng/mL	Major hepatic first-pass effect	No	35 (fecal) 60 (renal) <1 (unchanged in urine)
Nicardipine SR	30–60 mg bid		20 min		28–50 ng/mL			35 (fecal) 60 (renal) <1 (unchanged in urine)
Nimodipine	60 mg q4h	<30 min			7 ng/mL	Major hepatic first-pass effect Oxidation	No	
Amlodipine	5–10 mg q24h		90–120 min in vitro		5–15 ng/mL	Extensive but slow hepatic metabolism	No	20–25 (fecal) 60 (renal) 10 (unchanged in urine)

716

	Dose	$t_{1/2}$	PC	Metabolism	Active metabolite	Elimination (%)
Isradipine	2.5–10 mg q12h	120 min		Hepatic deesterification and aromatization	No	30 (fecal) 70 (renal) 0 (unchanged in urine)
Felodipine ER	2.5–10 mg q24h	2–5 h	4–6 nmol/L	Hepatic microsomal P450 system oxidation Major hepatic first-pass effect	No	10 (fecal) 60–70 (renal) <0.5 (unchanged in urine and feces)
Bepridil	200–400 mg q24h	1–2 h	1200–3500 ng/mL	Hepatic	No	70 (renal) 20 (fecal)
Nisoldipine ER	20–40 mg q24h			Hepatic hydroxylation	Yes	80 (renal) <1 (unchanged in urine) 12 (fecal)

KEY: PC, plasma concentrations; bid, twice daily; tid, thrice daily; SL, sublingual; nd, no data.
SOURCE: From Frishman WH, Sonnenblick WH, Sonnenblick EH. Calcium-channel blockers. In: Schlant RC, Alexander RW, O'Rourke RA, Roberts R, Sonnenblick EM (eds): *Hurst's The Heart*, 8th ed. New York, McGraw-Hill, 1993:1291–1308.

TABLE 48-3

CARDIOVASCULAR USES
OF CALCIUM CHANNEL BLOCKERS

Angina pectoris*
  Effort angina
  Rest angina
  Prinzmetal's variant
Arrhythmia treatment and prophylaxis* (acute and
  chronic)—verapamil, diltiazem
Systemic hypertension—verapamil, diltiazem, nifedi-
  pine, amlodipine, isradipine, felodipine, nicardipine,
  nisoldipine, mibefradil
Hypertensive emergencies*—nicardipine
Perioperative hypertension—nicardipine
Hypertrophic cardiomyopathy—verapamil, diltiazem
Congestive heart failure—amlodipine, felodipine,
  nicardipine, mibefradil
Myocardial infarction (containing size of infarct)
Myocardial infarction (preventing Q-wave infarcts in
  non-Q-wave infarction)—diltiazem, verapamil
Primary pulmonary hypertension—nifedipine
Peripheral vascular disease
  Raynaud's phenomenon
  Intermittent claudication
  Cerebral arterial spasm* (subarachnoid hemor-
    rhage)—nimodipine
  Stroke—nimodipine
  Mesenteric insufficiency
  Migraine headache prophylaxis—nimodipine
Deacceleration of atherosclerosis
Prevention of cardiomyopathy

*Approved by the Food and Drug Administration.

question about the origin of the rhythm, it should be treated as ventricular in origin. A number of adverse outcomes have been reported when verapamil was used for presumptive supraventricular tachycardia with aberrancy that turned out, in fact, to be ventricular tachycardia.

Calcium channel blockers are effective in treating systemic hypertension and can be used once or twice daily. Calcium antagonists are effective in reducing both systolic and diastolic blood pressures, and have been found to be effective in black patients and in the elderly. They have also been shown to cause regression of left ventricular hypertrophy.

Other conditions in which calcium channel blockers have been used include hypertensive emergencies; silent myocardial ischemia, and hypertrophic cardiomyopathy, although FDA approval has not been granted for these indications. Calcium antagonists also have been studied in use against congestive heart failure and primary pulmonary hypertension; however, current data do not suggest that they should be used as routine first-line therapy in these conditions.

ADVERSE EFFECTS
(TABLE 48-4)

Besides their expected effects on the cardiovascular system, these drugs have a wide array of side effects. Nifedipine has a fairly high incidence of minor side effects; major side effects are far less common. The most frequent adverse effects reported with nifedipine include headache, pedal edema, flushing, parasthesias, and dizziness. More serious side effects include exacerbation of ischemic symptoms and hypotension. These appear to be less common with the new long-acting form of nifedipine—the nifedipine gastrointestinal therapeutic system (GITS). Diltiazem and verapamil can potentiate conduction system dysfunction. The most common adverse effect with verapamil is constipation, which can be particularly difficult in older patients. Exacerbation of congestive heart failure also can be seen with verapamil, especially with concomitant beta-blocker therapy. *Combinations of verapamil and diltiazem with beta blockers should be used only with extreme caution.* Diltiazem can cause headache and gastrointestinal complaints in addition to cardiovascular side effects. Bepridil can cause diarrhea, constipation, and torsades de pointes. The calcium antagonists have the potential for exacerbating heart failure.

Drug Withdrawal and Overdose

Although serious problems have been reported with abrupt withdrawal of beta blockers, experience to date with the calcium antagonists does not suggest rebound or overshoot phenomena. There may be a return of symptoms, however, when therapy is discontinued.

TABLE 48-4
ADVERSE EFFECTS OF THE CALCIUM CHANNEL BLOCKERS

	Overall (%)	Head-ache	Dizzi-ness	GI	Flushing	Paresth-esia	↓SA/AV Conduc-tion	CHF	Hypo-tension	Pedal Edema	Worsen-ing of Angina	Palp.
Diltiazem	≈5	+	+	+	+	0	3+	+	+	+	0	+
Diltiazem SR	≈5	+	+	+	+	0	3+	+	+	+	0	+
Diltiazem CD	≈5	+	+	+	0	+	+	+	+	+	0	+
Dilacor XR	≈5	+	+	+	+	+	+	+	+	+		+
Diltiazem ER	≈5	+	+	+	+	+	+	+	+	+	+	+
Verapamil	8	+	+	3+	0	0	3+	2+	+	+	0	+
Verapamil SR	~8	+	+	3+	0	0	3+	2+	+	+	0	+
Verapamil Coer 24	≈8	+	+	3+	0	0	3+	2+	+	+	0	+
Bepridil	15	0	2+	3+	0	0	+	+	0	0	0	
Amlodipine	~15	2+	+	+	+	+	0	0	+	2+	0	+
Isradipine	~15	2+	2+	+	+	+	0	0	+	2+	0	+

720

Isradipine GITS	≈15	2+	+	+	0	0	+	2+	0	+
Nifedipine	~20	3+	+	+	3+	+	+	2+	+	2+
Nifedipine GITS	~10	+	+	+	+	+	0	+	0	+
Nifedipine ER	≈15	+	+	+	+	+	+	+	0	+
Nicardipine	~20	3+	+	+	3+	+	0	2+	+	2+
Nicardipine SR	≈10	+	+	+	+	+	0	+	0	+
Nicardipine IV		+	0	0	0	0	0	0	+	+
Nimodipine	15	+	+	+	0	+	0	0+	+	
Felodipine ER	≈20	2+	2+	2+	+	0	0	2+	0	+
Nisoldipine ER	≈15	2+	+	+	+	0	0	2+	0	+
Mibefradil	≈10	+	+	+	0	3+	0	+	0	+

KEY: GI, gastrointestinal; SA, sinoatrial node; AV, atrioventricular node; CHF, congestive heart failure; Palp, palpitations; 0, no report; +, rare; 2+, occasional; 3+, frequent; SR, sustained release; ER, extended release; GITS, gastrointestinal therapeutic system.

SOURCES: Adapted from Frishman WH, Stroh JA, Greenberg SM, Suarez T, Karp A, Peled HB: Calcium channel blockers in systemic hypertension. *Med Clin North Am* 1988; 72:449–499; and Frishman WH, Sonnenblick WH: Beta-adrenergic blocking drugs and calcium channel blockers. In: Alexander RW, Schlant RC. Fuster V, O'Rourke RA, Roberts R. Sonnenblick EH (eds): *Hurst's The Heart*, 9th ed. New York, McGraw-Hill, 1998:1583–1618.

The effects seen with calcium channel blocker overdosage are those expected, based on the physiological action of the drugs. Hypotension, heart block, and left ventricular dysfunction are among the major complications.

DRUG INTERACTIONS

A number of drug interactions have been reported with the available calcium antagonists. Both nifedipine and verapamil have been reported to increase serum digoxin levels, verapamil by as much as 70 percent through decreased renal and nonrenal clearance as well as volume of distribution. Nifedipine causes smaller increases. Verapamil bioavailability is decreased by rifampin. The latter causes an increase in first-pass hepatic metabolism. Mibefradil, a selective T-channel antagonist, can potentiate toxic effects of lovastatin and simvastatin, leading to rhabdomyolysis, and was removed from the market because of this and other serious drug-drug interactions. Verapamil, diltiazem, and mibefradil may have additive effects on atrioventricular conduction. Combinations of beta blockers with nifedipine or verapamil have been studied in patients with angina pectoris; care must be used when employing these combinations because of the potential for impairment of left ventricular function, especially in patients with abnormal function at baseline.

CONCLUSION

Although calcium antagonists are linked by the common thread of inhibition of calcium transport into the cell, their exact mechanisms differ. This results in different actions at the various target organs. These differences allow the clinician to select the agent most appropriate to each clinical situation.

SUGGESTED READING

Frishman WH (ed): Calcium antagonists. In: *Current Cardiovascular Drugs*, 2d ed. Philadelphia, Current Medicine, 1995:129–148.

Frishman WH: Mibefradil: a new selective T-channel calcium antagonist for hypertension and angina pectoris. *J Cardiovasc Pharmacol Ther* 1997; 2:321–330.

Frishman WH, Skolnick AE: Secondary prevention post infarction: the role of β-adrenergic blockers, calcium-channel blockers and aspirin. In: Gersh BJ, Rahimtoola SH (eds): *Acute Myocardial Infarction*. New York, Chapman & Hall, 1996:766–796.

Frishman WH: Calcium channel blockers. In: Frishman WH, Sonnenblick EH (eds): *Cardiovascular Pharmacotherapeutics*. New York, McGraw-Hill, 1997:101–130.

Frishman WH: In: Frishman WH, Sonnenblick EH (eds): Calcium channel blockers. *Cardiovascular Pharmacotherapeutics Companion Handbook*. New York, McGraw-Hill, 1998:73–106.

Frishman WH, Sonnenblick EH: Beta-adrenergic blocking drugs and calcium-channel blockers. In: Alexander RW, Schlant RC, Fuster V, O'Rourke RA, Roberts R, Sonnenblick EH (eds): *Hurst's The Heart*, 9th ed. New York, McGraw-Hill, 1998:1583–1618.

Katz B, Rosenberg A, Frishman WH: Controlled-release drug delivery systems in cardiovascular medicine. *Am Heart J* 1995; 19:359–368.

Opie LH, Frishman WH, Thadani U: Calcium channel antagonists. In: Opie LH et al (eds): *Drugs for the Heart*, 4th ed. Philadelphia, Saunders, 1995:50–82.

Packer M, Frishman WH (eds): *Clinical Pharmacology of the Calcium-Entry Blocking Drugs*. Norwalk, CT, Appleton-Century Crofts, 1984.

INDEX

Note: Numbers followed by an "f" or a "t" refer to figures or tables, respectively.

725

ISBN 0-07-001024-2

90000

9 780070 010246

HURST: THE HEART HANDBOOK